D1702966

Chemical and Functional Properties of Food Saccharides

Chemical and Functional Properties of Food Components Series

SERIES EDITOR
Zdzisław E. Sikorski

Chemical and Functional Properties of Food Proteins
Edited by Zdzisław E. Sikorski

Chemical and Functional Properties of Food Components, Second Edition
Edited by Zdzisław E. Sikorski

Chemical and Functional Properties of Food Lipids
Edited by Zdzisław E. Sikorski and Anna Kołakowska

Chemical and Functional Properties of Food Saccharides
Edited by Piotr Tomasik

Chemical and Functional Properties of Food Saccharides

EDITED BY Piotr Tomasik

CRC

CRC PRESS

Boca Raton   London   New York   Washington, D.C.

ΠC ⁄124⁄ 2006

Library of Congress Cataloging-in-Publication Data

Chemical and functional properites of food saccharides / edited by Piotr Tomasik.
   p. cm. — (Chemical and functional properites of food components series ; 5)
   Includes bibliographical references and index.
   ISBN 0-8493-1486-0 (alk. paper)
   1. Sweeteners. I. Tomasik, Piotr. II. Title. III. Series.
TP421.C44 2003
664—dc21
                                       2003053186

Visit the CRC Press Web site at www.crcpress.com

© 2004 by CRC Press LLC

No claim to original U.S. Government works
International Standard Book Number 0-8493-1486-0
Library of Congress Card Number 2003053186
Printed in the United States of America 1 2 3 4 5 6 7 8 9 0
Printed on acid-free paper

Preface

In preparing this new book on food carbohydrates, I have looked for a formula to distinguish it from titles already published. The team of contributors to this volume presents modern developments in the field of food carbohydrates. The book is a guide to the chemistry, biochemistry, and technology of food carbohydrates, and because of the abundant scope of the data included, it can serve as a handbook of carbohydrate chemistry.

As editor of this volume, I have tried to present food carbohydrates within a wider context. Therefore, I invited as contributors not only food chemists and technologists but also organic chemists, biochemists, a chemist theoretician, and one M.D., an expert in human nutrition.

The contents of this book can be divided into the following areas:

Information on structure and reactivity of carbohydrates (Chapters 1, 2, 9, and 10). Chapter 1 presents the comprehensive structural chemistry of saccharides. It includes a method of interconversion of open-chain structures into cyclic structures as well as a collection of over 100 structures of mono-, oligo-, and polysaccharides which are the most common in food chemistry and technology. Chapter 2 introduces the reader to the chemistry of mono- and disaccharides, revealing the simplicity of apparently unusual reaction pathways of carbohydrates. Only those reactions which are essential in food chemistry and technology are covered in this chapter. Chapters 9 and 10 describe chemical reactivity and enzymatic conversions of high-molecular carbohydrates, respectively.

Low-molecular carbohydrates: sources, modifications, and applications. The production of sugar from sugar beet and sugar cane are discussed in Chapter 4, together with applications of sucrose in nutrition. In Chapter 6, this presentation is extended to honey. Currently, considerable attention is being paid to nonnutritional applications of carbohydrates, including obtaining carbohydrates from nonagricultural sources, such as seaweeds, for use in the food industry. Chapter 3 presents chemical modifications of lower saccharides useful in food processing and as an important source for designing products for nonnutritional applications. A similar treatment of polysaccharides is presented in Chapter 19. Chapter 21 focuses on carbohydrates as a source of energy.

Carbohydrates and functional properties of food. The role of carbohydrates in controlling the texture of food is discussed in Chapter 11. Carbohydrates' roles in the flavor and aroma of food is presented in detail in Chapter 18. Sweetening is one of the most common applications of carbohydrates. A modern theory of sweetening is developed in Chapter 5.

Selected problems of the carbohydrate analytics. These are limited to polysaccharides. Molecular characteristics of polysaccharides are discussed in Chapters 22 and 23.

Health and nutritional problems. These issues are discussed by an M.D. in Chapter 20.

Sources, properties, and applications of particular groups of carbohydrates. This includes cyclodextrins (Chapter 17), carbohydrates of animal tissues (Chapter 16), starch (Chapters 7 and 8), pectic polysaccharides (Chapter 12), fructans (Chapter 13), chitosan (Chapter 14), and plant and algal gums and mucilages (Chapter 15).

Natural and synthetic nonsaccharide sweeteners. These are discussed in Chapter 24. Natural and synthetic sweeteners compete with saccharide sweeteners for a variety of reasons. Technological problems encountered during food production and issues related to nutrition have created interest in their applications in food products jointly with carbohydrates. In addition, several essentially nonsaccharide sweeteners carry in their structures saccharide aglycones. These factors warrant the inclusion of this chapter.

Throughout the book, the chapters summarize the current state of knowledge and provide prospects for the development of research and technology in a given area. I hope that our joint efforts have brought interesting results for our readers.

Editor

Piotr Tomasik earned his M.Sc. and Ph.D. degrees from the Technical University of Wrocław and his D.Sc. from the Institute of Organic Chemistry of the Polish Academy of Sciences in Warsaw. He is professor and head of the Department of Chemistry at the Agricultural University in Cracow. He spent 1 year as a postdoctoral fellow at the University of Alabama and subsequently was a visiting researcher and visiting professor. He has worked at Clemson University, South Carolina; Ames Laboratory, Iowa State University; Saginaw Valley State University; Academia Sinica in Taipei, Taiwan; and the Institute of Industrial Research, Tokyo University. His major interest is organic chemistry. He is active in heterocyclic chemistry, chemistry of polysaccharides, inorganic chemistry, food chemistry, and nanotechnology. Dr. Tomasik has published 21 monographs, chapters, and extended reviews; 19 other reviews; 3 student textbooks; and almost 300 full research papers. He holds 56 patents and has supervised 26 Ph.D. theses.

Contributors

Wolfgang Bergthaller
Federal Center for Cereal, Potato, and
  Lipid Research
Detmold, Germany

Stanisław Bielecki
Institute of Technical Biochemistry
Technical University of Łódź
Łódź, Poland

Ewa Cieślik
Institute for Human Nutrition
Agricultural University
Cracow, Poland

Jiri Davidek
Department of Food Chemistry and
  Analysis
Institute of Chemical Technology
Prague, Czech Republic

Tomas Davidek
Friskies R&D Center
Aubigny, France

Aslak Einbu
Norwegian Biopolymer Laboratory
Department of Biotechnology
Norwegian University of Science and
  Technology
Trondheim, Norway

Jan Grabka
Institute of Chemical Technology of
  Food
Technical University of Łódź
Łódź, Poland

Anton Huber
Institute of Chemistry, Colloids, and
  Polymers
Karl-Franzens University
Graz, Austria

Jay-lin Jane
Department of Food Science and
  Human Nutrition
Iowa State University
Ames, Iowa

Tadeusz Kołczak
Department of Animal Products
  Technology
University of Agriculture
Cracow, Poland

Vivian M.-F. Lai
Department of Food and Nutrition
Providence University
Taichung, Taiwan

Cheng-yi Lii
Institute of Chemistry
Academia Sinica, Nankang
Taipei, Taiwan

Alistair J. MacDougall
BBSRC Institute of Food Research
Norwich, United Kingdom

Władysław Pietrzycki
Department of Chemistry
University of Agriculture
Cracow, Poland

Werner Praznik
Institute of Chemistry
University of Agriculture
Vienna, Austria

Lawrence Ramsden
Department of Botany
University of Hong Kong
Hong Kong, China

Stephen G. Ring
Institute of Food Research
Norwich Laboratory
Norwich, United Kingdom

Heinz Th. K. Ruck
University of Wuppertal
Wuppertal, Germany

Helena Rybak-Chmielewska
Research Institute of Pomology and
  Floriculture
Apiculture Division
Pulawy, Poland

Jeroen J.G. van Soest
ATO B.V.
Wageningen, The Netherlands

Wiesław Szeja
Department of Chemistry
Silesian Technical University
Gliwice, Poland

József Szejtli
CYCLOLAB
Cyclodextrin R&D Laboratory Ltd.
Budapest, Hungary

Piotr Tomasik
Department of Chemistry
University of Agriculture
Cracow, Poland

Przemysław Jan Tomasik
Department of Clinical Biochemistry
Collegium Medicum
Jagiellonian University
Cracow, Poland

Michal Uher
Faculty of Chemical and Food
  Technology
Slovak University of Technology
Bratislava, Slovakia

Kjell M. Vårum
Norwegian Biopolymer Laboratory
Department of Biotechnology
Norwegian University of Science and
  Technology
Trondheim, Norway

Ya-Jane Wang
Department of Food Science
University of Arkansas
Fayetteville, Arkansas

Table of Contents

Chapter 1   Saccharides and Polysaccharides: An Introduction 1
Piotr Tomasik

Chapter 2   Structure and Reactivity of Saccharides 19
Wiesław Szeja

Chapter 3   Saccharides: Modifications and Applications 35
Ya-Jane Wang

Chapter 4   Production of Saccharides 47
Jan Grabka

Chapter 5   Saccharide Sweeteners and the Theory of Sweetness 57
Władysław Pietrzycki

Chapter 6   Honey 73
Helena Rybak-Chmielewska

Chapter 7   Starch: Structure and Properties 81
Jay-lin Jane

Chapter 8   Starch World Markets and Isolation of Starch 103
Wolfgang Bergthaller

Chapter 9   Chemical Modifications of Polysaccharides 123
Piotr Tomasik

Chapter 10 Enzymatic Conversions of Carbohydrates 131
Stanisław Bielecki

Chapter 11 Role of Saccharides in Texturization and Functional Properties
                        of Foodstuffs 159
Vivian M.F. Lai and Cheng-yi Lii

Chapter 12 Pectic Polysaccharides 181
Alistair J. MacDougall and Stephen G. Ring

Chapter 13 Fructans: Occurrence and Application in Food 197
Werner Praznik, Ewa Cieślik, and Anton Huber

Chapter 14 Structure–Property Relationships in Chitosan 217
Aslak Einbu and Kjell M. Vårum

Chapter 15 Plant and Algal Gums and Mucilages 231
Lawrence Ramsden

Chapter 16 Carbohydrates of Animal Tissues 255
Tadeusz Kołczak

Chapter 17 Cyclodextrins 271
József Szejtli

Chapter 18 Chemistry of the Maillard Reaction in Foods 291
Tomas Davidek and Jiri Davidek

Chapter 19 Nonnutritional Applications of Saccharides
and Polysaccharides 315
Piotr Tomasik

Chapter 20 Carbohydrates: Nutritional Value and Health Problems 325
Przemysław Jan Tomasik

Chapter 21 Glucose — Our Lasting Source of Energy 335
Heinz Th. K. Ruck

Chapter 22 Analysis of Molecular Characteristics of Starch
Polysaccharides 349
Anton Huber and Werner Praznik

Chapter 23 Spectroscopy of Polysaccharides 371
Jeroen J.G. van Soest

Chapter 24 Natural and Synthetic Nonsaccharide Sweeteners 387
Michal Uher

Index 405

1 Saccharides and Polysaccharides: An Introduction

Piotr Tomasik

CONTENTS

1.1 Occurrence and Significance 1
1.2 Carbohydrates as a Class of Compounds 2
1.3 Further Remarks on the Structure 3
1.4 Food Carbohydrates 5
References.. 6

1.1 OCCURRENCE AND SIGNIFICANCE

Carbohydrates, one of the most common groups of natural products, are construction materials for plant (cellulose, hemicelluloses, pectins) and animal tissues. They are the sources of energy and storage material in flora (starch, inulin) and fauna (glycogen).

Several specific carbohydrates are utilized in nature, for instance, as water-maintaining hydrocolloids, sex attractants (pheromones), and others formed by photosynthesis and associated processes. Some of them are subsequently utilized by plants, which transform them into other products; for instance, some plants may turn D-galactose into ascorbic acid, or some carbohydrates reside in plants as final products of photosynthesis and metabolism. Several of them are interesting for their pharmacological properties.

The list of known natural carbohydrates is continuously growing, owing to new discoveries in animal and, particularly, plant material. Recently, two novel polysaccharides have been isolated from deep-sea hydrothermal microorganisms.[1,2] Their unique thickening and gelling properties make them suitable for future commercialization.

The number of synthetic and semisynthetic carbohydrates is also growing. Several interesting features of the biological activity and functional properties of natural carbohydrates are the driving force for designing products of novel structure with improved properties. For example, syntheses of carbohydrate derivatives with heparin-like properties are being continuously developed. There are also a number of

0-8493-1486-0/04/$0.00+$1.50

natural, modified carbohydrates, such as aminated (chitin, chitosan), esterified with inorganic acids such as sulfuric acid (heparin, carrageenans) and phosphoric acid (potato amylopectin), oxidized (pectins, hyaluronic acid), alkylated (glycosides), and reduced (deoxy sugars such as fucose, deoxyribose).

Several natural substances composed of carbohydrate and protein moieties (glycoproteins) and lipid (glycolipids) constitute a group of so-called gangliosides. Monomeric carbohydrates combine to form oligomers and polymers. Carbohydrates are also called *saccharides* (a term originally common for sugars), because the first-isolated compounds of this class were sweet. Research in this area has shown that only selected saccharides are sweet, and a number of compounds have a sweet taste but their structures do not even resemble those of saccharides.

1.2 CARBOHYDRATES AS A CLASS OF COMPOUNDS

Carbohydrates as a class of compounds received their name based on results of elemental analysis of first-characterized compounds of this class. The analysis showed that formally they are a combination of n carbon atoms and n molecules of water always in a 1:1 proportion. As per this criterion, metanal (formaldehyde) can be considered the first member of the series. It follows that subsequent members of the series must be either hydroxyaldehydes [glyoxal ($CHO–CH_2OH$), glyceric aldehyde ($CHO–CHOH–CH_2OH$) and so on], or hydroxyketones [hydroxyacetone ($CH_2OH–CO–CH_2OH$) and so on]. To include hydroxyaldehydes and hydroxyketones among saccharides, an additional condition — the presence of at least one asymmetric carbon atom — has to be met.. Thus, the series of saccharides beginning with hydroxyaldehydes (*oses*) is that containing a three-carbon glyceric aldehyde called, therefore, triose, or more precisely, aldotriose. The series of saccharides beginning with hydroxyketones (*uloses*) is 1,3,4-trihydroxy-2-butanon (erythrulose), a tetrulose generally called a ketotetrose. Because of the asymmetric center localized at the C-2 atom (starred in the structures shown), two possible configurations of the hydrogen atom and hydroxyl group at this atom can be distinguished in a glyceric aldehyde [(1.1) and (1.2), Figure 1.1].

Chirality is a natural consequence of asymmetry although, for various reasons, not all compounds with asymmetric centers are chiral. Thus a glyceric aldehyde has two optically active (chiral) isomers called enantiomers, each a mirror image of the

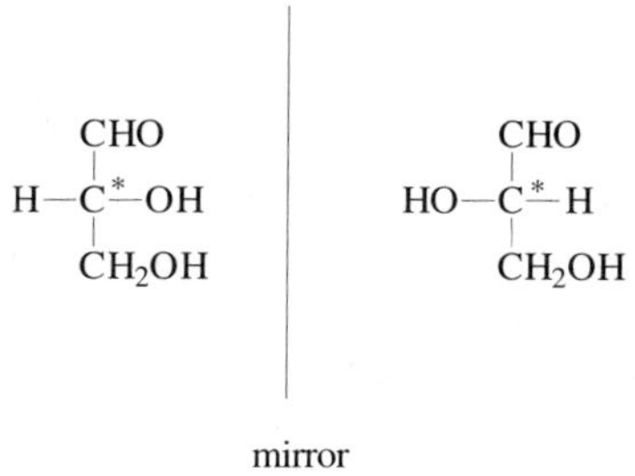

$$
\begin{array}{c|c}
\begin{array}{c}
\text{CHO} \\
| \\
\text{H}-\overset{*}{\text{C}}-\text{OH} \\
| \\
\text{CH}_2\text{OH}
\end{array}
&
\begin{array}{c}
\text{CHO} \\
| \\
\text{HO}-\overset{*}{\text{C}}-\text{H} \\
| \\
\text{CH}_2\text{OH}
\end{array}
\end{array}
$$

mirror

FIGURE 1.1 Enantiomeric D- and L-glyceric aldehydes.

other. The aqueous solution of one enantiomer twists the plane of passing polarized light to the left and that of the other twists it to the right. Because of the environment of the H–2C–OH moiety (a rotaphore), the angle of the twist in both directions is identical. The angle of the twist in other chiral carbohydrates, including those with higher number of rotaphores, results from the additivity of the chiral effects specific for every rotaphore present in a given molecule.

In his fundamental studies, Emil Fischer assigned the direction of optical rotation in (1.1) and (1.2) to the structure of its rotaphore. Thus, (1.1) appeared to be a right-twisting compound [(D) from *dextra*] and (1.2) a left-twisting compound [(L) from *levo*]. The following principle linked the steric configuration of (1.1) and (1.2) to experimental chirality. The one that takes position at C-2 of a given saccharide looking straight toward the H-atom of that rotaphore has to turn right in order to see the most oxidized group of the molecule (the aldehyde or keto group) and belongs to the D-family of saccharides. Because this condition is met in (1.1), it is the mother compound of the D-family, and for the same reason (1.2), the L-molecule, opens the L-family of saccharides.

Subsequent members of the families of oses are formally formed by inserting one, two, or more H–C–OH or HO–C–H moieties between the aldehyde group and the before last H–C*–OH group in (1.1) and the before last HO–C*–H group in (1.2) and subsequently formed family members. Structures of two tetroses, four pentoses, eight hexoses, and so on are designed in such a manner in each group. Structures (1.1) and (1.2) are mirror images, because of which they are called enantiomers. It is easy to check that designed structures of higher oses within a given D- or L-family cannot be enantiomers with respect to one another. Because they are isomeric but not enantiomeric, they are named diastereoisomers. It is also easy to check that enantiomers of particular members of the D-family can be found among members of the L-family. Such cases also exist within the families of D- and L-uloses. The operations start from inserting the H–C–OH or HO–C–H fragments, or both, between the C=O and before last H–C*–OH and HO–C*–H groups of D- and L-tetruloses, respectively.

The Fischer link of absolute configuration to direction of the experimental chirality is valid only for trioses and tetruloses. Thus, for (1.1) and (1.2) one may write D(+)-glyceric aldehyde and L(−)-glyceric aldehyde, where the notations (+) and (−) are related to experimental chirality of right and left, respectively. Saccharides being D(−) and L(+) are common.

1.3 FURTHER REMARKS ON THE STRUCTURE

One should note that chain structures designed in the manner presented above reflect only to a limited extent the real structure of these saccharides. Open-chain structures are the only structures of trioses and tetroses, logically, tetruloses. Pentoses, hexoses, pentuloses, and hexuloses take cyclic structures. Such cyclization is common in chemistry because it is energetically beneficial. The energy of five- and six-membered cyclic saccharides is lower than that of corresponding open-chain isomers. Cyclization in open-chain structures involves the electron gap at the carbonyl carbon atom and a lone electron pair orbital of the oxygen atom of this hydroxyl group, nucleophilic attack of which on the carbonyl carbon atom provides ring closure. Such cyclization is reversible, and equilibria specific for a given carbohydrate are established between two molecules of cyclic structures and one open-chain structure. Positions of such equilibria are specific for each saccharide and are dependent on concentration and temperature. Figure 1.2 presents such equilibria, so-called mutarotation, for a pentulose (1.4) and a hexose (1.7). A tendency to the six-membered ring closure is common among higher saccharides such as heptoses, octoses, nanoses, and related uloses. Saccharides with five-membered rings are called furanoses (*f*) and those with six-membered rings are called pyranoses (*p*).

For D-glucose at 20°C, the concentration of the open structure (1.7) reaches 0.003%. Thus, open-chain structures of saccharides are of minor importance. Despite this, open-chain structures are commonly presented in many textbooks and mono-graphs. Simple rules of conversion of open-chain structures into cyclic structures are shown in Figure 1.3. Resulting cyclic structures are, in fact, hemiacetals and hemiketals of oses and uloses, respectively. As usual, the C-1 atom (formerly the carbonyl carbon atom, now called the anomeric carbon atom) behaves as the hemia-cetal (hemiketal) carbon atom, that is, it has an electron gap available for an intra- or intermolecular nucleophilic attack. Such attack may produce substitution products with the introduced substituent oriented in two opposite directions. Thus, two iso-mers can be formed.

Based on the orientation of the substituent at the anomeric carbon atom, two families of saccharides can be distinguished: the family with orientation of the substituent such as in (1.3) and (1.6), and the family with orientation of the substit-uent as in (1.5) and (1.8). The orientation of the substituents at the C-1 atom affects several properties of saccharides and extends to polysaccharides also (compare properties of starch and cellulose, both polymers of D-glucose). Every hydroxyl group in the monomeric saccharide can act as the nucleophile attacking the anomeric carbon atoms of another saccharide. In this manner dimerization, oligomerization,

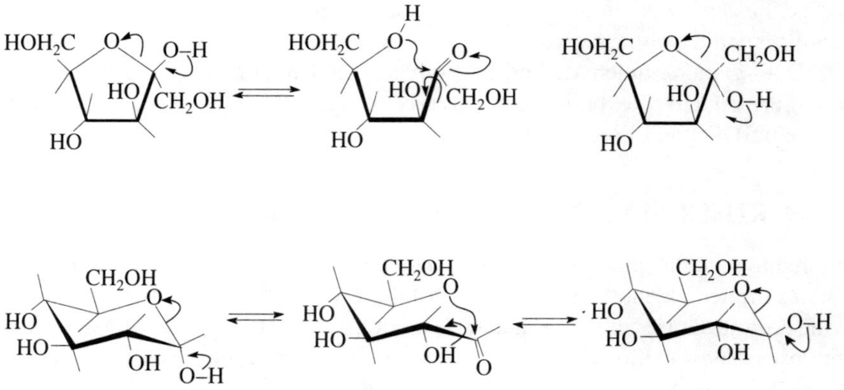

FIGURE 1.2 Mutarotation in pentuloses and hexoses.

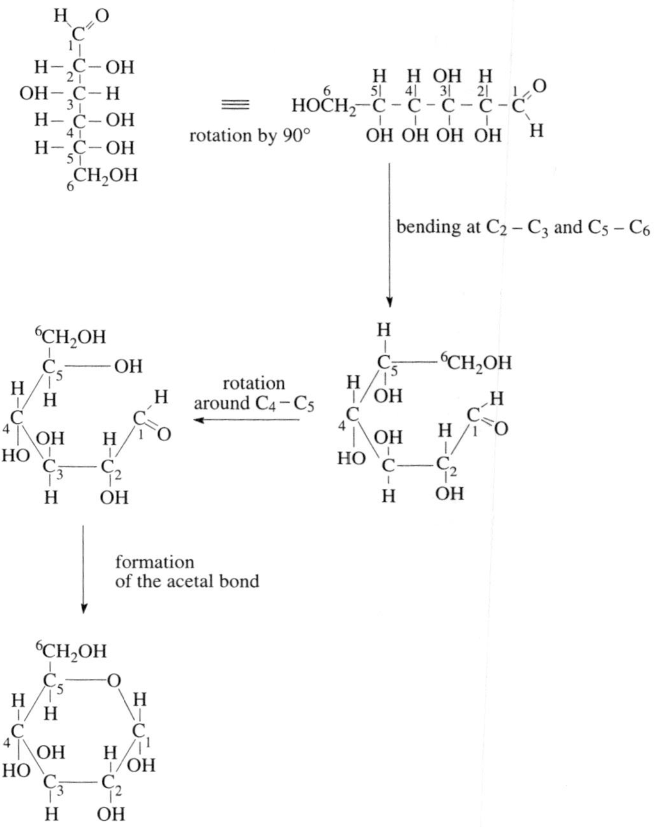

FIGURE 1.3 Conversion of open-chain structures into cyclic structures.

and polymerization occur. In resulting products, particular monomeric residues are bound through the properties of the glycosidic bond, which resemble those of hemiacetal bonds.

1.4 FOOD CARBOHYDRATES

Among the abundant number of carbohydrates, only approximately 100 were either identified in food as components of several natural foodstuffs (ingredients) or as additives. Figure 1.4 gives the structures of 109 saccharides, oligosaccharides, and polysaccharides present in food or essential in food technology as additives.

REFERENCES

1. Rougeaux, H. et al., Novel bacterial exopolysaccharides from deep-sea hydrothermal vents, *Carbohydr. Polym.* 31, 237, 1996.
2. Raguenes, G. et al., Description of a new polymer-secreting bacterium from a deep-sea hydrothermal vent, *Alteromonas macleodii* subsp. *fijiensis*, and preliminary characterization of the polymer, *Appl. Environ. Microbiol.,* 62, 67, 1996.

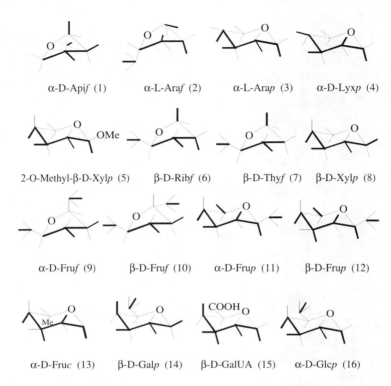

α-D-Api*f* (1)   α-L-Ara*f* (2)   α-L-Ara*p* (3)   α-D-Lyx*p* (4)

2-O-Methyl-β-D-Xyl*p* (5)   β-D-Rib*f* (6)   β-D-Thy*f* (7)   β-D-Xyl*p* (8)

α-D-Fru*f* (9)   β-D-Fru*f* (10)   α-D-Fru*p* (11)   β-D-Fru*p* (12)

α-D-Fru*c* (13)   β-D-Gal*p* (14)   β-D-GalUA (15)   α-D-Glc*p* (16)

FIGURE 1.4A   Structures of food carbohydrates. Bold valences directed outside the ring should be understood as –OH groups; *f* and *p*: furanose and pyranose rings, respectively.

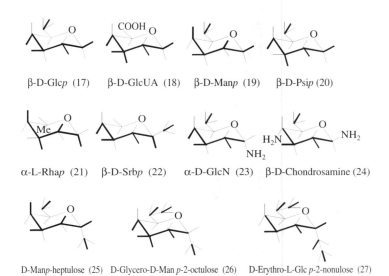

β-D-Glc*p* (17) β-D-GlcUA (18) β-D-Man*p* (19) β-D-Psi*p* (20)

α-L-Rha*p* (21) β-D-Srb*p* (22) α-D-GlcN (23) β-D-Chondrosamine (24)

D-Man*p*-heptulose (25) D-Glycero-D-Man *p*-2-octulose (26) D-Erythro-L-Glc *p*-2-nonulose (27)

FIGURE 1.4B Structures of food carbohydrates. Bold valences directed outside the ring should be understood as –OH groups; *f* and *p*: furanose and pyranose rings, respectively.

Disaccharides (28-43: Oses; 44 and 45: Osuloses)

β-D-Glcp-(1→4)-α-D-Glcp	β-D-Glcp-(1→6)-α-D-Glcp	α-D-Glcp-(1→6)-α-D-Glcp
Cellobiose (28)	Gentobiose (29)	Isomaltose (30)
α-D-Glcp-(1→2)-α-D-Glcp	β-D-Galp-(1→4)-α-D-Glcp	β-D-Glcp-(1→3)-α-D-Glcp
Kojibiose (31)	Lactose (32)	Laminaribiose (33)
α-D-Glcp-(1→4)-α-D-Glcp	α-D-Galp-(1→6)-α-D-Glcp	α-L-Rhap-(1→2)-α-D-Glcp
Maltose (34)	Mellibiose (35)	Neohesperidose (36)
α-D-Glcp-(1→1)-β-D-Glcp	α-D-Glcp-(1→3)-α-D-Glcp	α-D-Glcp-(1→6)-β-D-Fruf
Neotrehalose(37)	Nigerose (38)	Palatinose (39)
α-L-Rhap-(1→6)-α-D-Glcp	β-D-Fruf-(2→1)-α-D-Glcp	α-D-Glcp-(1→1)-α-D-Glcp
Rutinose (40)	Sucrose (41)	Trehalose (42)
α-D-Glcp-(1→3)-β-D-Fruf	β-D-Galp-(1→4)-β-D-Fruf	α-D-Glcp-(1→4)-β-D-Fruf
Turanose (43)	Lactulose (44)	Maltulose (45)

Trisaccharides

α-D-Glcp-(1→4)-α-D-Glcp-(1→2)-β-D-Fruf	α-D-Fucp-(1→2)-β-D-Galp-(1→4)-β-D-Galp
Erlose (46)	Fucosidolactose (47)
β-D-Glcp-(1→6)-α-D-Glcp-(1→2)-β-D-Fruf	α-D-Glcp-(1→4)-α-D-Glcp-(1→2)-α-D-Glcp
Gentianose (48)	Gentose (49)
O-α-D-Glcp-(1→2)-O-α-D-Fruf-(1→2)-β-D-Fruf	O-α-D-Glcp-(1→6)-O-α-D-Glcp-(1→3)-α-D-Glcp
Isokestose (1-Kestose) (50)	3-α-Isomaltosylglucose (51)
O-αD-Glcp-(1→6)-O-α-D-Glcp-(1→6)-α-D-Glcp	O-α-D-Glcp-(1→4)-O-α-D-Glcp-(1→6)-α-D-Glcp
Isomaltotriose (52)	Isopanose (53)
O-α-D-Glcp-(1→2)-O-β-D-Fruf-(6→2)-β-D-Fruf	O-α-D-Glcp-(1→6)-O-α-D-Glcp-(1→6)-α-D-Galp
Kestose (6-Kestose) (54)	Lycotriose (55)
O-α-D-Glcp-(1→4)-O-α-D-Glcp-(1→4)-α-D-Glcp	O-α-D-Galp-(1→6)-O-α-D-Galp-(1→6)-α-D-Galp
Maltotriose (56)	Manninotriose (57)
O-αD-Glcp-(1→3)-O-β-D-Fruf-(2→1)-α-D-Glcp	O-β-D-Fruf-(2→6)-O-α-D-Glcp-(1→2)-β-D-Fruf
Melesitose (58)	Neokestose (59)
O-α-D-Glcp-(1→6)-O-α-D-Glcp-(1→4)-α-D-Glcp	O-α-D-Galp-(1→6)-O-α-D-Glcp-(1→2)-β-D-Fruf
Panose (60)	Raffinose (61)
O-β-D-Glcp-(1→3)-O-β-D-Galp-(2→1)-O-α-L-Rhap	O-α-D-Glcp-(1→6)-α-D-Glcp-(1→2)-β-D-Fruf
Solatriose (62)	Theanderose (63)

O-α-D-Galp-(1→2)-O-α-D-Glcp-(1→2)-β-D-Fruf

Umbelliferose (64)

FIGURE 1.4 C,D Structures of food carbohydrates. *f* and *p*: furanose and pyranose rings, respectively.

Tetrasaccharides

[*O*-α-D-Glc*p*-(1→6)]₄   [*O*-α–D-Glc*p*-(1→4)]₄

Isomaltotetrose (65)   Maltotetrose (66)

[*O*-α-D-Gal*p*-(1→6)]₂ -*O*-α-D-Glc*p*-(1→2)-β-D-Fru*f*

Stachyose (67)

Pentasaccharides

[*O*-α-D-Glc*p*-(1→6)]₅   [*O*-α-D-Glc*p*-(1→4)]₅

Isomaltopentaose (68)   Maltopentaose (69)

Hexa– to Octa-saccharides

[-*O*-αD-Glc*p*-(1→4)]ₙ

Cyclodextrins: α: n = 6; β: n = 7; γ: n = 8   (70-72)

FIGURE 1.4E Structures of food carbohydrates. *f* and *p*: furanose and pyranose rings, respectively.

Polysaccharides

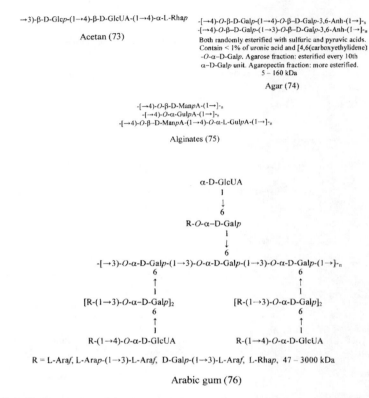

→3)-β-D-Glc*p*-(1→4)-β-D-GlcUA-(1→4)-α-L-Rha*p*

Acetan (73)

-[→4)-*O*-β-D-Gal*p*-(1→4)-*O*-β-D-Gal*p*-3,6-Anh-(1→]-ₙ
-[→4)-*O*-β-D-Gal*p*-(1→3)-*O*-β-D-Gal*p*-3,6-Anh-(1→]-ₙ
Both randomly esterified with sulfuric and pyruvic acids.
Contain < 1% of uronic acid and [4,6(carboxyethylidene)
-*O*-α-D-Gal*p*. Agarose fraction: esterified every 10th
α–D-Gal*p* unit. Agaropectin fraction: more esterified.
5 – 160 kDa

Agar (74)

-[→4)-*O*-β-D-Man*p*A-(1→]-ₙ
-[→4)-*O*-α-Gul*p*A-(1→]-ₙ
-[→4)-*O*-β-D-Man*p*A-(1→4)-*O*-α-L-Gul*p*A-(1→]-ₙ

Alginates (75)

α-D-GlcUA
1
↓
6
R-*O*-α-D-Gal*p*
1
↓
6
-[→3)-*O*-α-D-Gal*p*-(1→3)-*O*-α-D-Gal*p*-(1→3)-*O*-α-D-Gal*p*-(1→]-ₙ

6   6
↑   ↑
1   1
[R-(1→3)-*O*-α-D-Gal*p*]₂   [R-(1→3)-*O*-α-D-Gal*p*]₂
6   6
↑   ↑
1   1
R-(1→4)-*O*-α-D-GlcUA   R-(1→4)-*O*-α-D-GlcUA

R = L-Ara*f*, L-Ara*p*-(1→3)-L-Ara*f*, D-Gal*p*-(1→3)-L-Ara*f*, L-Rha*p*, 47 – 3000 kDa

Arabic gum (76)

FIGURE 1.4F Structures of food carbohydrates. *f* and *p*: furanose and pyranose rings, respectively.

Carrageenans:

-[→3)-*O*-β-D-Gal*p*-4-OSO₃H-(1→4)-*O*-β-D-Gal*p*-3,6-Anh-2-OSO₃H- (1→]-ₙ

ι-Carrageenan (77)

-[→3)-*O*-β-D-Gal*p*-4-OSO₃H-(1→4)-*O*-β-D-Gal*p*-3,6-Anh-(1→]-ₙ

κ-Carrageenan (78)

-{→3)-*O*-β-D-Gal*p*-4-OSO₃H-(1→4)-*O*-β-D-Gal*p*-3,6-Anh-2,6-di-OSO₃H-(1→]-ₙ

λ-Carrageenan (79)

-[→3)-*O*-β-D-Gal*p*-4-OSO₃H-(1→4)-*O*-β-D-Gal*p*-6-OSO₃H-(1→4)-*O*-β-D-Gal*p*-3,6-Anh-(1→]-ₙ

μ-Carrageenan (80)

-[→3)-*O*-β-D-Gal*p*-4-OSO₃H-(1→4)-*O*-β-D-Gal*p*-2,6-di-OSO₃H-(1→4)-*O*-β-D-Gal*p*-3,6-Anh-(1→]-ₙ

ν-Carrageenan (81)

FIGURE 1.4G Structures of food carbohydrates. *f* and *p*: furanose and pyranose rings, respectively.

-[→4)-*O*-β-D-Glc*p*-(1→]-ₙ   -[→4)-*O*-β-D-Glc*p*-2-NHAc-(1→]-ₙ

Cellulose (82)   Chitin (83)

-[→4)-*O*-β-D-GlcUA-(1→4)-*O*-α-D-Glc*p*—2-NHAc-6-OSO₃⁻ - (1→]-ₙ

Chondroitin sulphates A, B, and C (84)

-[→6)-*O*-α-D-Glc*p*-(1→6)-*O*-α-D-Glc*p*-(1→]-ₙ
                      4(3)(2)
                        ↑
                        1
                O-α-D-Glc*p*

Dextran (85)

-[→4)-*O*-β-D-Gal*p*-(1→4)-*O*-β-D-Gal*p*-2-SO₃⁻ -(1→4)-*O*-β-D-Gal*p*-4-SO₃⁻ ——┐
┌───┘
(1→4)-*O*-β-D-Gal*p*-6-SO₃⁻ -(1→4)-*O*-β-D-Gal*p*-3,6-Anh-(1→]-ₙ   20 – 80 kDa

Furcellaran (86)

FIGURE 1.4H Structures of food carbohydrates. *f* and *p*: furanose and pyranose rings, respectively.

$$R$$
$$|$$
$$\uparrow$$
$$3$$

$$R$$
$$|$$
$$\uparrow$$
$$6$$

$$R$$
$$|$$
$$\uparrow$$
$$6$$

-[→4)-*O*-β-D-GlcUA-(1→6)-*O*-β-D-Gal*p*-(1→]-ₙ

-[→4)-*O*-α-D-GlcUA-(1→2)-*O*-β-D-Man*p*-(1→2)-*O*-β-D-Man*p*-(1→]-ₙ

$$3$$
$$\downarrow$$
$$1$$
$$R$$

$$R$$
$$|$$
$$\uparrow$$
$$4$$

-[→6)-*O*-α-D-Gal*p*-(1→6)-*O*-α-D-Gal*p*-(1→3)-*O*-α-L-Ara*f*-(1→]-ₙ

$$3$$
$$\downarrow$$
$$1$$
$$R$$

R = L-Ara*f*, L-Ara*p*-(1→3)-*O*-α-Ara*f*, L-Rha*p*

Gatti gum (87)

-[→3)-*O*-deoxy-β-D-Glc*p*-(1→4)-*O*-β-D-GlcUA-(1→4)-*O*-β-D-Glc*p*-(1→4)-*O*-α-L-Rha*p*-(1→]-ₙ

Gellan (88)

-[→4)-*O*-α-D-Glc*p*-(1→]-ₙ

Glycogen (89)

-[→4)-*O*-β-D-Man*p*-(1→4)-*O*-β-D-Man*p*-(1→4)-*O*-β-D-Man*p*-(1→]-ₙ

$$6$$
$$\downarrow$$
$$1$$
α-D-Gal*p*

Guar gum (90)

FIGURE 1.41 Structures of food carbohydrates. *f* and *p*: furanose and pyranose rings, respectively.

Hemicelluloses:

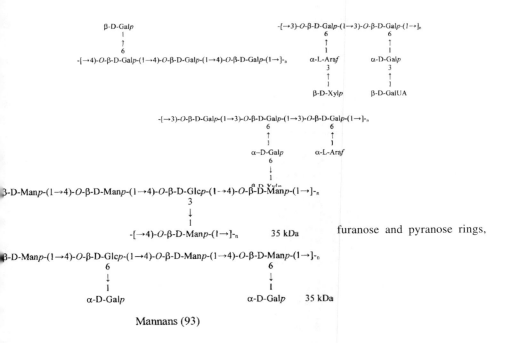

Arabinogalactan (91)

Mannans (93)

FIGURE 1.4K Structures of food carbohydrates. *f* and *p*: furanose and pyranose rings, respectively.

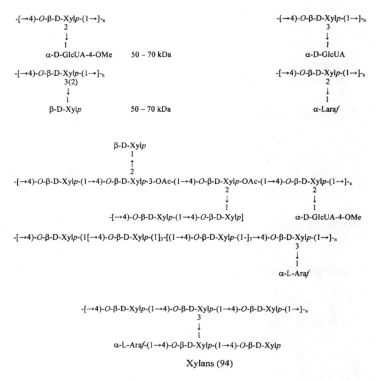

-[→4)-*O*-β-D-Xyl*p*-(1→]-ₙ   -[→4)-*O*-β-D-Xyl*p*-(1→]-ₙ
     2   3
     ↓   ↓
     1   1
α-D-GlcUA-4-OMe   50 – 70 kDa   α-D-GlcUA

-[→4)-*O*-β-D-Xyl*p*-(1→]-ₙ   -[→4)-*O*-β-D-Xyl*p*-(1→]-ₙ
   3(2)   2
    ↓   ↓
    1   1
β-D-Xyl*p*   50 – 70 kDa   α-L-Ara*f*

β-D-Xyl*p*
1
↑
2
-[→4)-*O*-β-D-Xyl*p*-(1→4)-*O*-β-D-Xyl*p*-3-OAc-(1→4)-*O*-β-D-Xyl*p*-OAc-(1→4)-*O*-β-D-Xyl*p*-(1→]-ₙ
                2   2
                ↓   ↓
                1   1
-[→4)-*O*-β-D-Xyl*p*-(1→4)-*O*-β-D-Xyl*p*]   α-D-GlcUA-4-OMe

-[→4)-*O*-β-D-Xyl*p*-(1[→4)-*O*-β-D-Xyl*p*-(1]₃-[(1→4)-*O*-β-D-Xyl*p*-(1-]₇→4)-*O*-β-D-Xyl*p*-(1→]-ₙ
                                       3
                                       ↓
                                       1
                                  α-L-Ara*f*

-[→4)-*O*-β-D-Xyl*p*-(1→4)-*O*-β-D-Xyl*p*-(1→4)-*O*-β-D-Xyl*p*-(1→]-ₙ
                          3
                          ↓
                          1
α-L-Ara*f*-(1→4)-*O*-β-D-Xyl*p*-(1→4)-*O*-β-D-Xyl*p*

Xylans (94)

FIGURE 1.4L Structures of food carbohydrates. *f* and *p*: furanose and pyranose rings, respectively.

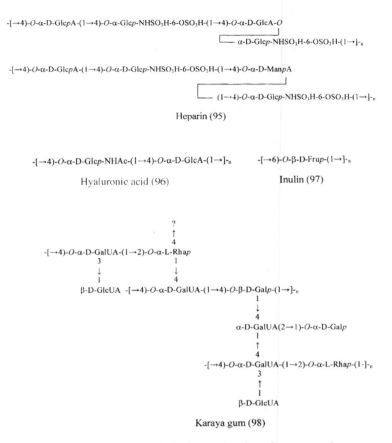

-[→4)-O-α-D-GlcpA-(1→4)-O-α-D-Glcp-NHSO₃H-6-OSO₃H-(1→4)-O-α-D-GlcA-O
└── α-D-Glcp-NHSO₃H-6-OSO₃H-(1→]-ₙ

-[→4)-O-α-D-GlcpA-(1→4)-O-α-D-Glcp-NHSO₃H-6-OSO₃H-(1→4)-O-α-D-ManpA
└── (1→4)-O-α-D-Glcp-NHSO₃H-6-OSO₃H-(1→]-ₙ

Heparin (95)

-[→4)-O-α-D-Glcp-NHAc-(1→4)-O-α-D-GlcA-(1→]-ₙ   -[→6)-O-β-D-Frup-(1→]-ₙ

Hyaluronic acid (96)   Inulin (97)

?
↑
4
-[→4)-O-α-D-GalUA-(1→2)-O-α-L-Rhap
3   1
↓   ↓
1   4
β-D-GlcUA -[→4)-O-α-D-GalUA-(1→4)-O-β-D-Galp-(1→]-ₙ
1
↓
4
α-D-GalUA(2→1)-O-α-D-Galp
1
↑
4
-[→4)-O-α-D-GalUA-(1→2)-O-α-L-Rhap-(1-]-ₙ
3
↑
1
β-D-GlcUA

Karaya gum (98)

FIGURE 1.4M Structures of food carbohydrates. *f* and *p*: furanose and pyranose rings, respectively.

-[→2)-O-β-D-Fru$f$-(1→]-$_n$ -{→4)-[O-β-D-Man$p$-(1→4)]$_3$-O-β-D-Man$p$-(1→}-$_n$
                                                                                6
        Levan (99)   ↓
                                                                                α-D-Gal$p$

Locust beam gum (100)

-[→4)-O-αD-GalUA-(1→]-$_n$

The carboxyl groups are in part
methylated and in pierce. sunflower.
beet. and potato pectins the hydroxyl
groups are partly acetylated. 20 – 150 kDa

Pectins (101)

-[→4)-O-α-D-GalUA-(1→2)-O-α-L-Rha$p$-(1→4)-[O-α-D-GalUA]$_n$
                                                      └─────(1→4)-O-α-L-Rha$p$-(1→]-$_n$
                                                                              2
                                                                              ↓
                                                                              1
                                                                        α-G-Gal$p$

Protopectin (102)

–[→4)-α-D-Glc$p$-(1→4)]$_2$-O-α-D-Glc$p$
                                          1
                                          ↓
                                          6
D-Glc$p$-(1→[→4-O-α-D-Glc$p$-(1→]$_m$→4)-0-α-D-Glc$p$
    6
    ↑
    1
α-D-Glc$p$-(1→]-$_n$ m = 1 –2

Pullulan (103)

FIGURE 1.4N Structures of food carbohydrates. f and p: furanose and pyranose rings, respectively.

Starch:

-[→4)-O-α-D-Glc$p$-(1→]-$_n$
  10-50 kDa

Amylose (104)

-{→4)-[-O-α-D-Glc$p$-(1→]$_n$→4)-O-α-D-Glc$p$-(1→4)-O-α-D-Glc$p$-(1→}-
                                                    6
                                                    ↑
                                                    1
                          [α-D-Glc$p$-(4→]$_{15-30}$ 100 – 200 kDa

In potato amylopectin every 12 – 200th glucose unit is esterified with phosphoric acid

Amylopectin (105)

-[→4)-O-β-D-Glc$p$-(1→4)-O-β-D-Glc$p$-(1→4)-O-β-D-Glc$p$-(1→4)-O-β-D-Glc$p$-(1→]-$_n$
    6   6   6
    ↑   ↑   ↑
    1   1   1
  β–D-Xyl$p$ β–D-Xyl$p$ α-L-Ara$f$
                               2
                               ↓
                               1
                             β-D-Gal$p$

Tamarind flour (106)

Tragacant gum (107 and 108):

α-L--Ara$f$ + αD-Gal$p$ + α-D-GalUA + α--Rha$p$

Bassarin (107)

FIGURE 1.4O Structures of food carbohydrates. f and p: furanose and pyranose rings, respectively.

-[→4)-O-α-D-GalUA-(1→4)-O-α-D-GalUA-(1→4)-O-α-D-GalUA-(1→4)-O-α-D-GalUA-(1→]-$_n$
    3   3   3
    ↑   ↑   ↑
    1   1   1
  β-D-Xyl$p$ β-D-Xyl$p$ β-D-Xyl$p$
                              2   2
                              ↑   ↑
                              1   1
                            α-D-Fuc$p$ α-D-Gal$p$

Tragacantin (108)

-[→4)-O-β-D-Glc$p$-(1→4)-O-β-Galcp-(1→]-$_n$
                           3
                           ↑
                           1
       α-D-ManUA(2→1)-O-α-D-Glc$p$-(4→1)-O-β-D-Man$p$-4-OAc

Xanthan gum (109)

FIGURE 1.4P Structures of food carbohydrates. f and p: furanose and pyranose rings, respectively.

2 Structure and Reactivity of Saccharides

Wiesław Szeja

CONTENTS

2.1 Stereochemistry of Monosaccharides 19
2.2 The Complex Formation of Carbohydrates with Cations 21
2.3 Chemical Reactions of Saccharides 22
    2.3.1 Reduction 22
    2.3.2 Oxidation.. 23
    2.3.3 Alkylation 25
    2.3.4 Acylation 27
    2.3.5 Glycosylation 28
    2.3.6 Biological Significance of the Efficiency of O-Glycosylation 29
    2.3.7 Reactions with Amino Compounds 31
References 32

2.1 STEREOCHEMISTRY OF MONOSACCHARIDES

The physical, biological, and chemical properties of organic compounds depend, among other things, on the nature and steric arrangement of the chemical bonds. Stereoisomeric hexoses, which differ in configuration of their molecules, are examples of a relationship between three-dimensional structure and biological activity of compounds. The free rotation around a single bond produces a different three-dimensional arrangement of the atoms in space; that is, different molecule conformations. Conformations rapidly convert into one another and are therefore nonseparable. In a vast majority of compounds with an alicyclic six-membered ring, the molecules, in most cases, take the chair conformation. One bond on each carbon atom is directed perpendicularly to the plane of the ring (axially) and the other resides in the plane of the ring (the equatorial bond), denoted in (2.1) as *a* and *e,* respectively.

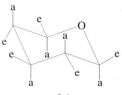

2.1

Substituted cyclohexanes with equatorial substituents, and also corresponding the six-membered rings with heteroatoms, are usually more stable. Conformational analysis of the pyranose sugars accepts identical geometry for the pyranose and cyclohexane rings. Thus, the substituted pyranose ring may take one of two chair conformations denoted as $^4C_1$ (2.2) or $^1C_4$ (2.3). The superscript corresponds to the number of the carbon atom in the upper position of the chair and the subscript to that in the lower position. In an aqueous solution these conformations reside in equilibrium, but usually one of conformers preponderates. The preponderance of either one or the other chair conformation can be estimated by conformational analysis.

2.2   **2.3**

In monosaccharides, being pyranoses, substituents usually reside in the equatorial position. These preferences can be rationalized in terms of the energy of interactions between the axial and equatorial substituents (Table 2.1). Apart from the nonbonding interactions, the so-called anomeric effect has to be taken into account. Thus, a polar group at the anomeric carbon atom favors the axial position. The magnitude of the anomeric effect depends on the configuration of the hydroxyl group on C-2. In an aqueous solution, the energy associated with the anomeric effect of the hydroxyl group is estimated at 2.30 and 4.18 kJ mole^{-1} for the equatorial and axial group, respectively. In organic solvents of lower dielectric constant, an increase in the energy of the anomeric effect has been observed. Calculated energies fit experimental values.

The data in Table 2.1 show that a conformation of the aldohexopyranoses in aqueous solution depends to a great extent on the orientation of the hydroxymethyl group. Stability of the conformation is reduced by the bulky axial CH$_2$OH group, and most of the β-D-anomers take predominantly the $^4C_1$ form, because the $^1C_4$ form involves a strong interaction (10.45 kJ mole^{-1}) between the axial hydroxymethyl group and the axial hydroxyl group.

Changes in free energy associated with the transformation of equatorial into the axial groups point to the priority for the equatorial alkyl groups at the anomeric carbon atom, but for polar groups (e.g., the hydroxyl group), the axial position is preferable. The same reason rationalizes a higher stability of methyl-α-D-glucoside over its β-isomer. Estimated free energy differences between equatorially and axially substituted pyranose rings help determine the proportions of stereoisomers. Mutarotation of D-glucose (Chapter 1) is an excellent example of the application of

TABLE 2.1
Free Energy of Nonbonding Interaction in the Pyranose Ring

Interaction	Energy (kJ mole^{-1})
Axial–Axial	
H–O	1.88
O–O	3.76
O–(CH$_3$, CH$_2$OH)	6.27
Equatorial–Equatorial	
O–O	10.45
O–(CH$_3$, CH$_2$OH)	1.46
Equatorial–Axial	
O–O	1.88
O–(CH$_3$, CH$_2$OH)	1.46

Note: In aqueous solution at room temperature.

conformational analysis to monosaccharides. The relative stability of the α-(2.4) and β-(2.5) anomers and the α to β ratio in the equilibrium can be predicted for this equilibrium process.

2.4 **2.5**

$_{conf}$ = 10.03 kJ/mol E_{conf} = 8.57 kJ/mol

$\alpha : \beta$ =36 : 64 (experimented and calculated)

2.2 THE COMPLEX FORMATION OF CARBOHYDRATES WITH CATIONS

The hydroxyl groups of carbohydrates can coordinate to metal cations. The ability to form complexes and the stability of the relevant complexes of neutral carbohydrates strongly depend on the conformational orientation of the neighboring hydroxyl groups. Several carbohydrates with the axial–equatorial–axial sequence of hydroxyl groups form stable complexes with the Na$^+$, Mg^{2+}, Zn^{2+}, Ba^{2+}, Sr^{2+}, and Ca^{2+} ions. Thus, α-D-gulopyranose forms a tridentate complex (2.6) with calcium chloride in solution (stability constant $K = 3.7$ M^{-1}) whereas β-D-gulopyranose (2.7) cannot form such complex.

2.6   2.7

The complex formation of some methyl glycosides with the Na^+, Ca^{2+}, and La^{3+} ions has been demonstrated but when the anomeric hydroxyl group was replaced by the methoxyl group, the formed complexes were weaker. The electrophenic mobility of a number of carbohydrates coordinated to the metal cations revealed that pyranoses with *cis-* and *trans-*oriented two consecutive hydroxyl groups provided weak complexes.

Oligo- and polysaccharides can also form complexes with cations. The solubility is lower at low salt concentration and rises as the salt concentration increases. Precipitation of saccharides in the form of complexes after admixture of salts, for instance, cupric acetate, is an important method of purifying sugars. The stability of the complexes depends on the ionic radius of the ions. The more stable complexes are formed with Na^+, Ca^{2+}, and La^{3+} cations (ionic radius 10 to11 Å). Smaller cations such as Li^+, Mg^{2+}, and Y^{3+} produce weak complexes.

The aldonic and alduronic acids, and also a number of α-hydroxy acids, form much stronger complexes with cations than do neutral sugars. Tridentate complexes of gluconic acids coordinated through the O-1, O-2, and O-3 atoms have been frequently found (e.g., calcium lactobionate, D-arabinate).

The coordination to bivalent cations such as Mg^{2+} and Ca^{2+} has important practical applications. A complex formation is involved in the formation of gels from alginic acid (Chapter 15) and pectins (Chapter 12). The complexation of sucrose to Ca^{2+} ion is commonly used in sucrose production (see Chapter 6). A ferric complex of D-fructose is soluble in water and might appear suitable in delivery of metabolized form of iron in case of anemia.

2.3 CHEMICAL REACTIONS OF SACCHARIDES

2.3.1 REDUCTION

Reduction of the carbonyl group in monosaccharides produces cyclic polyhydric alcohols (alditols) (see also Chapter 3). The catalytic reduction of monosaccharides, mainly D-glucose, with hydrogen under pressure has been widely applied in industrial scale production of alditols. A nickel catalyst with various promoters is most frequently used. The reduction of monosaccharides with sodium borohydride or lithium aluminum hydride offers the most convenient means to synthesize alditols on a laboratory scale.

2.3.2 OXIDATION

Commonly, primary alcohols (2.8) may be converted to aldehydes (2.9) and secondary alcohols (2.10) to ketones (2.11).

$$RCH_2OH \rightarrow RCHO \qquad \text{and} \qquad RCH(OH)R^1 \rightarrow RCOR^1$$
$$\textbf{2.8} \qquad \textbf{2.9} \qquad\qquad\qquad \textbf{2.10} \qquad \textbf{2.11}$$

The hydroxyl group can be easily oxidized with a number of strong oxidizing agents, the most common being $KMnO_4$, Br_2, $K_2Cr_2O_7$, and $Pb(CH_3COO)_4$. Permanganate and dichromate oxidize the primary hydroxyl group to the carboxyl group rather than to the aldehyde group. Periodic acid, its salts, and lead tetracetate cleave 1,2-glycols (2.12) under mild conditions and with a good yield. Two molecules of either aldehyde (2.13a) ($R^1 = R^4 = H$) or ketone (2.13b) of the structure dependent on the substrate are the products of the oxidation.

$$R^1R^2C(OH)-C(OH)R^3R^4 \rightarrow R^1COR^2 + R^3COR^4$$
$$\textbf{2.12} \qquad\qquad\qquad \textbf{2.13 a, b}$$

The reaction serves to identify the *cis*-diol structure.

Results of applying any of these oxidants to aldoses and ketoses are formally the same, except when furanoses and pyranoses are oxidized either with periodates or lead tetracetate, in which case one molecule of dialdehyde is formed as only the C-2–C-3 bond in the ring is cleaved.

Halogen molecules and their oxy compounds are widely used to oxidize the aldehyde group to the carboxylic group. It is a way to prepare aldonic acids from aldoses and aldaric acids from glucuronic acids. Ketoses are usually resistant to bromine. Cyclic aldoses (2.14) in acid solution are oxidized directly to aldonic acids (2.15). In aqueous solutions, they reside in equilibrium with their lactones (2.16) and (2.17). Pyranoses and furanoses produce 1,5-(2.16) and 1,4-(2.17) lactones, respectively. The higher oxidation rate for β-D-glucopyranose is probably because

its equatorial anomeric hydroxyl group is sterically not hindered. Catalytic oxidation of aldoses with oxygen leads to corresponding aldonic acids. It readily proceeds over the platinum catalyst in an alkaline solution. Also, microbial oxidation of reducing sugars provides a high yield of aldonic acids (Chapter 10). Aldonic acids and lactones are stable toward acids. By treatment in alkaline solution, the configuration at C-2 of aldonic acids can be altered. These acids easily form crystalline salts, esters, amides, and hydrazides. These derivatives have been helpful to identify aldonic acids and sugars.

D-Gluconic acid is commercially important. A number of its salts exhibit biological activity. Ferrous D-gluconate is used to treat iron-deficiency anemia. Calcium deficiency is treated with calcium D-gluconate. Because of the sequestering ability in alkaline solutions, D-gluconic acid is added to various cleaning compositions. In the dairy industry, it prevents milk scale; similarly, in breweries, it prevents deposition of beer scale.

Oxidation of the primary hydroxyl group of aldoses to the carboxyl group results in the formation of uronic acids (aldehydocarboxylic acids), which, similar to aldoses, take two possible pyranose forms as shown for D-galacturonic (2.18) and D-mannuronic (2.19) acids. Depending on their configuration, uronic acids can also

form lactone rings. Prior to oxidation of the primary hydroxyl groups, the secondary hydroxyl groups should be protected from oxidation to the carbonyl group. The isopropylidene group appeared to be the most suitable. It is stable under neutral and alkaline conditions and after oxidation can be removed by acid hydrolysis. Thus, D-glucose (2.20) is reacted with acetone under proton catalysis. Protection involves the hydroxyl groups at C-1 and C-2 and transformation of pyranose to furanose. In consequence, 1,2-O-isopropylidene-α-D-glucofuranose (2.21) is formed, which is subsequently oxidized over platinum at C-6 with a high yield of acid (2.22), which is then hydrolyzed to D-glucuronic acid (2.23).

Uronic acids are common constituents of polysaccharides (Chapter 1 and Chapter 9). They play an essential role in developing texture of foodstuffs (Chapter 11). D-Glucuronic acid serves in veterinary medicine as a detoxifying agent for mammals.

Aldaric acids (called also saccharic acids) are hydroxydicarboxylic acids, HOOC(CHOH)4–COOH. Their systematic names are formed by replacing the suffix *ose* in the name of aldose with *aric acid*. These acids are produced from saccharides when strong oxidizing agents are applied. D-Glucaric acid (2.24) is available from D-glucose (2.20) either oxidized with concentrated nitric acid or catalytically oxidized over the platinum catalyst. The latter method additionally oxidizes D-gluconic acid into D-glucaric acid and is an alternative route of synthesis of aldaric acids from uronic acids. The two carboxyl groups in glycaric acids allow formation of numerous types of mono- (2.17) and di- (2.25) lactones.

2.3.3 ALKYLATION

Williamson reaction, the alkylation of alkoxides (2.26) with alkyl halides (2.27), is the most efficient general method of preparing unsymmetrical ethers (2.28).

$$R'O^- + RX \rightarrow ROR' + X^-$$
$$\mathbf{2.26} \quad \mathbf{2.27} \quad \mathbf{2.28}$$
$$X: Cl, Br, J, RSO_3, R^3Si$$

Reactive alkyl halides such as triphenyl methyl chloride $(C_6H_5)_3$ CCl (trityl chloride) may react directly with alcohols in the presence of tertiary organic amines. This is the most common method for preparing trityl saccharide ethers. The preparation of the methyl ethers is a key step in the gas chromatographic quantitative determination of sugars, especially in multicomponent mixtures resulting from the hydrolysis of polysaccharides. Mono- and oligosaccharides are nonvolatile and thermally unstable. Therefore, prior to the analysis, their conversion into volatile derivatives, usually to *O*-methyl ethers or *O*-trimethysilyl ethers, is indispensable. Methylation is useful in elucidation of the structure of oligo- and polysaccharides. Thus, after the methylation of trehalose and hydrolysis of *O*-permethylated sugar, tetra-*O*-methyl glucopyranose has to be separated and analyzed quantitatively.

Thus, a disaccharide (2.29) with two D-glucose units linked with one another with the 1,1-glycosidic bond is *O*-permethylated into ether (2.30), which is then split into a tetramethylated monosaccharide (2.31). Usually, methylation is per-

a) CH₃J, NaOH, (CH₃)₂SO
b) H₂O, H⊕

2.29   **2.30**   **2.31**

formed in a dry dimethyl sulfoxide solution with methyl iodide, in the presence of either sodium hydride (Hakamori method) or powdered sodium hydroxide (Ciucanu–Kerek procedure). The permethylation method has gained importance in the gas chromatographic/mass spectrometric structure elucidation of oligosaccharides with up to 11 monosaccharide units. Conversion of the polyhydroxy compounds into their more volatile *O*-trimethylsilyl ethers (TMS-ethers) allows their gas chromatographic separation. A simple method for such quantitative derivatization involves the reaction of saccharide (2.32) with either trimethylchlorisilane in dry pyridine or trimethylchlorosilane–hexamethyldisilazone mixture. Two products are formed: one is the product of pertrimethylsilylation (2.33) and the other results from monoetherifcation of the 1-hydroxyl group (2.34).

Other ethers such as carboxymethyl (R–O–CH₂COOH) and 2-(diethylamino)ethyl (R–O–CH₂CH₂NEt₂) ethers are important commercial products. *O*-(Carboxymethyl) cellulose (CH-cellulose), *O*-(2-diethylaminoethyl) cellulose (DEAE-Cellulose), and *O*-(2-diethylaminoethyl)-Sephadex (DEAE-Sephadex) are used to separate proteins and complex polysaccharides according to their molecular weight.

a) TMSCl, pyridine

2.32   **2.33**   **2.34**

Trityl ethers of sugars are important substrates in synthetic carbohydrate chemistry. Alkylation of the less sterically crowded hydroxyl group proceeds much more readily, providing a selective substitution of the primary hydroxyl groups. Pentoses that normally take the pyranose structure may react as furanoses having a more reactive primary hydroxyl group.

2.3.4 ACYLATION

Under laboratory conditions and in industry, esters (2.37) are available from acids (2.35) and alcohols (2.36) refluxed in the presence of an acid catalyst. To increase

$$R^1\text{-COOH} + R^2OH \rightleftharpoons R^1COOR^2 + H_2O$$

2.35   **2.36**   **2.37**

the yield of ester in this reversible reaction, either ester or water should be continuously evacuated from the reaction mixture. In the presence of water, the hydrolysis of esters predominates. Because carbohydrates are unstable in the acidic medium at elevated temperature, this method does not apply to synthesis of esters of saccharides (acylated saccharides). The reactions either with acyl chlorides in the presence of tertiary amine (pyridine, triethylamine) or with acyl anhydrides in the presence of acid catalyst (zinc chloride, strong mineral acid, acidic ion-exchange resin) are the most suitable routes of acylation of saccharides.

Reducing sugars in solution form a tautomeric equilibrium. The position of the acylation depends on the catalyst applied and temperature. The peracylation to penta-O-acetyl-β-D-glucopyranose is achieved at higher temperature with sodium acetate as the catalyst.

Acylated aldopyranoses readily undergo acid-catalyzed anomerization. Because of the anomeric effect, the axial acetyl group is preferred, and therefore α-isomer of aldohexopyranose peracetate (2.38) is the principal product of the reaction performed on the tautomeric mixture. However, change in the catalyst applied can result in the domination of the β-isomer (2.39).

A wide variety of the acyl groups have been introduced to saccharides and their derivatives. Exhaustively benzoylated derivatives are usually obtained by treating saccharides with benzyl chloride in pyridine. Saccharide esters of higher alkanoic (fatty) acids can be prepared by transesterification with other esters in the presence

a) CH$_3$COONa
b) ZnCl$_2$

of either base or acid catalyst. Polymerizing sugar esters can be prepared with methacrylic anhydride in pyridine

2.3.5 GLYCOSYLATION

When aldehydes or ketones (2.40) are treated with alcohols (2.41) in the presence of an acid catalyst, acetals and ketals (2.42) are formed, respectively. This reaction

2.40a,b   2.41   2.42a,b

is reversible, and acetals and ketals may hydrolyze into carbonyl compounds. In absence of acids, particularly in alkaline solutions, acetals are stable. Hydroxy aldehydes and ketones (2.43), in reaction with alcohols, can be converted into cyclic products, derivatives of furane or pyrane (2.44) with $n = 2$ and 3, respectively. Aldoses and ketoses belonging to the same family of hydroxy carbonyl compounds undergo a similar reaction, leading to cyclic saccharides (glycosides) with the hemi-acetal hydroxyl group replaced by an alkoxy group, an aglycone. The saccharide residues are either glucofuranosyl or glucopyranosyl groups.

Reaction of monosaccharides with alcohols but not with phenols (Fischer syn-thesis) is particularly useful in preparing alkyl pyranosides and furanosides. The reaction produces a complex mixture. The α/β isomer ratio depends on the reaction conditions. Presumably, the initial reaction of xylose (2.45) involves the open-chain structure (2.46) from which an intermediate ion (2.47) is generated. The first stage of that reaction providing furanosides (2.48) proceeds under kinetic control. The α-isomer preponderates, with $\alpha/\beta = 2$. The α- and β-xylofuranosides formed initially anomerize to give an equilibrium mixture with $\alpha/\beta = 0.2$. Several mechanisms for the formation and anomerization of furanosides have been proposed. The furanosides undergo a ring expansion to α- and β-pyranosides formed in the α/β ratio of 0.6.

Pyranosides (2.45) anomerize into (2.50) and (2.51). An intermediary cyclic carboxonium ion (2.49) is postulated. The α/β ratio can be predicted taking into account the interaction between substituents in the tetrahydropyranose ring (Table 2.1). Alcohols as solvents and the OCH_3 group cooperate with the anomeric effect. Thus, the α-pyranoside structure becomes favorable.

Glycosides and acetals readily hydrolyze in acid media. The rate of hydrolysis strongly depends on the structure of the aglycone and sugar. Examination of the hydrolysis of methyl glucosides, mannosides, galactosides, and xylosides reveals that furanosides hydrolyze more readily than pyranosides. The rate of hydrolysis of β-D-glycosides with the equatorial methoxy group is higher than that of α-anomers with the axial methoxy group. The hydrolysis of methyl aldofuranosides with the 1,2-*cis* configuration of the hydroxyl group at C-2 and aglycone proceeds at higher rate, and hydrolysis of glucopyranosides and glucofuranosides resembles that of simple acetals. It requires protonation of the acetal oxygen atom followed by substitution at the anomeric carbon atom. At room temperature, in contrast to aryl glycosides, alkyl glycosides are stable on hydrolysis in dilute alkali. The hydrolysis of aryl glycosides (2.52) involves a 1,2-epoxide intermediate (2.53), which is attacked by the C-6 hydroxyl group to give a 1,6-anhydride (2.54).

Glycoside formation and degradation are fundamental biochemical processes. Several natural biologically active compounds are complex glycosides (flavanoids, cardiac glycosides, antibiotics). Such glucoconjugates play essential roles in several molecular processes, such as hormone activity, induction of protective antibody response, control the development and defense mechanism of plants, and cell proliferation.

2.3.6 BIOLOGICAL SIGNIFICANCE OF THE EFFICIENCY OF O-GLYCOSYLATION

There are no general methods for glycoside synthesis. An interglycosidic bond is formed in a reaction of suitably O-protected glucosyl donor bearing a leaving group, for instance the bromine atom in (2.55), on its anomeric carbon atom with a suitable nucleophile. For example, phenolate anion resulting from the dissociation of potassium phenolate (2.56) provides phenylglycoside (2.57).

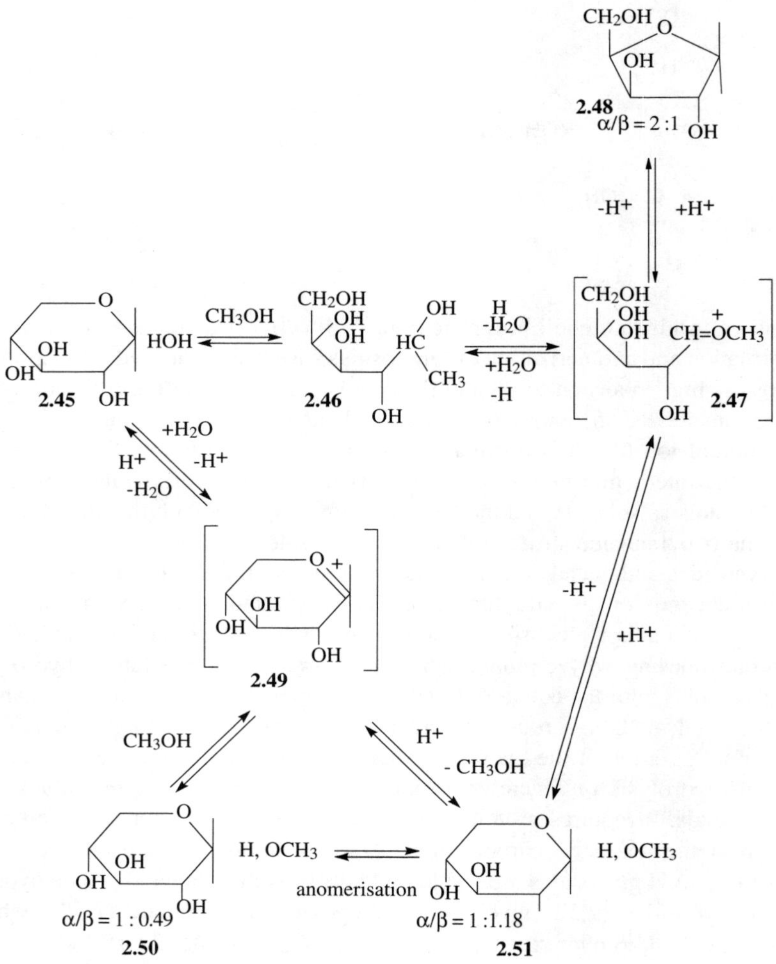
2.45
2.46
2.47
2.48
α/β = 2 : 1
2.49
2.50
α/β = 1 : 0.49
2.51
α/β = 1 :1.18
CH3OH
+H2O
-H2O
-H+
+H+
H+
- CH3OH
H, OCH3
anomerisation

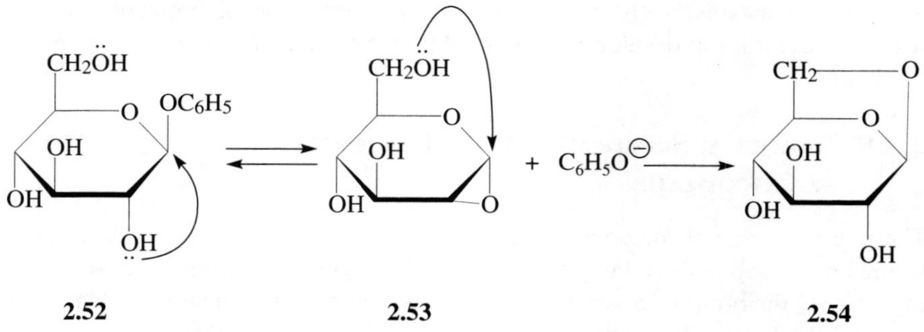
2.52
2.53
2.54
OC6H5
C6H5O

CH$_2$OAc ... O ... OAc ... OAc ... Br ... OAc + C$_6$H$_5$OK $\longrightarrow$ CH$_2$OAc ... O ... OC$_6$H$_5$ ... OAc ... OAc ... OAc

2.55   **2.56**   **2.57**

2.3.7 REACTIONS WITH AMINO COMPOUNDS

Primary amines (2.58) can add to aldehydes (2.59a) and ketones (2.59b). The initial *N*-substituted hemiaminals (2.60a, b) lose water to give more stable Schiff bases (2.61a, b). Addition of alcohols to the C=N double bond gives alkoxyamines (2.62a, b). Amino sugars are commonly available from reducing aldohexoses and aldopentoses (e.g., D-glucose) (2.63) treated with primary or secondary amines (2.64) to give *N*-alkylaldosylamines (2.65). The reactive compounds readily undergo either hydrolysis or rearrangements.

$$R^1\text{-}NH_2 + R^2\text{-}C\!\!=\!\!O \longrightarrow \underset{(H, R^3)}{R^2\text{-}\overset{OH}{\underset{|}{C}}\text{—}NHR^1} \xrightarrow{H_2O} \underset{(H, R^3)}{R^2\text{-}C = NR^1} \xrightarrow{R^4OH} \underset{(H, R^3)}{R^2\text{-}\overset{OR^4}{\underset{|}{C}}\!=\!NHR}$$

$$\underset{(H, R^3)}{}$$

2.58   **2.59(a,b)**   **2.60(a,b)**   **2.61(a,b)**   **2.62(a,b)**

CH$_2$OH ... O ... OH ... OH ... OH ... H, OH + RNH$_2$ $\rightleftharpoons$ CH$_2$OH ... O ... OH ... OH ... OH ... H, NHR + H$_2$O

2.63   **2.64**   **2.65**

Reaction of saccharides (2.66) with urea (2.67) requires more vigorous conditions. On heating their equimolar reaction mixture in aqueous sulfuric acid *N*-β-D-glucopyranosylurea (2.68) is formed. Synthesis of glycosylamines derivatives of sugars and amino acids or peptides has been thoroughly examined because similar compounds, glycoproteins, occur in nature. Transglycosylation, an exchange of substituent in the anomeric carbon atom, is observed when glycosylamine (2.69) is treated with amine (2.70). In such acid-catalyzed reactions under thermodynamic control, the cyclic structure of reacting saccharides is retained. Thus, Schiff bases

CH₂OH and structural diagrams

$$2.66 + H, OH + NH_2CONH_2 \rightleftharpoons 2.68 + H_2O$$

2.66   **2.67**   **2.68**

typical for reactions of common aldehydes do not form. Proportions of anomeric mixtures of glycosylamines (2.71), which are formed instead, depend on the reaction conditions and the structure of amine. In the reaction with bulky amines, reaction of the equatorial isomer is strongly favored for steric reasons.

In aqueous solution, glycosylamines undergo mutarotation and hydrolyze. Formation of an open-chain immonium structure was postulated in the intermediary step followed by cyclization (mutarotation) or substitution (hydrolysis). There is a strong link between these reactions and the Maillard reaction (see Chapter 18).

Several monographs[1–8] are recommended for further reading.

REFERENCES

1. Lehninger, A.L., *Principles of Biochemistry,* Worth Publishers, New York, 1982.
2. Armstrong, E.F., and Armstrong, K.F., *The Glycosides,* Longmans & Green, New York, 1982.
3. *The Carbohydrates, Chemistry and Biochemistry,* Pipman, W. and Horton, D., Eds., Academic Press, New York, 1972.
4. Schers, H. and Bonn, G., *Analytical Chemistry of Carbohydrates,* Georg Thierne Verlag, Stuttgart, 1998.
5. Driquez, H., and Thiem, J. (Eds), *Glycoscience,* Springer Verlag, Berlin, 1997.
6. Eliel, E.L., Wilen, S.H., and Mander, L.N., *Stereochemistry of Organic Compounds,* Wiley–Interscience, New York, 1994.
7. Stoddart, J.F., *Stereochemistry of Carbohydrates,* Wiley–Interscience, New York, 1971.
8. Eliel, E.L. et al., *Conformational Analysis,* John Wiley & Sons, New York, 1965.

TABLE 2.2
Structure and Occurrence of Alditol

Name	Occurrence and Application
D-Ribitol	Occurs in the plants, constituent of riboflavin (vitamin B2), teichoic acids
D-Xylitol	Isolated from field mushrooms; occurs in pentosans that accumulate as agricultural waste products (e.g., corncobs); sweetening properties similar to that of sucrose
D-Glucitol (sorbitol, sorbite)	Found in apples, pears, cherries, apricots; by-product in the synthesis of vitamin C; physiological function: conversion into liver glycogen; sugar substitute in dietetic food formulation; softener, crystallization inhibitor, additive in lyophilized food preparation
D-Mannitol	Found in exudates of plants, olive, plane trees, marine alga, mushrooms, onion; physiological function: energy storage

3 Saccharides: Modifications and Applications

Ya-Jane Wang

CONTENTS

3.1 Introduction 35
3.2 Sugar Alcohols and Polydextrose 36
3.3 Caramels 38
3.4 Saccharide Ethers 39
3.5 Fatty Acid Esters 39
3.6 Fatty Acid Polyesters 40
3.7 Physicochemical Properties and Applications 41
    3.7.1 Sugar Alcohols 41
    3.7.2 Polydextrose 41
    3.7.3 Saccharide Ethers 42
    3.7.4 Fatty Acid Esters 42
        3.7.4.1 Emulsifying Properties 42
        3.7.4.2 Complex Formation of Sucrose Esters with Starch
                and Proteins 43
        3.7.4.3 Antimicrobial Properties of Sucrose Fatty Acid Esters 43
    3.7.5 Fatty Acid Polyesters 44
References 44

3.1 INTRODUCTION

The massive availability of saccharides, mainly sucrose and D-glucose, has made them attractive sources for many products. Natural biodegradability and versatility of saccharides and health awareness were principal factors stimulating consumer attempts to reduce uptake of the calorie food. Response from industry resulted in reduction of energy in processed food products. Several technologies were developed, which were based on the processing of sucrose and D-glucose into sugar substitutes (e.g., sugar alcohols), food flavoring ingredients and colorants (e.g., caramels), surfactants, emulsifiers, and fat replacers (saccharide ethers with long-chain alkyl groups, fatty acid esters, and polyesters of saccharides). The processing involves either biochemical (see Chapter 10) or chemical methods.

Among saccharides, sucrose is the only organic compound that is pure, low-cost, renewable, and available worldwide on a scale of million tons a year. Sucrose being a hydrophilic molecule can be converted into a product having both hydrophilic and lipophilic groups by an esterification reaction with a fatty acid molecule. Sucrose esters have been transformed into nonionic surfactants from sucrose. Eight hydroxyl groups in sucrose can be substituted with fatty acids, but selective esterification of sucrose has been difficult because of its insolubility in most common organic solvents and its tendency to caramelize at high temperatures. D-Glucose and D-fructose for nutritive and nonnutritive purposes are available, among others, in the form of so-called invert from sucrose. When sucrose is hydrolyzed by dilute aqueous acid or enzyme invertase, it yields equal amounts of D-glucose and D-fructose on acidic hydrolysis. This 1:1 mixture of glucose and fructose is often referred to as invert sugar, because the sign of optical rotation changes (inverts) during the hydrolysis from sucrose ($[\alpha]_D = +66.5°$) to a glucose–fructose mixture ($[\alpha]_D = -22.0°$).

3.2 SUGAR ALCOHOLS AND POLYDEXTROSE

Sugar alcohols (3.1) to (3.7), also referred to as polyols, polyalcohols or polyhydric alcohols, are naturally present in many fruits and vegetables to contribute to sweetness. The majority of sugar alcohols are produced industrially by hydrogenation in the presence of Raney-nickel as a catalyst from their parent reducing sugars, except that erythritol (3.1) is commercially produced from a fermentation process. The reaction is exothermic and the reaction temperature and pH have to be well controlled.

Erythritol is a four-carbon symmetrical polyol and exists only in the meso form. Unlike other polyols of the hydrogenation process, the industrial production of erythritol uses a fermentation process with an osmophilic yeast (*Moniliella* sp. and *Trichosporonoides* sp.) or a fungus (*Aureobasidium* sp.).[1] D-Glucose resulting from hydrolyzed starch is fermented with the osmotolerant microorganisms. A mixture of erythritol and minor amounts of glycerol and ribitol (3.2) is formed.[2–4] In general, the microorganisms have the characteristics of tolerating high sugar concentrations and giving high erythritol yields. Conversion yields of 40 to 50% have been achieved with the fermentation process, and erythritol is crystallized at over 99% purity from the filtered and concentrated fermentation broth.

Xylitol (3.3) is a pentitol and is produced from hydrogenation of D-xylose, a hydrolyzed product of xylan from wood pulp of paper processing and other agricultural by-products such as sugar cane bagasse, corncob, and wheat straw. The hydrogenation is conducted at 80 to 140°C and hydrogen pressure up to 5 MPa, and the solution after hydrogenation requires further processing, including chromatographic fractionation, concentration, and crystallization,[5,6] which are expensive and increase the cost of xylitol production. About 50 to 60% of the initial xylose is converted into xylitol. Because of the high cost of hydrolysis, purification, and separation involved in catalytic hydrogenation of xylose,[7] efforts have been directed to developing biotechnological processes to produce economic sources of xylitol

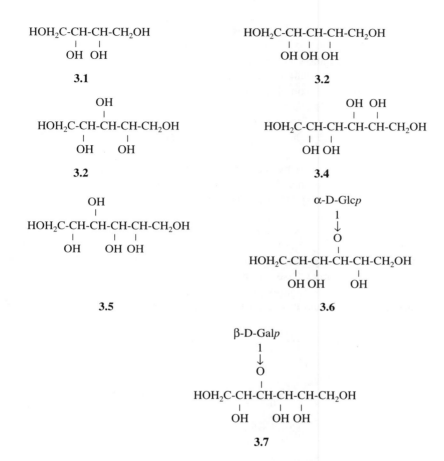

and fermentation processes by using bacteria, fungi, yeasts, mixed cultures, and enzymes. An enzymatic redox system is believed to be responsible for the conversion of xylose to xylitol in bacteria, fungi, and yeasts.[8]

Because mannose is not available industrially, mannitol (3.4) is obtained from hydrogenation of fructose, which is from either sucrose or starch. Sucrose, after hydrolysis, can be hydrogenated to produce a mixture of sorbitol (3.5) and mannitol. Dextrose from hydrolyzed starch can be isomerized to fructose, which is then reduced to sorbitol and mannitol. Mannitol is isolated by fractional crystallization due to its lower solubility in water compared with sorbitol, but the yield does not exceed 15 to 18%.[9]

Sorbitol is a reduced form of dextrose and the most available polyol. Sorbitol can be produced from electrochemical reduction of dextrose in an alkaline medium, which leads to the production of a considerable amount of mannitol due to alkali-catalyzed epimerization of dextrose to fructose, which is reduced more readily than dextrose. Sorbitol is industrially produced either by catalytic hydrogenation of dextrose, which produces less than 2% mannitol, or by catalytic hydrogenation of sucrose as a mixture with mannitol.

Maltitol (3.6), a 1,4-linked molecule of glucose and sorbitol, is produced by hydrogenation of high-maltose syrups from enzymatic hydrolysis of starch and subsequent purification and concentration. Maltitol, liquid or crystalline, contains a small amount of sorbitol, maltotriitol, and hydrogenated starch hydrolysates.

Unlike other polyols, isomalt is produced in two steps from sucrose. The first step involves enzymatic transglucosidation to form the intermediate isomaltulose, which is then hydrogenated to isomalt.[10,11] Isomalt is a mixture of two isomers: 1-O-α-D-glucopyranosyl-D-mannitol dihydrate and 6-O-α-D-glucopyranosyl-D-sorbitol.

Lactitol (3.7) is obtained by reducing the glucose moiety of lactose. A 30 to 40% lactose solution is subjected to a pressure of 40^6 Pa or more in the presence of the Raney-nickel catalyst at 100°C. After purification, lactitol is crystallized in a mono- or dihydrate form.

Sugar alcohols are widely used in many food products because of their unique functional and nutritional properties, such as reduced calories, cooling effect, sweetness, and noncariogenicity. Recently, sugar alcohols appeared to be suitable binders and plasticizers in ceramics based on nanometric metal oxide powders.[12]

D-Glucose and sugar alcohols provide an interesting product — polydextrose (3.8). Polydextrose is a randomly crosslinked glucose polymer prepared by vacuum

Glcp-(1→6)-Glcp-(1→6)-Glcp-(1→6)-Glcp-(1→3)-Glcp-(1→6)-Glcp-1-OR
```
                 4              2
                 ↓              ↓
                 1              1
               Glcp           Glcp-(4→1)-Glcp-(6→1)-Glcp
                 6
                 ↓
            1-Glcp-(3→1)-Glcp
```

3.8

thermal polymerization of glucose with sorbitol as a plasticizer and in the presence of low levels of some acids (citric acid or phosphoric acid) as a catalyst. The D-glucose units in polydextrose are nonuniformly α and β. Both sorbitol and admixed acids are bound covalently to the polymer. Because of the nature of the random reaction, polydextrose contains all types of glucosidic linkages, including α- and β-1→2, 1→3, 1→4, and 1→6, with 1→6 linkage being predominant. The average molecular weight of polydextrose is 2000–2500, with an average degree of polymerization (DP) of 12–15,[13] although the range of molecular weight is 162 to about 20,000. The addition of sorbitol is to control the upper molecular weight limit and to prevent the formation of large-molecular-weight molecules that are water insoluble.

3.3 CARAMELS[14]

When saccharides are melted or heated in the presence of acidic or alkaline catalysts, brown-colored products with a typical caramel aroma are obtained. The thermal degradation (caramelization) of saccharides leads to the formation of volatiles (caramel aroma) and brown-colored products (caramel colors). The process is acid or base catalyzed and generally requires high temperatures > 200°C at pH < 3 or pH > 9 and low water/high sugar content. The generation of flavors and colors in thermally induced caramelization requires that sugars, normally monosaccharide structures, first undergo intramolecular rearrangements. The reaction causes the release of H[+]. Thus, the pH of the solution undergoing caramelization falls with time, eventually into the slightly acidic region of 4 to 5. Isomerization, water elimination, and oxidation occur during caramelization, which leads to the formation of nice color and flavors in many foods, such as coffee and beverage colors, but can also cause undesirable flavors and colors, such as a burnt sugar smell.

3.4 SACCHARIDE ETHERS

Preparation of saccharide ethers involves treating saccharides with hydrophobic fatty alcohols (C-8 to C-16) under acid catalysis. Because of the difficulties involved in removing these high-boiling alcohols, a high temperature and prolonged reaction time is required, which results in the formation of a complex mixture of mono-, di-, tri-, and oligoglycosides. Typically, the reaction proceeds on a suspension of saccharide in 2 to 6 moles of alcohol heated at 100 to 120°C in the presence of an acid catalyst. Water formed is continuously removed under vacuum.[15] After the reaction, the acid catalyst is neutralized with a suitable base, followed by removal of excess alcohol under vacuum.

3.5 FATTY ACID ESTERS

Fatty acid esters of saccharides constitute an important group of nonionic surfactants with a degree of substitution (DS), defined as the number of hydroxyl groups esterified with fatty acids, of 1 to 3. Fatty acid esters are hydrophilic, absorbable, and digestible, and are integral parts of many food processes. On the other hand, fatty acid polyesters have a DS > 3 and are lipophilic, nondigestible, or partially digestible, with characteristics of conventional fats and oils.

Sucrose fatty acid esters are available[16] through transesterification of sucrose with the corresponding fatty acid methyl esters in the presence of an alkaline catalyst, potassium carbonate, and sufficient solvent such as dimethylformamide (DMF) or dimethylsulfoxide (DMSO) at 90–95°C under 80–100 mmHg pressure for 9–12 h. This process is called the Snell process. The reaction follows first-order kinetics and the rate of reaction is independent of the sucrose concentration and the molecular weight of fatty acid. Only the three primary hydroxyl groups at C-6 of glucose and C-1' and C-6' of fructose can be substituted to yield mono-, di-, and triesters. The yield of sucrose esters exceeds 90%. The monosubstitution in sucrose takes place preferably at C-6 of the D-glucose unit, with the ratio of the substitution to the glucose to fructose moiety being 4:1.[17]

Because those solvents are not approved in food technology, propylene glycol or water was employed as a solvent in the Snell process. Soap or the product itself acted as an emulsifier to conduct the reaction in a microemulsion system with potassium carbonate as the catalyst.[18] Reported yield is 85% for the sucrose monoester and 15% for the diester after purification. Feuge et al.[19] developed a solvent-free interesterification, which carried out the reaction between molten sucrose (melting point 185°C) and fatty acid methyl esters at 170 to 187°C, with lithium, potassium, and sodium soaps as catalysts and solubilizers. Lithium palmitate produced mostly sucrose polyesters with DS > 3, and lower esters were best produced with combinations of lithium oleate with sodium or potassium oleate at a level of 25% based on sucrose weight. Vegetable oil was a solvent as well as a fatty acid source to carry out the reaction at 110 to 140°C with potassium carbonate as a catalyst.[20] The product can be used without further purification. In both solvent and solvent-free processes, distillation was often needed to remove the nonreacted reactants and by-products. As the molar ratio of sucrose to fatty acid increased, the yield of sucrose esters was higher than that of sucrose polyesters.

Inorganic carriers commonly used in biotransformation with immobilized enzymes, such as Celite and disodium hydrogen phosphate, catalyze the sucrose acylation with vinyl esters of fatty acids. 2-O-Lauroylsucrose is the principal product.[21] The chemical acylation of sucrose takes place with the preference decreasing in the order of 6-OH ≥ 6'-OH ≥ 1'-OH > secondary-OH. However, under a general-base catalysis, the most acid 2-hydroxyl group of sucrose was activated for a more nucleophilic alkoxide. A simple chemical reaction in dimethyl sulfoxide at 40°C with disodium hydrogen phosphate as the catalyst for selective synthesis of sucrose monolaurate was reported by the same group, with 60% conversion to monoester and 30% to diester.[22]

3.6 FATTY ACID POLYESTERS

Fatty acid polyesters of saccharides are nondigestible, nonabsorbable, fat-like molecules with physical and chemical properties of conventional fats and oils. Olestra (generic name) or Olean® (brand name), a sucrose polyester (DS 4 to 8) is a mixture of sucrose esters with six to eight fatty acids.[23] This polyester was approved by the U.S. Food and Drug Administration (FDA) in 1996 for limited uses.

Also, sucrose polyesters are available by transesterification and interesterification. Transesterification is a two-stage process. Potassium soaps provide a homogenous melt of sucrose and fatty acid methyl esters (FAME). Then, at 130 to 150°C, FAME and NaH are added in excess. The interesterification is a simple ester–ester interchange reaction involving reaction of alkyl ester of sucrose with FAME in the presence of sodium methoxide (NaOCH₃) or sodium or potassium metal as catalyst.[24] There are preferences for the sucrose ester formation when the molar ratio of sucrose to fatty acid increases. Sucrose polyesters are the predominant derivatives when the molar ratio of sucrose to fatty acid decreases.[24]

Sorbitol[20] and other carbohydrates, including trehalose,[25] raffinose,[26] stachyose,[27] and alkyl glycoside,[27] have been esterified. Up to six hydroxyl groups in sorbitol can be esterified (interesterification or transesterification). Trehalose, raffinose, and stachyose are nonreducing saccharides with 2, 3, or 4 sugar units and consist of 8, 11, or 14 available OH groups for esterification, respectively. They can be esterified with FAME catalyzed by Na metal to produce highly substituted polyesters at lower temperatures because they are nonreducing sugars. Glycosylation converts the reducing C-1 anomeric center to nonreducing, the alkyl glycosides become substrates for transesterification with FAME, and the remaining hydroxyl groups can be esterified as described for sucrose polyester to produce alkyl glycoside fatty acid polyesters.

Fatty acid esters and ethers of saccharides are nonionic surfactants commonly used as wetting, dispersing, or emulsifying agents and as an antimicrobial and protective coating for fruits. Fatty acid polyesters of saccharides function as fat replacers and provide the characteristic functional and physical properties of fat but have lower caloric contents because of reduced digestion and absorption. Their properties depend on the type and number of sugar units in carbohydrate moiety as well as the number, type, and length of attached alkyl chain or fatty acid residue.

3.7 PHYSICOCHEMICAL PROPERTIES AND APPLICATIONS

3.7.1 SUGAR ALCOHOLS

All polyols exhibit negative heats of solution, resulting in a cooling of the solution and consequently a cooling sensation when the dry powder of polyols is dissolved in the mouth. The intensity of the cooling effect depends on the magnitude of the heat of the dissolution. For example, maltitol and isomalt show no cooling effect and xylitol imparts a strong cooling sensation. This cooling property has found applications in food and pharmaceutical products in which a cooling sensation is desired.

Because of the absence of carbonyl group, polyols exhibit a good stability against the exposure to heat, alkali, and acid, and do not take part in nonenzymatic browning (Maillard) reactions (Chapter 18). They exert freezing point depression because of their small molecular weight. They have low hygroscopicity with, in increasing order, mannitol, isomalt, lactitol, maltitol, xylitol, and sorbitol (Table 3.1).

Polyols are low-calorie bulk sweeteners with applications where calorie reduction is desired, such as diabetic products, table-top sweeteners, chocolate, hard-boiled candies, jellies and jams, chewing gum, and ice creams. Some polyols may require the addition of intensive sweeteners to provide the right sweetness to certain applications. Polyols also function as humectants and emulsifiers to provide freshness and extend shelf life to bakery products and confectionery where products tend to dry or harden during storage. Because of their small molecular weight, they also have a plasticizing effect to improve the texture of the products. Sugar alcohol syrups can function as antifreeze to preserve the functionality of products such as fish protein. The selection of polyols depends on the sweetness profile, cooling effect, and aftertaste, and combinations of polyols often have a synergistic effect.

3.7.2 POLYDEXTROSE

The highly branched structure of polydextrose renders it water soluble and gives it reduced-calorie (1 cal/g) properties. It has no sweetness and its viscosity is slightly

TABLE 3.1
Properties of Sugar Alcohols

Sugar	Sweetness (Sucrose = 1)	Heat of Solution (kJ/kg)	Cooling effect	Hygroscopicity	Caloric Value (kJ/g)
Sucrose	1.00	−18	None	Medium	16.74
Erythritol	0.53–0.70	−180	Cool	Very low	1.67
Xylitol	0.87–1.00	−153	Very cool	High	10.05
Mannitol	0.50–0.52	−121	Cool	Low	6.70
Sorbitol	0.60–0.70	−111	Cool	Medium	10.88
Maltitol	0.74–0.95	−79	None	Medium	8.79
Isomalt	0.35–0.60	−39	None	Low	8.37
Lactitol	0.35–0.40	−53	Slightly cool	Medium	8.37

From Bornet, F.J.R., *Am. J. Clin. Nutr.*, 59S, 763, 1994. With modification.

greater than that of sugar solutions of equal concentrations. It can function as a humectant to prevent or slow undesirable changes (either gain or loss) in the moisture content of foods. It also has higher water solubility than other common bulking agents, which is advantageous in certain foods.

The U.S. FDA has accepted polydextrose as an additive for 13 food categories: (1) baked goods and baking mixes (restricted to fruit, custard, and pudding-filled pies); (2) cakes; (3) cookies and similar baked products; (4) chewing gum; (5) confections and frostings; (6) dressings for salads; (7) frozen dairy desserts and mixes; (8) fruit spreads; (9) gelatins, puddings, and fillings; (10) hard and soft candies; (11) peanut spread; (12) sweet sauces, toppings, and syrups; and (13) tablespreads (21 CFR 172.841). Polydextrose conforms to the Japanese definition of a dietary fiber and is used in beverages as a dietary fiber source. Polydextrose provides the foods with the functions of bulking agent, formulation aid, humectant, and texturizer.

3.7.3 SACCHARIDE ETHERS

Saccharide ethers with long hydrocarbon chains exhibit low solubility both in water and in nonpolar solvents and tend to crowd the interface, resulting in high interfacial activity. They are moderate foamers and their foam ability is a function of chain length; that is, the foam volume decreases with an increase in chain length. Saccharide ethers can potentially replace the more commonly used nonionic surfactants. They exhibit better stability to temperature changes and excellent surfactance.

3.7.4 FATTY ACID ESTERS

3.7.4.1 Emulsifying Properties

Sucrose esters are nonionic surfactants with a wide range of hydrophilic–lipophilic balance (HLB) values. Their surface-active properties are derived from the original hydrophilic group of sucrose and the original lypophilic group of fatty acids. When sucrose esters are added to an oil–water mixture, they are adsorbed at the interface between water and oil and orient themselves so that the hydrophilic portions toward water and the lipophilic portions toward oil, thereby reducing surface tension.

Sucrose fatty acid esters from lauric (C12:0) to docosanoic acid (C22:0) have surfactant properties to reduce surface tension of water. Lower fatty acid esters than lauric do not possess significant surfactant properties. The monoesters are soluble in many organic solvents but only slightly in water. Because of toxicological clearance, the U.S. FDA has approved only sucrose esters blended with mono-, di-, and triglycerides of palmitic and stearic acids. They are completely absorbed and metabolized surfactants. Surfactant properties depend on the number of the esterified groups and the length and degree of saturation of the fatty acid chain. Sucrose esters with a shorter, more unsaturated fatty acid and fewer esterified groups show a greater hydrophilic function.[29]

Little research compares the emulsification properties among different sucrose ester isomers. Husband et al.[30] reported that both pure sucrose 6-monolaurate and sucrose 6'-monolaurate exhibited similar foaming and interfacial properties, and sucrose dilaurate displayed inferior foaming properties and higher surface tension than did sucrose monolaurates.

3.7.4.2 Complex Formation of Sucrose Esters with Starch and Proteins

Osman et al.[31] observed a reduced iodine affinity of amylose, indicative of complex formation, in the presence of sucrose esters and suggested that the ability of sucrose esters to form complexes was related to the percentage of the fatty acid portion. Later, Matsunaga and Kainuma[32] reported that sucrose esters had the ability to prevent starch retrogradation by forming a helical complex with amylose. The Brabender peak viscosity, peak time, and maximum set back viscosity of maize, tapioca, or wheat starch were higher in the presence of sucrose esters (HLB9-16), supporting the inclusion complex formation, which retarded the migration of amylose from the starch granule.[33]

It has been suggested that sucrose esters function as dough conditioners by interacting with flour protein. Results of the study by Hoseney et al.[34] suggested that sucrose simultaneously bound to glutenin by hydrophobic bonding and to gliadin by hydrogen bonding. Krog[35] proposed two possible mechanisms for sucrose esters to function as dough conditioners: (1) hydrophobic and/or hydrophilic binding with gluten protein, or (2) interaction in bulk form with the water phase of the dough. It is likely that both mechanisms are involved because considerable evidence supports both hypotheses.

3.7.4.3 Antimicrobial Properties of Sucrose Fatty Acid Esters

The antimicrobial activity of sucrose esters comes from the interaction of esters with cell membranes of bacteria, which causes autolysis. The lytic action is assumed to be due to stimulation of autolytic enzymes rather than to actual solubilization of cell membranes of bacteria. Both gram-positive and gram-negative bacteria are affected, but the susceptibility of very similar organisms to comparable sucrose esters varies significantly.[36] Results have demonstrated that the antimicrobial ability of sucrose esters is determined by the structure of the esterified fatty acids. Monoesters are more potent than polyesters. Sucrose esters exhibit substantial antimycotic activity against some toxinogenic and spoilage molds but little or no inhibitory activity against yeast.

Sucrose esters were approved as food additives in the U.S. in 1983 (21 CFR, 172.859) and in many other countries. Because sucrose has eight hydroxyl groups to be esterified, a variety of sucrose esters can be manufactured ranging from low HLB to high HLB values by controlling the DS. At present, sucrose esters with a low HLB value are not permitted for typical W/O emulsified food such as margarine, except in Japan. The main users of sucrose esters are the baking and confectionery industries. Sucrose esters are used in bread to improve mixing tolerance, water absorption, and gas retention, resulting in improved loaf volume and texture. Their interactions with amylose also reduce starch retrogradation and maintain softness. Sucrose esters also improve cookie spread; volume and softness of high-ratio white layer cakes; and tenderness of spongy cakes, pastries, biscuits, shortbreads, and shortcakes. They can also be incorporated as components of protective coating for fresh fruits to retard ripening and spoilage. Sucrose esters with high monoester content are effective in dispersing and stabilizing oil-soluble vitamins and calcium in beverages.[37]

3.7.5 FATTY ACID POLYESTERS

When fatty acid ester groups are four or more, sucrose polyesters behave as fats and have physical and organoleptic properties similar to those of conventional fats. However, they are absorbed and digested to a lesser extent and thus have a reduced calorie property. Sucrose polyesters are not hydrolyzed by pancreatic lipase and not absorbed in the small intestine. They show characteristics similar to those of conventional oil such as soybean or cotton oil but do not contribute any significant calories. They are not good surfactants of O/W emulsions but are excellent stabilizers of W/O emulsions.[38]

In 1996, the U.S. FDA approved Olestra, a sucrose polyester for limited use in savory snacks (chips, curls, and crackers). It has been shown that carbohydrate polyesters have the potential of lowering cholesterol levels in certain lipid disorders by selective partitioning of cholesterol in the nonabsorbable Olestra phase. Carbohydrate polyesters may also benefit persons at high risk of diabetes, coronary heart disease, colon cancer, and obesity. They can be used as a frying or cooking medium or incorporated into products to replace the oil such as salad dressing, margarine, or dairy or meat products.

REFERENCES

1. Kasumi, T., Fermentative production of polyols and utilization for food and other products in Japan, *Jpn. Agric. Res. Q.*, 29, 49, 1995.

2. Sasaki, T., Production of erythritol, *J. Agric. Chem. Soc. Jpn.*, 63, 1130, 1989.
3. Ishizuka, H., et al., Breeding of a mutant of *Aureobasidium* sp. with high erythritol production, *J. Ferm. Biotechnol.*, 68, 310, 1989.
4. Aoki, M.A.Y., Pastore, G.M., and Park, Y.K., Microbial transformation of sucrose and glucose to erythritol, *Biotechnol. Lett.*, 15, 383, 1993.
5. Melaja, A. and Hämäläïnen, L., U.S. Patent 4,008,285, 1977.
6. Hyvönen, L., Koivistoinen, P., and Voirol, F., Food technological evaluation of xylitol, *Adv. Food Res.*, 28, 373, 1982.
7. Nigam, P., and Singh, D., Processes for fermentative production of xylitol — a sugar substitute, *Process Biochem.*, 30, 117, 1995.
8. Parajó, J. C., Domínguez, H., and Domínguez, J. M., Biotechnological production of xylitol. Part 1: Interest of xylitol and fundamentals of its biosynthesis, *Bioresource Technol.*, 65, 191, 1998.
9. Sicard, P.J. and Leroy, P., Mannitol, sorbitol, and Lycasin: Properties and food applications, in *Developments in Sweeteners*, Grenby, T.H., Ed., Applied Science Publishers, London, 1987, p. 1.
10. Schiweck, H., Isomalt (Palatinit®), A versatile alternative sweetener — production, properties and uses, in *Carbohydrates in Industrial Synthesis*, Clarke, M.A., Ed., Verlag Dr. Herbert Bartens, Berlin, 1992, p. 37.
11. Kunz, M., Sugar alcohols, in *Ullmann's Encyclopedia of Industrial Chemistry,* 5th ed., Vol. A 25, Elvers, B., Hawkins, S., and Russey, W., Eds., Weinheim, VCH Verlagsgesellschaft, Weinheim, 1994, pp. 426, 436.
12. Schilling, C.H. et al., Rheology of alumina-nanoparticle suspensions: Effects of lower saccharides and sugar alcohols, *J. Eur. Ceram. Soc.*, 22, 917, 2002.
13. Mitchell, H., Auerbach, M.H., and Moppett, F.K., Polydextrose, in *Alternative Sweeteners, Food Science and Technology*, Vol. 112, Nabors, L.O., Ed., Marcel Dekker, New York, 2001, p. 499.
14. Tomasik, P., Palasiński, M., and Wiejak, S., The thermolysis of saccharides. Part I. *Adv. Carbohydr. Chem. Biochem.*, 47, 203, 1989.
15. Koeltzow, D.E., and Urfer, A.D. Preparation and properties of pure alkyl glucosides, maltosides and maltotriosides, *J. Am. Oil Chem. Soc.*, 61, 1651, 1984.
16. Osipow, L. et al., Methods of preparation fatty acid esters of sucrose, *Ind. Eng. Chem.*, 48, 1459, 1956.
17. York, W.C. et al., Structural studies on sucrose monolaurate, *J. Am. Oil Chem. Soc.*, 33, 424, 1956.
18. Osipow, L. and Rosenblatt, W., Micro-emulsion process for the preparation of sucrose esters, *J. Am. Oil Chem. Soc.*, 44, 307, 1967.
19. Feuge, R.O. et al., Preparation of sucrose esters by interesterification, *J. Am. Oil Chem. Soc.*, 47, 56, 1970.
20. Khan, R.A. and Mufti, K.S., Ger. Offen. DE 2,412,374, 1974.
21. Plou, F.J. et al., Acylation of sucrose with vinyl esters using immobilized hydrolases: Demonstration that chemical catalysis may interfere with enzymatic catalysis, *Biotechnol. Lett.*, 21, 635, 1999.
22. Ferrer, M. et al., Chemical versus enzymatic catalysis for the regioselective synthesis of sucrose esters of fatty acids, *Stud. Surf. Sci. Catal.*, 130A, 509, 2000.
23. Mattson, F.H., Healthy, M., and Volpenhein, R.A., U.S. Patent 3,600,186, 1971.
24. Akoh, C.C. and Swanson, B.G., Optimization of sucrose polyester synthesis: Comparison of properties of sucrose polyesters, raffinose polyesters, and salad oils, *J. Food Sci.*, 55, 236, 1990.

25. Akoh, C.C. and Swanson, B.G., Preparation of trehalose and sorbitol fatty acid polyesters by interesterification, *J. Am. Oil Chem. Soc.*, 66, 158, 1989.

26. Akoh, C.C., and Swanson, B.G., One-step synthesis of raffinose fatty acid esters, *J. Food Sci.*, 52, 1570, 1987.

27. Akoh, C.C. and Swanson, B.G., Synthesis and properties of alkyl glycoside and stachyose fatty acid polyesters, *J. Am. Oil Chem. Soc.*, 66, 1295, 1989.

28. Bornet, F.J.R.,Undigestible sugars in food products, *Am. J. Clin. Nutr.*, 59S, 763, 1994.

29. Bhatt, S. and Shukla, R.P., Sucrose fatty acid esters — Compounds of divergent properties and applications, *Bharatiya Sugar*, 1, 75, 1991.

30. Husband, F.A. et al., Comparison of foaming and interfacial properties of pure sucrose monolaurates, dilaurate and commercial preparations, *Food Hydrocol.*, 12, 237, 1998.

31. Osman, E.M., Leith, S.J., and Fles, M., Complexes of amylose with surfactants, *Cereal Chem.*, 38, 449, 1961.

32. Matsunaga, A. and Kainuma, K., Studies on the retrogradation of starch in starchy foods. Part 3: Effect of the additive of sucrose fatty acid ester on the retrogradation of corn starch, *Starch/Staerke* 38, 1, 1986.

33. Deffenbaugh, L.B. and Walker, C.E., Use of the rapid visco-analyzer to measure starch pasting properties, *Starch/Staerke*, 42, 89, 1990.

34. Hoseney, R.C., Finney, K.F., and Pomeranz, Y., Functional (bread-making) and biochemical properties of wheat flour components. VI. Gliadin-lipid-glutenin interaction in wheat gluten, *Cereal Chem.*, 47, 135, 1970.

35. Krog, N., Theoretical aspects of surfactants in relation to their use in breadmaking, *Cereal Chem.*, 58, 158, 1981.

36. Marshall, D.L. and Bullerman, L.B., Antimicrobial properties of sucrose fatty acid esters, in *Carbohydrate Polyesters as Fat Substitutes*, Akoh, C.C., and Swanson, B.G., Eds., Marcel Dekker, New York, 1994, p. 149.

37. Nakamura, S., Applications of sucrose fatty acid esters as food emulsifiers, in *Industrial Aapplications of Surfactants IV*, Royal Society of Chemistry, London, 1999, p. 73.

38. Akoh, C.C., Emulsification properties of polyesters and sucrose ester blends I. Carbohydrate fatty acid polyesters, *J. Am. Oil Chem. Soc.*, 69, 9, 1992.

4 Production of Saccharides

Jan Grabka

CONTENTS

4.1 Introduction 47
4.2 Production of Beet Sugar 48
    4.2.1 Beet Preparation Processes 49
    4.2.2 Extraction 49
    4.2.3 Juice Purification 50
    4.2.4 Evaporation 51
    4.2.5 Crystallization 52
    4.2.6 Commercial Sugar Varieties 53
    4.2.7 By-Products and Waste 54
4.3 Production of Cane Sugar 54
    4.3.1 By-Products and Waste 55
4.4 Production of Other Saccharides Essential in Food Technology 55
References 56

4.1 INTRODUCTION

Sugar beets and sugar cane are the two most important and competing source materials for production of saccharides in the sugar industry.[1,2] The composition of sugar beets and sugar cane varies widely, depending on the genetic strain, agronomic factors, soil and weather conditions during growth, plant diseases, and treatment between harvesting and slicing. The content of sugar in mature beets ranges from 14–20% of the beet mass, whereas in mature cane this value ranges from 8–16% of the cane mass. Table 4.1 presents the typical chemical composition of sugar beet. Table 4.2 outlines the average composition of dry substance in sugar cane juice.

Production of other sweeteners such as glucose syrups (corn or potato syrups), dextrose (glucose), izoglucose (high-fructose corn syrup), and fructose is developing rapidly. Sweetener consumption is highest in the industrialized countries, though limited to some degree by health and economic considerations. Sugar cane and sugar beets are among the plants producing the highest yields of carbohydrates per hectare and are considered ideal raw materials in the production of ethanol, which can be used as an ecologically more acceptable motor fuel.

0-8493-1486-0/04/$0.00+$1.50
© 2004 by CRC Press LLC

TABLE 4.1
Chemical Composition of Sugar Beet in g/100 g Beet

Component	Content
Water	73.0–76.5
Dry substance	23.5–27.0
Sucrose	14.0–20.0
Nonsucrose	7.0–9.5
Water-insoluble compounds (cellulose, pectin, lignin, proteins, saponins, lipids, ash)	4.5–5.0
Soluble compounds (monosaccharides, raffinose, organic acids)	2.5
Nitrogenous compounds (betaine, amino acids, amides, NH_4 salts)	1.0–1.2
Inorganic compounds (K^+, Na^+, Ca^{2+}, Mg^{2+}, Cl^-, SO_4^{2-}, PO_4^{3-}, Fe^{3+}/Al^{3+}, SiO_2)	0.4–0.5

TABLE 4.2
General Composition of Sugarcane Juice in g/100 g Soluble Dry Substance

Component	Content
Sugars	75.0–94.0
Sucrose	70.0–90.0
Glucose	2.0–4.0
Fructose	2.0–4.0
Oligosaccharides	0.001–0.05
Salts	2.5–7.5
Of inorganic acids	1.5–4.5
Of organic acids	1.0–3.0
Organic Acids	1.5–5.5
Carboxylic acids	1.1–3.0
Amino acids	0.5–2.5
Other Organic Nonsugars	0.54–1.45
Protein	0.5–0.6
Starch	0.001–0.18
Soluble polysaccharides	0.03–0.5
Waxes, fat, phosphatides	0.01–0.15

4.2 PRODUCTION OF BEET SUGAR

The fundamentals of production of beet sugar are separated into five operations: (1) beet preparation, (2) diffusion or extraction, (3) juice purification, (4) evaporation, and (5) crystallization.

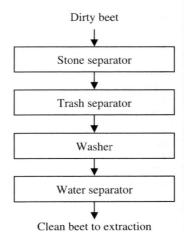

Clean beet to extraction

FIGURE 4.1 Beet preparation process.

4.2.1 Beet Preparation Processes

Figure 4.1 schematically presents the preparation process. Preparation of beets for processing takes place after mechanical harvesting and before slicing. Beets delivered to a sugar plant are contaminated with soil, stones, and vegetable matter such as leaves and beet tails. Washing is the final and most important cleaning step in sugar beet preparation. Cleaned beets must not contain more than 0.1 to 0.2% of soil. Fluming water must be removed from the beets between the final washing station and the slicing machines. Use of coagulants is the new method to separate colloidal precipitates from flume water,[3] though biological anaerobic or aerobic fermenters have some advantages.

4.2.2 Extraction

Sugar beets are sliced into long, thin strips or cossettes by the slicers and then weighed and transported to continuous extractors. The extraction plant (Figure 4.2) comprises (1) a heating unit where cossettes are warmed to 70–80°C to achieve denaturation or scalding, and (2) countercurrent transport, heat exchange, and mass transfer between the cossettes and extraction liquid. The raw juice (the diffusion juice) contains approximately 13–16% sugar of approximately 86–89% purity. About 98% of the sugar in the beet is extracted. Residual juice is removed by pressing the pulp.

The desugared cossettes leaving the upper end of the extractor, known as wet pulp, are delivered to pulp presses, where the moisture content is reduced from 95 to 76–84%. The pressed wet pulp, or dried pulp, with or without added molasses has a ready market as animal feed.

The extraction process theory is based on Fick's law, derived from the general diffusion law (Equation 4.1):

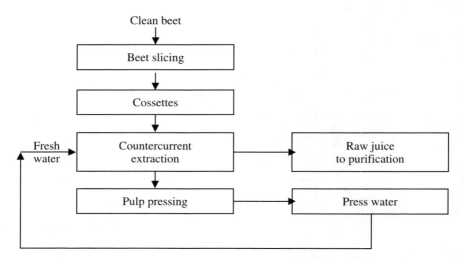

FIGURE 4.2 Beet extraction process.

$$ds = D_f A \frac{dc}{dx} d\tau \tag{4.1}$$

where ds is the weight of sugar diffusing through the area A in time $d\tau$; dc/dx the concentration gradient of the solute; and D_f the diffusion coefficient, which depends on temperature according to the Einstein and Smoluchowski correlation (Equation 4.2):

$$D_f = \frac{kT}{\eta} \tag{4.2}$$

where k is the constant for the dissolved substance, T the absolute temperature, and η the viscosity of the solution.

The juice after extraction has a considerable quantity of colloidal matter and a large number of fine pulp particles. They must be eliminated prior to juice purification.[4] The nonsucrose content of the raw juice is determined by beet quality, conditions under which sugar is extracted, quality of cossettes, draft, pH value, temperature, time, and bacterial control.

4.2.3 JUICE PURIFICATION

Figure 4.3 shows the fundamental options of raw juice purification. Juice purification in the lime/carbonic acid treatment can be considered as a sequence of the following operations: (1) precipitation and flocculation of colloidal components such as pectins and proteins; (2) precipitation of phosphates, citrates, oxalates, and other insoluble salts; (3) alkaline degradation of invert sugar and amides into organic acids, providing thermostability of the juice on evaporation; (4) reduction of the calcium salt

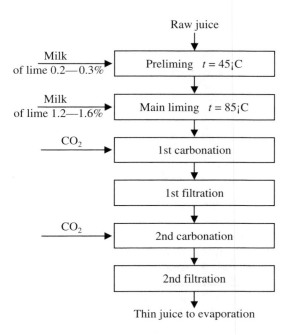

FIGURE 4.3 Beet juice purification.

concentration in thin juice after a second carbonation to a minimum required to achieve optimum alkalinity. The pH is reduced to 9–9.5.

Normally, fresh and healthy sugar beet contains below 0.1 g invert sugar/100 g beet. The quantity varies considerably, depending on plant growing condition, maturity, variety, climate, and conditions and duration of storage. The level of invert sugar rises during storage of frozen beets because thawing increases the activity of the enzyme invertase.

Recently proposed have been the application of a slurry activated with lime milk[5] and the use of flocculants for preliming in order to reduce lime consumption, possibly down to 0.2–0.3%, a level available in the sugar cane industry.

4.2.4 EVAPORATION

Thin juice is warmed to 130–135°C and submitted to evaporators. Multiple-effect evaporators usually consist of five individual bodies. Such a system allows multiple use of the same heat energy. It provides a decrease in pressure and temperature when the juice proceeds through the bodies. The evaporation results in concentration of dissolved solids in the juice from 12–16% to 60–70%. The juice leaving the last segment of the evaporator is called thick juice.

The steam boiler in beet sugar production is fired by a variety of fuels such as coal, natural gas, or oil, and possibly pressed and dried pulp in the future. In the cane sugar industry, bagasse is used almost exclusively.

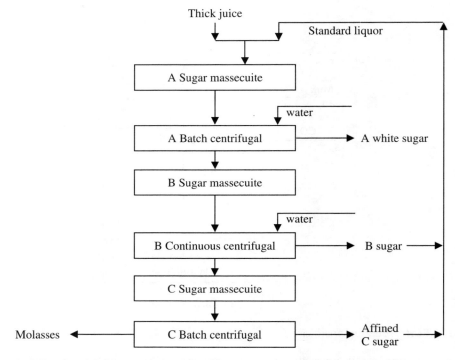

FIGURE 4.4 Crystallization scheme for the production of a standard white sugar.

4.2.5 CRYSTALLIZATION

Recrystallization leads to a standard white sugar (Figure 4.4). It proceeds as a three-stage process. In the first crystallization stage, sugar massecuite (A), the standard liquor (syrup), is obtained by dissolving washed raw sugar (B) and affined C sugar [the so-called after-product (C)] in the thick juice.

Sugar crystallizes when boiled in vacuum pans. To avoid inversion and caramelization, it is boiled at low temperatures (see Figure 4.5). In this manner, an undersaturated solution (1) turns into a supersaturated solution (2). Alternatively, a decrease in temperature (3, Figure 4.5) provides standard liquor subsequently boiled in vacuum pans until the syrup becomes sufficiently supersaturated. The supersaturation coefficient y_p is available by estimating the separation of sucrose between water and solution as a ratio of the mass fractions of sucrose in a saturated solution at a given temperature.

The liquor is then either "shocked" to initiate the crystal formation (in the intermediate supersaturated zone $1.2 < y_p < 1.3$) by the addition of a small amount of powdered sugar or it is seeded at a lower supersaturation (metastable zone $y_p > 1.2$) by adding finely milled sugar.[6] The size of seed crystals exceeds 15 to 20 μm. The crystals are then carefully grown under controlled supersaturation. The control involves vacuum, temperature, feed-liquor additions, and steam. Equation 4.3 expresses the rate of the crystal growth on boiling:

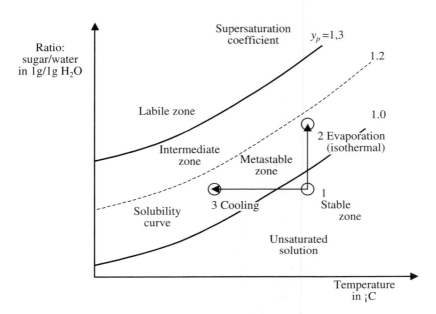

FIGURE 4.5 Dry substance and sugar/water ratio of pure saturated ($y_p = 1$) and supersaturated ($y_p = 1.20$, $y_p = 1.30$) sucrose solution as a function at temperature.

$$\frac{dm}{Adt} = B_k \, (C - C_0) \qquad (4.3)$$

where B_k is the coefficient of mass transfer of sucrose from the solution phase to the solid phase and $(C - C_0)$ expresses the supersaturation of solution. This equation is valid for the growth rate in all the surface area of the crystal slurry.

When the desired size of crystals (above 0.5 to 0.7 mm) of the commercial A white sugar, the desired number of crystals (125×10^9 in the massecuite volume of 35 m³), and the amount of mother liquor known as A massecuite (starting from 10 tons and ending on 50 tons) are achieved, the crude product passes from the vacuum pan into the mixer. The massecuite from the mixer is then sent to A centrifugals batch. Following one or two washings with hot water, the wet A white sugar crystals are discharged from the centrifugal basket and sent to the dryer or granulator and to the cooler. The granulated A white sugar is then screened and stored in bulk delivery or packaging.

4.2.6 COMMERCIAL SUGAR VARIETIES

To satisfy a variety of consumer demands, a variety of sucrose products is available on the market. One of the differentiations, for example, is the energy value. Thus, sugar compounded with other sweeteners such as D-fructose, D-glucose, sugar alcohols (Chapter 3), and nonsaccharide sweeteners (Chapter 24) are common in several countries as sweeteners of low-calorie value; commercial sucrose with different size and shape of crystals used for texturizing and decorating of food

(Chapter 11) is another commercial form of the product. The normal crystalline form, the fine, granulated sucrose powder (frequently stabilized with approximately 0.3% admixture of starch to prevent aggregation caused by the hygroscopicity), is used as fondant and icing sugar. Esthetic factors may also play a part in creating demand for icing sugar or sugar cubes, or, for more exalted tastes, aromatized or colored sugar crystals. Instant sugar and sugar syrups are produced for those who value convenience. Brown sugars (golden brown and old fashioned brown) are used by those who value their specific flavor.

4.2.7 BY-PRODUCTS AND WASTE

Essentially, there are three by-products of beet sugar manufacture: molasses, pulp, and lime mud. Molasses, a dark-brown viscous liquid, contains 48 to 52 wt% of noncrystallizing sucrose. It is rich in potassium salts; salts of sodium, magnesium, and calcium are also present, but in one order lower concentration.

Apart from glutamic acid, some sucrose can be recovered from molasses. It is used mainly as animal fodder and as raw material in the production of ethanol and yeast. Lime mud is used as alkalizing fertilizer and pesticide protecting stored beets prior to their processing. Recovery of calcium oxide is also possible. Pulp is generally used as cattle fodder; however, it can also be dried and used as fuel in either the original or palletized form. Methods of turning either wet or dry pulp into biodegradable plastics have been recently patented.

4.3 PRODUCTION OF CANE SUGAR

Cane harvesting and loading can be manual or mechanical. Harvesting machines cannot keep the sugar cane as clean and as undamaged as field workers can. Cane transport and unloading is a material-handling problem. Whenever the cane is chopper-harvested, unloading it becomes important as its condition can shorten the cut-to-crush period. Delays always lead to losses of sugar and a number of difficulties in all production stages.

Extraction of sugar from cane by crushing or grinding is carried out in a milling plant comprised of a series of roller mills. The separation of sugar involves a volumetric reduction on dilution of residual juice by counterflow washing. For best results, the imbibition liquor applied to the feed of the final mill is usually at 85°C. The juice collected from the first and second mill, the so-called mixed juice, is returned to the process. The whole milling process usually takes about 20 min. Final bagasse serves as fuel.

A simple defecation of raw juice is the most common purification method used in cane sugar factories. Preliming and main liming by adding 1.2 to 1.6% lime milk to beets is used. The purification process consists of the following steps: (1) addition of milk (milk of lime) to pH 7, (0.2% CaO on juice), (2) warming to about 103°C, and (3) addition of a flocculating agent at the entrance to the clarifier. The clarification time varies between 1 and 8 h and depends on flocculating agent used, the cane quality, and the type of clarifier. The scum from clarifiers (precipitate) is mixed with the final bagasse, filtered, and desweetened on a rotary vacuum filter.

Either sulfitation (addition of SO_2) of cane juice prior to the liming or liming prior to the sulfitation is commonly used to achieve mill white sugar. Better sedimentation of slurry is achieved when the sulfitation precedes liming. Calcium sulfite is precipitated in the process and the clarified juice with 12–16% solids is passed through a multiple-effect evaporator, producing a thick juice containing 60–70% of dry substance.

Cane juice, after concentration in evaporators, is of lower purity than that of the thick beet juice. It contains more reducing sugars (5–13%) and high-molecular components (0.4–3%). Commercial cane sugar is crystallized in three stages. The C massecuite is centrifuged in continuous centrifuges to yield C sugar and final molasses. The C sugar of about 85% purity is mixed into a magma, which is used as a seed for the A and B crystallization stages.

4.3.1 By-Products and Waste

Bagasse is the major by-product in cane sugar production. It is used, first of all, as the source of energy available from combustion. Methane can also be produced from bagasse. Other uses of bagasse are composting for fertilizer for cane plantation, paper and board manufacture, and cattle fodder. Methods of utilization of molasses and lime mud are common for both technologies of sucrose production.

4.4 PRODUCTION OF OTHER SACCHARIDES ESSENTIAL IN FOOD TECHNOLOGY

Apart from sucrose, D-fructose, D-glucose, and lactose have found wide application in food technology. Importance of D-fructose arises, most importantly, from its insulin-independent metabolism, nonpromoting formation of cavities, and hygroscopicity, the latter frequently employed to preserve moisture in foodstuffs. Its sweetness, higher than that of sucrose (Chapter 5), is an additional advantage of that saccharide. D-Glucose is most commonly used as the source of the "fast" energy in the recovery stage after various medical treatments or strenuous activity. Lactose is frequently used as a carrier of other sweeteners and as an additive to improve the taste of dairy products and the appearance of microwave-heated foodstuffs.

D-Fructose and D-glucose are currently produced in biotechnological processes, preferably from enzymatically saccharified starch (Chapter 10). In the first part of the last century, considerable attention was paid to saccharification of wood in the acid-catalyzed processes as the source of D-glucose.

Lactose is moderately soluble in water (ca. 20 wt% at 20°C). It is most readily available from whey (almost the 5% content) by careful evaporation in order to achieve crystals in the preferred β form. Fast evaporation results in the crystalline α form of lactose which, when present in food, produces sandiness.

REFERENCES

1. McGinnis, R.A, *Beet-Sugar Technology*, 3rd ed., Beet Sugar Development Foundation, Fort Collins, Colorado, 1982.
2. van der Poel, P.W., Schiweck, H., and Schwartz T., *Sugar Technology: Beet and Cane Sugar Manufacture*, Verlag Dr. Albert Bartens KG, Berlin, 1998.
3. Grabka, J., Separation of colloidal precipitates from flume water using coagulants and flocculants, in *1999 CITS Antwerpen Proceedings,* Verlag Dr. Albert Bartens, Berlin, 1999, p. 375.
4. Grabka, J., and Baryga, A., Effect of removing sugar beets pulp during raw juice purification, *Int. Sugar J.*, 103, 310, 2001.
5. Grabka, J. and Baryga, A., Purification of raw juice with activated defeco-carbonation deposit, in *1999 CITS Antwerpen Proc*eedings, Verlag Dr. Albert Bartens, Berlin, 1999, p. 379.
6. Grabka, J., Eine neue Slurry zum Saatimpfen von Kohlmassen, *Z. Zuckerind.*, 114, 467, 1989.

5 Saccharide Sweeteners and the Theory of Sweetness

Władysław Pietrzycki

CONTENTS

5.1 Introduction 57
5.2 AH,B Shallenberger–Acree Glycophore 58
5.3 RSj: Relative Sweetness of Saccharides 58
5.4 AH,B,X and AH,B,γ Triangle Sweeteners 58
5.5 B,AH,XH,G1–G4 Heptagonal Sweetener for Saccharides 61
5.6 Sugar Electron Structure–Sweetness Relationships [Quantitative
    Structure Activity Relationships (QSAR) Linking Sweetness and Quantum
    Chemical Parameters] 63
5.7 Domination of the B1,XH1 Glycophore in Aldopyranoses 65
5.8 Activity of E1, E4 Electron Donors and G1, G4 Dispersion Sweeteners
    in the Sucrose/Galactosucrose System 66
5.9 Biochemistry of Sweet-Taste Transduction 69
References 70

5.1 INTRODUCTION

Sensory chemistry, especially the chemistry of taste and smell, has become a common field of food chemistry. However, in the past, problems of sensory impressions belonged to the sphere of philosophy. Sensualism often occurred as an element of epistemology. Later, sensory impressions became a subject of medical investigations. Sensory physiology and "chemical senses"[1] opened wide the door to taste and olfactory chemistry. Developments in molecular biology and biochemistry have provided the background for sweet-taste chemistry, making it one of the most modern and attractive domains of bioorganic as well as food chemistries.

5.2 AH,B SHALLENBERGER–ACREE GLYCOPHORE[2]

It was observed at the turn of the 20th century that pairs of functional groups such as hydroxyl groups, amino groups, and the oxygen atom in ethers were usually present in the structure of sweet-tasting compounds. They were called glycophores.[2,3] The sweet-taste-eliciting group in sugars was the glycol (–CHOH—CHOH–) unit. Shallenberger marked the sweetener site pair of glycophores as AH and B, an acid and base site, respectively. Birch and Lee[4] have accepted that in aldopyranoses, the hydrogen atom of the 4-OH hydroxyl group is the AH site, whereas B is the O-3 atom of the 3-OH group. The AH and B sites were differently localized in ketopyranoses. The AH and B sites were recognized as the hydrogen atom of the 2-OH group and the O-1 atom, respectively. Figure 5.1 presents the localization of the AH,B Shallenberger glycophore in the most common aldo- and ketopyranoses. The sweet taste is generated when the *ah,b* dipoles of a sweet receptor interacted with the AH,B glycophore, resulting in the formation of two hydrogen bonds (Figure 5.2).

5.3 *RS*$_j$: RELATIVE SWEETNESS OF SACCHARIDES

The experimental relative sweetness [RS_j (%/%), or RS_j] usually expresses the sweetness of a 10% aqueous solution of a *j*th sugar measured with respect to the sweetness of a 10% aqueous solution of sucrose (the standard). Such a method is very useful when the molecular structure and weight of a given sugar are unknown. Table 5.1 lists the RS_j (%/%) values for several saccharides.

The concentration of solutions in molecular theories is frequently expressed in mole/dcm^3. In such cases, the RS_j (mole/mole) sweetness attributed to equimolar solutions is considered. Transformation of RS_j (%/%) into RS_j (mole/mole) involves Equation 5.1:

$$RS_j\left(\frac{mole}{mole}\right) = \frac{d_0 M_j}{d_j M_0} RS_j\left(\frac{\%}{\%}\right) \tag{5.1}$$

where M_j and M_0 are molecular weights of the *j*th and standard sugar (sucrose), respectively, and d_j and d_0 the densities of the *j*th sugar and sucrose solution, respectively. Frequently, especially for diluted solutions, $dj = d_0$.

5.4 AH,B,X AND AH,B,γ TRIANGLE SWEETENERS[5,6]

A two-point *ah,b* receptor cannot be stereospecific with respect to the AH,B glycophore. Thus, it does not explain the sweetness of D-leucine and lack of the sweet taste of L-leucine. Kier[5] has modified the two-point glycophore approach into the AH,B,X triangle model. In a triangle model, the AH,B glycophore retains its meaning and role, and the third, additional X point interacts with the receptor by dispersion or hydrophobic interactions, or both. The A–B, A–X, and B–X distances in the AH,B,X glycophore (Figure 5.3b) are accepted as ~0.26 nm, ~0.35 nm, and ~0.55 nm, respectively. The presence of such a triangle in many amino acids explains their

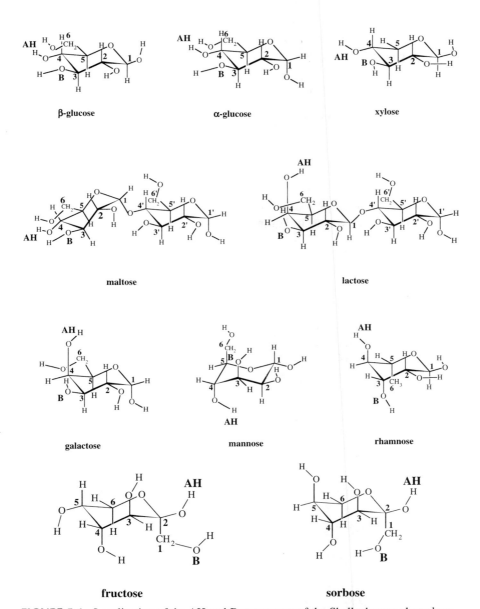

FIGURE 5.1 Localization of the AH and B sweeteners of the Shallenberger glycophore in aldo- and ketopyranoses.

sweet taste. Figure 5.3a shows the AH,B,X glycophore of D-phenylalanine. A triangle sweetener theory formulated for amino acids also contributes to the theory of sweet taste of saccharides. It may explain the high sweetness of sucrose ($RS_j = 1.000$ mole/mole) with respect to α-D-glucopyranose ($RS_j = 0.368$ mole/mole) and maltose ($RS_j = 0.330$ mole/mole). In terms of the two-point AH,B theory, these three sugars

```
     A–H                 b            K          A–H----- b
      |          +        |      ------------>     |        |
      B                  h–a      <------------     B-------h–a

  Saccharide          Sweet taste              Glycophore-receptor
  glycophore           receptor                    complex
```

FIGURE 5.2 The glycophore–receptor complex according to the Shallenberger concept. K is the equilibrium constant of this taste-generating reaction. The equilibrium constant for the analogous reaction of the standard saccharide is K_0.

TABLE 5.1
Relative Sweetness, RS_j (%/%), of the Most Common Mono- and Disaccharides Measured in Equipercent Solutions

Saccharide	Abbreviation	RS_j^{expl} (%/%)
1′,4,6′- Trichloro-1′,4,6′-trideoxygalactosucrose	1′,4,6′- TriClG	2000
1′,4,6,6′-Tetrachloro-1′,4,6,6′-tetradeoxygalactosucrose	1′,4,6,6′-TetraClG	1000
1′,4-Dichloro-1′,4-dideoxygalactosucrose	1′,4,6′- TriClG	600
1′,6′-Dichloro-1′,6′-dideoxysucrose	1′,6′-DiClS	500
1′,4,6,6′-Tetrachloro-1′,4,6,6′-tetradeoxysucrose	1,4,6,6′-TetraClS	200
6′- Chloro-6′-deoxysucrose	6′- ClS	20
1′-Chloro-1′-deoxysucrose	1′-ClS	20
4-Chloro-4-deoxysucrose	4-ClS	5
β-D-Fructopyranose	β-D-Fructose	1.72
2-β-D-Fructofuranosyl-α-D-glucopyranoside	Sucrose	1.00
α-L-Sorbopyranose	α-L-Sorbose	0.86
β-D-Glucopyranose	β-D-Glucose	0.80
α-D-Glucopyranose	α-D-Glucose	0.70
β-D-Xylopyranose	β-D-Xylose	0.40
4′-O-(α-D-Glucopyranosyl)-α-D-glucopyranose	Maltose	0.33
6-Deoxy-α-L-mannopyranose	Rhamnose	0.33
α-D-Galactopyranose	α-D-Galactose	0.32
α-D-Mannopyranose	α-D-Mannose	0.30
4′-O-(β-D-Galactopyranosyl)-α-D-glucopyranose	Lactose	0.20
2-β-D-Fructofuranosyl-α-D-galactopyranoside	Galactosucrose	~0.20

Note: A 10% aqueous sucrose solution is used as the standard

should indicate almost the same sweetness because of the same geometry of the AH,B glycophore. Sucrose (Figure 5.3c) apart from this glycophore carries, additionally, the 2-OH–O1′–O6′ triangle as the AH,B,X Kier sweetener.

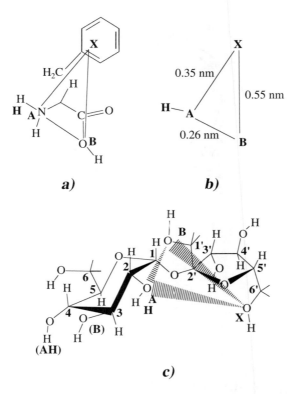

FIGURE 5.3 AH,B,X Kier triangle sweetener. (a) In D-phenylalanine. (b) Typical dimensions. (c) In sucrose.

If a sweet molecule is not planar, two configurations of AH,B,X are likely, of which only one precisely fits the *ah,b,x* stereospecific receptor.

Subsequently, Shallenberger introduced the three-point AH,B,γ glycophore[6] in which the γ point had a mixed lipophilic and hydrophobic character. In this theory, dimension of a triangle is variable. The A-γ and B-γ distances for dipeptides are higher than for the sweet α-amino acids.

5.5 B,AH,XH,G1–G4 HEPTAGONAL SWEETENER FOR SACCHARIDES[7]

Nofre and Tinti[7] formulated the multipoint attachment theory (MPA), which might explain a binding of sweet ligands by the receptor in a transmembrane (TM) pocket. A saccharide sweetener has seven interaction sites: B, AH, XH, G1, G2, G3, and G4. Each site is composed of two interaction points, subsites (B1,B2), (AH1,AH2), (XH1,XH2), (E1,G1), (E2,G2), (E3,G3) and (E4,G4), respectively. Figure 5.4 presents the structure of this sweetener in the sweet-taste receptor pocket.

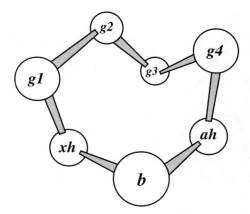

FIGURE 5.4 Localization of all the 14 subsites of Nofre–Tinti saccharide sweetener in a sweet-taste receptor pocket.

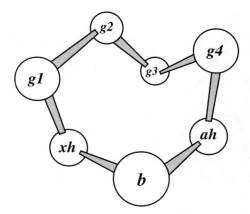

FIGURE 5.5 Nofre–Tinti sweetener subsites in the sucrose molecule. (From Nofre, C., and Tinti, J.M., *Food Chem.*, 56, 263, 1996. With permission.)

Figure 5.5 shows the localization of all 14 subsites in sucrose. The AH1, AH2, XH1, and XH2 glucopyranose points (the hydrogen atoms in the 3-OH, 2-OH, 4-OH, and 6-OH groups, respectively) have positive atomic net charges. They either interact via electrostatic forces or may be a part of acceptors of the *n*-electron pairs (the hydrogen bond donors). The B1 and B2 subsites are the oxygen atoms of the 4-OH and 3-OH glucopyranose groups, respectively. They carry negative atomic net charges. They either interact with the sweet-taste receptor via electrostatic forces or they can be donors of the *n*-electron pairs (the hydrogen bond acceptors). The hydroxyl oxygen atoms of the fructopyranose moiety are the E1–E4 subsites. They

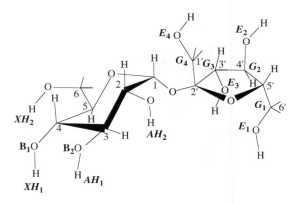

FIGURE 5.6 Structure of the heptagonal saccharide receptor. Approximate distance between recognition sites (b–ah, ah–g4, g4–g3, …) is 0.65 nm. (From Pietrzycki, W., *Pol. J. Chem.*, 76, 875, 2002. With permission.)

are the *n*-electron donors, exclusively. In turn, the G1–G4 subsites represent small, weakly polar groups of atoms, such as the CH_2 and CH groups of the fructofuranose moiety. The latter subsites dispose with small atomic net charges and, therefore, cannot interact with the receptor by electrostatic forces and are unable to form the electron donor–electron acceptor centers. The G1–G4 points are assumed to interact with receptor recognition sites via dispersion forces. Saccharides do not possess the eighth sweetener site, D. In some cyano compounds, for instance, D is attributed to the 4-cyanophenyl and other groups with the CN moiety.

The heptagonal saccharide receptor (Figure 5.6) is composed of seven amino acid residues, which form receptor dipoles and recognition sites, *b,ah,xh,g1–g4*, corresponding to the sweetener sites. The lysine residue in Figure 5.5 binds the B site of the sweetener, and either the aspartic acid or glutamic acid residues bind the AH and XH saccharide sites. In turn, the threonine residue interacts with the E1–E4 and G1–G4 sweetener subsites. In the octagonal nonsaccharide sweetener, the side chain of serine (CH_2OH) or threonine (CH_2CHOH) residues interacts with D sweetener sites.

5.6 SUGAR ELECTRON STRUCTURE–SWEETNESS RELATIONSHIPS[8,9] [QUANTITATIVE STRUCTURE ACTIVITY RELATIONSHIPS (QSAR) LINKING SWEETNESS AND QUANTUM CHEMICAL PARAMETERS]

Höltje and Kier[10] found that experimental $\log RS_j^{expl}$ for a set of sweet 1-X_j-2-amino-4-nitrobenzenes linearly correlated with calculated X_j-glycophore–receptor intermolecular interaction energies, $E_{int}(j)$.

$$\log\left(RS_j^{\mathrm{expl}}\right) = b_1 E_{\mathrm{int}}(j) + b_0 \qquad (5.2)$$

Thus, $\log(K_j/K_0)$ in a usual QSAR correlation equation is equivalent to $\log RS_j^{\mathrm{expl}}$ for an arbitrary jth sweetener. Simulation of the $E_{\mathrm{int}}(j)$ interaction energies in Equation 5.2 is formally based on the perturbation calculus with the electron exchange.[11,12] Assuming a small overlap of molecular wave functions, $E_{\mathrm{int}}(j)$, the total glycophore–receptor interaction energy, is a sum of $E_{elst}^{(2)}$ electrostatic, $E_{CT}^{(2)}$ charge-transfer, and $E_{disp}^{(2)}$ dispersion components [Equation 5.3]:

$$E_{\mathrm{int}}(j) = E_{elst}^{(1)}(j) + E_{CT}^{(2)}(j) + E_{disp}^{(2)}(j) \qquad (5.3)$$

for $j = 1, 2, 3, \ldots n$ sugars in a set.

Combination of Equation 5.3 with Equation 5.2 and inclusion of intermolecular interaction integrals into C-type coefficients (interaction integrals are multiplied by the b_1 coefficient in Equation 5.2) lead to correlation Equation 5.4:

$$\log\left(RS_j^{\mathrm{expl}}\right) = C_0 + \sum_{a\varepsilon Gly} C_a Q_a(j) + \sum_{i\varepsilon Gly}^{occ} \frac{C_i'}{\varepsilon_{LU}(Rec) - \varepsilon_i(j)} + \sum_{x\varepsilon Gly}^{unocc} \frac{C_x'}{\varepsilon_x(j) - \varepsilon_{HO}(Rec)} +$$

$$+ \sum_{i\varepsilon Gly}^{occ} \sum_{x\varepsilon Gly}^{unocc} \frac{C_{ix}''}{\varepsilon_x(j) - \varepsilon_i(j) + \varepsilon_{LU}(Rec) - \varepsilon_{HO}(Rec)}, \qquad j = 1, 2, 3, \ldots n \quad (5.4)$$

In Equation 5.4, $\varepsilon(j)$ and $\varepsilon(Rec)$ represent the j-saccharide and receptor molecular energies, respectively. This equation correlates $\log RS_j^{\mathrm{expl}}$ with the m quantum chemical parameters, that is, $Q_a(j)$ atomic net charges, $[\varepsilon_{LU}(Rec) - \varepsilon_i(j)] - 1$ and $[\varepsilon_x(j)\varepsilon\varepsilon_{HO}(Rec)] - 1$ charge-transfer orbital and $[\varepsilon_x(j) - \varepsilon_i(j) + \varepsilon_{LU}(Rec) - \varepsilon_{HO}(Rec)] - 1$ dispersion orbital parameters for the jth saccharide in a set. These parameters are available from a quantum population analysis for the closed-shell configurations of the isolated, attributed to a receptor recognition site, the jth saccharide, and amino acid molecules in their ground states. The PM3 semiempirical quantum method in HyperChem-5 standard is subsequently applied. Initial geometry of the molecules is estimated by the *Add H & Build Model* function from the HyperChem *Build* menu and then followed by the MM+ procedure. The *Gly* index represents a sugar molecule.

All C_0, C_a, C', and C'' coefficients are dependent on the receptor geometry and include glycophore–receptor intermolecular interaction integrals. The C_a coefficients are attributed to the electrostatic terms and the C_i' and C_x' coefficients represent charge-transfer interaction integrals associated with the $i(Gly) \rightarrow$ LUMO(Rec) and HOMO(Rec) $\rightarrow x$ (Gly) transitions, respectively, where LUMO and HOMO are related to the lowest unoccupied and the highest occupied molecular orbitals, respectively. In turn, the C'' coefficients represent dispersion interaction integrals attributed to the coupling of the sugar $i \rightarrow x$ MOs electron excitations with the receptor HOMO $\rightarrow$ LUMO excitation. Because the exact receptor geometry is unknown, these coefficients are calculated by mathematical statistics according to Equation 5.4. It is accepted that Equation 5.4 is a multiple regression equation. It applies to similar sugars, for example, aldopyranoses, which react with the same receptor. In such cases, the resulting glycophore–receptor complexes have a similar geometry for all sugars in the set and the C_0, C_a, C'_i, and C''_{ix} unknown coefficients in Equation 5.4 are common and constant for all the sugars in the set of $j = 1,2,3, \ldots n$. These coefficients can be found by solving Equation 5.4 as a system of n linear equations with m unknown quantities, $m < n$, using the least-square procedure. If the R coefficient of such multiple correlation exceeds 0.95 with $m \approx \frac{1}{2} n$, it points to a strong dependence of sweetness on the electron structure of the saccharides.

The method presented permits finding a moiety of a sugar molecule responsible for an interaction with the sweet-taste receptor and describes a general mechanism of the saccharide–receptor interaction. It provides the recognition of the chemoreception properties together with a semiempirical expression useful in determination of RS_j of an arbitrary, unknown saccharide, because its structure is similar to that of sugars from the set under consideration.

5.7 DOMINATION OF THE B1,XH1 GLYCOPHORE IN ALDOPYRANOSES[8]

Atomic net charges on the O-4 oxygen and hydrogen atoms in the 4-OH group of aldopyranoses (Figure 5.1) excellently correlate with $\log RS_j^{expl}$ (mole/mole) (Table 5.2). The corresponding correlation coefficients, Rs, are 0.981 and 0.964, respectively. These atoms are identified as B1 and XH1 sweetener subsites in the Nofre–Tinti heptagonal sweetener. The relations B1 = A and XH1 = H are valid for the AH,B Shallenberger glycophore (Figure 5.2). However, generally, XH1 is identified as AH. Because of the above-mentioned property, QSAR multiple correlation Equation (5.4) reduces to the two-parameter Equation (5.5):

$$\log\left(RS_j\right) = -6.8368 \cdot Q_j(\text{B1}) + 2.5937 \cdot Q_j(\text{XH1}) - 3.3754 \qquad (5.5)$$

with $R = 0.9832$ and $S_2 = 0.0015$.

Three constants, −6.8368, 2.5937, and −3.3754, satisfactorily estimate the RS_j values for all the eight aldopyranoses (Table 5.2). The QSAR equation leads to the following conclusions: (1) The B1 and XH1 sweetener subsites situated on the 4-OH group in aldopyranoses have pure electrostatic centers and they determine the sweetness of these compounds. (2) The B2 subsite, equivalent to the B

TABLE 5.2
QSAR Correlation Analysis of Sweetness for Aldopyranoses, RS_j, by Equation 5.5[a]

jth Saccharide	$Q_{XH1}(j)$	$Q_{B1}(j)$	RS_j (mole/mole)		RS_j (%/%)	
			Calc.	Expl.	Calc.	Expl.
β-D-Glucose	0.23087	−0.34304	0.370	0.421	0.704	0.800
α-D-Glucose	0.22910	−0.34333	0.368	0.368	0.700	0.700
α-Maltose	0.22885	−0.34170	0.358	0.330	0.358	0.330
α-Lactose	0.21109	−0.31812	0.222	0.200	0.222	0.200
β-D-Xylose	0.20607	−0.30594	0.178	0.175	0.406	0.400
α-D-Galactose	0.19458	−0.30438	0.162	0.168	0.308	0.320
Rhamnose	0.19108	−0.30507	0.161	0.158	0.335	0.330
α-D-Mannose	0.19600	−0.29832	0.149	0.158	0.283	0.300
R	0.9644[b]	0.9812[c]	0.9832[d]			

[a] Atomic net charges on the XH1 and B1 glycophore atoms, $Q(j)$ $Q(j)$, were calculated by the PM3 quantum method.
[b] Correlation coefficient for the single correlation of $\log RS_j^{expl}$ against $Q_{XH1}(j)$.
[c] For the single correlation of $\log RS_j^{expl}$ against $Q_{B1}(j)$.
[d] Correlation coefficient for the multiple correlation of $\log RS_j^{expl}$ against either $Q_{XH1}(j)$ or $Q_{B1}(j)$ parameters, Equation 5.5

Shallenberger sweetener, is inactive in aldopyranoses. This subsite becomes active in ketopyranoses,[8] but it indicates pure *n*-electron donor character. (3) Sweetness of maltose and lactose depends on the 4-OH group of one nonreducing terminal aldopyranose moiety.

5.8 ACTIVITY OF E1, E4 ELECTRON DONORS AND G1, G4 DISPERSION SWEETENERS IN THE SUCROSE/GALACTOSUCROSE SYSTEM[9]

Sucrose, galactosucrose, and their eight deoxychloro derivatives are considered (Figure 5.7). To illustrate the nature of the sweetener–receptor interaction, QSAR correlations (Equation 5.4) are carried out for m = 5. These constants are calculated by the least-square procedure. Simultaneously, the numerically overlapped R = max condition allows one to find the character of four quantum parameters. The following QSAR correlation equation results:

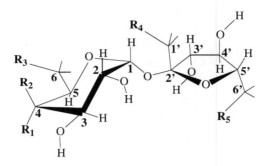

FIGURE 5.7 Structure of sucrose and galactosucrose deoxychloro derivatives. The following is an enumeration of the carbon atoms (1-6, 1'-6'), an arrangement of the Nofre–Tinti sweetener subsites (XH$_1$–G$_4$), and situation of the R substituents that form the following derivatives:

	R_1	R_2	R_3	R_4	R_5	
1.	H	Cl	OH	Cl	Cl	1',4,6'-Trichloro-1',4,6'-trideoxygalactosucrose
2.	H	Cl	Cl	Cl	Cl	1',4,6,6'-Tetrachloro-1',4,6,6'-tetradeoxygalactosucrose
3.	H	Cl	OH	OH	OH	1',4-Dichloro-1',4-dideoxygalactosucrose
4.	OH	H	OH	Cl	Cl	1',6'-Dichloro-1',6'-dideoxysucrose
5.	Cl	H	Cl	Cl	Cl	1',4,6,6'-Tetrachloro-1',4,6,6'-tetradeoxysucrose
6.	OH	H	OH	OH	Cl	6'-Chloro-6'-deoxysucrose
7.	OH	H	OH	Cl	OH	1'-Chloro-1'-deoxysucrose
8.	Cl	H	OH	OH	OH	4-Chloro-4-deoxysucrose
9.	OH	H	OH	OH	OH	Sucrose
10.	H	OH	OH	OH	OH	Galactosucrose

(From Pietrzycki, W., *Pol. J. Chem.*, 76, 875, 2002. With permission.)

$$\log\left(RS_j\right) = -67.387 + 2.9473 \cdot Q_{XH2}(j) + \frac{526.705}{\varepsilon_{LU}^{Rec} - \varepsilon_5(j)} +$$

$$+ \frac{1556.877}{\varepsilon_2^{\#}(j) - \varepsilon_2(j) + \varepsilon_{LU}^{Rec} - \varepsilon_{HO}^{Rec}} - \frac{1038.128}{\varepsilon_2^{\#}(j) - \varepsilon_7(j) + \varepsilon_{LU}^{Rec} - \varepsilon_{HO}^{Rec}}$$

(5.6)

with $R = 0.9996$ and $S_2 = 0.0815$. The sign # denotes an unoccupied orbital. The theoretical relative sweetness (RS_j) values, calculated from Equation 5.6, are listed in Table 5.3.

Equation 5.6 describes a nature of the saccharide–receptor interaction denoted as {XH2, 5, 2 → 2#, 7 → 2#}. Details of this interaction for individual sugars are presented in the last four columns of Table 5.3. They are calculated in relation to the corresponding energies of galactosucrose (G). *ELST* electrostatic, *CT* charge-transfer, and *DISP* dispersion contributions to the interaction energies are calculated from Equation 5.6 as:

$$E_{elst}^{(1)}(j) = -2.303 \cdot RT \times 2.9473 \cdot Q_{XH2}(j) , \quad ELST_j = E_{elst}^{(1)}(j) - E_{elst}^{(1)}(G) \quad (5.6a)$$

TABLE 5.3
QSAR Correlation Analysis of Relative Sweetness, RS_j, of
Sucrose/Galactosucrose Chlorodeoxy Derivatives by Equation 5.6

*j*th Saccharide	RS_j (mole/mole)		Sugar–Receptor Interaction Energy (kJ/mole)			
	Calculated		ELST	CT	DISP	
	(Eq. 5.6)	Expl.	(Eq. 5.6a)	(Eq. 5.6b)	(Eq. 5.6c)	E_{tot}
1', 4, 6'-TriClG	2656.06	2323.26	−0.013	−16.389	−6.786	−23.188
1',4,6,6'-TetraClG	1230.95	1215.52	3.180	−15.732	−8.749	−21.301
1',4-DiClG	556.59	664.65	−0.059	−12.439	−6.899	−19.397
1',6'-DiClS	537.92	553.87	−0.142	−15.539	−3.594	−19.276
1',4,6,6'-TetraClS	239.14	243.10	3.180	−15.698	−4.761	−17.280
6'ClS	20.05	21.08	−0.155	−9.498	−1.548	−11.201
1'-ClS	23.15	21.08	−0.201	−8.581	−2.778	−11.560
4-ClS	5.45	5.27	0.013	−4.845	−3.167	−8.000
Sucrose	0.94	1.00	−0.184	−3.146	−0.372	−3.703
Galactosucrose	0.21	0.20	0.000	0.000	0.000	0.000
R	0.9996[a]					
S^2	0.0816[b]					

[a] Correlation coefficient of the multiple regression, Equation 5.6.
[b] Remainder variance of the above multiple regression.

$$E_{CT}^{(2)}(j) = -2.303RT \cdot \left(\frac{526.705}{\varepsilon_{LU}^{Rec} - \varepsilon_s(j)} \right), \quad CT_j = E_{CT}^{(2)}(j) - E_{CT}^{(2)}(G) \qquad (5.6b)$$

$$E_{disp}^{(2)}(j) = -2.303RT \cdot \left(\frac{1556.877}{\varepsilon_2^{\#}(j) - \varepsilon_2(j) + \varepsilon_{LU}^{Rec} - \varepsilon_{HO}^{Rec}} - \frac{1038.128}{\varepsilon_2^{\#}(j) - \varepsilon_7(j) + \varepsilon_{LU}^{Rec} - \varepsilon_{HO}^{Rec}} \right)$$

$$DISP_j = E_{disp}^{(2)}(j) - E_{disp}^{(2)}(G) \qquad (5.6c)$$

Primary sweetness effects belong to the *CT* sugar–receptor interactions. They are included in this term as the 526.705 coefficient, which describes the *n*-electron transfer from the fifth occupied saccharide orbital to the receptor LUMO. All five, the most sweet sugars (RS_j = 200–2000), have a high *CT* energy constituting about 60% contribution to the total, E_{tot}, energy (Table 5.3). The LCAO (linear combination of atomic orbitals) structure of the 5th occupied orbital indicates a significant domination of the fructofuranose Cl-1' and Cl-6' atoms, identified as the E4 and E1 sweetener subsites, respectively.

Secondary sweetness effects originate from the saccharide–receptor dispersion interactions. They are represented by two 2→2# and 7→2# sugar molecular orbitals (MO) excitations coupled with the HOMO → LUMO receptor transition (terms with 1556.877 and 1038.128 coefficients in Equation 5.6, respectively). For five of the sweetest chlorosaccharides, the absolute values of the *DISP* energies are about 2.42 kJ/mole higher than the *DISP* energies for the remaining derivatives (Table 5.3). The LCAO structure of the 2# empty orbital indicates a higher contribution of fructofuranose $1'$-CH$_2$ and $6'$-CH$_2$ groups, which are identified as the G4 and G1 sweetener subsites (Figure 5.5).

The tertiary effects, abbreviated XH2, belong to the electrostatic interactions of the *Q*XH2 net charges. It is observed that deoxychlorination at the 6th position of the glucopyranose moiety reduces sweetness. Thus, the less chlorinated $1',4,6'$-triClG is twice as sweet as the more chlorinated $1',4,6,6'$-tetraClG. Analogously, $1',6'$-diClS is twice as sweet as $1'4,6,6'$-tetraClS. The substitution of the 6-OH group with the weakly charged Cl-6 chlorine atom ($Q \approx -0.050$) reduces the *ELST* energy component by about 3 kJ/mole in $1',4,6,6'$-tetraClG.

5.9 BIOCHEMISTRY OF SWEET-TASTE TRANSDUCTION[13]

The formation of a sugar–receptor complex is the most important phase in the first stage of the mechanism of sweet-taste development. At this stage, energy stimulus is produced in the receptor cell, which initiates a chain of biochemical reactions essential for neurotransmitter release. Adsorption of neurotransmitter molecules on sensory nerve fibers is a biochemical equivalent of the sweet-taste sensation.

A progress in the sweet-taste theory after 1990 was followed by a turning point in the development of biochemistry of receptors and molecular biology with cloning receptor DNA techniques.[13] Martin Rodbell and Alfred Gilman won the Nobel Prize in 1994 for discovering *G*-coupled proteins, which fulfil an important part of the transduction pathway of the sweet stimulus. Structure of the sweet-taste receptor is considered similar to the structure of other *G*-protein receptors.[14]

The receptor shows a polypeptide chain distinguished by seven transmembrane domain segments, TM I to TM VII helices, forming a pocket in which the sweet ligands are bound. The energy of formation of the ligand–receptor complex (about 21 kJ/mole for hydrogen bonds) is the origin of the stimulus for signal transduction. Figure 5.8 presents a schematic representation of the transduction mechanism for sweet taste. The sugar falls into a receptor box, which is coupled with a heterotrimetric GTP-binding regulatory protein of the Gs type. Consequently, the α-subunit of this Gs-protein presumably activates adenylyl cyclase (AC), which acting on ATP increases concentration of the intracellular second messenger, cyclic AMP (Figure 5.9).

The cAMP messenger may then stimulate a protein kinase A (PKA), which phosphorylates ion channels, causing their depolarization and opens a Ca^{2+} ionic channel. The intracellular calcium ion (Ca^{2+}) activity increases, leading ultimately to neurotransmitter release (Figure 5.8).

Nonsugar sweet substances have another transduction path mechanism also. Artificial sweeteners induce the production of inositol 1,4,5-triphosphate (IP3).[15] In

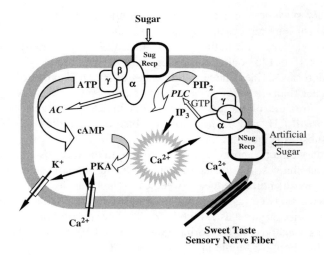

FIGURE 5.8
The biochemical transduction path of the sweet stimulus. GTP — guanosine triphosphate coupled to G-regulatory protein; α, β, and γ — G protein subunits; ATP — adenosine triphosphate; cAMP — cyclic adenosine monophosphate; AC — adenylyl cyclase; PLC — phospholipase C; PIP_2 — phosphatidylinositol biphosphate; IP_3 — inositol triphosphate.

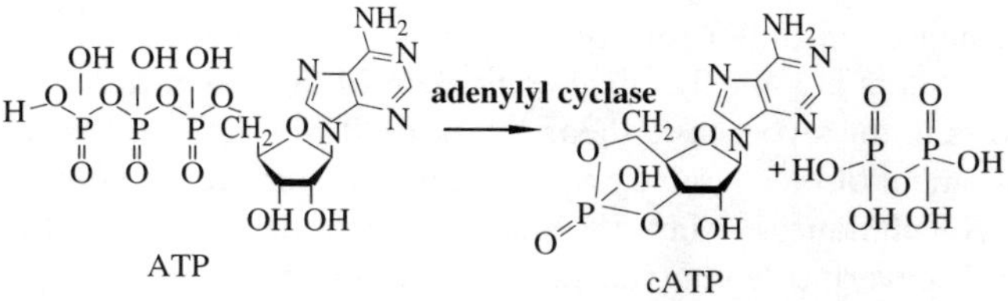

FIGURE 5.9 Transformation of adenosine triphosphate into cyclic adenosine monophosphate (cAMP) under the influence of adenylyl cyclase.

such a case, the artificial sweetener is adsorbed in a nonsugar receptor box. The arising stimulus is transferred to G-type protein, which activates phospholipase C (PLC). The latter, acting on phosphatidylinositol biphosphate (PIP_2), produces IP3 as a second messenger, which phosphorylates the intracellular plasma and produces an increase in the activity of Ca^{2+} ions in the receptor cell.[16] Intracellular Ca^{2+} ions activate sweet-taste nerve fibers, leading to neurotransmitter release.

REFERENCES

1. Acree, T.E., Chemical senses, Part 2, in *Handbook of Sensory Physiology,* Vol. 4, Beidler, L.M., Ed., Springer-Verlag, Berlin, 1971.
2. Shallenberger, R.S. and Acree, T.E., Molecular theory of sweet taste, *Nature,* 216, 480, 1967.

3. Shallenberger, R.S., Sweetness theory and its application in the food industry, *Food Technol.*, 52, 72, 1998.

4. Birch, G.G. and Lee, C.K., The theory of sweetness, *Dev. Sweeteners,* 1, 165, 1979.

5. Kier, L.B., A molecular theory of sweet taste, *J. Pharm. Sci.*, 61, 1394, 1972.

6. Shallenberger, R.S., *Sensory Properties of Foods*, Birch, C.G. Ed., Applied Science Publishers, London 1977, 91.

7. Nofre, C. and Tinti, J.M., Sweetness reception in man: multipoint attachment theory, *Food Chem.*, 56, 263, 1996.

8. Pietrzycki, W., QSAR computational model of Nofre–Tinti theory on sweetness of mono- and disaccharides composed by pyranose units, *Pol. J. Chem.*, 75, 1569, 2001.

9. Pietrzycki, W., Nofre–Tinti E1 and E4 sweetener subsites activity on high sweetness of chlorodeoxy derivatives of sucrose and galactosucrose, *Pol. J. Chem.*, 76, 875, 2002.

10. Höltje, H.D. and Kier, L.B., Sweet taste receptor studies using model interaction energy calculations, *J. Pharm. Sci.*, 63, 1722, 1974.

11. Sokalski, A. and Chojnacki, H., Approximate exchange perturbation study of intermolecular interactions in molecular complexes, *Int. J. Quant. Chem.*, 13, 679, 1978._

12. Pietrzycki, W., Noniterative analytical solution in the exchange perturbation theory of intermolecular interaction with the nonsymmetric zeroth-order hamiltonian, *Acta Phys. Polon.*, A76, 421, 1989.

13. Brand, J.G. and Feigin, A.M., Biochemistry of sweet taste transduction, *Food Chem.*, 56, 199, 1996.

14. Ruffolo, R.R., Jr. et al., Structure and function of α-adrenoceptors, *Pharmacol. Rev.*, 43, 475, 1991.

15. Bernhardt S.J. et al., Changes in IP3 and cytosolic Ca^{2+} in response to sugars and non-sugar sweeteners in transduction of sweet taste in the rat, *J. Physiol.*, 490, 325, 1996.

16. Eggers, S.C., Acree T.E., and Shallenberger R.S., Sweetness chemoreception theory and sweetness transduction, *Food Chem.*, 68, 45, 2000.

6 Honey

Helena Rybak-Chmielewska

CONTENTS

6.1 Introduction 73
6.2 Physical Properties of Honey 74
6.3 Chemical Composition of Bee Honey...74
    6.3.1 Saccharides 74
    6.3.2 Nonsaccharide Honey Components 75
        6.3.2.1 Proteins and Amino Acids 75
        6.3.2.2 Organic Acids 77
        6.3.2.3 Mineral Substances 77
        6.3.2.4 Dyes and Other Components 77
6.4 Honey Adulteration and Possibilities of Detection 77
References 78

6.1 INTRODUCTION

Honey is a natural sweet substance produced by honey bees from the nectar of plants (blossom honey), secretions of living parts of plants, or excretions of plant-sucking insects (*Hemiptera*) (honeydew honey). Bees collect honey, transform it by combining it with their specific substances, deposit, dehydrate, store, and leave in the honey comb to ripen and mature.

Honey consists essentially of various sugars, predominantly D-fructose and D-glucose, as well as other compounds and substances such as organic acids, enzymes, and solid particles collected by bees. The appearance of honey varies from nearly colorless to dark brown. It may be fluid, viscous, or solid. Its flavor and aroma vary depending on the plant origin. Honey varieties can be identified by their color, taste, flavor, and manner of crystallization. Under exceptional circumstances, the honey sediment is analyzed for the content of pollen grains. Alternatively, in honeydew honey varieties, other components characteristic of them, such as algae, spores, mycelium fragments, or leaf fragments, are determined. Other characteristics helpful in identifying the honey type include specific conductivity, variety-specific flavor compounds, and saccharide makeup.

6.2 PHYSICAL PROPERTIES OF HONEY

Density of honey at 20°C ranges from 1.38 to 1.45 g/cm³. Fresh honey is a dense liquid of substantial viscosity. The viscosity at 25°C varies between 1816.9 and 256.0 Pa/sec and is determined to some extent by the honey botanical provenance, but more substantially by its water content and temperature during examination. The variety-dependent differences in the viscosity cease with the increase in temperature. Above 40°C, water content within the range of 16.4 and 20% has no significant effect on viscosity. Viscosity depends not only on sugar concentration (water content) and temperature but also on content of dextrin, trisaccharides, and proteins. Rheological properties of heather honey, which is richer in protein and can turn into a gelatinous substance (a tixotropy), are of particular interest.

Duration of the liquid phase in honey depends on many external and intrinsic factors such as water content, ratio of D-fructose to D-glucose, ratio of D-fructose to nonsugars, and the dextrin content. High total sugar and D-glucose contents or low dextrin content favor crystallization. Temperature is the major ambient factor affecting crystallization. Formation of crystal nuclei in the first phase of crystallization, manifested by a subtle haze, occurs at between 5 and 8°C. The growth rate of the crystals is the highest at 13 to 17°C. Storage temperature affects not only the liquid phase but also the manner of crystallization. For instance, rapeseed honey, in which crystal nuclei are readily formed and crystallization proceeds very fast, solidifies at low temperature into a hard, fine-crystalline solid, whereas at higher temperatures (14 to 18°C) it forms a mass of lard-like consistency. Honey with fewer crystal nuclei crystallizes more slowly, forming larger crystals that tend to fall to the bottom of the container (buckwheat and honeydew honeys), resulting in a stratified structure. When honey is liquefied, its recrystallization is usually not uniform. Large crystal aggregates gradually concentrate throughout the container, leaving bulky domains of the liquid phase. The higher the degree of crystal dissolution during recrystallization, the less appealing is the appearance of recrystallized honey.

6.3 CHEMICAL COMPOSITION OF BEE HONEY

6.3.1 SACCHARIDES

The sugar composition of honey depends on the content of saccharides in the nectar or honeydew. The quantitative sugar composition and pollen analysis can be suitable indicators of the origin of honey.[2] For instance, a large amount of D-fructose in the nectar of *Labiatae* and *Epilobium* and a high glucose content in the nectar of rape (*Brassica napus*), forget-me-not (*Myosotis* sp.), and lime-tree (*Tilia cordata*) is reflected by a high content of those sugars in the corresponding honey varieties. Generally, the ratio of fructose to glucose in honey is close to unity, with D-fructose being the prevalent sugar.

The sucrose content in honeys (Table 6.1) randomly exceeds 1% of the total saccharide content. The level of maltose is often three times higher than that of sucrose. In nectar honeys the concentration of oligosaccharides reaches approximately 2% and is higher in honeydew honeys.

TABLE 6.1
Composition of Nectar Honey Saccharides

Saccharide	Content (%)
D-Fructose	35.7–41.7
D-Glucose	29.7–34.9
Maltose	1.6–3.8
Isomaltose	0.3–0.9
Turanose	0.9–1.7
Sucrose	0.4–1.4
Trehalose	0.6–1.1
Erlose	0.2–0.9
Maltotriose	0.04–0.14
Melecitose	0.25–1.1
Raffinose	0–0.4

D-Glucose and D-fructose can originate from the nectar or honeydew and from the enzymatic hydrolysis of sucrose and other sugars residing in the honey. Raffinose and mellecitose originate from the nectar or from honeydew. Other di- and trisaccharides result from the action of honey enzymes. Honey also contains certain amounts of dextrin (3 to 10%). Honey dextrins have a lower molecular weight than those designed from starch. They are not blue-stained with iodine and do not precipitate on an admixture of alcohol. They form glutinous colloidal solutions and are strongly dextrorotatory. They are more abundant in honeydew than in nectar honeys.

6.3.2 NONSACCHARIDE HONEY COMPONENTS

6.3.2.1 Proteins and Amino Acids

Table 6.2 presents the nonsaccharide components of honey. There is approximately 175 mg of free amino acids (from 27 to 875 mg) in 100 g of nectar honey. In honeydew honey varieties, the average content of free amino acids is 178 mg (from 54 to 269 mg). The major free amino acid, proline, constitutes 49 and 59% of the total free amino acid content of the nectar and honeydew honeys, respectively.[3]

Honey contains a number of specific enzymes: invertase (3.2.1.26; fructohydrolase-D-fructofuranoside), α- and β-amylase, glucose oxidase, catalase, and phosphatase. Together with invertase that hydrolyzes glycosidic bonds of sucrose and maltose, honey contains a specific enzyme, maltase [3.2.1.20; glucoside hydrolases (3.2.1) and under the name of glucohydrolase-D-glucoside, able to hydrolyze sucrose and maltose].[4] Maltase has a transglucosidase character against those sugars. It splits sucrose and maltose to monosugars through an intermediary stage, erlose and maltotriose, respectively. Invertase and maltase are produced in the bee organism as pharyngeal gland enzymes and midgut enzymes. Apart from invertase and maltase, honey also contains diastatic enzymes, α- and β-amylase. The designations α and

TABLE 6.2
Nonsugar Honey Components

Major Groups of Compounds Nitrogen Compounds	Content
Total proteins (mg/100 g)	50–1000
Free proline (mg/100 g)	20–300
Other free amino acids (mg/100 g)	30–700
Acids (gluconic, citric, lactic, malic, succinic, butyric, propionic, and other) (mg/100g)	10–300
Ash (Mn, Co, Fe, and others) (mg/100 g)	70–900
Essential oils (in fresh honey) (mg/100 g)	30–200
Dyes (carotenoids, anthocyanines, flavones) (µg/100g)	1.5–180
Vitamins and other active substances (mg/100 g)	0–0.1

β do not refer here to the hydrolyzed glycosidic bond configuration. Enzymes of both types catalyze hydrolysis of the α-(1→4) bond. α-Amylase, by cleaving some α-(1→4) bonds of polysaccharides, forms combinations of several monosugar molecules, and is hence sometimes called the dextrinogenic enzyme (3.2.1.1; α-1,4-glucane 4-glucanohydrolase). β-Amylase (3.2.1.2; α-1,4-glucane maltohydrolase) catalyzes hydrolysis of amylose into maltose.

As other amylases, honey amylases are activated by chloride ions. At a 0.01 M concentration of Cl^- ions, enzyme activity increases by 170%. The optimum pH for honey amylases is 5.4 whereas human saliva amylases show highest activity at pH 7. Honey amylases originate exclusively from the bee organism.

Activities of both enzymes decrease during storage of honey and heating accompanying the decrystallization; therefore, the diastase number (the α-amylase activity) determination was used for a long time to evaluate potential overheating of honey. Currently, the overheating is evaluated by determining the level of 5-hydroxymethyl-furfural (HMF) in honey.[5] HMF results from dehydration (loss of three water molecules) of either D-glucose or D-fructose. Because of overheating, the HMF level can increase from commonly occurring ~1.20 mg/100 g up to 20 mg/100 g, although a sucrose invert made by hydrolysis with citric acid may contain 170–650 mg HMF/100 g. Such a difference can be useful in detecting and evaluating honey adulteration by an admixture of invert. Honey containing over 20 mg HMF/100 g should be recognized as a product adulterated with invert made by acid hydrolysis (see Section 6.4).

Several biochemical reactions make the composition of honey labile. During one-year storage at 20°C, the activities of α-amylase and invertase decrease by 30–50% and 10%, respectively. Simultaneously, honey acidity rises by several mval/kg. The sugar composition also changes.[6]

6.3.2.2 Organic Acids

Organic acids contribute substantially to the characteristic flavor of honey. They enrich and diversify the taste of honey varieties. Experienced honey tasters can detect the adulteration even with a 20% admixture of bee-inverted sucrose.[7] Butyric, acetic, formic, lactic, succinic, folic, malic, citric and gluconic acids have been identified in honey, the last two being the main acids. Gluconic acid is the product of specific catalytic oxidation of D-glucopyranose with glucose oxidase, a honey flavoprotein enzyme. Glucolactone, which results from the oxidation, readily hydrolyzes into gluconic acid.[8] In this oxidase-mediated oxidation, a strongly antibacterial hydrogen peroxide is formed.

6.3.2.3 Mineral Substances

Potassium, magnesium, sodium, calcium, phosphorus, iron, manganese, cobalt, copper and some other elements have been identified in honey by spectral analysis. Potassium was found to be a major element in honey, the content of which exceeded that of other elements by several orders. A high correlation was also found between potassium and magnesium content.[9] Mineral salts, organic acids, and amino acids in honey dissociate, making honey an electrolyte. The standardization requires measurements for 20% honey solutions (on dry weight basis) at 20°C. The unit of specific conductivity is millisiemens per centimeter (mS/cm).[10,11] Typical conductivity of honey ranges from 0.09 to 1.4 mS/cm, but for chestnut honey a value of 2.07 mS/cm has been found.[12]

6.3.2.4 Dyes and Other Components

Honey dyes belong to carotenoids, flavones, and anthocyanines (Table 6.2). The levels of the β-carotenoids (μg/100 g) in particular honey varieties are as follows: rapeseed honey, 4.18–8.45; linden honey, 19.25–183.07; buckwheat honey, 1.49–7.34; multifloral honey, 1.49–10.44; honeydew honey, 3.44–13.06.

About 80 aromatic compounds have been detected in honey, including carboxylic acids, aldehydes, ketones, alcohols, hydrocarbons, and phenols. They also contribute to the organoleptic properties of honey.[13]

Honey is poor in vitamins. Only some varieties such as heather honey and honeydew honey contain traces of vitamins A, B2, C, B6, and PP.

6.4 HONEY ADULTERATION AND POSSIBILITIES OF DETECTION

Despite improved methods to determine the quality of honey, any evaluation of various forms of honey adulteration presents a difficult task. The ratio of natural carbon isotopes, ^{13}C and ^{12}C, in the nectar of honey plants and in sugars of maple and sugar cane syrup is entirely different.[14] It allows the detection of honey adulteration by feeding those syrups to bees. A modification of this method allows detection of a 7% admixture of enzymatically hydrolyzed sucrose; however, this detection is only if the sucrose originates from cane, because the sugar beet sucrose has a $^{13}C/^{12}C$ ratio similar to that for sugars from honey plants.[14] There is still no simple method to detect adulteration of honey with bee-inverted sucrose, especially after storage. The sucrose-adulterated fresh honey contains a detectable excess of sucrose. However, because sucrose content decreases on storage, other indicators of honey quality are necessary (see Section 6.3.2.1). Specifically, high proline content in honey (usually 10 orders higher)[6] appeared promising in this respect. A study showed fresh honey samples from the U.S. to contain, on average, 48.3 mg proline/100 g of honey.[15] After a 20-year storage at 4°C, this value changed to 51.5 mg proline/100 g of honey, with 14.8 mg proline/100 g of honey being the lowest value ever recorded. Assays based on proline content can therefore be helpful to assess the degree of honey adulteration with bee-inverted sucrose.[16] The erlose assay is also helpful to assess honey integrity.[16] Erlose is detected even when the concentration of bee-inverted sucrose is as low as a few percent. However, erlose occurs in honey from the nectars with a high natural sucrose content.

REFERENCES

1. Rybak-Chmielewska, H., and Szczęsna, T., Viscosity of honey, *Pszczeln. Zesz. Nauk.*, 43, 209, 1999.
2. Maurizio, A., Das Zuckerbild Blütenreiner Sortenhonige, *Ann. Abeille*, 7, 289, 1964.
3. Bosi, G., and Battaglini, M., Gas-chromatographic analysis of free and protein amino acids in some unifloral honeys, *J. Apic. Res.,* 17, 152, 1978.
4. Takenaka, T., An α-glucosidase from honey, *Honeybee Sci.*, 1, 13, 1980.
5. White, J. W., Jr., Hydroxymethylfurfural content of honey as an indicator of its adulteration with invert sugars, *Bee World*, 61, 29, 1980.
6. Rybak, H., and Achremowicz, B., Changes in chemical composition of natural and adulterated with inverted by bees sucrose honeys, during storage (in Polish), *Pszczeln. Zesz. Nauk.*, 30, 19, 1986.
7. Simpson, J., Maxley, E., and Greenwood, S. P., Can honey from sugar-fed bees be distinguished from natural honey by its flavour? *Bee World*, 55, 10, 1975.
8. White, J. W., Jr., Subers, M. H., and Schepartz, A. J., The identification of inhibine, the antibacterial factor in honey, as hydrogen peroxide and its origin in a honey glucose-oxidase system, *Biochem. Biophys. Acta*, 73, 57, 1963.
9. McLellan, A. R., Calcium, magnesium, potassium and sodium in honey and in nectar secretion, *J. Apic. Res.,* 14, 57, 1975.
10. Council Directive 2001/110/EC of 20 December 2001 relating to honey, *Off. J. Eur. Commun.*, L 10, 47, 2002.
11. Codex Alimentarius Commission, 24th Session, July 2001, adopting the draft revised standard for honey. *Alinorm* 01/25, Appendix II, pp. 22-24.
12. Bogdanov, S., Honey quality, methods of analysis and international regulatory standards: review of the work of the International Honey Commission, *Mitt. Gebiete Lebensm. Hyg.*, 90, 108, 1999.
13. Steeg, E., and Montag, A., Minorbestandteile des Honigs mit Aroma-Relevanz, *Dtsch. Lebensm. Rund.*, 84, 147, 1988.
14. White, J. W., Jr., and Doner, L.W., The $^{13}C/^{12}C$ ratio in honey, *J. Apic. Res.*, 17, 94, 1978.

15. White, J. W., Jr., and Rudyj, O. N., Proline content of United States honeys, *J. Apic. Res.*, 17, 89, 1978.
16. Bogdanov, S., and Martin, P., Honey authenticity: a review, *Mitt. Gebiete Lebensm. Hyg.*, 93, 232, 2002.

Starch: Structure and Properties

7

Jay-lin Jane

CONTENTS

7.1 Introduction 81
7.2 Chemical Structures and Starch Molecules 82
    7.2.1 Amylose 82
    7.2.2 Amylopectin 83
    7.2.3 Minor Components of Starch 87
7.3 Organization of Starch Granules 89
7.4 Properties of Starch 90
    7.4.1 Starch Gelatinization 91
    7.4.2 Pasting of Starch 92
    7.4.3 Starch Retrogradation 94
    7.4.4 Glass-Transition Temperature of Starch 95
References 96

7.1 INTRODUCTION

Starch is produced by higher plants for energy storage and is the second largest biomass produced on earth, next to cellulose. Starch is also the major energy source in human and animal diets. Starch granules, consisting of highly branched amylopectin and primarily linear amylase (Chapter 1), are synthesized by apposition in amyloplasts of plants. Amylopectin and amylose molecules are organized in semicrystalline starch granules, and the outer chains of amylopectin molecules are arranged in double-helical crystalline structures.[1] The native granular structure of starch,[2,3] with a specific density about 1.5 g/cm^3, economizes the space for energy storage in seeds, grains, and tubers until it is ready to be utilized during germination. Transit starch in leaves, which is synthesized using sunlight and carbon dioxide during the daytime and hydrolyzed to glucose to be transported to storage organs at night, is in flat, small granule forms.[4] The granular form of starch present in grains, tubers, and roots facilitates its isolation by gravity and by centrifugation during wet milling. The granular structure of starch also makes it possible to conduct chemical reactions followed by washing to produce chemically modified starches that are used as commodity products for food and industrial applications.[5] Major markets of starch applications are in corn syrups, high-fructose corn syrup, paper, textile, and food industries. In recent years, there is increasing demand in alcohol fuel produced from starch fermentation to partially replace gasoline and to increase the octane number of the fuel. Alcohol containing gasoline has been reported to reduce air pollution. Readers are encouraged to refer to other chapters for information on starch structures[1,6] and properties.[7]

7.2 CHEMICAL STRUCTURES AND STARCH MOLECULES

Starch consists of two major types of molecules, primarily linear amylose and highly branched amylopectin. Normal starch consists of about 75% amylopectin and 25% amylose, waxy starches consist of mainly amylopectin and 0–8% amylose, and high-amylose starches consist of 40–70% amylose. A newly developed genetically modified starch has more than 90% amylose content.[8,9] In addition to amylose and amylopectin, most cereal normal starches also contain lipids and phospholipids,[10–13] which have profound impacts on the pasting property of the starch.[14] Most tuber and root starches and some cereal starches consist of phosphate monoester derivatives that are found exclusively on amylopectin molecules.[15,16] Sugary-1 starch consists of phytoglycogen that is a water-soluble glucan with a highly branched structure and substantially shorter branch chains. The presence of phytoglycogen is a result of the lack of starch-debranching enzymes.[17,18] Many starches, such as high-amylose maize starches[19-22] and sugary-2 starches,[23] also contain intermediate components that are branched molecules with smaller molecular weights and longer branch-chain lengths than does amylopectin.[24] Structures and properties of these components are discussed in the following sections.

7.2.1 AMYLOSE

Amylose is a primarily linear polymeric molecule, consisting of α 1-4 linked D-glucopyranose with a few branches.[25,26] Molecular weight of amylose varies from ca. 500 anhydroglucose units (AGU) of high-amylose maize starch[19] to more than 6000 AGU of potato starch.[27] It is known that amylose molecules cannot be totally hydrolyzed by β-amylase to produce maltose.[27] Hizukuri and coworkers[25] demonstrated that some amylose molecules consist of a few branches, and the number of branches seems proportional to the molecular size of the molecule. The same authors propose that amylose contains some clusters of branches that resemble clusters of amylopectin molecules.[28] The structure may be a result of transferring a cluster from an amylopectin molecule to the amylose molecule catalyzed by the branching enzyme.

The α 1-4 glycosidic bonds of amylose molecules, differing from the β 1-4 glycosidic bonds of cellulose, give random coil conformations to the molecule. The hydrocarbon and the hydroxyl moieties of the anhydroglucose unit prompt the formation of double helices by folding two linear chains of starch and having the hydrocarbon moiety of the chains folded inside of the helix, away from a polar, aqueous medium, to reach a low energy state. The two chains in the double helix can be arranged in either parallel or antiparallel orientation.[2,3,29] This recrystallization or aging process is known as retrogradation, which is responsible for bread staling and syneresis of gravy. Double helices of amylose chains are packed in a hexagonal unit that consists of six double helices and an empty channel at the center of the packing unit and has ca. 36 water molecules in each unit. This type of unit packing displays a B-type x-ray pattern.[2,3,30,31] Another type polymorph, the A-type x-ray pattern, is not found in retrograded native amylose.

In the presence of many chemicals that consist of a hydrophobic moiety, such as alcohols and fatty acids (known as complexing agents), amylose molecules can form helical complexes with these chemicals by having the chemicals located inside of the hydrophobic cavity of the helix as inclusions.[32] Depending on the size of the cross-section of the complexing agent, amylose can form single helices of different sizes to accommodate the diameters of the complexing agents. It has been reported that sizes of amylose single helices of six, seven, and eight glucose units per turn are found when complexed with n-butyl alcohol (linear molecule), isobutyl alcohol (branched molecule), and α-naphthol (bulky molecule), respectively, by using electron microscopy and x-ray diffraction[33] and by enzymatic method.[34] X-ray diffraction of amylose single-helical complexes gives V-type patterns.[30]

7.2.2 AMYLOPECTIN

Amylopectin is a highly branched molecule, consisting of α 1-4 linked D-glucopyranose chains that are connected by α 1-6 branch linkages. Amylopectin has a very high molecular weight, varying between hundreds of thousands to tens of millions anhydroglucose units (Table 7.1).[35] Figure 7.1 shows a plot of the relationship between molecular weight and gyration radius of the amylopectin molecule. These analytical results show that molecular weights of waxy starch amylopectins are higher than those of normal starch counterparts.

Branch chains of amylopectin molecules can be categorized to the A chains that are chains whose reducing ends attach to other B or C chains but do not carry any other chains; the B chains have reducing ends attached to other B or C chains and also carry other A or B chains; the C chain is the only chain of the molecule carrying a reducing end. The A chains are generally short and extend within one cluster. The B chains have different chain lengths, B_1 chains have lengths that extend within one cluster, B_2 chains extend through two clusters, B_3 chains through three clusters, and so on.[36] The bimodal branch-chain length distribution of amylopectin differs from the single modal distribution of glycogen[37] and phytoglycogen.[38] Branch chain lengths of amylopectin vary with botanical sources of the starch, which control the polymorphic forms of crystalline structure,[39] gelatinization, pasting, and retrogradation properties of the starch.[40]

Branch chains of amylopectin are arranged in clusters.[1] After native granular waxy maize starch was hydrolyzed by acid at room temperature, Yamaguchi et al.,[41] using transmission electron microscopy, observed worm-like cluster structures that resisted acid hydrolysis. The worm-like cluster is attributed to the crystalline structure of outer chains of amylopectin molecules. The amorphous structure of the waxy

TABLE 7.1
Amylopectin Molecular Weights and Gyration Radii of Selected Starches[a]

Starch	M_w $(\times 10^8)$[b]	R_z (nm)[c]	ρ (g/mol/nm^3)[d]
A-Type Starches			
Normal maize	4.9 (0.8)[e]	312 (23)	16.1
Waxy maize	8.3 (0.2)	372 (11)	16.1
du wx Maize	4.9 (0.5)	312 (13)	16.1
Normal rice	26.8 (2.9)	581 (41)	13.7
Waxy rice	56.8 (9.3)	782 (36)	11.9
Sweet rice	13.9 (1.0)	486 (5)	12.1
Normal wheat	3.1 (0.3)	302 (3)	11.3
Waxy wheat	5.2 (0.4)	328 (6)	14.7
Barley	1.3 (0.1)	201 (8)	16.0
Waxy barley	6.8 (0.1)	341 (3)	17.1
Cattail millet	2.7 (0.2)	278 (6)	12.6
Mung bean	3.8 (0.2)	312 (3)	12.5
Chinese taro	12.6 (3.6)	560 (15)	7.2
Tapioca	0.7 (0.1)	191 (25)	10.0
B-Type Starches			
ae wx Maize	3.2 (0.2)	306 (8)	11.2
Amylomaize V	2.4 (0.0)	357 (24)	5.3
Amylomaize VII	1.7 (0.0)	389 (57)	2.9
Potato	1.7 (0.2)	356 (36)	3.8
Waxy potato	2.0 (0.2)	344 (37)	4.9
Green leaf canna	3.4 (2.2)	436 (85)	4.1
C-Type Starches			
Lotus root	1.5 (0.4)	280 (57)	6.8
Water chestnut	7.1 (1.5)	230 (25)	58.4
Green banana	1.9 (0.8)	286 (29)	8.1
Glycogen			
Cyanobacterial glycogen[f]	0.2 (0.0)	55 (4)	99.2

[a] Data are averages of at least two injections.
[b] Weight-average molecular weight.
[c] z-Average radius of gyration.
[d] Density $(\rho) = M_w/R_z^3$
[e] Standard deviation.
[f] Glycogen was isolated from *Synechocystis* sp. PCC6803 in our laboratory.

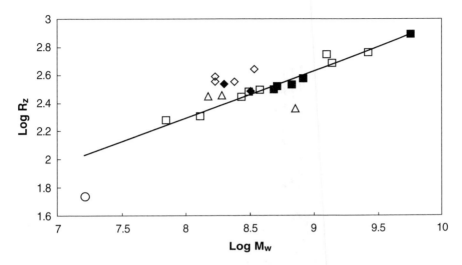

FIGURE 7.1 Relationships between the weight-average molecular weight (M_w) and z-average gyration radius (R_z) of amylopectin. Data are plotted on Log-Log scale; normal A-type (£), B-type (Ø), C-type (r) amylopectin; waxy A-type (¢), B-type (ø) amylopectin; and cyanobacterial glycogen (Å). The linear regression line on the graph comprises data of the A-type amylopectin, $r = 0.98$ ($P < 0.05$).

maize starch, mainly consisting of branch linkages, is hydrolyzed during the acid hydrolysis. The alternating crystalline and amorphous structure of amylopectin is also observed after enzyme hydrolysis of starch. Amylose of normal starch is present at an amorphous form in the starch granule.

Starches that consist of amylopectin with long average branch-chain lengths, such as potato, high-amylose maize, and ae waxy maize starch, display the B-type x-ray pattern.[39] Starches that consist of amylopectin with shorter average branch-chain lengths, such as maize, wheat, barley, rice and tapioca starch, display the A-type x-ray pattern.[39] These are consistent with amylodextrins of shorter chain length (degree of polymerization, DP, 10 to 12) that develop to the A-type crystalline structure and amylodextrins of longer chains length (DP > 13) that develop to the B-type crystalline structure.[42] X-ray diffraction signals and space distances calculated from the signals reveal that the A-type polymorphic form of starch crystal has a monoclinic packing unit, which consists of seven sets of double helices closely packed in the unit and contains only two water molecules in each unit.[2,3] The B-type crystalline packing is the same as that for amylose crystals. The sizes of crystalline lamellae of A- and the B-type starch granules are similar, for example, 9.0 nm for maize and 9.2 nm for potato, determined by x-ray diffraction[31] and the small angle electron scattering method.[43]

Structures of Naegeli dextrins prepared from starches of A- and B-type polymorphisms are quite distinct. The Naegeli dextrins of B-type polymorphic starches consist of a small number of branched molecules, whereas those of A-type starches consist of large proportions of branched molecules (Figure 7.2). These distinct

A

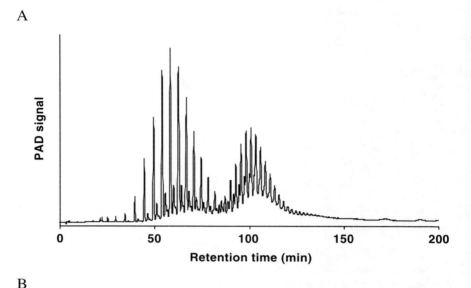

B

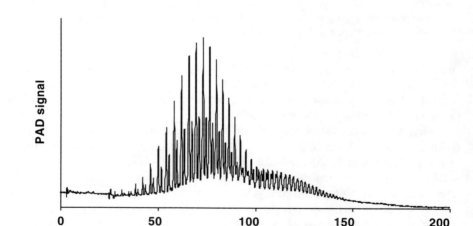

FIGURE 7.2 Anion-exchange chromatograms of Naegeli dextrins prepared from an A-type crystalline starch, waxy maize (A), and a B-type crystalline starch ae-waxy maize (B). The Naegeli dextrins are prepared and analyzed according to the methods reported in Jane, J., Wong, K.-S., and McPherson, A.E., *Carbohydr. Res.*, 300, 219, 1997.

structures reflect that many branch linkages, carrying short chains, are present within the crystalline region of amylopectin clusters in A-type starch. In contrast, there are few branch linkages carrying short branch chains present in B-type starches.[44] These observed branch structures of amylopectin molecules are in good agreement with the fact that A-type starches consist of more short chains than do B-type starches.

It is also intriguing to attribute A-type close packing polymorphism to the short branch chains derived in the middle of the crystalline region, which can be more easily arranged into the monoclinic packing unit.

Extra-long branch chains are found in amylopectin of many cereal normal starches, such as normal wheat, rice, barley, and maize starches, but are not found in cereal waxy starch and potato starch.[14,45,46] The extra-long branch chains do not have highly branched structures such as normal amylopectin molecules but have amylose-like structures. The extra-long branch chains are responsible for giving a higher blue value and iodine affinity of amylopectin isolated from cereal normal starch than those from waxy starch counterparts. The molecular size of extra-long branch chains in normal wheat starch is ca. 720 AGU.[14] The concentration of extra-long branch chains of normal wheat starch amylopectin is proportional to dosage of the waxy gene encoding granular-bound starch synthase that is primarily responsible for amylose biosynthesis. Thus, it indicates the extra-long branch chains are likely synthesized by the granular-bound starch synthase in a possessive elongation reaction pattern.

7.2.3 MINOR COMPONENTS OF STARCH

Minor components, such as lipids, phospholipids, and phosphate monoester derivatives, are found in starch and have profound effects on the properties of starch. Lipids and phospholipids are found in cereal starches. Cereal normal starches contain up to 1% lipids, and the level of lipid content is proportional to the amylose content of the normal starch.[47] Starches of different botanical origins consist of different species of lipids. For example, normal maize starch consists of mainly free fatty acids, glycerides, and little phospholipids; normal rice starch contains substantial amount of phospholipids and some free fatty acids; and wheat, barley, rye, and triticale starches consist of exclusively phospholipids.[12,13,47] Cereal waxy starches contain few lipids, whereas high-amylose starches contain substantially more lipids. Root and tuber starches contain very little lipids and no detectable phospholipids.[48]

Using solid ^{13}C cross-polarization/magic angle spinning-NMR, Morrison et al.[49] reported that 43% amylose in native nonwaxy rice starch, 33% in oat starch, and 22% in normal maize starch are complexed with lipids in single-helical conformation. The remaining amylose is present in a random coil conformation. Fatty acids, fatty acid monoglycerides and diglycerides, and phospholipids are good complexing agents with amylose and long branch chains of amylopectin. Differential scanning calorimetry (DSC) thermograms show melting peaks of the amylose–lipid complex at temperatures higher than the starch gelatinization temperature (above 90°C). Unlike the gelatinization peak of normal and waxy starches, the thermal transition peak of the amylose–lipid complex reappears on immediate rescan, reflecting an instant reaction of an amylose–lipid complex formation. The amylose–lipid complex in native starch is amorphous (Type 1), which does not give a V-type x-ray pattern and has a lower melting temperature (94 to 100°C).[50,51] On annealing or after cooking, the amylose–lipid complex is organized to a more orderly, semicrystalline and lamellar structure (Type 2), which displays the V-type x-ray pattern and has a higher melting temperature (100 to 125°C).

Amylose–lipid or amylose–phospholipid complex formed in starch granules or during gelatinization is known to entangle the amylose and amylopectin molecules and thus restrict swelling of granules. Consequently, starches containing lipids or phospholipids display higher pasting temperatures and lower paste viscosities. Phospholipids, the primary lipids found in normal wheat starch, have more profound impact to starch pasting behavior than do free fatty acids present in maize starch.[14] The pasting temperature of normal wheat starch is 28.1°C higher than that of waxy wheat starch, whereas the pasting temperature of normal maize is only 11.7°C higher than that of waxy maize starch.[14] The viscosity difference between waxy and normal wheat starches is also significantly greater than that between waxy and normal maize starches. Amylose–lipid and amylose–phospholipid complexes also render starch pastes opaque in color.

In contrast to phospholipids, phosphate monoester derivatives found in potato and other tuber and root starches enhance the swelling of starch granules through charge repelling of the negatively charged phosphate derivatives. Phosphate monoester derivatives are covalently attached to long branch chains (average DP 42) of amylopectin molecules.[15] Potato starch displays low gelatinization and pasting temperatures and a very high peak viscosity.[40] In addition, potato starch produces a crystal clear paste. In the presence of less than 0.001% NaCl in the solution, the viscosity of potato starch can be reduced by 50%. This is attributed to masking of negative charges of the phosphate monoesters.

Quantitative analyses of phospholipids and phosphate monoester derivative contents of starch have been a challenge, particularly because of their distinct impacts on starch properties. ^{31}P-NMR spectroscopy has been demonstrated to be a precise and unambiguous analytical method for qualitative and quantitative analyses of phospholipids and phosphate monoester contents of starch.[12,13] Phospholipids, being phosphate diesters, display chemical shifts of phosphorus between 0 and 1 ppm at pH 8, whereas phosphate monoesters display chemical shifts between 4 and 5 ppm at pH 8. Quantitative results are obtained when adequate acquisition times are programmed to allow spin relaxation.

Intermediate material is found in many mutants of starches, such as high-amylose maize starches, the ae-gene-containing maize double mutants,[19–22] and sugary-2 maize starch.[23,52] The intermediate material has branched structures but has lower molecular weights than does amylopectin. After fractionation of starch by complexing agents such as n-butyl alcohol to precipitate amylose, the intermediate material remains in the supernatant with amylopectin. Gel-permeation chromatograms of the fraction collected from the supernatant show that high-molecular-weight amylopectin is eluted at the void volume and the lower-molecular-weight intermediate material is eluted later at larger eluting volumes.[19,20] The chromatogram shows an additional peak of the soluble fraction of high-amylose maize starch containing primarily linear small molecules.[19,20,24] The linear small molecules can result from deficiency of the branching enzyme IIb in high-amylose maize starch.[38]

Phytoglycogen is another water-soluble glucan and is present in sugary-1 mutants of maize, rice, and other species deficient in the starch-debranching enzyme.[53,54] Phytoglycogen has lower molecular weights and shorter average branch chains (DP 10.3) than does amylopectin (DP 18.5 for waxy maize amylopectin).

The branch-chain lengths of phytoglycogen display a single-modal chain-length distribution, which is different from the bimodal distribution of amylopectin.

7.3 ORGANIZATION OF STARCH GRANULES

Starch is synthesized in granules inside amyloplasts. Starch granules isolated from different botanical sources and different organs display different shapes and sizes.[55] Diameters of starch granules vary from submicrons to more than 100 μm. Examples of diameters of starch granules are potato starch 15 to 75 μm, wheat A-granules 18 to 33 μm and B-granules 2 to 5 μm, maize starch 5 to 20 μm, rice starch 3 to 8 μm, and amaranth starch 0.5 to 2 μm.[55] Starch granules display spherical, oval, disk, polygonal, elongated, kidney, and lobe shapes. Leaf starch has flat-shaped small granules that are submicron in diameter.[4] Wheat, barley, rye, and triticale starches display bimodal granule size distributions: the disk-shaped, large A-granules and the spherical, small B-granules.[55] Rice and oats starches are known as compound starches, which are defined as multiple granules synthesized within a single amyloplast.[55] Thus, starch granules are tightly packed together and develop into polygonal irregular shapes.

When viewed under a polarized light-microscope, starch granules display birefringence, known as the Maltese cross. The Maltese cross birefringence reflects the radial arrangement of starch molecules in the starch granule. The center of the Maltese cross, the hilum, is the organic center of the granule where biosynthesis of the starch granule initiates. The hilum is not necessarily at the geometric center of the granule; some can be located close to the end of the granule, reflecting the initiating site of the starch biosynthesis.

Many starch varieties, particularly those having the A-type x-ray pattern and short average branch-chain length (e.g., sorghum and maize), display pinholes on the surface of starch granules. The pinholes are open ends of serpentine-like channels penetrating into the granule[56] and are attributed to enzyme hydrolysis of starch granules. Scanning electron micrographs of cracked corn kernels shows enzyme-degraded starch granules located around germs, indicating that enzyme hydrolysis of starch initiates at the location surrounding the germ to generate energy for germination.[57]

Surface gelatinization of starch granules by saturated neutral salt solutions, such as LiCl and $CaCl_2$, enables separation of starch molecules by layers from the periphery of the granule. Therefore, one can investigate structures of starch molecules at different radial locations of a granule.[58,59] The amylose content is higher at the periphery of the granule than at the hilum. This difference in the amylose content is consistent with a known fact that amylose content of starch increases with size of starch granules. The same studies also reveal that amylopectin molecules at the hilum have longer branch chains than do those at the periphery.[58,59]

Structures of amylopectin molecules isolated from disk-shaped, large A-granules and from spherical, small B-granules of barley, wheat, and other starches have been analyzed. Large disk-shaped granules consist of amylopectin that has fewer short branch chains than does that of small spherical granules.[57] Differences in branch structures determine the shape of amylopectin molecules, which, in turn, governs the geometric packing of amylopectin molecules in the granule and results in different shapes of granules. Amylopectin molecules consisting of fewer short branch chains and more middle-size chains are likely to pack in a parallel arrangement and result in disk-shaped granules. In contrast, amylopectin molecules consisting of a larger proportion of short branch chains tend to pack into a spherical shaped granule. Different chain-length distributions of amylopectins in disk-shaped A-granules and spherical B-granules may be related to the branching enzyme SBE 1c.[60]

When intact native starch granules are subjected to chemical cross-linking reactions, amylose molecules are cross-linked to amylopectin molecules but are not cross-linked to other amylose molecules.[61,62] On the basis of these results, it is postulated that amylose molecules are interspersed among amylopectin molecules in the starch granule instead of being present in bundles among themselves. Starch chains can be elongated on the surface of the starch granule, which indicates that there is no membrane present on the surface of the granule.[63] The apparent shell or ghost on the surface of starch granules[64] can be attributed to highly crystalline starch molecules, possibly associated with amylose, to give a strongly entangled layer.

7.4 PROPERTIES OF STARCH

Starch is known to go through transformation and gives diverse physical structures and properties (Figure 7.3). Native granular starch, after being soaked in water, can absorb up to 30%, by weight, of moisture. The water absorbed is present in the amorphous region of the starch granule. This process is reversible, and water can be evaporated on drying at the ambient temperature or at a temperature below that for gelatinization. When starch granules are heated in the presence of water, the granules eventually lose the double-helical crystalline structure and the Maltese cross. This process is known as gelatinization. The gelatinization of starch is an irreversible process, and each starch has its own characteristic gelatinization temperature. When the gelatinized starch is continuously heated in excess water, the starch granules swell, develop viscosity, and become a paste. This is known as

Starch Functionalities

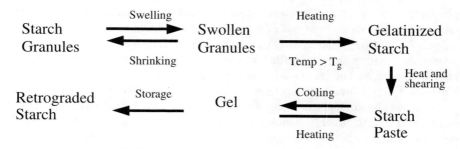

FIGURE 7.3 Transformation of starch structures.

pasting. On cooling, viscosity increases as temperature decreases, and starch molecules in the paste develop a network and gel. After an extended storage period or repeated freeze and thaw cycles, starch molecules in the paste or gel crystallize, and the process is known as retrogradation.

The versatile physical structures of starch facilitate a wide variety of applications; for example, granular starch is used for facial powder; large A-granules of wheat starch are used in carbonless copy paper, and small particle starch[65] or small granular amaranth starch (diameter 0.5 to 2 μm, resembling that of a fat micelle) is used for fat substitutes.[66] Starch pastes are used as sizing agents in paper and textile industries and as thickening agents for soup and canned foods in the food industry. Starch gel is used to make tapioca pudding, desserts, and fillings. Retrograded or crystalline high-amylose maize starch is used as resistant starch that has limited digestibility by humans and is desirable for low-caloric diet foods.

7.4.1 STARCH GELATINIZATION

Starches of different botanical sources and genetic backgrounds display different gelatinization properties, such as gelatinization temperature, enthalpy change, and melting of the amylose–lipid complex. Gelatinization temperature of starch can be determined by a light microscope equipped with a hot stage. The temperature at which starch granules lose the Maltese cross in the presence of excess water is the gelatinization temperature. The most common and reliable method for analysis of starch gelatinization temperature and enthalpy change is by a differential scanning calorimeter (DSC). Starch gelatinization is an endothermic reaction and requires the presence of water or other plasticizers such as glycerol. Without water or other proper plasticizers, starch will not gelatinize until it decomposes. Thus, water is critical for starch gelatinization. Two times or more, by weight, of water to starch is required to assure a constant gelatinization temperature. With insufficient water content, starch gelatinization temperature increases and the range of gelatinization temperature also broadens.[67]

The gelatinization properties of starches vary substantially. For example, onset gelatinization temperatures vary from 47.8 (sugary-2 maize starch) to 71.5°C (ae waxy maize starch), ranges of gelatinization temperatures vary from 6.6 (barley starch) to 58.8°C (high-amylose maize VII starch), enthalpy changes of starch gelatinization vary from 10 (barley starch) to 22 J/g (ae-waxy maize starch), and percentage retrogradation of gelatinized starch after being stored at 5°C for 7 days varies from 4.3 (sweet rice) to 80.8% (high-amylose maize V starch).[40] The gelatinization temperature of starch is highly correlated to the branch-chain length of amylopectin.[22,68-71] Starches that consist of amylopectin with more long branch chains, such as high-amylose maize starches and ae-waxy maize starch, display higher gelatinization temperatures and greater enthalpy changes, gelatinization temperature ranges, and percentage retrogradation. Potato starch amylopectin also has more long branch chains but displays a very low gelatinization temperature ($T_0 = 58.2°C$). This may be attributed to amylopectin of potato starch consisting of a large concentration of phosphate monoester derivatives. The negative charges of the phosphate derivatives repel one another and destabilize the granular structure. Thus, potato starch has unique gelatinization properties, a low gelatinization temperature with a narrow range, and a slow retrogradation rate.

The gelatinization temperature of Naegeli dextrin prepared from normal maize starch is higher than that of native starch,[23] but the gelatinization temperature of Naegeli dextrin prepared from waxy maize starch is lower than that of native starch. The difference indicates that amylose in the normal maize starch may have retrograded during acid hydrolysis; the retrograded amylose causes increase in the gelatinization temperature of normal maize Naegeli dextrin.

Crystalline spherulites of A- and B-type polymorphisms, prepared from a short chain amylose (DP 15), show different melting behaviors.[72,73] The melting temperature of the B-type crystallite is about 20°C lower than that of the A-type crystallite when measured at a constant water content. This indicates that the A-type structure is the more stable of the two polymorphs.[74,75]

Gelatinization of starch can also be achieved by using aqueous alkaline solutions (e.g., NaOH, KOH); dimethyl sulfoxide; and various neutral salt solutions such as $CaCl_2$, LiCl, KI, and KSCN.[76,77] Saturated LiCl and $CaCl_2$ solutions are known to gelatinize starch at the periphery of the granule.[58,59,78] Thus, the solutions can be used for controlled surface gelatinization.[58,59,79] KI and KSCN are known to have a water structure-breaking effect. Consequently, aqueous solutions of KI and KSCN are less viscose and can easily penetrate into starch granules and gelatinize starch from the hilum.[77]

Sodium sulfate is commonly used to stabilize starch granules during chemical reactions at an alkaline pH for manufacturing chemically modified starches.[5] Sodium sulfate increases the gelatinization temperature of starch granules through its structure-making effect and its large negative charge density.[77] In the presence of a high concentration of sucrose, starch gelatinization temperature increases substantially.[80] As a result, cakes containing large concentration of sugar sometimes collapse when removed from the oven after baking. This is attributed to starch in the cake batter not being fully gelatinized during baking and failure to develop the structure that holds the volume of the cake. Chlorine-treated flour containing oxidized starch is preferred for cake batter because oxidized starch has a lower gelatinization temperature and is more prompt to gelatinize to develop the structure, which prevents cake collapsing. Starch in sugar cookies is frequently found ungelatinized, resulting from large concentrations of sugar and lipids and little moisture content in the cookie dough.

7.4.2 PASTING OF STARCH

Pasting is an important function for many applications of starch, such as thickening agents and sizing agents. Different starches display different pasting properties.[40] Pasting properties of starch depend on the amylose and lipid contents and branch-chain length of the amylopectin. Cereal normal starches, consisting of amylose and lipids, display higher pasting temperatures and less peak viscosity and less shear thinning than do their waxy starch counterparts.[40] Normal starches, containing amylose, display greater set back viscosity. Species and contents of lipids of the starch significantly affect pasting properties. For example, normal wheat starch consists of a large concentration phospholipid (0.06%), whereas normal maize starch consists of mainly free fatty acids, some glycerides, and little phospholipids (~0.01%).[10,12,13] The difference in pasting temperature between normal wheat and waxy wheat starches, measured by a Rapid Visco Amylograph at a heating rate of 6°C/min, is 28.1°C, whereas the difference between normal maize and waxy maize starches is only 11.7°C. The peak viscosity of waxy wheat starch is 134 RVU (rapid viscosity units) greater than that of normal wheat starch, compared with the 41 RVU difference between waxy and normal maize starch (Figure 7.4).[14] The extremely high pasting temperature and low peak viscosity of normal wheat starch are attributed to its high phospholipid content.[14] Phospholipids form the helical complex with amylose and with the long branch chains of amylopectin. The helical complex entangles with amylopectin molecules and prohibits swelling of starch granules. However, waxy starches consist of little or no amylose and have little lipids, thus display lower pasting temperatures and unrestricted swelling to a greater peak viscosity. After ultrahigh-pressure treatments, starch pasting properties change, depending on lipid content, molecular structure, and crystalline structures.[81,82]

Waxy starches consist of 92 to 100% amylopectin, and amylopectin molecules in starch granules are primarily responsible for the swelling power of starch.[83] Higher amylopectin content of waxy starches also contributes to greater viscosity of waxy starch pastes. High-amylose starches, such as high-amylose maize[40] and high-amylose barley,[84] display little viscosity, which can be attributed to their lower amylopectin contents and higher amylose and lipids contents.

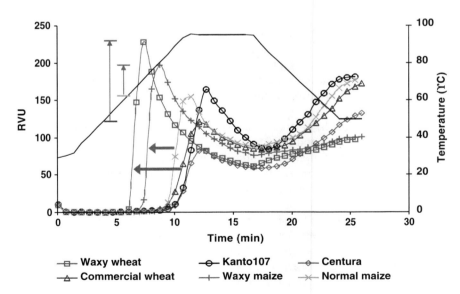

FIGURE 7.4 Pasting profiles of waxy wheat starch, semiwaxy wheat starch (Kanto 107), and normal wheat starches (Centura and Commercial wheat) compared with those of normal and waxy maize starches. Arrows stand for differences between normal and waxy starch counterparts.

Root and tuber starches, such as potato and tapioca, consist of little lipids and thus display lower pasting temperatures and greater peak viscosities.[48] Potato starch has an exceptionally low pasting temperature and high peak viscosity, which can be attributed to its large content (~0.08%) of phosphate monoester derivatives and also its large granule size (diameter up to 75 μm). Phosphate monoester derivatives carry negative charges that generate repulsion and facilitate swelling and pasting of starch during cooking, resulting in a low pasting temperature and a high paste viscosity. In presence of salts, negative charges of phosphate groups are masked and viscosity of the starch paste is substantially reduced.

Pastes of normal cereal starch, such as normal wheat and normal maize, are more opaque than those of waxy and tuber starches.[85] This is because of light reflected by the dense, limited swollen granules of normal cereal starches. Normal cereal starch is known for consisting of amylose and lipids/phospholipids and has restricted swelling.[85] Pastes of waxy starch and tapioca starch display greater clarity because the starches disperse to greater extents, and the highly dispersed starches do not reflect light. Potato starch and chemically modified starches that carry charge groups display the greatest clarity, resulting from charge repelling.

7.4.3 STARCH RETROGRADATION

Dispersed amorphous starch in pastes, gel, or solutions gradually develops double-helical crystalline structures and loses its water-binding capacity. This process is known as starch retrogradation. Retrograded starch that consists of large crystalline size, such as retrograded amylose having 31 AGU in the crystalline region, is highly resistant to enzyme hydrolysis.[34] Resistant starch is produced commercially from high-amylose maize starch[86–89] for bulking agent and is used in low-caloric diet food products.

Amylose molecules having linear structures develop double-helical crystallites faster. Amylopectin molecules with branched structures, in general, crystallize more slowly. The rate of retrogradation or crystallization of amylopectin depends on branch-chain length. Amylopectin molecules that have long branch chains, such as ae waxy maize starch, crystallize faster than those with short branch chains, such as waxy rice starch and sugary-2 maize starch.[23,40] Starches that consist of higher concentrations of lipids and phospholipids are also known to retrograde faster, which might be attributed to limited dispersion of starch during cooking. When starch chains are entangled and more clustered, they crystallize more promptly.

Studies have shown that linear-chain amylose molecules of DP 80 to 100 display maximal rate of retrogradation.[90,91] Linear amylose molecules with chain length longer or shorter than DP 80 to 100 display slower retrogradation rates. Amylose having chain length less than DP 110 precipitates from a solution (0.1%), and that having chain length more than DP 1100 forms predominantly gels. Amylose of DP 250 to 660 develops mixtures of precipitates and gel.[91] Chains of monodispersed molecular-weight amylose shorter than DP 10 do not retrograde. In a mixture with longer-chain linear molecules, amylose chains as short as DP 6 can cocrystallize with other longer chains.[91]

Studies on polymorphic forms of short amylose crystallization have shown that chain length of short amylose molecules[42,74] concentration of amylose solution,[74] temperature,[74] and presence of other chemicals such as salts[92] and alcohol[73] affect the type of polymorphism of retrograded amylose. Amylose of DP 10 to 12 crystallizes to develop A-type polymorphism, whereas that of DP 13 and longer develops B-type polymorphism.[42,74] High concentration of amylose (~50%), high crystallization temperature (30°C),[74] presence of salts of higher lyotropic number,[92] and presence of 30% ethanol (w/w)[73] favor A-type polymorphism. Most salts and those that have structure-breaking effects, such as NaCl, LiCl, KCl, RbCl, KI, KCN, KBr, favor A-type polymorphism. Sulfate salts, such as K_2SO_4, $(NH_4)_2SO_4$, and Na_2SO_4, however, favor the formation of B-type polymorphism. The trend of amylose chain length effect on the crystalline structure is in agreement with the polymorphism of starch that consists of amylopectin with different branch-chain lengths.[39] Starches having more long branch chains, such as potato and high-amylose maize, display B-type polymorphism, whereas starches with more short branch chains, such as rice, corn, taro, and wheat, display A-type polymorphism.

Potato amylose in an aqueous solution (3.5 mg/ml) displays the highest rate of retrogradation at 5°C, and retrogradation rate decreases with increase in temperature up to 45°C.[93] When incubation temperature reaches 45°C, only small amylose molecules (DP 110) crystallize, whereas amylose of larger molecular weight remains in the supernatant. Size of the crystallite in retrograded amylose increases with incubation temperature. After the amorphous region of retrograded amylose has been hydrolyzed and removed by α-amylase or by acid, chain length of the resistant crystalline region increases from DP 39 to DP 52 as incubation temperature increases from 5 to 45°C. Retrograded amylose displays exclusively B-type polymorphism, which does not change with incubation temperature. Repeated freeze–thaw treatments of amylose solutions enhance amylose retrogradation and result in a sponge-like structure.

Crystallization of short-chain amylose in an aqueous solution produces spherical particles.[73] Scanning electron micrographs of the spherical particles show diameters of 10 to 15 μm, and the particles are mostly clustered. Polarized light micrographs of the particles show Maltese cross, indicating a spherulitic morphology. Potato amylose retrogrades and develops to nodule-shaped particles when the solution is incubated at 5°C, but develops to a well-defined network when incubated at a higher temperature (e.g., 15 or 25°C).

7.4.4 GLASS-TRANSITION TEMPERATURE OF STARCH

Glass transition is the kinetic transition of amorphous polymers from a glassy state to a rubbery state. The glass-transition temperature (T_g) is specific to each material and depends on molecular weight of the polymer and presence of plasticizers. T_g of a homologous polymer increases with the increasing number-average molecular weight up to a plateau limit[94] and decreases with an increase of plasticizer concentration. Plasticizers are small chemicals compatible with the polymer and increase flexibility and extensibility. Water is known to be the most powerful plasticizer for starch and many hydrophilic biopolymers.[95,96] Glass-transition temperature of native wheat starch has been reported to decrease with the increase in moisture content.[97] When environmental temperature is above the glass-transition temperature, the polymer has molecular mobility and can go through conformational changes to crystallize and retrograde.

For frozen foods, the glass-transition temperature of a maximally freeze-concentrated polymer-dispersion (T_g') is critical for maintaining food quality during storage. The quality and stability of frozen foods are controlled by the temperature difference between freezer temperature and T_g' of the material.[98–101] Small sugars, such as fructose, glucose, and sucrose, have very low T_g's ($-42.5°C$, $-40.3°C$, and $-32.2°C$, respectively). In the presence of these small sugars, the T_g' of a starch dispersion decreases, and storage stability of the starch food decreases at freezing temperature ($-20°C$).

REFERENCES

1. French, D., Organization of starch granules, in *Starch Chemistry and Technology,* Whistler, R.L., BeMiller, J.N., and Paschall, E.F., Eds., Academic Press, Orlando, 1984, chap. 7.
2. Sarko, A., and Wu, H.-C.H., The crystal structures of A-, B- and C-polymorphs of amylose and starch, *Starch/Staerke,* 30, 73, 1978.
3. Imberty, A. et al., Recent advances in knowledge of starch structure, *Starch/Staerke,* 43, 375, 1991.
4. Zeeman, S.C. et al., A starch-accumulating mutant of *Arabidopsis thaliana* deficient in a chloroplastic starch-hydrolyzing enzyme, *Plant J.,* 5, 357, 1998.
5. Wurzburg, O.B., *Modified Starches: Properties and Uses,* Wurzburg, O.B., Ed., CRC Press, Boca Raton, 1986.
6. Hizukuri, S., Starch: analytical aspects, in *Carbohydrates in Food,* Eliasson, A.-C., Ed., Marcel Dekker, New York, 1996, chap. 9.
7. Eliasson, A.-C., Starch: physicochemical and functional aspects, in *Carbohydrates in Food,* Eliasson, A.-C., Ed., Marcel Dekker, New York, 1996, chap. 10.
8. Sidebottom, C. et al., Characterization of the difference of starch branching enzyme activities in normal and low-amylopectin maize during kernel development, *J. Cereal Sci.,* 27, 279, 1998.
9. Case, S.E. et al., Physical properties and gelation behavior of a low-amylopectin maize starch and other high-amylose maize starches, *J. Cereal Sci.,* 27, 301, 1998.
10. Morrison, W.R., Plant lipids, *Res. Food Sci. Nutr.,* 5, 247, 1984.
11. Medcalf, D.G., Youngs, V.L., and Gilles, K.A., Wheat starches. II. Effect of polar and nonpolar lipids fractions on pasting characteristics, *Cereal Chem.,* 45, 88, 1968.
12. Lim, S.-T., Kasemsuwan, T., and Jane, J. Characterization of phosphorus in starches using ^{31}P-NMR spectroscopy, *Cereal Chem.,* 71, 488, 1994.
13. Kasemsuwan, T. and Jane, J., A quantitative method for the analysis of phosphorus structures and contents in starch by P-31 nmr spectroscopy, *Cereal Chem.,* 73, 702, 1996.
14. Yoo, S.-H. and Jane, J., Structural and physical characteristics of waxy and other wheat starches, *Carbohydr. Polym.,* 49, 297, 2002.
15. Takeda, Y. and Hizukuri, S., Location of phosphate groups in potato amylopectin. *Carbohydr. Res.,* 102, 321, 1982.

16. Lim, S. and Seib, P.A., Location of phosphate esters in a wheat starch phosphate by 31-P NMR spectroscopy, *Cereal Chem.*, 70, 145, 1993.

17. Nakamura, Y. et al., Correlation between activities of starch debranching enzyme and α-polyglucan structure in endosperms of sugary-1 mutants of rice, *Plant J.*, 12, 143, 1997.

18. Colleoni, C. et al., Genetic and biochemical evidence for the involvement of α-1,4 glucanotransferases in amylopectin synthesis, *Plant Physiol.*, 120,993, 1999.

19. Jane, J. and Chen, J., Effects of amylose molecular size and amylopectin branch chain length on paste properties of starch, *Cereal Chem.*, 69, 60, 1992.

20. Kasemsuwan, K. et al., Characterization of dominant mutant amylose-extender (Ae-5180) maize starch, *Cereal Chem.*, 72, 457, 1995.

21. Inouchi, N. et al., Development changes in fine structure starches of several endosperm mutants of maize, *Starch/Staerke*, 35, 371, 1983.

22. Wang, Y.-J. et al., Characterization of amylopectin and intermediate materials in starches from mutant genotypes of the OH43 inbred line, *Cereal Chem.*, 70, 521, 1993.

23. Perera, C. et al., Comparison of physicochemical properties and structures of sugary-2 cornstarch with normal and waxy cultivars, *Cereal Chem.*, 78, 249, 2001.

24. Takeda, C., Takeda, Y., and Hizukuri, S., Structure of the amylopectin fraction of amylomaize, *Carbohydr. Res.*, 246, 273, 1993.

25. Takeda, Y., Shirasaka, K., and Hizukuri, S., Examination of the purity and structure of amylose by gel-permeation chromatography, *Carbohydr. Res.*, 132, 83, 1984.

26. Takeda, Y. and Hizukuri, S., Structure of branched molecules of amyloses of various origins, and molar fractions of branched and unbranched molecules, *Carbohydr. Res.*, 165, 139, 1987.

27. Hizukuri, S., Takeda, Y., and Yasuda, M., Multi-branched nature of amylose and the action of debranching enzymes, *Carbohydr. Res.*, 94, 205, 1981.

28. Takeda, Y., Shitaozono, T., and Hizukuri, S., Structures of sub-fractions of corn amylase, *Carbohydr. Res.*, 199, 207, 1990.

29. French, A.D. and Murphy. V.G., Computer modeling in the study of starch, *Cereal Foods World*, 22, 61, 1977.

30. Zobel, H.F., Starch crystal transformations and their industrial importance, *Starch/Staerke*, 40, 1, 1988.

31. Zobel, H.F., Molecules to granules: a comprehensive starch review, *Starch/Staerke*, 40, 1, 1988.

32. Kuge, T. and Takeo, K., Complexes of starchy materials with organic compounds. Part II. Complex formation in aqueous solution and fractionation of starch by 1-menthone, *Agric. Biol. Chem.*, 32, 1232, 1968.

33. Yamashita, Y., Ryugo, J., and Monobe, K., An electron microscopic study on crystals of amylose V complexes, *J. Electr. Microsc.*, 22, 19, 1973.

34. Jane, J. and Robyt, J.F., Structure studies of amylose studies of amylose-V complexes and retrograded amylose by action of α-amylases, and a new method of preparing amylodextrin, *Carbohydr. Res.*, 132, 105, 1984.

35. Yoo, S.-H. and Jane, J., Molecular weights and gyration radii of amylopectins determined by high-performance size-exclusion chromatography equipped with multi-angle laser light scattering and refractive index detection, *Carbohydr. Polym.*, 49, 307, 2002.

36. Hizukuri, S., Polymodal distribution of the chain lengths of amylopectins and it significance, *Carbohydr. Res.*, 147, 342, 1986.

37. Matsui, M., Kakuta, M., and Misaki, A., Fine structural features of oyster glycogen: mode of multiple branching, *Carbohydr. Polym.,* 31, 227, 1997.
38. Yoo, S.-H, Spalding, M.H., and Jane, J., Characterization of cyanobacterial glycogen isolated from the wild type and from a mutant lacking branching enzyme, *Carbohydr. Res.,* 337, 2195, 2002.
39. Hizukuri, S., Kaneko, T., and Takeda, Y., Measurement of the chain length of amylopectin and its relevance to origin of crystalline polymorphism of starch granules, *Biochem. Biophys. Acta,* 760, 188, 1983.
40. Jane, J. et al., Effects of amylopectin branch chain-length and amylose content on the gelatinization and pasting properties of starch, *Cereal Chem.,* 52, 555, 1999.
41. Yamaguchi, M., Kainuma, K., and French, D., Electron microscopic observations of waxy maize starch, *J. Ultrastruct. Res.,* 69, 249, 1979.
42. Pfannemuller, B., Influence of chain length of short monodisperse amyloses on the formation of A- and B-type x-ray diffraction patterns, *Int. J. Biol. Macromol.,* 9, 105, 1987.
43. Cameron, R.E. and Donald, A.M., A small-angle x-ray scattering study of the annealing the gelatinization starch, *Polymer,* 33, 2628, 1992.
44. Jane, J., Wong, K.-S., and McPherson, A.E., Branch-structure difference in starches of A- and B-type x-ray patterns revealed by their Naegeli dextrins, *Carbohydr. Res.,* 300, 219, 1997.
45. Shibanuma, Y., Takeda, Y., and Hizukuri, S., Molecular structures of some wheat starches, *Carbohydr. Polym.,* 25, 111, 1994.
46. Hizukuri, S. et al., Molecular structures of rice starches, *Carbohydr. Res.,* 189, 227, 1989.
47. Morrison, W.R., Starch lipids and how they relate to starch granule structure and functionality, *Cereal Foods World,* 40, 437, 1993.
48. McPherson, A.E. and Jane, J., Physicochemical properties of selected root and tuber starches, *Carbohydr. Polym.,* 40, 57, 1999.
49. Morrison, W.R., Law, R.V., and Snape, C.E., Evidence for inclusion complexes of lipids with V-amylose in maize, rice, and oat starches, *J. Cereal Sci.,* 18, 107, 1993.
50. Biliaderis, C.G. and Galloway, G., Crystallization behavior of amylose-V complexes: structure-property relationships, *Carbohydr. Res.,* 189, 31, 1989.
51. Tufvesson, F., Wahlgren, M., and Eliasson, A.-C., Formation of amylose-lipid complexes and effects of temperature treatment. Part. 1: Monoglycerides, *Starch/Staerke,* 55, 61, 2003.
52. Takeda, Y. and Preiss, J., Structures of B90 (sugary) and W64A (normal) maize starch, *Carbohydr. Res.,* 240, 265, 1993.
53. James, M.G., Robertson, D.S., and Myers, A.M., Characterization of the maize gene sugary 1, a determinant of starch composition in kernels, *Plant Cell,* 7, 417, 1995.
54. Wong, K.-S. et al., Structures and properties of amylopectin and phytoglycogen in the endosperm of sugary-1 mutants of rice, *J. Cereal Sci.,* 37, 139, 2003.
55. Jane, J. et al., Anthology of starch granule morphology by scanning electron microscopy, *Starch/Staerke,* 46, 121, 1994.
56. Fannon, J.E., Hauber, R.J., and Bemiller, J.N., Surface pores of starch granules, *Cereal Chem.,* 69, 284, 1992
57. Jane, J. et al., Structures of amylopectin and starch granules: how are they synthesized? *J. Appl. Glycosci.,* 50, 167, 2003.
58. Jane, J. and Shen, J.J., Internal structure of potato starch granule revealed by chemical gelatinization, *Carbohydr. Res.,* 247, 279, 1993.

59. Pan, D.D. and Jane, J., Internal structure of normal maize starch granules revealed by chemical surface gelatinization, *Biomacromolecules*, 1, 126, 2000.

60. Peng, M. et al., Starch branch enzymes preferentially associated with A-type starch granules in wheat endosperm, *Plant Physiol.*, 124, 265, 2000.

61. Jane, J. et al., Location of amylose in normal starch granules. I. Susceptibility of amylose and amylopectin to cross-linking reagents, *Cereal Chem.*, 69, 405, 1992.

62. Kasemsuwan, T. and Jane, J., Location of amylose in normal starch granules. II. Locations of phospho-diesters and -monoesters revealed by ^{31}P-NMR, *Cereal Chem.*, 71, 282, 1994.

63. Baba, T., Yoshii, M., and Kainuma, K., Incorporation of glucose to the outer chain of starch granule, *Starch/Staerke*, 39, 53, 1987.

64. Fannon, J.E. and BeMiller, J.N., Structure of corn starch paste and granule remnants revealed by low-temperature scanning electron microscopy after cryopreparation, *Cereal Chem.*, 69, 456, 1992.

65. Jane, J. et al., Preparation and properties of small particle corn starch, *Cereal Chem.*, 69, 280, 1992.

66. Hanson, L.F., Preparation and properties of phosphorylated amaranth starch for use in low-fat mayonnaise, M.S. Thesis, Iowa State University, Ames, Iowa, USA, 1998.

67. Donovan, J.W., Phase transitions of the starch-water system, *Biopolymers*, 18, 263, 1979.

68. Franco, C.M.L. et al., Structural and functional characteristics of selected soft wheat starches, *Cereal Chem.*, 79, 243, 2002.

69. Yuan, R.C., Thompson, D.B., and Boyer, C.D., Fine structure of amylopectin in relation to gelatinization and retrogradation behavior of maize starches from three wx-containing genotypes in two inbred lines, *Cereal Chem.*, 70, 81, 1993.

70. Shi, Y.-C. and Seib, P.A., The structure of four waxy starches related to gelatinization and retrogradation, *Carbohydr. Res.*, 227, 131, 1992.

71. Shi, Y.-C. and Seib, P.A., Fine structure of maize starches from four wx-containing genotypes of the W64A inbred line in relation to gelatinization and retrogradation, *Carbohydr. Polym.*, 26, 141, 1995.

72. Whittam, M.A., Noel, T.R., and Ring, S.G., Melting behaviour of A- and B-type crystalline starch, *Int. J. Biol. Macromol.*, 12, 359, 1990.

73. Ring, S.G. et al., Spherulitic crystallization of short chain amylase, *Int. J. Biol. Macromol.*, 9, 159, 1987.

74. Gidley, M.J. and Bulpin, P.V., Crystallisation of malto-oligosaccharides as models of the crystalline forms of starch: minimum chain-length requirement for the formation of double helices, *Carbohydr. Res.*, 161, 291, 1987.

75. Gidley, M.J., Factors affecting the crystalline type (A-C) of native starches and model compounds: a rationalization of observed effects in terms of polymorphic structures, *Carbohydr. Res.*, 161, 301, 1987.

76. Evans, I.D. and Haisman, D.R., The effect of solutes on the gelatinization temperature range of potato starch, *Starch/Staerke*, 34, 224, 1982.

77. Jane, J., Mechanism of starch gelatinization in neutral salt solutions, *Starch/Staerke*, 45, 161, 1993.

78. Gough, B.M. and Pybus, J. N., Effect of metal cations on the swelling and gelatinization behavior of large wheat starch granules, *Starch/Staerke*, 25, 123, 1973.

79. Koch, K. and Jane, J., Morphological changes of granules of different starches by surface gelatinization with calcium chloride, *Cereal Chem.*, 77, 115, 2000.

80. Ahmad, F.B. and Williams, P.A., Effect of sugars on the thermal and rheological properties of sago starch, *Biopolymers*, 50, 401, 1999.
81. Kudla, E. and Tomasik, P., The modification of starch by high pressure. Part II. Compression of starch with additives, *Starch/Staerke,* 44, 253, 1992.
82. Katopo, H., Song, Y., and Jane, J., Effect and mechanism of ultrahigh-hydrostatic pressure on the structure and properties of starches, *Carbohydr. Polym.*, 47, 233, 2002.
83. Tester, R.F. and Morrison, W.R., Swelling and gelatinization of cereal starches. I. Effects of amylopectin, amylose, and lipids, *Cereal Chem.*, 67, 551, 1990.
84. Song, Y. and Jane, J., Characterization of barley starches from waxy, normal and high amylose varieties, *Carbohydr. Polym.*, 41, 365, 2000.
85. Craig, S.A.S. et al., Starch paste clarity, *Cereal Chem.*, 66, 173, 1989.
86. Sievert, D. and Pomeranz, Y., Enzyme-resistant starch. I. Characterization and evaluation by enzymatic, thermoanalytical, and microscopic methods, *Cereal Chem.*, 66, 342, 1989.
87. Sievert, D. and Pomeranz, Y., Enzyme-resistant starch. II. Differential scanning calorimetry studies on heat-treated starches and enzyme-resistant starch residues, *Cereal Chem.*, 67, 217, 1990.
88. Sievert, D., Czuchajowska, Z., and Pomeranz, Y., Enzyme-resistant starch. III. X-ray diffraction of autoclaved amylomaize VII starch and enzyme-resistant starch residues, *Cereal Chem.*, 68, 86, 1991.
89. Lin, P.Y., Czuchajowska, Z., and Pomeranz, Y., Enzyme-resistant starch in yellow layer cake, *Cereal Chem.*, 71, 69, 1994.
90. Pfannemuller, B., Mayerhoffer, H., and Schulz, R.C., Conformation of amylose in aqueous solution: optical rotatory dispersion and circular dichroism of amylose-iodine complexes and dependence on chain length of retrogradation of amylase, *Biopolymers,* 10, 243, 1971.
91. Gidley, M.J. and Bulpin, P.V., Aggregation of amylose in aqueous systems: the effect of chain length on phase behavior and aggregation kinetics, *Macromolecules,* 22, 341, 1989.
92. Hizukuri, S., Fujii, M., and Nikuni, J., Effect of inorganic ions on the crystallization of amylodextrin, *Biochim. et Biophys. Acta*, 40, 346, 1960.
93. Lu, T.-J., Jane, J., and Keeling, P., Temperature effect on the amylose retrogradation rate and crystalline structure, *Carbohydr. Polym.*, 33, 19, 1997.
94. Billmeyer, F.W., Polymer structure and physical properties, in *Textbook of Polymer Science,* Wiley-Interscience, New York, 1984, p. 330.
95. Slade, L. and Levine, H., Water relationships in starch transitions, *Carbohydr. Polym.,* 21, 105, 1993.
96. Slade, L. and Levine, H., Glass transitions and water-food structure interactions, *Adv. Food Nutr. Res.,* 38, 103, 1995.
97. Zeleznak, K.J. and Hoseney, R.C., The glass transition in starch, *Cereal Chem.,* 64, 121, 1987.
98. Jouppila, K. and Roos, Y.H., The physical state of amorphous corn starch and its impact on crystallization, *Carbohydr. Polym.*, 32, 95, 1997.
99. Lievonen, S.M. and Roos, Y.H., Water sorption of food models for studies of glass transition and reaction kinetics, *J. Food Sci.*, 67, 1758, 2002.
100. Roos, Y.H. and Karel, M., Water and molecular weight effects on glass transition in amorphous carbohydrates and carbohydrate solutions, *J. Food Sci.*, 56, 1676, 1991.

101. Wang, Y.-J. and Jane, J., Correlation between glass transition temperature and starch retrogradation in the presence of sugars and maltodextrins, *Cereal Chem.*, 71, 527, 1994.

8 Starch World Markets and Isolation of Starch

Wolfgang Bergthaller

CONTENTS

8.1 Production and Markets 104
    8.1.1 Globally Used Substrates 104
    8.1.2 World Starch Production 104
8.2 Corn (Maize) Starch 105
    8.2.1 General Remarks 105
    8.2.2 Substrates 105
    8.2.3 Processes 106
        8.2.3.1 Steeping and Grinding 107
        8.2.3.2 Fiber Separation 107
        8.2.3.3 Starch–Gluten Separation 107
        8.2.3.4 Starch Refinement — Washing 108
8.3 Wheat Starch 109
    8.3.1 General Remarks 109
    8.3.2 Substrates 110
    8.3.3 Modified Martin Process 110
        8.3.3.1 General Remarks 110
        8.3.3.2 Dough Preparation and Gluten–Starch Separation 111
        8.3.3.3 Starch Purification, Dewatering, and Drying 111
    8.3.4 Westfalia Process 112
        8.3.4.1 General Remarks 112
        8.3.4.2 Flour Slurry Preparation and Gluten Agglomeration 112
        8.3.4.3 A-Starch Purification and Gluten Recovery 113
8.4 Potato Starch 114
    8.4.1 General Remarks 114
    8.4.2 Substrates 114
    8.4.3 Processes and Alternative Process Steps 114
        8.4.3.1 General Remarks 114
        8.4.3.2 Tuber Disintegration — Rasping 116
        8.4.3.3 Fruit Water Separation 116
        8.4.3.4 Starch Extraction and Purification 117
        8.4.3.5 Starch Dewatering and Drying 118

0-8493-1486-0/04/$0.00+$1.50
© 2004 by CRC Press LLC

           8.4.3.6   Protein Recovery — By-Products 118
8.5   Cassava (Tapioca, Manioc) Starch 118
       8.5.1   General Remarks 118
       8.5.2   Processes 119
           8.5.2.1   Chopping and Rasping 120
           8.5.2.2   Starch Extraction and Purification 120
           8.5.2.3   Starch Dewatering and Drying 120
References 120

8.1 PRODUCTION AND MARKETS

8.1.1 GLOBALLY USED SUBSTRATES

Starch is deposited in different storage organs of numerous plants in the green environment and functions as an essential energy reservoir. Thus, starch, together with other storage compounds, enables plants to start a new cycle in plant reproduction. Several cereal grains as well as some tubers and roots are cultivated for being utilized as substrates for starch production.

Maize (corn) is the major source of starch worldwide, even in many countries that use rather small quantities in production and are therefore frequently indicated in the category "other countries." Over 80% of the starch produced globally is extracted from maize, generally named corn in the U.S. Wheat comes second as raw material because the European Union (E.U.) relies greatly on this crop. However, not more than ca. 8% of global starch production is derived from wheat. Potato takes the third position, and again the E.U. dominates the production of this starch type by historical tradition. The group of the most important plants used as raw materials for starch production is completed with a tropical root known by different botanical synonyms, such as cassava, tapioca, or manioc. The production of starch from this root contributes to over 5% of the world's production and is concentrated in Southeast Asia and South America. In addition to these substrates, only a few other plants are raw materials, although numerous others have promising property profiles. Rice, barley, oats, sweet potatoes, and sago trunks are locally used in substantial quantities.

The domination of maize can be easily explained by the cost competitiveness in regional markets and the scale and value of by-products that contribute to cost calculation. Here again, maize and wheat offer several advantages, with valuable products such as vegetable oil provided by maize germs or vital gluten recovered from wheat. In contrast, with potatoes and cassava as starch plants, there is no significant additional profit because commercialization of by-products, proteins, and fibers covers only expenses. In the potato protein, however, a generally low sensorial quality allows it to be used as feed only, but it can eventually be transformed into a food-grade product by additional but costly purification measures.

8.1.2 WORLD STARCH PRODUCTION

A recently published E.U. study estimated the world starch market at 48.5 mt (million tonnes) in 2000.[1] This figure includes the total amount of starch used for production of modified starches as well as the share converted into syrups of varied composition, such as isoglucose or very highly converted glucose syrups for fermentation into organic chemicals, including ethanol. The output of this branch of agro-based industry is U.S. $16 billion (€ 15 billion) per year. The U.S. has the largest starch industry, with maize being the main substrate and wheat supplementing by 10% only. The U.S. production comes to a share of about 51%, whereas the E.U., standing second, contributes to more than 17% of the total share. Maize is still the main substrate for starch but is increasingly replaced by wheat. A third of the world production (32%) comes from many other countries that also have maize as the dominant raw material. Prominent shares come from Japan, and also from South Africa, China, and Russia. However, a substantial contribution to world starch production comes from cassava (also known as tapioca), the major producers being Thailand in Southeast Asia and Brazil in South America.[1]

8.2 CORN (MAIZE) STARCH

8.2.1 General Remarks

As explained previously, kernels of the botanical species *Zea mays* L. are a major source of starch. For the grain, two synonyms are used in parallel, but with different local distributions worldwide. Coming from the main production areas, the Midwest of the U.S., the popular name is corn. In other traditional production areas such as Europe, the expression maize is preferred.

In wet milling of corn/maize, almost 100% of the processed material is recovered in products, contributing considerably to commercial success. Technologically speaking, the separated components form the spectrum of products derived from corn/maize: starch, germs, and feed products are differently recombined from gluten, fibers, and steep liquor isolated during the process as distinct components. Process effluents and wastewater load are not critical issues in maize starch production. Wastewater streams (reduced to 1.5 to 3.0 m^3 t^{-1}) are circulated mainly as process water and only condensates from evaporation [specific biological oxygen demand (BOD) 0.6–0.8 kg t^{-1} and specific chemical oxygen demand (COD) 1.0–1.3 kg t^{-1}] and process water (specific BOD 5.2–11.0 kg t^{-1} and specific COD 7.9–16.7 kg t^{-1}) surplus are sent to sewage.[2]

8.2.2 Substrates

In the past, the standard and worldwide-accepted grain quality for maize starch production was known as No. 2 yellow dent corn or U.S. Yellow No. 2.[3,4] But maize from other origins, for example, from Argentina (preferred white-colored cultivars) or South Africa, was also used because of specific whiteness of the starch. Table 8.1 gives the composition of U.S. Yellow No. 2. Efforts of agricultural scientists to improve cultivars grown in different areas of Europe and France, in particular, and to adapt their produce to requirements of starch production helped remove U.S. Yellow No. 2 from the European market. Success was achieved by the installing suitable drying equipment and adapting drying procedures that resulted in most of

TABLE 8.1
Proximate Analysis of No. 2 Yellow Dent Corn and Maize of French Origin

Component	Yellow Dent Corn (% Dry Basis)	Maize of French Origin (% Dry Basis)
Moisture	15.0 ± 1.0	10.1
Starch	71.8 ± 1.5	71.8
Protein	9.6 ± 1.1	10.2
Lipids	4.6 ± 0.5	4.6
Crude fiber	2.9 ± 0.5	1.93
Minerals	1.4 ± 0.2	n.d.[a]
Soluble Carbohydrates	2.0 ± 0.4	n.d.

[a] n.d.: Not determined.

Source: For No. 2 yellow dent corn from Blanchard, P.H., *Technology of Corn Wet Milling and Associated Processes*, Elsevier Science, Amsterdam, 1992, chaps. 1 and 3 and for maize of French origin from Kempf, W., *Starch/Staerke*, 24, 269, 1972

the grain being undamaged and properly dried. Because of the much higher kernel moisture content under European harvesting conditions (a maximum of 40% moisture at harvest), only carefully controlled drying conditions, in particular avoiding damage by heated air drying, provide undamaged starch and protein bodies with proper density differences, which are prerequisites for successful separation in the gravity field during centrifugal separation.[5,6]

8.2.3 PROCESSES

Typical in wet milling of corn/maize is the physical breakdown of kernel firmness after chemical and biological treatments. Kernels enter processing by an initial steeping procedure that is designed to soften the kernel tissue texture by soaking to optimum hydration. Furthermore, low-molecular-weight compounds, such as carbohydrates, peptides and amino acids, and mineral substances, are solved in this state. Addition of sulfur dioxide (SO_2) or sulfurous acid solution supports chemical and biological processes in this controlled environment. The dominant reaction of reduction of sulfur bridges loosens the protein matrix within endosperm cells, where at the beginning starch granules lie tightly packed and glued to one another. The differences in intensity of packing can be seen in kernel sections having horny or floury endosperm. Both the starch granules and protein matrix contribute to the specific property profiles of these sections. The horny endosperm represents a structure formed mainly by small-size polygonal starch granules embedded in a protein matrix that becomes fortified by protein corpuscles. The floury endosperm, in contrast, contains the large and more spherical starch granules. Its protein matrix is less dense because of enclosed air and the lack of protein corpuscles.[3–5]

8.2.3.1   Steeping and Grinding

Steeping is the key operation in preparing kernels for effective separation and splitting of their components.[3] As previously addressed, water diffusion, chemical reactions, and removal of solved compounds together with fermentative processes (in particular, a lactic acid fermentation) produce tissue structures adequately softened for mechanical separation operations.[4] Understeeping results in high-protein starch and starch losses with by-products, whereas oversteeping damages starch quality and gluten losses by excessive solubilization.

Cleaned corn/maize is soaked in process water from starch refining to which 0.1 to 0.2% SO_2 is added to produce steep acid. The steeping battery is managed in a complex system of countercurrent flow where steep acid enters the battery section with oldest corn (approximately 45% moisture) and leaves the section with fresh corn as light steep water. The temperature applied during steeping is set by tolerance of lactic acid bacteria and ranges from 49 to 53°C. Steeping time may vary from 30 to 50 h; however, effective steeping times are 40 to 42 h. After reaching the correct point of treatment, steeped corn is drained (about 50% moisture) and fed to degerminating mills (attrition mills) to tear open kernels and leave the intact germ free for the following separation. In general, degermination takes place in two stages of a first and second grind, with intermediate removal of germs via hydrocyclones 152 mm in diameter. Ground germ-free material passes 50-µm screen bends (120°, grit screens) to separate finely split starch and gluten (40 to 50% of the stream) from coarse material, which is fed to the third grind by refiner or impact mills for completed disintegration. Separated starch-free germs are dried. They have a lipid content of 50 to 58% and are high-grade material for vegetable oil production.

8.2.3.2   Fiber Separation

Fibers made flexible by steeping become resistant to further fragmentation by milling and are removed by screening from the slurry consisting mainly of starch and gluten. The material leaving the grit screens for the third grinding passes the fiber-washing station, a system of five or six stages of screen bends in series. Screen bends used in the fiber-washing station resemble the 120° screens used as grit screens. The aperture of the screens is also 50 µm. While starch and gluten are transported to the first wash screen, fibers leave the last screen for dewatering and drying. Coarse fiber is discharged after optimum washing with maximum 15% bound and 5% free starch (dry substance basis).

8.2.3.3   Starch–Gluten Separation

The stream coming from grit screens and the first section of the fiber-washing screens is designated as mill starch and contains separated starch, insoluble proteins (zein, glutelins), and soluble impurities, mainly proteins and minerals. Mill starch produced with a density of 6 to 8° Bé first passes degritting hydrocyclones to remove sand and fine debris that potentially may damage separators and hydrocyclone stations used for starch–gluten separation and starch refinement. Debris formed by the action of centrifugal forces and high-speed movement applied there can cause significant wear. Oversized materials are also removed from mill starch thickeners that adjust feed for primary separation to a density of 10 to 12° Bé preferred in starch–gluten separation.

Splitting of both components relies on the difference of densities in suspension. For starch, the maximum density is about 1.5 kg l^{-1} whereas intact maize protein corpuscles have a maximum density of about 1.3 kg l^{-1}. For utilization of European maize, it is essential that drying procedures for maize harvested as a moist produce do not compensate this density difference.[5] Under appropriate conditions, starch is separated from protein in a primary separator, which is a nozzle separator equipped with wash-water feed close to nozzle concentration. Starch is discharged at a concentration of about 19° Bé and a protein content of 2 to 4%, whereas gluten is discharged via the overflow and gets concentrated in a subsequently installed separator (gluten thickener), followed by dewatering to a gluten cake by a rotary drum filter (40 to 43% dry substance). With a decanter centrifuge, 30 to 40% or even 44% dry substance can be achieved. Finally, dewatered maize gluten with a protein content of 72 to 75% in dry substance is dried in tubular bundle dryers, flash dryers, or disc dryers[4,8] or mixed with fiber fractions to result in a feed mixture.

8.2.3.4   Starch Refinement — Washing

The starch suspension resulting from primary separation and having a protein concentration of 2 to 4% is then washed countercurrent in a multistage hydrocyclone washing station in nozzle separators or even mixed systems. Fresh water enters the process at the last stage of the washing station and transports soluble and insoluble impurities to previous sections. Because water quality is decisive for starch quality, softened water is used, in general, together with good-quality vapor condensates or filtrates from starch dewatering.

Particle size of maize starch requires hydrocyclones small in size, with an inner diameter of 10 mm and the overall length about 150 mm.[9] Capacity of these systems is low (3.8 l min^{-1}), in principle, but can be simply adapted to needs in total feed capacity by parallel adjustment of high numbers (up to several hundred hydrocyclones) in one housing (multicyclone). Size of feed inlet, vortex finder (low-density outlet), and apex opening (concentrate outlet) define the operation characteristics of hydrocyclones, but there are differences in the apex opening. Hydrocyclone washing is most effective at elevated temperatures, but limited by the sensitivity of starch. Water temperature is therefore kept at 41 to 43°C.

Impurities are displaced from starch by countercurrent dilution with wash water. Usually, over 99% of soluble components need to be removed. Depending on the number of stages available, specific figures for the wash ratio (ratio of weight of wash water to weight of starch solids in the underflow of the last stage) are recommended. A nine-stage hydrocyclone system should have a ratio of 2.5 to 2.8. A disadvantage of hydrocyclones is the loss of small-size starch granules in the light phase, which go back in increasing amounts to the middling clarifier. Alternatively, three-phase nozzle separators equipped with wash-water feed close to concentration sections in the front of nozzles are also common for different sections of maize starch production. Their specific construction principles and unique separation effect allow them to successfully replace two-phase separators in primary separation or as middling separators. As a result, separation and refining steps can be performed more effectively at a reduced cost.

The purified starch slurry leaves the washing section with about 23°Bé (equal to about 40% dry substance) to intermediate storage for modification or saccharification or finally to dewatering in discontinuously working peeler centrifuges or vacuum drum filters. The minimum water content of starch cakes produced there varies from 36 to 40%. To achieve storable starch powder with a maximum moisture content of 14%, dewatered starch is dried in a pneumatic dryer, preferably a flash dryer. Before being packed, dried starch is sifted in horizontally working sifters to remove gritty particles that are ground for packing or readded to starch cakes.

8.3 WHEAT STARCH

8.3.1 GENERAL REMARKS

Wheat starch production was based both on wheat grains and wheat flours for a long time. As with other starches, potential uses of by-products were decisive in selecting the finally preferred technology.[10,11] The values of specific components in wheat grain proteins are important, in particular those that constitute most of the water-insoluble wheat protein fractions, the so-called gluten, and contribute to the extraordinary property profile essential for forming a satisfying crumb in baking wheat bread. When extraction and drying procedures for wheat gluten were developed into technologies that maintained the valuable viscoelastic properties in the dried product, wet milling processes were established exclusively in industrial practice, which produced typical wheat starch together with high yields of the vital wheat gluten. The principal characteristic of this gluten is to recover instantaneously viscoelastic behavior when hydrated.[12] From an economical point of view, the valuable, and as compared with the commodity product starch, high-priced by-product was responsible for selecting the best-suited technique.

For a long time, the flour-based dough system was the most suitable substrate for washing out starch. Because of the high flour quality needed for this process, vital high-quality gluten was recovered at high levels too. Purification of the received starch stream resulted then in the main product consisting of highly purified A-starch and another rather impure starch fraction derived from tailings, called B-starch. Other by-products such as fibers and pentosans, together with bran coming from the milling process, are now mostly used as feed.

The dough-based Martin process dominated production for a long time.[13] The main principle of this process was the formation of a dough similar to baker's dough, from which starch was washed out with rather high quantities of water. The enormous water consumption of approximately 15 m^3 per tonne flour was the main reason for the disuse of this process and replacement by procedures that consumed less water. Later, however, a process was derived from the Martin process. This technique, adapted to modern requirements and reestablished with modern installations such as hydrocyclone systems, was called the modified Martin process, and reduced water consumption to approximately 6 m^3 and even lesser volumes in up-to-date production plants.[13] An effective water regime achieved by the use of hydrocyclones in starch purification was another step forward in process engineering. More savings in water consumption, finally down to 3 to 2 m³, were achieved with the switch from dough systems to batters or even more diluted flour water systems, represented, for example, by the Westfalia three-phase decanter process, and, following this, the separation of starch, gluten, pentosans, and soluble components by systems based on centrifugal separation. A similar process design called the Flottweg tricanter process has also been developed.[13] It should be noted that wastewater load of wheat starch plant effluents is very high and may reach a specific BOD of 45 kg t^{-1} and a specific COD of 120 kg t^{-1} based on a specific wastewater volume of 2.8 to 3.5 kg t^{-1} wheat flour.[2]

8.3.2 SUBSTRATES

Although in the past specifications for wheat were frequently published as grain characteristics as, for example, by Witt,[14] industrial practice increasingly considered flour characteristics only. The standards indicated figures for moisture and protein content as well as some qualitative and descriptive terms for endosperm hardness, falling number, and amylograph consistency. Also, principal figures were presented for proximate composition, comprising moisture, protein, and starch content as the main components. Minor components such as minerals, lipids, and fibers were also mentioned. This catalog of requirements does not refer fully to the actually applied processes, in particular, the modified Martin process and the Westfalia process.

In the search for technologically important characteristics, wet gluten quantity as well as the characteristics mentioned previously for grain were still of interest. New insights established the starch potential and the limit for small-size granules as additional characteristics in an extended catalog for modern wheat starch manufacture (Table 8.2).[15]

Introduction of centrifugal separation techniques, wheat flours of reduced protein content, and, to a limited extent, reduced protein quality have been discussed for application using practical conditions. Even wheat cultivars having a more pronounced soft endosperm structure have been successfully used in wheat starch factories.[16]

8.3.3 MODIFIED MARTIN PROCESS

8.3.3.1 General Remarks

The modified Martin process[13] is the modern variant of the traditional procedure. After introducing the necessary improvements, different modifications are still applied in wheat starch plants worldwide. The improvements include reduction in fresh water consumption by expanded process water recycling and substitution of equipment directed toward more efficient starch–gluten separation.

TABLE 8.2
**Catalog of Wheat Flour Requirements for Modern Starch Production for
100 g Flour on a Fresh Weight Basis**

Characteristics	Limit
Moisture (%)	~15.0
Protein (*N* conversion factor: 5.7) (% d.b.)	~12.0
Minerals (% d.b.)	~0.63
Lipids (% d.b.)	~1.5
Fibers (% d.b.)	~1.5
Starch (% d.b.)	~80.0
Wet gluten (g)	~28.0
Amylogram peak viscosity (BU)	~500
Falling number (s)	~250
Starch potential[a] (%)	min. 70
Starch granules < 10 μm (%)	max. 30

[a] Determined by the mixer test/wash test.

Source: From Bergthaller, W.J., Witt, W., and Seiler, M., *Proceedings of the 2nd International Wheat
　　Quality Conference,* Manhattan, KS, May 20–24, 2001, in press.

d.b. = dry basis
BU = Brakender units
s = second
g = gram

8.3.3.2　Dough Preparation and Gluten–Starch Separation

Wheat flour is mixed continuously with water at 32°C to form a stiff dough and to
develop gluten. The cohesive dough formed is allowed to rest to complete hydration
of flour particles and, in particular, gluten proteins. The dough is then mixed vigor-
ously with additional water to accelerate segregation of gluten from starch. Further
separation of gluten aggregates from starch suspension occurs while pumping the
mixture into the gluten washer, a long, slanted rotating cylinder, the wall of which
is equipped with 40–mesh–plate screens. During passage, water is sprayed via
nozzles on gluten and screens to wash away starch and to avoid plugging of screen
slits. While the mixture of starch, tailings, and fibers is rinsed to an intermediate
tank, gluten is further purified with excess water in a gluten washer. Then gluten
gets dewatered, split up into small pieces, and is simultaneously remixed with dry
gluten powder to prepare for drying in a flash dryer. This final treatment allows
gluten to adapt to drying conditions in a way that allows maximum gluten vitality.
By this process, a dried gluten product results that can be reconstituted into the
original material and is highly functional.

8.3.3.3　Starch Purification, Dewatering, and Drying

The suspension leaving gluten washers is removed of fibers by sieving and purification with centrifugation and hydrocyclone washing. The first step is the separation
of residual small gluten particles by rotating sieves, shaker sieves, or sieve bends.

Residual bran particles, fibers, and endosperm pieces that have not been disintegrated are removed the same way. The clarified starch suspension is refined further in nozzle centrifuges with a central wash–water device by removal of small-size starch, pentosans, sugars, low-molecular-weight proteins, and minerals. In this manner, A-starch, concentrated to a suspension of approximately 35% dry matter, leaves the separator via nozzles. For final purification and concentration, multistage hydrocyclone systems are applied in a countercurrent washing. The resulting starch milk of 23°Bé (equal to approximately 40% dry substance) is then dewatered over a vacuum drum filter or peeler centrifuge. Finally, starch cakes resulting from dewatering are dried in flash dryers similar to those for maize starch.

With wheat starch, however, optimum dewatering is hardly ever achieved when small-size granules are present in large amounts. In such cases, dewatering can be accomplished alternatively by pressure filtration. With rotary pressure filters, the water content can be reduced to 33 to 35%, and the handling of the resulting starch cakes is less problematic. Also, drying costs can be significantly reduced.[17] Pressure filtration has been found to be effective even under practical conditions, wherein a pressure level of 3 bars, a drum rotation speed of 1.2 to 1.6 rotations/min[-1], and a specific throughput of 650 to 800 kg m[-2] h[-1] can be used.[18]

8.3.4 WESTFALIA PROCESS

8.3.4.1 General Remarks

The Westfalia three-phase decanter process was first designed in the 1980s and has undergone various improvements to achieve its present design, presented in Figure 8.1. The principal design feature of this process is the use of centrifugal forces to separate a diluted mixture of wheat flour and water into its main components — starch, gluten, pentosans, and process water. Prior to separation, the undiluted mixture undergoes shearing treatment, by the energy of which flour particles become fully hydrated and protein bodies are transformed into agglomerated gluten.

8.3.4.2 Flour Slurry Preparation and Gluten Agglomeration

The flour coming from suitable dosing systems is mixed with water (30 to 40°C) in a 1:0.85–0.95 ratio to produce a lump-free and fully hydrated slurry, in which flour proteins are allowed to form agglomerated gluten. Gluten development is frequently supported by a homogenization step with subsequent maturation. Depending on the applied pressure (up to 80 bars) and maturation time, the treatment results in improved differentiation of gluten and pentosans when centrifuged. Also, the relative amount of A-starch increases whereas gluten quantity reduces. However, gluten produced is more pure and offers a more clear separation from the superincumbent pentosans, and a distinct process water layer is developed. Prior to decanter separation, the pretreated flour–water mixture is further diluted by adding process or fresh water, or both. The generally used water quantity of 0.3 to 0.9 m[3] provides acceptable segregation of flour components in the decanter bowl according to their density. Because starch has the highest density of ca. 1.5 kg l[-1],

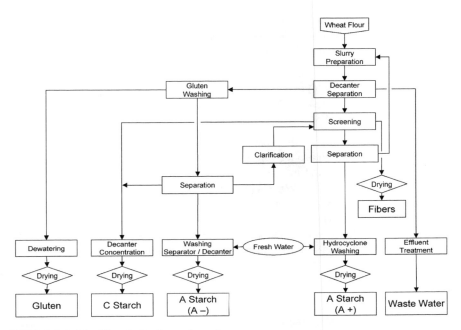

FIGURE 8.1 The Westphalia three-phase decanter process.

it settles as a sediment around the bowl wall and gets transported to the concentrate side by the decanter screw operating with a differential speed of approximately 60 rotation/min^{-1}. Depending on the decanter design, gluten, fibers, and process water with pentosans are removed together with the overflow (two-phase decanter), or process water and pentosans form the light phase and gluten and B starch the middle phase (three-phase decanter).

8.3.4.3 A-Starch Purification and Gluten Recovery

As described previously, the recovered concentrate is further treated by screening, multiple washing, and concentration. The A–starch, the starch fraction made up of the big-size and more pure granules, is either washed in a combination of a centrifuge and a multiple-stage hydrocyclone unit or a sequence of three-phase separators. The commercial starch usually has a protein content below 0.3% dry basis (d.b.).

Gluten is split off the middle phase and washed and dried as described for the modified Martin process. The filtrate resulting after removal of gluten and its wash water contains some small starch granules, a few gluten particles, and pentosans. As far as possible, the small starch gets purified in a separate process line and is finally added to the A-starch stream again to improve the starch recovery rate.[15] The run-offs that contain fibers, B-starch, and pentosans are concentrated, dewatered, and finally dried in application-oriented drying procedures. Pregelatinization via drum drying is a preferred method in handling this highly unstable, and from a microbiological point of view, critically loaded product used predominately as feed.

The organic matter removed with the light phase is used in liquid or dried form as feed. Wheat bran, for example, is mixed with such effluents and concentrated process water to prepare high-value feed, the wheat gluten feed.

8.4 POTATO STARCH

8.4.1 GENERAL REMARKS

Although potato (*Solanum tuberosum*) is grown on different continents, the utilization of potatoes for starch production is concentrated in the European countries. The main producers are situated in central Europe, in particular The Netherlands, Germany, and France. Poland and Russia are producers too. As mentioned earlier, starch production from potatoes cannot compete with that from other raw materials as regards the economic support required by coproducts and well paid in food and other industries. Because of alternatives in some regions of middle and eastern Europe, the production is still subsidized but also limited by a politically fixed system of quotas.[19,20]

8.4.2 SUBSTRATES

Potato starch factories buy potatoes on the basis of weight and starch content.[21] Both figures are determined for each delivered load immediately after entering the plant. In general, in-plant installed automatic sampling systems take samples from each lorry, at least two independent probes of up to 50 kg of tubers. The tubers are washed in automatic washing systems and inspected for general tuber quality. Finally, the tuber starch content is determined by underwater weighing. In terms of long–term selection, the starch content of potatoes preferred for starch production ranges from 17 to 19%, although tables established for regulating payment on the basis of starch content allow use of potatoes having a starch content from 13 to 23%. Annually established E.U. tables, for instance, describe the minimum price as well as subsidies based on starch content within the described framework.[21]

8.4.3 PROCESSES AND ALTERNATIVE PROCESS STEPS

8.4.3.1 General Remarks

From an engineering point of view, potatoes are a relatively uncomplicated product for starch extraction. Because potatoes are generated by secondary growth of subterranean shoots, their tissue structure is relatively soft in contrast to that of root tissue. Deposition of starch granules occurs in cells filled with fruit water, and the overall composition of potato starch *in situ* is uncomplicated. These facts make the starch accessible by simply disintegrating tubers by small mechanical forces, followed by washing out the interior content of cells, separating the starch, washing, and drying. A stable, but not valuable, commodity product of high purity and versatility can be recovered this way. Although the process itself seems simple, the production of potato starch has to overcome several serious obstacles that burden its economic success. The first and main problem is the limited stability of potatoes in the climatic conditions of the main production areas. The high water content in tubers and the connected transportation costs justify only limited distances between the production area and processing plant. Also, depending on the harvesting conditions, a considerable load of dirt gets transported to factories and must be removed and handled in initial washing steps.

Compared with wet milling of maize and wheat, the production of potato starch is rather simple (Figure 8.2). Some well-defined processes, however, differ in requirements of water consumption and, in particular, the techniques used to recover proteins solved in fruit water. The presence of partially recovered proteins leads to high economic costs in unalterable but necessary process water treatments. One of the modern-day processes therefore works toward the early removal of proteins or separation of high–molecular–weight proteins by isoelectric precipitation and denaturation from concentrated fruit water (reduction in about 50% of the dissolved

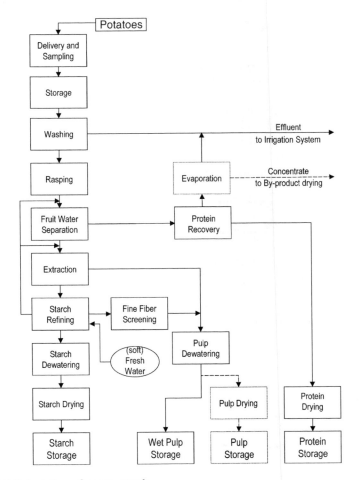

FIGURE 8.2 Isolation of potato starch.

protein) and the concentration of residual fruit water or diluted process water streams.[21] Alternative processes, however, operate either by partial removal or do not have separation of fruit water.[22]

Process water coming from internal water circulation (after initial fruit water removal) has a specific BOD of 6.3–8.2 kg t^{-1} and a specific COD of 7.4–9.6 kg t^{-1}. The specific wastewater volume is set to 1 to 1.3 m^3 t^{-1}. Although tuber transportation and wash water must be considered additionally (BOD 1.3–2.6 kg t^{-1} and COD 1.5–2.9 kg t^{-1}), the presented figures, even when multiplied by a factor of 5 to achieve comparable dry substance levels, are far from those representative of wheat starch production.[2] The presented figures do not include residual fruit water and fruit water.

8.4.3.2 Tuber Disintegration — Rasping

Because potato starch yield depends greatly on disintegration of potato tissue, rupture as complete as possible is significant when working with specific rasping or disintegration systems. If modern rasping machines operating at 2100 rotations/min^{-1} operate efficiently, starch extraction rates can reach up to 96 to 98%. Even in the yields given, starch bound to rasps/gratings can reach 25 to 30%. The rasping systems (such as Ultra Rasper) used allow throughputs between 20 and 30 t h^{-1}.

Depending on age of potatoes, oxidation of polyphenolic compounds of rasps plays a critical role following tuber disintegration, particularly when concentrated potato juice is separated and used for protein recovery. To minimize enzymatic discoloration, SO_2 is added either as sodium hydrogen sulfite (500–600 g t^{-1} = 120–144 g SO_2) or as aqueous sodium disulfite solution (52–34 g SO_2) during rasping. Encapsulated systems allow grinding under vacuum.

8.4.3.3 Fruit Water Separation

As stated previously, early and full separation of fruit water produces a concentrated protein-containing liquid favorably used for isoelectric precipitation and heat denaturation. A suitable protein concentration is required for optimum process design.[23] For separation of fruit water, horizontally oriented centrifuges, called decanters, are used. In the one-stage decanting system, undiluted fruit water is recovered at separation rates of 62 to 65%.[21,22] The rate can be significantly increased (up to 95%) when process water coming from starch refining is used to dilute rasps prior to decanting.

Alternatively, two-stage decanting systems can be used. Here, the rasps are diluted with overflow water from the second decanter before entering the first in the sequence. The concentrate from the first decanter is diluted, however, with process water coming from starch refinement. The quantity of process water used affects fruit water separation rate, which can reach over 92%. Because of early fruit water removal, total fresh water supply (including water supply for downstream starch extraction and refining) — water quality should preferably fulfil requirements for soft-water quality — can be reduced to 0.4 to 0.5 m^3 per tonne of potatoes.

8.4.3.4 Starch Extraction and Purification

When fruit water separation is not performed, rasped potatoes are extracted immediately after rasping in a four-stage screening centrifuge system. The screening centrifuges equipped with rotating conical sieves (centrisieves) ensure extraction of starch from rasps and dewater partially the residual fibrous material, called pulp. At every stage, pulp is transported from the small side of the sieve basket to the open side (basket diameter 850 mm) while being washed intensively via nozzles mounted inside and outside (called back washing or cleaning of sieves). Rotating sieve baskets are now equipped with etched slit plates or mechanically pressed perforation plate screens (125 μm for initial extraction and 60 to 80 μm for fine-fiber removal), which provide highest performance. The resulting starch suspension passes the screening system countercurrent and leaves the system with a density of 4 to 5 Bé, and washed pulp (5% d.s.) is dewatered in a decanter to a dry substance content of 17 to 22%.[21,22] For removal of insoluble (sand, peel fragments as black spots, fine fibers) and soluble (minerals, proteins, SO_2) impurities, mill starch first passes hydrocyclones for degritting and after concentration to 15 to 16 Bé a nozzle centrifuge (in large factories) or a four-stage multicyclone concentrator (in small factories) as the starch washing system. For starch washing, at most two systems exist, either nozzle centrifuges in two-phase or three-phase setups or a series of multicyclone stations.

Today in nozzle centrifuge washing of degritted mill starch, preferably three-phase nozzle centrifuges equipped with internal wash-water supply are used for displacement washing. A minimum 50% exchange with demineralized and softened wash water provides high-quality purified starch milk. Washing with three-phase separators offers the advantage of separation of fine fibers and small-size starch granules in the middle phase, which is further divided via fine-fiber screening (65 to 80 μm). The whole is then further separated in a recovery separator that recycles, in particular, small-size starch granules to the main stream of purified coarse starch. Finally, purified starch milk reaches a concentration of 21 to 22 Bé and fine fiber is removed to 98%. In the starch washing line, the light phase is used as wash water for the next stage and finally becomes process water for fruit water separation or extraction when leaving the first-stage separator.[21]

When multicyclones are used for purification and washing,[22] degritted mill starch gets purified first in a ten-stage washing section with cyclones of 15-mm inner diameter. The section enables classification into a 22 Bé underflow stream of coarse starch and an overflow (light phase) containing small-size starch and fibers. Fibers are removed by a screening centrifuge and fed back to the pulp. The very diluted flow of small-size starch having a concentration of 2 to 3 Bé only gets concentrated in a three-stage multicyclone concentrator and is further purified in an eight-stage multicyclone section equipped with cyclones of 10-mm diameters. Both purified streams, of coarse and of small-size starch, are combined finally for dewatering and drying. For the described hydrocyclone refining, a system lower in investment and energy costs has been proposed by feeding back concentrated fine starch after the third stage of the modified ten-stage washing section. This modification consists of expansion by one washing stage, and, depending on starch size characteristics, application of either 10 or 15 mm hydrocyclones from the fourth to the eleventh stage. Extra washing of fine starch is not performed.

8.4.3.5 Starch Dewatering and Drying

Because of the excellent filtering characteristics of potato starch, rotating vacuum drum filters that use filter cloth are standard equipment in dewatering.[22,24] They make available a minimum water content of 38%. The water content is further reduced to equilibrium relative moisture content of approximately 20% by flash drying. Drying of starch granules is performed rapidly without much damage in functionality. Maximum moisture level of commercially available potato starch is 21%.

8.4.3.6 Protein Recovery — By-Products

As stated previously, approximately 50% of the crude proteins solved in potato fruit water are high-molecular-weight proteins. This fraction can be separated by isoelectric precipitation and heat coagulation. The desired efficiency of these measures depends on concentration of the respective solute components. A process developed by GEA Westfalia Separator Industries GmbH, a machinery supplier, uses undiluted fruit water with a dry substance content of 4.5 to 5.2% or even diluted fruit water with 4% dry substance. Potential tuber protein recoveries range from 83 to 85%.

Technically, fruit water effluents are adjusted to a pH of 4.8 to 5.2 after preheating. Then, optimized precipitation is supported by steam injection and an induced heat shock of 110 to 120°C. The coagulated proteins are then separated by decanting and result in protein cakes with 38 to 43% dry matter. Dry and stable proteins having a protein content of 83 to 85% are recovered from air drying in ring dryers. For utilization, for example, as animal feed, the content of potato glycoalkaloids, solanines and chaconines, is critical. Provided suitable extraction or washing measures are applied, potato protein isolates with considerably reduced content of glycoalkaloids and of improved sensorial quality might serve valuable and palatable protein sources for human consumption.[25,26]

After removing the high-molecular-weight protein fraction, the remaining deproteinized fruit water contains a considerable amount of low-molecular-weight organic compounds such as amino acids, organic acids, and sugars, as well as minerals, predominantly potassium. The effluents produced mainly in a period of reduced plant growth get concentrated and as such are well-accepted storable fertilizers providing nitrogen, phosphorous, and potassium. Evaporation or membrane filtration is a suitable technique to prepare residual fruit water concentrates for future farm uses.[21,27–29]

8.5 CASSAVA (TAPIOCA, MANIOC) STARCH

8.5.1 General Remarks

Cassava, also known as tapioca or manioc, is a major tuber crop of tropical countries. Cultivation is reported in more than 80 countries, with a diversity of species belonging botanically to *Manihot utilissima* Pohl or *Manihot esculenta* Crantz and divers names such as yucca (Central America), mandioca (South America), cassava (English-speaking countries of Africa), and manioc (Madagaskar and French-speaking countries of Africa). Brazil, Thailand, Malaysia, Indonesia, and Nigeria are the major producers of cassava starch.[30,31] For starch production, the rootstock and, in particular, the big-size tuberous roots of this moderate, low-input semishrubbery perennial are used. Agricultural efforts, however, led finally to production yields of approximately 50 t ha^{-1}. The major constraint in utilization of roots is their storage. After harvest, roots rapidly deteriorate as a result of enzymatic decay or microbial infestation. To avoid decay and reduction in starch yield in cultivating areas, roots are processed within 24 h of harvest. Another way of preservation is dehydration in tropical counties by sun drying of roots chopped into small pieces or chips. Another alternative of preserving the economic value of cassava is by pellet production, which delivers uniform pieces of high density. Utilization of dehydrated root material for starch production is limited by starch quality. Even with specifically employed treatments, the starch extracted from chips or pellets does not reach the criteria of starch made from fresh and unspoiled roots.[32]

Cassava roots are rich in carbohydrates (30 to 35%), the main fraction of which is starch. Typical for roots, 3 to 5% of the carbohydrate fraction consists of fibers. About 80% of the fraction is made of starch. Calculated on this basis, the carbohydrates consist of fibers and about 80% of starch. The overall starch content may vary from 25 and 30% depending on variety. On the other hand, roots are poor in proteins (1 to 2%) and low in lipids, vitamins, and minerals. A major constraint in utilization is the content of cyanogenic glycosides, linamarin and lotaustralin, in bitter-tasting varieties of *Manihot utilissima*. For human consumption, these toxic compounds can be easily destroyed by simple measures such as drying, soaking, boiling, and fermentation.[33]

8.5.2 PROCESSES

A unique feature of starch production from cassava is that because of its simplicity, starch can be produced on all potential levels, from small-size handicrafts at the farm level to the industry. The scheme of the specific steps in starch manufacture resembles that of starch production from potato tubers. The main difference is in a peeling procedure performed immediately after washing, wherein the corky region at root ends and the protective epidermis are removed.

As regards the technological level achieved (cottage and small and large industrial scales) in cassava starch production, equipment and techniques differ mainly in precision of rasping or disintegration of roots (manual scraping of roots in contrast to perforated plates, perforated rotating drums, or saw tooth blades arranged longitudinally around the circumference of drums and, finally, modern rasping devices) as well as techniques used in dewatering. Further steps are connected with screening or extraction of starch from ruptured cell wall material, purification, and drying to a moisture content of less than 12% to prevent microbial growth, preparation of dried starch, and packing. Production capacities for modern cassava starch plants during the harvesting season are 200 to 300 tonnes per day.[33]

8.5.2.1   Chopping and Rasping

Prior to rasping by equipment of divers standards, roots are carefully washed to remove soil material and then reduced in size by chopping (30- to 50-mm slices) because root dimensions can be extreme. For instance, lengths can surpass 1.0 m and diameters may range up to 20 cm. Modern rasping machines do not differ from machines used for potatoes; however, because of the more fibrous structure of roots, the capacity, in general, is half. Quality of rasping is judged by the rasping effect, which is controlled by the amount of starch set free in rasping.

As with potatoes, fruit water can be removed from rasped cassava root material by decanter separation prior to starch extraction. Fruit water is separated with a dry matter content of about 5% whereas rasp dry matter can reach a dry matter content of ca. 45%.

8.5.2.2   Starch Extraction and Purification

The modern practice in separating starch from rasps resuspended with countercurrent wash water is to use curved sieve bends or conical jet extractors equipped with plate sieves instead of stainless steel wires. However, wire or rod type sieves are still in use. The slot width is 75 µm. Such screens work in groups of three to six stages, and fibers leave the system finally as pulp with a maximum amount of 4% free starch when these screening packages are operated with high efficiency. On a dry basis, cassava pulp may contain 56% starch, 36% fibers, 3.5% protein, and 2.7% minerals.[34] Dewatered pulp is used as fresh feed in regions surrounding the factory or gets sun dried for storage.

As regards industrial levels and size of plants, existing processes show a broad range of techniques applied in starch concentration and washing, ranging from old-fashioned settling tanks or semicontinuously working settling tables to use of three-phase nozzle separators or hydrocyclone washing in sophisticated setups. The purification steps of potato starch washing in all modern factories are very similar. Sulfuric acid is used to improve whiteness and control microbial status. The effect of this application is controlled by the pasting characteristics of isolated starch to which the acid was added at the concentration of 0.001 ml l⁻¹ starch milk or in other arrangements with 0.2% sulfuric acid at rates of 6 ml kg⁻¹ starch.

8.5.2.3   Starch Dewatering and Drying

Final removal of water directly before drying is done in a similar manner to that for potato starch production. Vacuum drum filters equipped with filter cloth are commonly used, and in small-scale plants drying is often done in tray driers whereas modern factories successfully use conventional flash driers.

REFERENCES

1. Anon., Evaluation of the community policy for starch and starch products, Report of the, European Commission, Brussels, 2002.

2. Seyfried, C.F. and Rosenwinkel, K.H.: Abwässer der Stäerkeindustrie. Grundlage zur Erabeitung der Mindestanforderungen nach 7 a (1) WHG. *Forschungsbericht* Nr. 10206031/05 *im Auftrag des Umweltbundesamtes*, 1981.

3. Blanchard, P.H., *Technology of Corn Wet Milling and Associated Processes*, Elsevier Science, Amsterdam, 1992, chaps. 1 and 3.

4. Watson, S.A., Corn and sorghum starches: production, in *Starch: Chemistry and Technology*, Whistler, R.L., BeMiller, J.N., and Paschall, E.F., Eds., Academic Press, New York, 1984, chap. 12.

5. Lasseran, J.C., Chemical and physical changes in maize components affecting quality for the wet-milling industry, in *Uniformity by 2000 — An International Workshop on Maize and Soybean Quality*, Hill, L.D., Ed., Scherer Communications, Urbana, 1991, chap. 14.

6. Eckhoff, S.R., Quality factors and measurement in the wet-milling industry, in *Uniformity by 2000 — An International Workshop on Maize and Soybean Quality*, Hill, L.D., Ed., Scherer Communications, Urbana, 1991, p. 419.

7. Kempf, W., Verarbeitungstechnische Eigenschaften von Inlandmais. 2. Stärkegewinnung, *Starch/Staerke*, 24, 269, 1972.

8. Bergthaller, W. and Lindhauer, M.G., Recent advances in the technology of starch production, in *Proceedings of the 6th Annual European Conference on Starch, Sugars and Sweeteners*, London, November 26, 1998.

9. van Esch, F., The efficiency of hydrocyclones for the separation of different starches, *Starch/Staerke*, 43, 427, 1991.

10. Kempf, W. and Röhrmann, C., Process for the industrial production of wheat starch from whole wheat, in *Wheat is Unique: Structure, Composition, Processing, End-Use Properties and Products*, Pomeranz, Y., Ed., American Association of Cereal Chemists, St. Paul, MN, 1989, p. 469.

11. Knight, B.W. and Olson, R.M., Wheat starch: production, modification, and uses, in *Starch: Chemistry and Technology*, Whistler, R.L., BeMiller, J.N., and Paschall, E.F., Eds., Academic Press, New York, 1984, chap. 15.

12. Dubois, D.K., History of vital wheat gluten, *Am. Soc. Bakery Eng. Bull.* 233, 991, 1996.

13. Maningat, C.C. and Bassi, S.D., Wheat starch production, in *Proceedings of the International Starch Technology Conference*, Tumbleson, M., Yang, P., and Eckhoff, S., Eds., University of Illinois, Urbana, 1999, p. 26.

14. Witt, W., Anforderungsprofil an den Rohstoff Weizen, in *Stärke im Nichtnahrungsbereich*, Anonymous, Landwirtschaftsverlag GmbH, Münster-Hiltrup, 1990, p. 125.

15. Bergthaller, W.J., Witt, W., and Seiler, M., Wheat quality and testing with regard to modern wet milling technology, *Proceedings of the 2nd International Wheat Quality Conference*, Manhattan, KS, May 20–24, 2001, in press.

16. Zwingelberg, H. and Lindhauer, M.G., Weizensorten für die Stärkeherstellung, *Getreide, Mehl, Brot*, 50, 303, 1996.

17. Meuser, F., New developments in the production of wheat starch and gluten, in *Starch 96 — The Book*, Van Doren, H. and van Swaij, N., Eds., Zestec b.v./Carbohydrate Research Foundation, The Hague, 1997, p. 65.

18. Gansäuer, H., Developments in dewatering of starch slurries, *Proceedings of the 48th Starch Convention*, Detmold, Germany, April 23–25, 1997.

19. De Baere, H., Starch policy in the European Community, *Starch/Staerke*, 51, 189, 1999.

20. Gordon, I., Starches from differing sources: supply, demand, price formation, *Starch/Staerke*, 51, 193, 1999.

21. Bergthaller, W., Witt, W., and Goldau, H.-P., Potato starch technology, *Starch/Staerke*, 51, 235, 1999.
22. van der Ham, W., Modern potato processing, *Proceedings of the Starch Experts Meeting*, Copenhagen, September 25–26, 2002, p. 6.
23. Meuser, F. and Köhler, F., Einsatz der Membranfiltrationstechnik zur Prozeβwasseraufbereitung in der Kartoffel- und Weizenstärkeindustrie. *Chem. Mikrobiol. Technol. Lebensm.*, 7, 51, 1981.
24. Tegge, G., *Staerke und Staerkederivate*, Behr's Verlag, Hamburg, 1988.
25. Knorr, D., Protein recovery from waste effluents of potato processing plants, *J. Food Technol.* 12, 563, 1977.
26. Wilhelm, E. and Kempf, W., Neue Erkenntnisse über die Gewinnung von Kartoffelprotein für die menschliche Ernaehrung, *Starch/Staerke* 29, 376, 1977.
27. von Tucher, T., Einsatz von Kartoffelfruchtwasser-Konzentrat als Düngemittel, *Kartoffelbau* 43, 278, 1992.
28. Meindersma, G.W., Application of membrane filtration in the potato starch industry, *Starch/Stärke* 32, 329, 1980.
29. Rüffer, H., Kremser, U., and Seekamp, M., Experiences with a reverse osmosis pilot plant for the concentration of potato fruit water in the potato starch industry, *Starch/Stärke* 49, 354, 1997.
30. Dziedzic, S.Z. and Kearsley, M.W., The technology of starch production, in *Handbook of Starch Hydrolysis Products and Their Derivatives*, Kearsley, M.W. and Dziedzic, S.Z., Eds., Blackie Academic & Professional, Glasgow, 1995, chap. 1.
31. Corbishley, D.A. and Miller, W., Tapioca, arrowroot, and sago starches: production, in *Starch: Chemistry and Technology*, Whistler, R.L., BeMiller, J.N., and Paschall, E.F., Eds., Academic Press, New York, 1984, chap. 13.
32. Meuser, F. et al., Comparison of starch extraction from tapioca chips, pellets and roots, *Starch/Stärke* 30, 299, 1978.
33. Rakshit, S.K., Cassava starch production technology, in *Proceedings of the International Starch Technology Conference*, Tumbleson, M., Yang, P., and Eckhoff, S., Eds., University of Illinois, Urbana, 1999, p. 41.
34. Balagopalan, C. et al., *Cassava in Food, Feed, and Industry*, CRC Press, Boca Raton, 1988.

9 Chemical Modifications of Polysaccharides

Piotr Tomasik

CONTENTS

9.1 Physical Modifications 123
    9.1.1 Introduction 123
    9.1.2 Dextrinization 124
    9.1.3 Pasting and Gelatinization 125
    9.1.4 Other Physical Modifications 125
    9.1.5 Complexation 126
9.2 Chemical Modifications 127
    9.2.1 Acid- and Base-Catalyzed Degradation 127
    9.2.2 Reduction and Oxidation 128
    9.2.3 Esterification 128
    9.2.4 Etherification 128
    9.2.5 Acetalation 129
    9.2.6 Halogenation 129
    9.2.7 Amination and Carbamoylation 129
    9.2.8 Graft Polymerization 129
    9.2.9 Metal Salts and Complexes 130
References 130

9.1 PHYSICAL MODIFICATIONS

9.1.1 INTRODUCTION

Among the polysaccharides, starch and cellulose are the most frequently and massively modified, followed by carrageenans. This chapter focuses on the modifications of starch. Modified starches are commonly used in food technology.

Generally, the methods and pathways used for modification of starch are also applicable for other polysaccharides. It should be emphasized that the result of any modification of starch depends on its botanical origin. Susceptibility of starch granules to swelling (water uptake) and penetration by chemicals are key but not the sole properties controlling the result of modification. Usually, small starch granules of a given botanical origin undergo modification to a lesser extent than do large granules.

0-8493-1486-0/04/$0.00+$1.50
© 2004 by CRC Press LLC

123

9.1.2 DEXTRINIZATION

Dextrinization is one of the two most common ways of starch modification. It is either thermal[1] or acid-catalyzed hydrolysis[2] of starch to oligosaccharides of much lower molecular weight. Apart from hydrolysis, there is scission of glycosidic bonds and rupture of inter- and intramolecular bonds. As a result, dextrins are much more readily soluble in water.

Treatment of starch with diluted hydrochloric acid at ambient temperature, called lintnerization, depending on the parameter applied, produces thick- or thin-boiled dextrins, which are slightly and strongly hydrolyzed materials, respectively. Thermal treatment of starch, usually from 240 to350°C, produces British gums or yellow dextrins, the color of which ranges from white to dark brown-black. Such dextrinization retains the granular form of starch, although granules may be severely damaged. Roasting of starch above 450°C, called pyrolysis, produces steam, CO_2, small amounts of CO, methane, ethane furan derivatives, and carbonizate. Acid-catalyzed dextrinization is an ionic process whereas thermolysis follows a free-radical mechanism. Thus, thermolysis of starch produces free radicals,[1,3] concentration of which depend on temperature and duration of roasting. Additives to starch such as mineral salts, proteins, low-molecular-weight saccharides, and lipids decrease temperature and shorten the time of generation of free radicals. Although temperature and time required to generate free radicals from pure starch exceed those usually needed for foodstuff blends of starch with previously mentioned additives and ingredients, free radicals may be generated under conditions applicable for foodstuff preparation. Free radicals of starch are stable and do not exhibit any mutagenicity.[4]

Frequently, elevated temperature and catalyst, the latter commonly being the proton of admixed acid, are used to prepare dextrins. In their physical and functional properties, these dextrins resemble dextrins from acid-catalyzed processes rather than British gums. Extended acid-catalyzed processes are useful for starch saccharification to maltose and glucose syrups.

Other methods besides conventional heating and acid-catalyzed hydrolysis provide energy for modification of starch.[5] Microwave heating of starch provides interesting results. Fast damage of granular starch has been noted, depending on the water content in starch.

Freezing is also a method of starch modification, particularly for moist starch granules changed within subsequent cycles of deep freezing in liquid nitrogen followed by thawing. The ice matrix probably causes mechanical deterioration of granules because water outside and inside starch granules has different abilities to freeze and expand its volume.[6]

Dextrins are commonly used as adhesives, thickeners, and sizes. Recently, dextrinization was developed in the presence of various additives such as amino acids, hydroxy acids, and carboxyamides such as urea in order to prepare dextrins for special uses, for instance, prebiotics, depressants for flotation, fodder for ruminants, or heavy-metal collectors. Microwave heating was found to be a convenient method for such modifications.[2]

9.1.3 Pasting and Gelatinization

Pasting of starch into gels is the most common method of starch modification. Starch granules are subjected to swelling followed by rupture of granule envelopes. To form a paste, starch (3 to 8 wt/wt%) is suspended in water and heated on agitation. Pasting carried out under standardized conditions (e.g., rate of temperature increase at constant velocity of a rotor followed by storage of gel under static conditions and controlled rate in the decrease of temperature) in a Brabender viscograph best characterizes specific properties of starches. Starches differ from one another in pasting temperatures, which for various starches range from 62 to 120°C. After reaching pasting temperature, viscosity of a suspension rapidly increases in order to reach the first maximum. Viscosity of pastes of tuber starches (potato, cassava, yams) slightly decreases with time and temperature whereas pastes of cereal starches do not exhibit such behavior. Further course of the viscogram recorded under standardized conditions characterizes functional properties of the paste and its ability to retrograde.

Heat–moisture–pressure treatment of starch, called extrusion cooking, specifically modifies starch. Pasting of starch at elevated temperature and pressure gives products with interesting functional properties. A variety of parameters may be controlled. The type of extruder plays a key role. A single-screw extruder offers a slightly modified starch pasted with minimum water, and this machine is used mainly for pasting starch and as a homogenizer to prepare starch inclusion complexes. A twin-screw extruder provides deeper modification of starch. This extruder may be used as a chemical reactor when fed with starch blended with a suitable reagent. Temperature gradient between feeding points throughout the barrel to the die, die diameter, feeding rate, expansion ratio, and moisture content of the feed are other parameters that effect extrusion cooking and influence properties of the product.

9.1.4 Other Physical Modifications

Besides dextrinization and pasting, several other methods of physical modification of starch are used to achieve the required functional properties and reactivity in various chemical processes. These include mechanical methods such as compression and sonication, and a wide range of electromagnetic radiation, from freezing and heating (also microwave radiation), exposure to low-pressure glow plasma, illumination with visible and ultraviolet nonpolarized and polarized light, to ionizing and neutron radiation. Such processing of starch results in reduction of the molecular weight of starch polysaccharides, that is, in dextrinization.

Compression delivers to starch energy in the form of heat and work. When a pressure of up to 1.2×10^9 Pa is applied to smashed granules of starch, small blocklets from their interior have been evacuated through cracks in granule envelops. When the water content in compressed granules is ca. 30%, a hard compressed gel is formed.[5] During sonication, when frequencies from 100 to 722,000 Hz are applied, cracks and other forms of erosion of granule envelopes develop. At 1 MHz, gelation of starch begins. Extent of dextrinization increases with time. Energy is applied to starch by silent discharges or potential gradient. Silent discharges in the form of low-pressure glow air, oxygen, hydrogen, and ammonia plasma act mainly on the surface of granules, causing their erosion. However, plasmas also penetrate granules to a certain extent and depolymerize the amylose and amylopectin of granules, depending on starch variety.[7] The processes involved in the potential gradient method are electrolysis and elelctrodialysis. Electrolysis results in hydrolysis of starch polysaccharides at the cathode in an acidic environment, providing a catalyzing proton, and a simultaneous oxidation at the anode. Electrodialysis separates amylose from amylopectin, but a reduction in molecular weight of polysacchrides, particularly amylopectin, also occurs by this process.

Illumination of starch with ultraviolet light, especially in the presence of sensibilizers (metal oxides, particularly ZnO), results in depolymerization of starch. Depolymerization to water and carbon dioxide is possible, but air is necessary to complete this reaction. Usually, photooxidation followed by photodecomposition takes place. Direct depolymerization of starch into formaldehyde is a novel method of utilizing starch as a novel and versatile renewable source for the chemical industry.[8] Air oxidation of starch in the presence of sensibilizers is also promoted by illuminating the reaction mixture with visible light. The role of illumination with visible light on amylolysis of starch has been recently discovered.[9] Polarized light causes depolymerization of starch in two ways. One method activates enzymes inside the starch granules. Thus, the method stimulates enzymatic decomposition. However, even when enzymes are not activated, polarized light, due to its better penetration of matter than visible nonpolarized light, passes its energy to the well-organized granules of starch, causing depolymerization of their interior. Prolonged illumination results in repolymerization of polysaccharides.[10]

High doses of energy carried by ionizing radiation cause fast and deep depolymerization of starch.[5] Neutrons have only a weak peptizing effect whereas x-rays and γ-rays have a powerful decomposing effect. Weak doses of radiation protect starch from germination and attack of microorganisms, with minimum decomposition producing formaldehyde and malonaldehyde. Depolymerization of starch with ionizing radiation follows the free radical mechanism. Free radicals formed are different in structure, stability (reactivity), and hence biological activity from radicals generated thermally. They are mutagenic. Extended exposure of starch to ionizing radiation produces monosaccharides, methanol, ethanol, ethanal, and hydroxy aldehydes, a variety of carboxylic acids and their esters, and oxaheterocycles such as hydroxymaltol and 5-(hydroxymethyl)-2-furalaldehyde. Irradiation of starch is also used as an activating pretreatment prior to its chemical modification.

9.1.5 COMPLEXATION

Frequently, starch is modified by the solvent effect.[5] Swelling of starch in water is the most common example. Starch granules soaked for a prolonged period in water swell, and exudations of amorphous content of the granules form on the surface. This is equivalent to the formation of inclusion complexes of starch, with water residing in empty domains inside granules. After evacuation of included water, microcapsules may act as natural microcapsules.[11] Inclusion complexes of granular starch are accompanied by surface and capillary complexes. Other solvents, even

nonpolar hydrocarbons, form such complexes.[12,13] Granular starch forms complexes
with a variety of mineral salts too, the structure of complexes depending on the salts.
Salts of metals from the first nontransition group of the periodic table interact with
starch in a specific way. Anions penetrate granules more than cations do,[15] whereas
salts of other metals form Werner-type complexes, that is, cations are ligated by
polysaccharides by the involvement of lone electron pair orbitals of oxygen atoms
of the hydroxyl groups.[15] Pasted starch also forms complexes. Complexes with lipids
or proteins are most common.[12,13] Such operations as agitation, kneading, mashing,
and beating, common in food preparation, lead to formation of complexes. Smaller
guest molecules such as lipids and other molecules carrying longer aliphatic hydrophobic chains may interact with amylose in such a manner that amylose coils around
such chains (helical complexes). Complexation protects guest molecules from evaporation, oxidation, polymerization, and other common transformations. Complexes
of starch (and some other polysaccharides) with proteins are promising as specific
nutrients as well as biodegradable materials. In the latter case, thermodynamic
compatibility of both partners is required. Complexation takes place when the
polysaccharide is anionic, that is, the polysaccharide contains groups capable ionization, after which a negative charge is left on this group (e.g., $COOH$, SO_3H,
PO_3H_2 and their salts). Among native starches, only potato starch with its phosphorylated amylopectin follows this condition and can be used without any modification
for production of such complexes.

9.2 CHEMICAL MODIFICATIONS

Chemical modification of starch[2] involves reactions common for saccharides.
Because of a number of glycosidic bonds, reaction of these bonds, a cleavage, and
acid-catalyzed hydrolysis is specific for polysaccharides. Reactions specific for 1-
and 4-OH groups and the anomeric carbon atom are limited. Additionally, conformational changes and steric hindrances as well as viscosity of the reaction mixtures
influence yield, rate, and degree of substitution of polysaccharide.

Chemical modification can be performed on granular as well as pasted starches.
Reaction conditions applied to granular starches may either retain starch granularity
or cause starch gelation. In the former case, the reaction depends on formation of
inclusion complexes (exudations) and penetrability of a reagent into the granules.
Only a limited number of chemical modifications are accepted in food technology.

9.2.1 ACID- AND BASE-CATALYZED DEGRADATION

Acid-catalyzed dextrinization and deep hydrolysis to maltose and glucose syrups
are the most frequent chemical modifications performed on starch. Various mineral
and organic, mainly carboxylic, acids as well as hydrolyzing salts of strong acids,
for instance, alum, cause hydrolytic scission of the glycosidic bonds, but the proton
from hydrochloric acid appears to be the most efficient catalyst. The catalytic
activities of sulfuric, nitric, and acetic acids are 50, 23, and 5% of the activity of
hydrochloric acid, respectively. There is very little preference for scission of peripheral 1→4 glycosidic bonds in the amylose chains and 1→6 glycosidic bonds in the
branching positions in amylopectin.

Diluted alkali and ammonium hydroxide cause swelling of granules. More concentrated alkali gelatinizes starch. Strong alkali forms complexes with starch. Degradation of starch in alkaline media also occurs. Various carboxylic acids (formic, acetic, glycolic, lactic, 2-hydroxybutyric, 2-hydroxyisobutyric, and 2-hydroxyvaleric acids) have been isolated from alkali-degraded starch. They constitute 41 to 46% of the weight of the starting material.

9.2.2 REDUCTION AND OXIDATION

Reduction, which provides saccharide alcohols, sweeteners for diabetes, and moisture-retaining food additives is performed on starch hydrolyzates. Sorbitol and maltitol are the main products. Reduction usually involves hydrogen over Raney-nickel catalyst.

Practically all known oxidants have been used for oxidation of starch, but two groups require particular attention. The first group is periodates, usually sodium periodate, which provide starch dialdehyde. This product, with various degrees of oxidation, received considerable interest but its use in food chemistry is illegal. The second group of oxidants is chlorine and bromine and their compounds, which give starch carboxylates on oxidation. Oxidation with chlorine in an alkaline medium, practically oxidation with sodium hypochlorite, is the sole oxidation method accepted in food technology. In this case too, degree of oxidation of starch has to be within the limits set by food regulations of various countries. The acceptable degree of oxidation never exceeds 0.1.

9.2.3 ESTERIFICATION

Starch can be esterified with inorganic and organic acids. Among inorganic acids, only phosphates are applied to food technology. They are known as gelating agents. Such esters are available in solid-state reactions of starch with various salts of phosphoric acids as well as with phosphorus pentoxide. Limits are set on degree of phosphorylation of esters by national food laws. Starch nitrates with a low degree of esterification, available by treating starch with a mixture of concentrated sulfuric and nitric acids, are considered film- and thin-foil forming biodegradable material. Starch sulfates (sulfated starch) resulting from the treatment of starch either with chlorosulfonic or sulfamic acids or with sulfur trioxide complexes with amines, preferably with pyridine, have received attention as heparine-like compounds. Ionic character of these starches makes their esters promising components of complexes with proteins. Esterification of starch with organic acids, mainly carboxylic acid, proceeds readily with acid anhydrides and chlorides. However, carboxylic acids readily esterify starch on heating. Thus far, only esters of acetic and adipic acids with a low degree of esterification have found their application in food technology.

9.2.4 ETHERIFICATION

Etherification offers a wide variety of starch derivatives. Among etherifying agents are compounds with the general structure X–R–Y. X represents active, that is, providing a nucleophilic substitution, F, Cl, Br, and I atoms and also substitutable

OSO_2H, OSO_2R, ONO groups. R represents an organic moiety providing the chemical activity of the X substituents. It is usually an aliphatic carbon chain carrying additionally a Y substituent. Many Ys, from the hydrogen atom to complex groups, have been tested. For instance, when Y = COOH, carboxylalkyl starches are formed. Etherification with compounds carrying amino, alkyl amino, and dialkylamino as well as quaternized dialkylamino groups leads to cationic starches, which are interesting additives to cellulose pulp and sizes. Etherification can be performed with ethylene oxide and aziridine and their derivatives too. When both X and Y functional groups are capable of nucleophilic substitution or esterification, cross-linking of starch is possible.

9.2.5 ACETALATION

The hydroxyl groups of starch glucose units are capable of forming hemiacetals with aldehydes. Cross-linking of starch occurs with dialdehydes.[2] Starch acetals are formed in reversible reactions. Hence, acetals are fairly unstable during storage and liberate acetalating aldehyde, which is usually unacceptable in food technology. However, acetals with such aldehydes as vanillin, eugenol, and other food-aromatizing aldehydes might have certain applications in food technology as semimicrocapsules.

9.2.6 HALOGENATION

Chlorination of cereals is sometimes used to improve functional properties of flours. Bleaching is a frequently used process but saccharides are not chlorinated in this manner. Chlorination can be achieved by reacting starch with such common chlorinating agents as PCl_3, PCl_5, $POCl_3$, or $SOCl_2$, by which the hydroxyl groups of glucose units are substituted with chlorine.

9.2.7 AMINATION AND CARBAMOYLATION

Amino starches are available by etherification of starch with compounds carrying amino groups. Treatment of starch with ammonia modifies starch and cereals, particularly, on thermolysis of starch under ammonia. There is no addition of nitrogen-containing functional groups to starch.[2] Reaction of starch with amides, particularly urea, produces carbamoylated starches, which as specific urea-enriched products are frequently used as a fodder for ruminants.

9.2.8 GRAFT POLYMERIZATION

Reaction of starch with vinyl monomers, that is, compounds with the general structure of $CH_2=CH$-X, where X = H, Cl, COOH, $CONH_2$, $COOCH_3$, $OCOCH_3$, and so on, results in formation of graft starch polymers. These reactions require initiation. Depending on initiators used, process follows a free radical or an ionic mechanism. The initiator used influences structure and properties of final products by either promoting grafting of starch with monomers over polymerization of vinyl monomers into chains subsequently grafted onto starch or, vice versa, promoting polymerization of vinyl monomers over grafting onto starch.

9.2.9 METAL SALTS AND COMPLEXES

Being a polyol, starch forms salts with certain metal hydroxides and alcoholates. Several metal derivatives of starch (metal starchates) have been reported [Na, K, Mg, Al, Tl(I), Sn(IV), As(III), Bi(III), Bi(V), V(III), Cr(III), Ti(IV), Mo(V), W(V), Fe(III)]. Depending on the metal they contain, they find use as synthons, heterogenic catalysts, rodent poisons, and carriers of biolements.

REFERENCES

1. Tomasik, P., Wiejak, S., and Pałasiński, M., The thermal decomposition of carbohydrates, Part II, *Adv. Carbohydr. Chem. Biochem.*, 47, 279, 1989.
2. Tomasik, P. and Schilling, C.H., Chemical modification of starch, *Adv. Carbohydr. Chem. Biochem.*, 53, 2003.
3. Ciesielski, W. and Tomasik, P., Starch radicals. Part I, *Carbohydr. Polym.* 31, 205, 1996.
4. Barabasz, W. et al., On the mutagenicity of caramels, *Starch/Stearke*, 42, 69, 1990.
5. Tomasik, P. and Zaranyika, M.F., Non-conventional modifications of starch., *Adv. Carbohydr. Chem. Biochem.*, 51, 243, 1995.
6. Szymońska, J., Krok, F., and Tomasik, P., Deep freezing of potato starch, *Int. J. Biol. Macromol.*, 27, 307, 2000.
7. Lii, C.Y et al., Behaviour of granular starches in air, low-pressure glow plasma. *Carbohydr. Polym.*, 49, 499, 2002.
8. Okkerse, C. and van Bekkum, H., Perspectives for a versatile raw material on the threshold of a new millennium, *Starch 96: The Book*, van Doren, H. and van Swaaj, N, Eds., Zestec bv/Carbohydrate Research Foundation, Nordwiijkhout, 1996, chap. 1.
9. Fiedorowicz, M. and Chaczatrian, G., Effect of illumination with the visible polarized and non-polarized light on alpha-amylolysis of starches of different botanical origin, *J. Agric. Food Chem.*, submitted.
10. Fiedorowicz, M., Tomasik, P., and Lii, C.Y., Degradation of starch by polarised light, *Carbohydr. Polym.*, 45, 7, 2001.
11. Korus, J., Lii, C.Y., and Tomasik, P., Potato starch granules as microcapsules. *J. Microencaps.*, 20, 47, 2003.
12. Tomasik, P. and Schilling, C.H., Starch complexes, Part I: Complexes with inorganic guests, *Adv. Carbohydr. Chem. Biochem.*, 53, 263, 1998.
13. Tomasik, P. and Schilling, C.H., Starch complexes, Part II: Complexes with organic guests, *Adv. Carbohydr.Chem. Biochem.*, 53, 345, 1998.
14. Lii, C.Y. et al., Revised look at starch interactions with electrolyte: interactions with salts of metals from the first non-transition group. *Food Hydrocoll.*, 16, 35, 2002.
15. Ciesielski, W. et al., Interaction of starch with metal ions from transition groups, *Carbohydr. Polym.*, 51, 47, 2003.

10 Enzymatic Conversions of Carbohydrates

Stanisław Bielecki

CONTENTS

10.1 Introduction 132
10.2 Enzymes Involved in Conversions of Carbohydrates 133
    10.2.1 Glycoside Hydrolases 133
    10.2.2 Glycosyltransferases 133
    10.2.3 Glycansucrases 135
    10.2.4 Mechanisms of Glycosidic Bond Hydrolysis
        and Transglycosylation 136
    10.2.5 Glycosynthases 138
    10.2.6 Other Sugar-Converting Enzymes 138
        10.2.6.1 Synthesis of Monosaccharides 140
        10.2.6.2 Selected Conversions Catalyzed by Oxidoreductases 140
10.3 Oligosaccharides 141
    10.3.1 Fructooligosaccharides 141
    10.3.2 Isomaltooligosaccharides 142
    10.3.3 Galactooligosachharides 142
10.4 Polysaccharides 143
    10.4.1 Starch 143
        10.4.1.1 Common Properties of the a-Amylase Family
            of Enzymes 145
        10.4.1.2 a-Amylases 146
        10.4.1.3 Exo-Acting Enzymes 147
        10.4.1.4 Debranching Enzymes 148
        10.4.1.5 Cyclodextrins and Cycloamyloses 148
        10.4.1.6 Trehalose 149
        10.4.1.7 Other Starch Conversions 149
    10.4.2 Cellulose, Chitin, and Chitosan 149
    10.4.3 Pectin 151
    10.4.4 Xylan 153
    10.4.5 Fructans 154
References 155

0-8493-1486-0/04/$0.00+$1.50
© 2004 by CRC Press LLC

10.1 INTRODUCTION

The diversity of possible structures of sugars corresponds to a multiplicity of their roles in nature and diversity of enzymes involved in their synthesis and decomposition. Practical applications of these enzymes range from hydrolysis of macromolecules such as starch to the production of monomers and oligomers with precisely defined structure to be used in cancer therapy.

Started in the 20th century, industrial conversions of polysaccharides involved processes that employed harsh conditions, consumed enormous portions of energy, and contributed to pollution of the natural environment. The increasing profitability of "green" technologies and growing public concern about the environment, resulting in stricter law regulations, have brought about the progressive dislodging of chemical technologies with environment-friendly production strategies that lead to a higher yield and quality of product and concomitantly reduce an input of energy and chemicals.

The main drawbacks of earlier biotechnologies resulted from relatively poor activity and stability of enzymes derived from natural sources. In some cases, enzymes showing sufficient activity and stability under adverse conditions were derived from extremophilic organisms, which thrive at extreme pH; very low or high temperature; or elevated pressure, salinity, or concentration of various compounds. Extremophiles such as thermo-, psychro-, baro-, acido- alkali-, or halophiles were shown to be the best natural sources of stable catalytic proteins and their genes. Studies on these organisms yielded an important insight into relationships between protein structure and stability, and many genes encoding their enzymes of industrial significance were cloned and expressed in hosts [usually generally regarded as safe (GRAS)]. At present, approximately 90% of all commercial enzymes are derived from genetically modified organisms (GMOs), due to advances in construction and selection of industrial mutant strains.

Tremendous advances in improvement of such properties of enzymes as catalytic efficiency, stereospecificity, and stability under diverse conditions and in various media, including organic solvents, supercritical fluids, or ionic liquids, make their industrial application more reasonable as compared to other technologies, and have contributed to implementing of biotechnologies in virtually all branches of industry, also those employing enzymes involved in conversions of saccharides. Immobilization of biocatalysts and construction of modern bioreactors have also enhanced production yield. An expansion of modified and improved enzymes or engineered whole cells gave rise to rapid development of glycobiotechnology, a research area focused on synthesis of various saccharides and their derivatives, with precisely determined structures. The range of possible applications of these compounds includes prevention of diverse infections, neutralization of toxins, and cancer immunotherapy. Controlled enzyme-catalyzed depolymerization, transglycosylation, isomerization, oxidation, and reduction of oligo- and polysaccharides lead to a variety of high-value-added products with improved functional properties. A benign effect can be also achieved by selective enzymatic degradation of some antinutritional oligosaccharides, consumption of which causes malfunction of the alimentary tract. The strictly enzymatic methods, termed as combinatorial biosynthesis,[1] have been applied to alter the structure of some natural compounds, including sugars. Mutations within gene clusters involved in carbohydrate synthesis and attachment to aglycon moieties have resulted in construction of recombinant microbial strains producing novel antibiotics. The wealth of possible enzymatic activities and specificities has dramatically increased due to cultivation-independent approaches also, based on direct DNA isolation from samples of various origin, giving rise to construction of a huge number of genomic libraries.

10.2 ENZYMES INVOLVED IN CONVERSIONS OF CARBOHYDRATES

Glycoside hydrolases (GHs) and glycosyltransferases (GTs) have been undoubtedly the most important biocatalysts exploited for conversions of sugars, independently on their scale, but recently oxidoreductases, isomerases, and lyases have also gained much interest. The basic information on GHs, GTs, as well as carbohydrate lyases and esterases, including their classification, structure and origin, is presented on the regularly updated CAZY web server.[2] A grouping of these enzymes into families and clans, based on primary-structure similarities and hydrophobic cluster analysis data, appears to be superior over commonly applied classification according to their substrate specificity.

10.2.1 GLYCOSIDE HYDROLASES

Glycosidases catalyze glycoside bond hydrolysis. However, under certain conditions they can also synthesize such linkages. Although GHs are stereospecific and precisely recognize the type of a bond (e.g., α-1,4, β-1,6, etc.), they are not always regiospecific, and produce mixtures of products. GHs are advantageous as catalysts of synthesis because of availability of their relatively inexpensive pure preparations and low prices of substrates. However, a poor synthesis yield is their main drawback. A satisfactory degree of glycosidic bond formation requires a large excess of substrates, preferably activated glycosyl donors (e.g., glycosyl fluorides) and acceptors, or organic solvents. The controlled reaction conditions can also reduce the number of product regioisomers. Transglycosylation activity is displayed by a number of hydrolases acting either in an exo or in an endo mode (Table 10.1). The application of glycosidases has been usually restricted to synthesis of short oligosaccharides. Ultrasounds carried out at high substrate concentrations improve the yield of hydrolysis catalyzed by α-amylase, glucoamylase, and invertase.[3]

10.2.2 GLYCOSYLTRANSFERASES

Glycosyltransferases catalyze the transfer of a monosaccharide unit from a donor to the saccharide acceptor. Based on the structure of the activating group at the anomeric center of the donor, they are categorized as Leloir (sugar-nucleotide-dependent) and non-Leloir (sugar-1P-dependent) enzymes.[2] Another criterion for differentiation of

TABLE 10.1
Selected Enzymes Used for Synthesis

Enzyme	EC Number	Systematic Name
β-1,4-Galactosyltransferase	EC 2.4.1.133	UDP-Galactose:O-β-D-xylosyl-protein 4-β-D-galactosyltransferase
α-2,3-Sialyltransferase	EC 2.4.99.6	CMP-N-Acetylneuraminate:β-D-galactosyl-1,4-N-acetyl-D-glucosaminyl-glycoprotein α-2,3-N-acetylneuraminyl-transferase
α-2,6-Sialyltransferase	EC 2.4.99.7	CMP-N-Acetylneuraminate:α-N-acetylneuraminyl-2,3-β-D-galactosyl-1,3)-N-acetyl-D-galactosaminide α-2,6-N-acetylneuraminyl-transferase
α-1,2-Mannosyltransferase	EC 2.4.1.131	GDP-Mannose:glycolipid 1,2-α-D-mannosyltransferase
α-1,3-Fucosyltransferase	EC 2.4.1.40	UDP-N-acetyl-β-Galactosamine:glycoprotein-α-L-fucosyl-(1,2)-D-galactose 3-N-acetyl-D-galactosaminyl-transferase
α-D-Glucosidase	EC 3.2.1.20	α-D-Glucoside glucohydrolase
α-D-Galactosidase	EC 3.2.1.22	α-D-Galactoside galactohydrolase
β-D-Galactosidase	EC 3.2.1.23	β-D-Galactoside galactohydrolase
α-D-Mannosidase	EC 3.2.1.25	α-D-Mannoside mannohydrolase
Invertase	EC 3.2.1.26	β-D-Fructan fructohydrolase
β-D-Xylosidase	EC 3.2.1.37	β-D-Xyloside xylohydrolase
Exo-N-acetyl-α-D-galactosaminidase	EC 3.2.1.49	α-N-Acetyl-D-galactosaminide N-acetylgalactosaminidase
α-L-Fucosidase	EC 3.2.1. 51	α-L-Fucoside fucohydrolase
Exo-β-1,4-Galactanase	EC not defined	
α-Amylase	EC 3.2.1.1	1,4-α-D-Glucan glucanohydrolase
Xylanase	EC 3.2.1.8	1,4-β-D-Xylan xylanohydrolase
Chitinase	EC 3.2.1.14	Poly(1,4-(N-acetyl-β-D-glucosaminide) glycanohydrolase
Lysozyme	EC 3.2.1.17	Peptidoglycan N-acetylmuramoyl-hydrolase
Endo-β-1,3-glucanase	EC 3.2.1.39	1,3-β-D-Glucan glucanohydrolase

GTs is the number of transferred sugar units. Nonprocessive GTs catalyze the transfer of a single sugar residue to the acceptor, whereas processive GTs, such as synthases of cellulose, chitin, and hyaluronan, attach more than one saccharide unit to respective acceptor molecules.[4] GTs display regio- and stereospecificity with respect to donor and acceptor structures as well as to the type of glycosidic linkage. However, some of them recognize more than one donor of a sugar residue and form more than one type of linkage. Non-Leloir GTs are said to be less efficient catalysts of synthesis than Leloir GTs. However, glycogen phosphorylase successfully provides synthesis of various not natural, tailor-made sugars with polymerization degrees up to 20, including 2-deoxyglucose derivatives and 2-chloro-4-nitrophenyl-maltooligosaccharides.[5]

In contrast to glycosidases and non-Leloir GTs, Leloir GTs provide high yield and regio- and stereoselectivity of synthesis. Therefore, they are captivating tools for preparative oligosaccharide synthesis, in particular, for synthesis of glycosidic components of glycopeptides, glycoproteins, proteoglycans, and glycolipids. Table 10.1 lists examples of Leloir GTs successfully tapped for synthetic purposes. Lack of inexpensive, pure commercial preparations of both these enzymes and sugar nucleotides, being the substrates of Leloir GTs, limit their practical significance. Because both chemical and enzymatic methods of synthesis of sugar nucleotides are laborious and expensive, some new approaches have been developed, including an *in situ* generation achieved by coupling of sugar synthesis catalyzed by a Leloir GT, with UTP synthesis from UDP and PEP catalyzed by pyruvate kinase.[6] Large-scale production of oligosaccharides can also exploit whole cells of GMOs bearing bacterial or mammalian GTs. Some enzymes from genetically engineered bacteria have been found to be specific toward nonnatural substrates.

GTs have also been applied for a solid-phase synthesis of oligosaccharides.[7] This approach facilitates product purification and synthesis of glycopeptides. The modified glycopeptide contains a chain substituted with a single sugar residue and attached to a soluble or insoluble support. Specific GTs elongate the carbohydrate chain, followed by the cleavage of the final glycopeptide from the support, either catalyzed by a specific protease or achieved by other methods.

10.2.3 Glycansucrases

Microbial polysaccharides such as dextran, inulin, and levan-type fructans, alternan, and 1,4-α-D-glucans are produced by transglycosylases (glycansucrases),[8] which are different from sugar-nucleotide-dependent GTs (Table 10.2). Glycansucrases derive the portion of energy necessary to catalyze the transfer of a given glycosyl group from the cleavage of the osidic bond of sucrose, which is their principal donor of glycosyl moieties. The best producers of extracellular glycansucrases are lactic acid bacteria of the genera *Leuconostoc*, *Streptococcus*, and *Lactobacillus*. Some of these bacterial enzymes have been harnessed for large-scale production of both polymers and oligosaccharides because, in the presence of suitable acceptor molecules (e.g., maltose) added to sucrose, glycansucrases preferably synthesize oligomers instead of high-molecular-mass sugars.

TABLE 10.2
Classified Glycansucrases

Common Name	EC Number	Systematic Name	Catalyzed Reaction
Amylosucrase	EC 2.1.4.4	Sucrose:1,4-α-D-glucan 4-α-D-glucosyltransferase	Sucrose + (1,4-α-D-glucosyl)$_n$ = D-fructose + (1,4-α-D-glucosyl)$_{n+1}$
Dextransucrase	EC 2.4.1.5	Sucrose:1,6-α-D-glucan 6-α-D-glucosyltransferase	Sucrose + (1,6-α-D-glucosyl)$_n$ = D-fructose + (1,6-α-D-glucosyl)$_{n+1}$
Inulosucrase	EC 2.4.1.9	Sucrose:2,1-β-D-fructan 1-β-D-fructosyltransferase	Sucrose + (2,1-β-D-fructosyl)$_n$ = D-glucose + (2,1-β-D-fructosyl)$_{n+1}$
Levansucrase	EC 2.4.1.10	Sucrose:2,6-β-D-fructan 6-β-D-fructosyltransferase	Sucrose + (2,6-β-D-fructosyl)$_n$ = D-glucose + (2,6-β-D-fructosyl)$_{n+1}$
Alternansucrase	EC 2.4.1.140	Sucrose:1,6(1,3)-α-D-glucan 6(3)-α-D-glucosyltransferase	Transfer of an α-D-glucosyl residue from sucrose alternatively to the 6-position and the 3-position of the nonreducing terminal residue of an α-D-glucan having alternatively α-1,6- and α-1,3 linkages

10.2.4 MECHANISMS OF GLYCOSIDIC BOND HYDROLYSIS AND TRANSGLYCOSYLATION

Enzymatic hydrolysis of the glycosidic bond is carried out via general acid/base catalysis that employs two crucial groups, such as a proton donor and a nucleophile/base. In most cases, these roles are played by carboxyl groups of either Asp or Glu residues.[2]

Due to two stereochemical outcomes received in reactions of glycoside linkage hydrolysis or transglycosylation — net retention and net inversion of anomeric configuration — GHs and GTs are grouped into either retaining or inverting enzymes. The mechanism of catalysis of inverting GTs and GHs are similar, whereas the function of retaining transferases still needs explanation.[9] Inverting GHs and GTs use a single-displacement reaction with base activation of the acceptor, and the retaining GHs follow the double-displacement mechanism with a glycosyl–enzyme intermediate. Both mechanisms of glycoside bond hydrolysis or transglycosylation (Figure 10.1.a and b) require formation of an oxocarbonium ion-like transition state as a prerequisite for glycoside bond cleavage.[10] The stereochemistry of catalysis is contingent on a spatial arrangement of two active-site-forming carboxyl groups in the active site of the enzyme. Groups in retaining GHs are approximately 0.55 nm from one another, whereas the distance between the relevant groups in inverting GHs is between 0.9 and 1.1 nm. The larger gap provides suitable localization of the nucleophilic water molecule between the base and the anomeric carbon atom base.

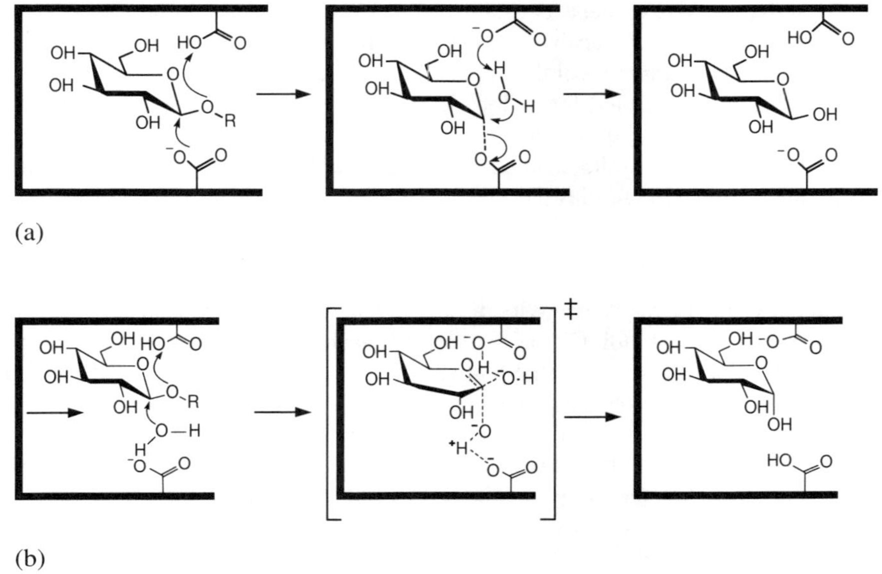

(a)

(b)

FIGURE 10.1 Mechanisms of glycosidic bond cleavage by retaining (a) and inverting (b) glycosidases. The retaining enzymes operate by a double-displacement mechanism, with two oxocarbonium ion-like transition states and a covalent glycosyl-enzyme intermediate to yield a product with the same anomeric configuration as that of the substrate. The inverting glycosidases apply a single-displacement mechanism with the oxocarbonium transition state, but without the covalent intermediate. (Withers, S.G., *Carbohydr. Polym.*, 44, 325, 2001. With permission.)

Apart from these two key groups involved in either cleavage or formation of the glycosidic bond, active sites of glycosidases contain a number of substrate-binding subsites responsible for proper accommodation of the saccharide molecules in the active-site cleft.[11] The subsites that bind sugar residues closer to the nonreducing end of the oligomer or polymer chain with respect to the scissile bond are marked with a negative sign, whereas those anchoring the substrate residues oriented toward its reducing end have a positive sign.

Retaining GTs probably follow the double-displacement strategy, but their glycosyl–enzyme intermediates have not been detected.[9] Furthermore, in certain retaining GT families, the Asp and Glu residues, playing the role of the acid and base in other GHs or GTs, are either not conserved or are not appropriately located in the active site. Some alternative nucleophiles, including enzyme amide groups or the substrate acetamide group, have also been detected.

Shape of the active-site cavities (pocket, cleft, and tunnel) dictates the manner of enzyme action.[12] "True" exoglycosidases, specifically liberating monomers or dimers from the polysaccharide chain ends, usually have their active sites located within a pocket, whose depth corresponds to the number of subsites and thus to the length of the reaction product. In contrast, an open cleft is characteristic of "true" endohydrolases that attack the whole polymer chain in a random manner. Tunnel-shaped active sites have been detected in "processive" or "multiple-attack" GHs active against fibrous polysaccharides with very few available chain ends, for example, cellulose. An entrapment of the polymer chain within the tunnel enables iterative acts of hydrolysis, as found for cellobiohydrolases. Many GHs and GTs appeared to be multimodular enzymes, containing separate carbohydrate-binding modules.[9] Similarity of their three-dimensional structures enable grouping of some of them into families and superfamilies.

10.2.5 GLYCOSYNTHASES

Insufficient yield and regioselectivity of oligosaccharide synthesis by using retaining glycosidases and non-Leloir GTs as well as poor availability of Leloir GTs and sugar nucleotides induced research on mutant enzymes catalyzing synthesis of the glycosidic bond. Glycosynthases are mutant enzymes with the active-site nucleophile (i.e., the residue of Asp or Glu) in a molecule of a retaining glycosidase replaced with a small nonpolar group such as Gly or Ala side chain[13] or hydroxyl-containing Ser.[14] This mutation makes the enzyme unable to catalyze glycosidic bond hydrolysis. However, under certain conditions, the mutant GH can efficiently synthesize "designer" oligosaccharides (Figure 10.2). Reactivation of the mutant protein as a catalyst is achieved by adding an external nucleophile, for instance, formate or azide anions, coupled with a highly activated glycosyl donor such as a glycosyl fluoride (α-1-F sugar) with a readily leaving aglycon group.[13] Very stable and reactive glycosyl fluorides have proved to be superior substrates for glycosynthases and wild-type GHs and GTs. Experiments on β-glucosidases reveal that the Glu–Ser mutation was superior because the hydroxyl group of Ser supported departure of the fluoride ion from glycosyl fluorides.[14]

10.2.6 OTHER SUGAR-CONVERTING ENZYMES

Apart from GHs and GTs, many enzymes of other classes catalyze conversions of saccharides. Some of them are applied in the analytics of sugars. For instance, common diagnostic kits for fast and accurate glucose assays, even in complex mixtures of various reducing carbohydrates, either use glucose oxidase coupled with

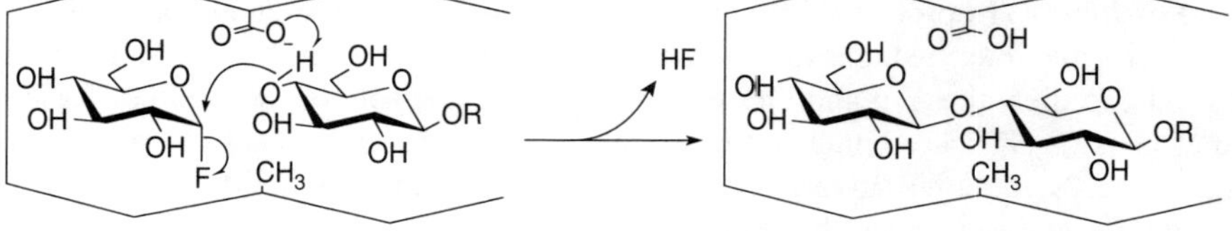

FIGURE 10.2 Mechanism of glycosidic bond synthesis by a glycosynthase. Glycosynthases are mutant-retaining glycosidases containing a small nonpolar group such as Gly or Ala side chain, instead of the active-site nucleophile. Glycosyl fluorides (α-1-F sugars) are preferred substrates of glycosynthases because they are highly activated glycosyl donors with a very readily leaving aglycon group. (Williams, S.J. and Withers, S.G., *Carbohydr. Res.*, 327, 27, 2000. With permission.)

peroxidase, or hexokinase and highly specific glucose-6-phosphate dehydrogenase. Oxidoreductases such as glucose dehydrogenase may also be used for large-scale production of sugar alcohols such as sorbitol and aldonic acids such as gluconic acid, applied in food, pharmaceutical, and chemical industries.[15] One of the principal industrial conversions of sorbitol is its selective oxidation by D-sorbitol dehydrogenase to L-sorbose, which is a chiral precursor of ascorbic acid. Similarly, as in many other processes, the latter reaction is run with whole microbial cells (*Gluconobacter* species), and not with purified enzymes. Other oxidoreductases useful in production of nonnatural carbohydrates and related compounds (Table 10.3) are pyranose 2-oxidase (P2O), aldose (xylose) reductase (ALR),[16] and cellobiose dehydrogenase, which oxidize soluble cellodextrins, mannodextrins, and lactose to their lactones. Cellobiose dehydrogenase seems to be applicable for conversion of lactose to lactobionic acid.

Xylose isomerase was the first technically applied preparation of sugar isomerase. It catalyzes the reversible isomerization of D-xylose to D-xylulose as well as conversion of glucose to fructose. The latter specificity has been industrially explored for production of high-fructose syrups. Also, sucrose isomerase is very useful in conversion of sucrose to isomaltulose, trehalulose, isomaltose, and isomelezitose. Products of the sucrose isomerization have attracted attention as potential acariogenic sweeteners (see Chapter 24).

Polysaccharide lyases and esterases are involved in conversion of pectin (see Chapter 12), xylan (see Chapter 13), and starch. Lyases cleave chains of polymers via a β-elimination mechanism that leads to the formation of a double bond at the newly formed nonreducing end. The most commercially important are pectate and pectin lyases, participating in depolymerization of pectins (Table 10.6). Another group of lyases that has gained increasing attention is aldolases, harnessed for synthesis of monosaccharides and their analogs in glycobiotechnology (see Section 10.2.6.1).

TABLE 10.3
Products of Conversion of Mono-and Disaccharides by Means of P2O and ALR

Substrate	Product of Oxidation with P2O[a]	Product of Further Reduction with ALR[b]
D-Glucose	2-Keto-D-glucose (D-glucosone)	D-Fructose
D-Galactose	2-Keto-D-galactose (D-galactosone)	D-Tagatose
Lactose	Lactosone	Lactulose
Gentiobiose	Gentiobiosone	Gentiobiulose
Allolactose	Allolactosone	Allolactulose
Isomaltose	Isomaltosone	Isomaltulose
Melibiose	Melibiosone	Melibiulose

[a] H_2O_2 formed as a by-product is degraded by catalase (to H_2O and O_2).
[b] NAD+ formed as a by-product is reduced by formate dehydrogenase, which catalyzes the reaction of formate oxidation to CO_2 ($HCOO^- + NAD^+ = CO_2 + NADH + H^+$).

Degradation of polysaccharides such as pectin and xylan also involves some carbohydrate esterases that catalyze the de-O or de-N-acetylation of substituted saccharides, and demethylation of pectin chains. The most common reaction mechanism of deacetylation involves a Ser–His–Asp triad, analogous to classical lipase- or serine-protease-catalyzed reactions, but other mechanisms such as catalysis by Zn^{2+} ions are also known.[2]

10.2.6.1 Synthesis of Monosaccharides

Nonnatural monosaccharides are potential inhibitors of glycosidases and therefore have certain therapeutic significance. Tailored monosaccharide synthesis, requiring formation of carbon–carbon bonds with defined stereochemical conformation, are catalyzed by aldolases, either dependent on pyruvate- or dihydroxyacetone phosphate (DHAP).[17]

These aldolases, which use pyruvate as the donor substrate, such as N-acetylneuraminate lyase, generate a new chiral center at C-4, yielding 3-deoxy-2-ketoulosonic acids, for example, N-acetylneuraminic acid. DHAP-dependent aldolases form a new C3–C4 bond. Fructose-1,6-diphosphate aldolase follows D-threo stereochemistry. These aldolases were exploited in synthesis of diverse sugar analogs, for example, azasaccharides, this is such with the nitrogen atom. Recently, a novel type of aldolases produced by *E. coli*, catalyzing the reversible synthesis of fructose-6-phosphate from dihydroxyacetone (and not its phosphate) and D-glyceraldehyde 3-phosphate, has been reported.[18]

Apart from their natural substrates, aldolases accept other compounds showing structural similarity.[18] For instance, 2-keto-3-deoxy-6-phosphogalactonate and 2-keto-3-deoxy-6-phosphogluconate aldolases have been harnessed for synthesis of derivatives of pyruvate and electrophilic aldehydes, different from D-glyceraldehyde-3-phosphate. D-threonine aldolase, uniquely forming amino alcohols from Gly and aldehydes, was employed for synthesis of some higher-molecular-weight compounds too. Coupling of fructose-1,6-diphosphate aldolase with triose isomerase and transketolase facilitated synthesis of D-xylulose-5-phosphate. Also some abzymes (antibodies with catalytic activity) catalyze the intermolecular or intramolecular reaction of aldol addition and accept several ketone donors and acceptors. 1-Deoxy-L-xylulose is one of the sugars obtained in this way.

The relatively broad substrate specificity of various aldolases and aldolase antibodies is promising for the synthesis of a variety of monosaccharides and their derivatives. Chemical synthesis of these compounds usually yields complex mixtures of isomers difficult for separation and extraction from natural sources.

10.2.6.2 Selected Conversions Catalyzed by Oxidoreductases

Fungal pyranose 2-oxidases (P2O), which in nature principally oxidize D-glucose and D-xylose to corresponding 2-keto analogs, have acquired interest as catalysts enabling conversion of a number of inexpensive sugars into chemically active dicarbonyl sugars.[16] The enzymes also catalyze simultaneous oxidation at C-2 and C-3, providing 2,3-diketo sugars. Monosaccharides, such as D-glucose, D-xylose, D-galactose, D-allose, D-glucono-1,5-lactone, and L-sorbose, and disaccharides, such as lactose, allolactose, gentiobiose, isomaltose, and melibiose, are substrates for P2O (see Table 10.3). Their lactones can be converted with aldose reductase (ALR) into corresponding keto analogs. Some of the disaccharides, for instance, lactulose, available in this manner are random in nature. Because of documented functional benefits, they can be tapped as low-calorie bulking sweeteners. Lactulose has been obtained mainly through the isomerization of lactose, catalyzed by sodium hydroxide, sodium hydroxide and boric acid, and sodium aluminate. Because of the toxicity of inorganic catalysts, enzyme-catalyzed lactulose production is riveting.

Tagatose, a ketohexose C-4 fructose epimer, is the lactose-derived noncariogenic monosaccharide. It can be produced by using either P2O and ALR (see Table 10.3) or in the sorbitol-dehydrogenase-catalyzed oxidation of D-galactitol, available from the reduction of D-galactose liberated from lactose by β-galactosidase. D-Galactitol is relatively inexpensive, and the process can be even more economically feasible if the oxidation reaction is conducted by whole cells of bacteria such as *Gluconobacter* species.

Advances in production of preparations of oxidoreductases and regeneration of their coenzymes will contribute to their common industrial applications. These enzymes are valuable biocatalysts, facilitating relatively simple synthesis of rare or nonnatural carbohydrates and sugar-based precursors (synthons) for further conversions.

10.3   OLIGOSACCHARIDES

Several oligosaccharides appeared to be biologically active compounds, and therefore their production became one of the most important areas in biotechnology. Some of them, such as nondigestible oligosaccharides (NDOs) have prebiotic properties (see Chapter 21). They are produced either by enzymatic transglycosylation of relatively inexpensive disaccharides such as sucrose and lactose, or by conversion of polysaccharides such as fructans or starch. Some other NDOs are available by a limited enzymatic digestion of pectin or hemicelluloses.[19]

10.3.1   FRUCTOOLIGOSACCHARIDES

The term *fructooligosaccharides* (FOs) is usually used for $1^F(1$-β-D-frucofuranosyl$)_n$-sucrose (GF$_n$; $n = 2$ to 10), though polyfructans and oligofructosides with another structure also exist (see Chapter 13).[20] 1-Kestose is the shortest fructan of the inulin-type and 6-kestose of the levan-type. The reported FOs producing enzymes are invertase and inulosucrase (also known as fructosyltransferase), whereas levansucrase converts sucrose to 2,6-β-D-fructans (Table 10.2). The majority of inulosucrases show high regiospecificity, and transfer the fructosyl moiety to C-1 hydroxyl group of terminal fructofuranosides to produce 1-kestose and its homologs, but some enzymes transfer the fructose to OH-6^F and OH-6^G to form 6-kestose or neokestose, respectively. Another transferase involved in the fructan formation, β-(2,1)-fructan:β-(2,1)-fructan 1-fructosyltransferase (FFT), transfers fructosyl groups between fructan chains. It catalyzes the reaction $GF_n + GF_m = GF_{n-1} + GF_{m+1}$, thus contributing to synthesis of higher fructans in nature.

The main bottleneck in FOs synthesis is that glucose, a by-product in reactions catalyzed by inulosucrase and levansucrase, negatively influences the degree of sucrose conversion, and, therefore, either its oxidation with glucose oxidase or polymerization with dextransucrase have been used to avoid the inhibiting effect.[21] Application of endodextranase coupled with dextransucrase facilitates concomitant production of isomaltooligosaccharides, apart from the fructose oligomers.

10.3.2 ISOMALTOOLIGOSACCHARIDES

Pure isomaltooligosaccharides (IMOs), made up by exclusively α-1,6-linked glucose residues, can be obtained from commercial dextrans synthesized from sucrose by dextransucrases (see Table 10.2), and digested with endodextranase.[22] Dextran can be also derived from linear 1,4-α-D-glucan by using dextrin dextranase, which transfers α-1,4-linked glucosyl units to the nonreducing end of 1,6-α-D-glucan. Dextransucrases usually form variable amounts of α-1,2-, α-1,3-, and α-1,4 glycosidic bonds, apart from the α-1,6 ones. Even dextrans produced by the best industrial *Leuconostoc mesenteroides* strains contain approximately 5% of branchings. Some *L. mesenteroides* dextransucrases form up to 65% of α-1,2 bonds, and the product has cosmetic and food applications. In the presence of efficient acceptor carbohydrates such as maltose and isomaltose, dextransucrases transfer glucopyranosyl units to the acceptor instead of to dextran, giving panose and related isomaltooligosaccharides, respectively. When weak acceptors such as fucose are used, only one type of acceptor products are formed, such as leucrose. The yield of the latter reaction reaches 90% when high concentrations of fructose (3.3 M) are used and the process is carried out below 0°C. Apart from sugars, some sugar derivatives such as alcohols can also act as glucose acceptors in reactions catalyzed by dextransucrases. Commercial preparations of IMOs are also derived from starch treated with α-amylase, neopullulanase, and α-glucosidase (see Section 10.4.1). They usually contain α-1,6- and α-1,4-linked glucose oligomers, because all three enzymes hydrolyze α-1,4 glycosidic bonds in starch, and α-glucosidase additionally synthesizes α-1,6 linkages.

10.3.3 GALACTOOLIGOSACHHARIDES

β-Galactooligosaccharides (β-GOS), also called transgalactooligosaccharides, display nutraceutical properties because they promote the growth of bifidobacteria. These beneficial dietary additives can be obtained by using either GTs or glycosidases. However, application of the latter enzymes is more economically feasible. β-GOS are produced from lactose through transgalactosylation catalyzed by β-galactosidases, which can be coupled with manufacturing of low-lactose milk for people with lactose intolerance, detected in nearly 70% of the human population.[23] An application of exo-β-1,4-galactanases for this purpose also seems possible.[24] Some β-galactosidases, which are retaining GHs, preferentially catalyze glycosidic bond formation, thus enabling large-scale synthesis of galactooligosaccharides and other compounds such as *para*-nitrophenyl galactosyl chromogens, various chiral sugar derivatives, and disaccharides such as galactosyl- xylose.[25]

Generally, the molecules of β-GOS can be presented as:

$$Gal^X (Gal)_n {}^Y Glc$$

where n is the number of β-1,4-linked galactose residues; x = β-1,6 > β-1,4, β-1,3; and y = β-1,2, β-1,6 > β-1,4. This formula indicates that apart from the dominating β-1,4 glycosidic bond, β-galactosidases synthesize some β-1,3 and β-1,6 linkages too.

Apart from galactooligosaccharides, lactulose, tagatose, and lactobionic acid can also be obtained from lactose, using relevant oxidoreductases (see Section 10.2.6). Because lactose is an attractive starting material, the list of its derivatives continues to grow. For example, hexyl glycosides have been prepared by lactose hydrolysis with β-galactosidase, followed by a treatment with thermostable β-glycosidases from hyperthermophiles.[26] It should be emphasized that their hydrolysis with α-galactosidase preparations enhances digestibility of certain plant foods.[27] α-Galactosidases are also applicable for synthesis of the so-called globo-oligosaccharides, having terminal α-D-galactose-1,4-α-D-galactose sequences and involved in interactions with pathogens and viruses.

10.4 POLYSACCHARIDES

10.4.1 STARCH

The application of enzymes in industrial starch processing dates back to the 1970s. Currently, amylolytic enzymes constitute approximately 30% of total market of biocatalysts. Starch-converting enzymes, listed in Table 10.4, comprise endo- and exoamylases, debranching enzymes, and transferases.[11] Their majority (except glucoamylase and β-amylase) belong to the α-amylase family, comprising numerous hydrolases and transferases, though some are active against substrates other than starch or dextrins. Enzymes of Family 13 of GHs, with a retaining mechanism of attack, dominate the α-amylase family, but some catalytic proteins assigned to the Family 70 (dextransucrase and alternansucrase) and Family 77 (4-α-glucanotransferase) GHs also share the structure of the catalytic domain [in the form of $(\beta/\alpha)_8$ barrel] and mechanism of the action with α-amylase.

Apart from amylolytic enzymes, some other enzymatic activities are also found to be necessary for a satisfactory degree of large-scale starch conversion. For instance, enzymatic starch processing has to face some problems related to the presence of amylose–lipid complexes formed either *in situ* in wheat starch granules or generated on gelatinization of containing lipids starch slurries. Lipases, mainly lysophospholipases, are applied together with amylolytic enzymes to improve the yield of processing.

TABLE 10.4
Enzymes Involved in Starch Conversions

Enzyme	EC Number	Systematic Name	Application
α-Amylase	EC 3.2.1.11	1,4-α-D-Glucan glucanohydrolase	Random endohydrolysis of α-1,4 bonds in starch and related glucans
Maltogenic amylase	EC 3.2.1.133	1,4-α-D-Glucan α-maltohydrolase	Successive hydrolysis of penultimate α-1,4 bonds to release maltose molecules from nonreducing ends of starch molecules
Maltotriohydrolase	EC 3.2.1.116	1,4-α-D-Glucan maltotriohydrolase	Hydrolysis of α-1,4 bonds and release of maltotriose molecules from nonreducing ends of starch chains
Maltotetraohydrolase	EC 3.2.1.60	1,4-α-D-Glucan maltotetraohydrolase	Release of maltotetraose molecules in a manner similar to the previous enzyme
Maltohexaosidase	EC 3.2.1.98	1,4-α-D-Glucan maltohexaohydrolase	Release of maltohexaose in a manner similar to the previous enzymes above
β-Amylase	EC 3.2.1.2	1,4-α-D-Glucan maltohydrolase	Successive hydrolysis of penultimate α-1,4 linkages and removal of β-maltose molecules from nonreducing termini of starch and related glucans
Glucoamylase	EC 3.2.1.3	1,4-α-D-Glucan glucohydrolase	Hydrolysis of α-1,4 and α-1,6 bonds and removal of α-D-glucose residues from nonreducing termini of starch, dextrins, and related glucans
α-Glucosidase	EC 3.2.1.20	α-D-Glucoside glucohydrolase	Hydrolysis of α-1,4 and α-1,6 bonds, and release of glucose from nonreducing termini of di- and oligosaccharides, and aryl-glucosides

-- *continued*

TABLE 10.4   (continued)
Enzymes Involved in Starch Conversions

Enzyme	EC Number	Systematic Name	Application
Oligo-1,6-Glucosidase	EC 3.2.1.10	Dextrin 6-α-D-glucanohydrolase	Release of terminal, α-1,6-linked glucose from isomaltose and dextrins produced from starch and glycogen by α-_amylase
Isoamylase	EC 3.2.1.68	Glycogen 6-glucanohydrolase	Hydrolysis of α-1,6 bonds in amylopectin, dextrins, and glycogen
Type I Pullulanase	EC 3.2.1.41	α-Dextrin 6-glucanohydrolase	Hydrolysis of α-1,6 bonds in pullulan, dextrins, amylopectin, and glycogen
Type II Pullulanase (amylo-pullulanase)	EC 3.2.1.1/41	α-Amylase-pullulanase	Hydrolysis of α-1,4 and α-1,6 bonds in starch, dextrins, and glycogen
Limit dextrinase	EC 3.2.1.142	Dextrin α-1,6-glucanohydrolase	Hydrolysis of α-1,6 bonds in dextrins
CGTase	EC 2.4.1.19	1,4-α-D-Glucan 4-α-D-(1,4-α-D-glucano)-transferase (cyclizing)	Conversion of starch to α-, β-, and γ-dextrins
Malto-oligosyl-trehalose synthase (MTS)	EC 5.4.99.15	1,4-α-D-Glucan 1-α-D-glucosyl-mutase	Conversion of the terminal α-1,4 bond from reducing end to α-1,1 bond
Malto-oligosyl-trehalose trehalohydrolase (MTH)	EC 3.2.1.141	4-α-D-((1,4)- α-D-Glucano) trehalose glucanohydrolase (trehalose-producing)	Releasing of trehalose molecules from dextrins formed by MTS
Exo-1,4-α-D-glucan lyase	EC 4.2.2.13	1,4-α-D-Glucan exo-4-lyase (1,5-anhydro-D-fructose-forming)	Conversion of starch, glycogen, and related oligo- and polymers of glucose to 1,5-anhydro-D-fructose

10.4.1.1   Common Properties of the α-Amylase Family of Enzymes

Some enzymes inactive against starch, for example, sucrose phosphorylase, dextran-sucrase, and alternansucrase, share significant structural similarities with α-amylases, and have been included in the α-amylase family.[11] Characteristic properties of the biocatalysts from this family are (a) the common, catalytic A-domain, having the $(\beta/\alpha)_8$ barrel structure, assisted by the small B-domain; (b) hydrolytic cleavage of α-1-4 or 1-6 glycosidic bonds, or both, with retention of α-configuration or formation of these linkages (the α-retaining double-displacement catalytic mechanism, Glu residue as the acid/base catalyst and Asp residue as the nucleophile); (c) presence of the second conserved Asp residue involved in substrate binding (hydrogen bonds with C2-OH and C3-OH groups of the substrate) and conversion, and two His residues crucial for transition-state stabilization, located in the cleft between Domains A and B; (d) presence of four highly conserved moieties in the amino acid sequence that contain both amino acid residues forming the catalytic site and those providing $(\beta/\alpha)_8$ barrel folding (within the second, fourth, fifth, and seventh β-sheets of the barrel), and the fifth conserved region responsible for calcium ion binding; (e) differences in substrate and product specificities resulting from the presence of other domains attached to the A-domain (up to 30 different reaction and product specificities detected within the α amylase family of enzymes); (f) folding of the C-terminal fragment of the molecule into the C-domain (found in the majority of family members), believed to stabilize the A-domain by shielding its hydrophobic residues from the polar environment; and (g) very high efficiency of catalysis (approximately 10^{15}-fold augmentation of the rate of the glycosidic bond hydrolysis).

10.4.1.2 α-Amylases

α-Amylases used for the starch liquefaction are endo-acting hydrolases, widely distributed in nature (produced by animals, plants, and numerous microorganisms). They randomly split in the α-1,4 glycosidic bonds in linear (amylose) and branched (amylopectin, only if not adjacent to the α-1,6 branchings) chains of starch and related glucans. They also catalyze oligosaccharide synthesis from soluble starch and some mono-, di-, and oligosaccharides.

The iterative attack of α-amylases ultimately converts starch to a mixture of glucose, maltose, oligosaccharides, and α-limit dextrins resistant to further hydrolysis. The attack is controlled by branching, formation of complexes with lipids, and other factors. The final pattern of digestion products, optimum hydrolysis conditions, and stability depend on α-amylase origin. α-Amylases from *Dictyoglomus thermophilum* and *Pyrococcus* species are classified in Family 57 of GHs[2]. They are retaining enzymes, and employ Glu residues as catalytic base, but the proton donor has not been identified to date.

Activity and stability of the majority of known α-amylases depend on the conserved and very tightly bound Ca^{2+} ion, located at the interface between A and B domains, and playing mainly the structural role.[28] Some α-amylases contain more than one calcium ion. Because the Ca^{2+} ions have to be removed by ion-exchange prior to the further conversion of liquefied starch, the protein-engineered *B. licheniformis* α-amylase, active and stable only at 5 ppm Ca^{2+}, has been obtained.[29]Another advantage of this mutant enzyme is optimum activity at pH 5.5, whereas bacterial α-amylases exhibit maximum activity usually at pH 6.0. An excess of Ca^{2+}, concentration of which depends on the enzyme origin, inhibits the activity of α-amylases because of the carboxyl groups in their active sites. Chloride ions, enhancing catalytic efficiency and affinity for the conserved Ca^{2+} ions, probably induce conformational changes around the active site. These ions have been detected in the active sites of several, mainly mammalian, α-amylases, but also in microbial α-amylases.[28] In contrast to the endo-acting α-amylases, some α-amylases such as maltogenic amylase, maltotriohydrolase, maltotetraohydrolase, and maltohexaohydrolase synthesized mainly by *Bacillus* strains (Table 10.4) preferentially attack one end of the glucan chain and liberate selected oligosaccharides only.

Mesophilic α-amylases, derived from *Bacillus* species and filamentous fungi (mainly *Aspergillus* species), which dominated industrial starch processing in its infancy, have been currently replaced with their thermophilic counterparts such as *B. stearothermophilus* or *B. licheniformis* enzymes. They provide satisfying liquefaction of the polysaccharide at 100 to 105°C, that is, under conditions causing rapid denaturation of mesophilic proteins. However, the most thermoactive α-amylases are those produced by the archaea *Pyrococcus woesei* and *P. furiosus,* active up to 130 and 120°C, respectively.[30]

Attention of technologists has been also drawn to raw starch-digesting α-amylases. They reduce costs of amylolysis because they degrade starch granules without prior energy-consuming gelatinization. These α-amylases posses an additional raw starch-binding E-domain, providing adsorption on glucan granules.[31] The enzymatic digestion of uncooked starch is much slower than that of gelatinized glucan, and light microscopy and scanning electron microscopy observations demonstrated that starch-adsorbable α-amylases and glucoamylases produce pits or pores, penetrating the granule surface toward its center. The diameter and depth of pores as well as the composition of water-soluble digestion products hinge on the enzyme used. The pitted starch granules remain insoluble in cold water, but are soluble in warm water and are more readily degradable. Such granules are useful in production of food, cosmetics, and pharmaceuticals.

10.4.1.3   Exo-Acting Enzymes

The exo-acting starch-digesting enzymes comprise β-amylases and glucoamylasesas well as enzymes of the α-amylase family, such as α-glucosidases, and oligo-1,6-glucosidases.[2] All these exo-hydrolases attack their substrates from the nonreducing ends.

α-Glucosidases hydrolyze the α-1,4 and, more slowly, also the α-1,6 glycosidic bonds and release glucose molecules from disaccharides, oligosaccharides, and aryl-glucosides. In nature, they participate in the final steps of starch hydrolysis into glucose. They are much less important in industrial starch processing. Oligo-1,6-glucosidases liberate terminal α-1,6-linked glucose from isomaltose and starch and glycogen dextrins. They degrade exclusively short-chain substrates.

In contrast to the α-amylase family enzymes, α- and gluco-amylases cause inversion of anomeric configuration on hydrolysis of the α-glycosidic bond. Active site of β-amylases (Family 14 of GH) is included as a part of the $(\alpha/\beta)_8$ barrel, whereas that of glucoamylase (Family 15 of GH) is localized in the $(\alpha/\alpha)_6$ fold. Some glucoamylases are also classified in Family 31 of GHs.

β-Amylases liberate maltose and split exclusively the α-1,4 linkages. High-molecular-weight limit dextrins are the by-products of starch digestion with β-amylases The latter can neither cleave nor omit the α-1,6 branchings. β-Amylases are widely distributed in plants but reside in fungi and bacteria too. Their concomitant use with starch-debranching enzymes such as pullulanse or isoamylase gives rise to significantly higher yield of maltose. Starch digestion with both α- and β-amylases provides relatively simple and fast production of certain branched maltooligosaccharides such as 6-α-maltosyl-maltotriose, whose chemical synthesis is laborious and difficult.

Glucoamylases, synthesized by filamentous fungi, yeast, eubacteria, and archaea,[2] chiefly saccharify liquefied starch to pure glucose. Glucoamylases prefer the α-1,4 glycosidic bonds, but, although more slowly, can also hydrolyze the α-1,6 and other α linkages between glucose residues. Grafting of amylose and amylopectin with sugar residues, for example, with fructofuranosyl groups, reduces the degree of the hydrolysis of the polymers by β- and gluco-amylase.

10.4.1.4 Debranching Enzymes

Type I isoamylases and pullulanases are debranching enzymes that cleave the α-1,6 glycosidic bonds in amylopectin, dextrins, and glycogen.[2] They are distinguished by the activity against pullulan, exclusively degraded by the pullulanases. Apart from this difference, isoamylases preferably attack long chains of amylopectin than shorter branched dextrins, the latter readily digested by pullulanases and limit dextrinases. For technological requirements, poorly thermostable mesophilic pullulanases have been replaced with thermoactive enzymes. The most heat-stable Type I pullulanases are synthesized by *Fervidobacterium pennavorans* and *Thermus caldophilus* GK-24, showing maximum activity at 85 and 75°C, respectively.[30]

Type II pullulanases, also termed as α-amylase-pullulanases, or amylopullulanases, randomly cleave the α-1,4 bonds apart from the α-1,6 linkages, and produce mainly glucose, maltose, and maltotriose from starch and dextrins. The most thermostable and thermoactive Type II pullulanases have been derived from *Pyrococcus woesei, P. furiosus,* and *Thermococcus litoralis,* showing optimum activity at 100°C.

Other pullulan-digesting enzymes, such as neopullulanases, which convert the polymer to panose, and isopullulanases, which degrade pullulan to isopanose, do not belong to starch-debranching enzymes.[2] The first exhibits transglycosylating activity and simultaneously hydrolyzes and forms α-1,4- and α-1,6 glycosidic bonds.

10.4.1.5 Cyclodextrins and Cycloamyloses

Cyclic glucose oligomers are synthesized from starch and dextrins by cyclodextrin glycosyltransferases (CGTases), classified in the α-amylase family (see Chapter 18).[11] Their function still needs to be improved because they usually produce a poorly separable mixture of α-, β-, and γ-cyclodextrins. CGTases were initially obtained from *Bacillus macerans*; however, their thermostable preparations from *Thermoanaerobacter* species, *Thermoanaerobacterium thermosulfurigenes*, and *Anaerobranca bogoriae* have attracted attention of cyclodextrins producers.

Apart from the production of cyclic glucose oligomers, the transglycosylating activity of CGTases has been also harnessed for synthesis of novel glycosylated substances such as acarbose analogs and monoesters of fatty acids and maltooligosaccharides.[32] The latter amphiphilic and bioactive compounds, used as "green" emulsifiers in food, pharmaceutical, and cosmetics production, are synthesized by lipases in organic media, from the fatty acids and mono- or disaccharides. An increase in the degree of polymerization of carbohydrate moieties is achieved in the second step by using the CGTase preparation, and cyclodextrins, maltooligosaccharides, or starch as donors.

CGTases and other transglycosylating enzymes of the α-amylase family also synthesize cyclic molecules of cycloamyloses consisting of over 20 glucose residues.[31] A low yield of synthesis, required low concentration of high-molecular-weight amylose, and high concentration of the enzymes make the larger-scale production of cycloamyloses economically not feasible.

10.4.1.6 Trehalose

Trehalose, a nonreducing disaccharide, containing two α-1,1-linked glucose residues, plays an important role in nature and in food, pharmaceutical, and cosmetics industries, and in tissue engineering (cryopreservation of mammalian cells). The enzymatic conversion of dextrins to trehalose is more profitable than its extraction from plants and bakery yeast. The enzymatic process involves concerted action of an isomerase called malto-oligosyltrehalose synthase (MTS) with malto-oligosyltrehalose trehalohydrolase (MTH).[33] The first of those biocatalysts (MTS) converts the first α-1,4 glycosidic linkage from the reducing end of the dextrin to the α-1,1 bond. This conformational alteration in the sugar molecule facilitates MTH hydrolysis of the neighboring α-1,4 linkage to the trehalose molecule. A mesophilic strain of *Arthrobacter* was the original source of MTS and MTH, but their thermostable counterparts from the thermophilic archaeon *Sulfolobus shibatae* appeared to be superior from the technological standpoint. Trehalose is also available from maltose by using trehalose synthase.[2]

10.4.1.7 Other Starch Conversions

Apart from GHs and GTs, starch processing employs other enzymes, including isomerases, mainly xylose isomerase for glucose conversion to fructose, and lyases.

The production of 1,5-anhydro-D-fructose by exo-1,4-α-D-glucan lyase from starch, glycogen, or related oligo- and polymers of D-glucose are examples of a potential large-scale application of a lyase in starch conversion.[2] Chemical synthesis of this precursor for antibiotics appeared to be extremely laborious, and therefore only the enzymatic process can contribute to elucidation of its metabolic role and potential uses in medicine.

10.4.2 Cellulose, Chitin, and Chitosan

Complete hydrolysis of cellulose, the most abundant natural biopolymer, composed of β-1,4-linked glucopyranose residues (degree of polymerization up to 15,000), produced by plants (approximately 10^{15} kg per year) and some bacteria, requires common,

TABLE 10.5
Cellulose-Digesting Enzymes

Enzyme	EC Number	Systematic Name	Application
Endocellulase	EC 3.2.1.4	1,4-β-D-Glucan glucanohydrolase	Random attack on β-1,4 bonds preferably in amorphous cellulose regions to release cellooligosaccharides, cellobiose, and some glucose
Cellobiohydrolase	EC 3.2.1.91	1,4-β-D-Glucan cellobiohydrolase	Liberates cellobiose molecules either from nonreducing or reducing cellulose chains, also in crystalline regions
Exoglucanase	EC 3.2.1.74	1,4-β-D-Glucan glucohydrolase	Releases glucose from nonreducing ends of cellulose and cellooligosaccharides
β-glucosidase	EC 3.2.1.21	β-D-Glucoside glucohydrolase	Hydrolyzes β-1,4 linkages in cellobiose and related disaccharides or glycosides

synergistic action of endoglucanases, cellobiohydrolases, and β-glucosidases, supported by exoglucanases. Basic information on these enzymes is compiled in Table 10.5.

Both endocellulases and cellobiohydrolases are either inverting or retaining GHs.[2] The latter enzymes hydrolyze crystalline cellulose more efficiently than the endo-acting cellulases due to shape of the active site, resembling a long (extends approximately 4 nm into the interior of the catalytic domain) tunnel and providing processive character of attack on crystalline cellulose.[34] The cellulose-binding domain is wedge-like with a flat and hydrophilic (three Tyr residues) face interacting with crystalline region of a cellulose microfibril. Cellobiohydrolases can be accounted to the enzymes attacking from the reducing or non-reducing ends.

Many fungal endocellulases derived from *Aspergillus*, *Acremonium*, and *Trichoderma* species show a relatively broad specificity toward other polysaccharides such as xylan and chitin. This property enables production of N-acetyl-D-glucosamine (GlcNAc) from chitin, the second most abundant natural polysaccharide, using these nonchitinase enzymes.[35] GlcNAc has been produced by reacetylation of D-glucosamine available from acidic hydrolysis of chitin. Serious drawbacks of the latter process are omitted by using inexpensive cellulase preparations releasing pure monosaccharide. The enzyme from *Acremonium cellulolyticus* is the most efficient.

Enzymatic hydrolysis of chitosan catalyzed by β-glucosidase or other hydrolases such as lysozyme, endocellulase, and endopolygalacturonase has recently attracted attention (see Chapter 14).

10.4.3 PECTIN

Degradation of pectic substances (see Chapter 12) involves enzymes presented in Table 10.6. The solubilization of pectin from protopectin is catalyzed by protopectinase (PPase), which also shows the classical endopolygalacturonase activity.[36] With respect to the site of attack against protopectin, PPases are specific for the polygalacturonic acid region of protopectin (A-type) and that combining with the polysaccharide chains connecting polygalacturonic acid with other cell wall components (B-type). PPases originate mainly from fungi and serve for extraction of pectin from citrus peels as well as maceration of vegetables.

Degradation of pectin by endopolygalacturonases (PGs) requires prior demethylation with pectinesterases. Due to the concomitant presence of some acetyl residues in pectin molecules, acetylesterases are also involved in their digestion.

Some fungal PGs are processive enzymes, able to multiple-attack on a single chain, whereas the others operate via a single-attack mechanism. Studies on *Aspergillus niger* endo- and exo-polygalacturonases (ExoPGs) revealed that they are inverting enzymes, unable to catalyze glycosidic bond synthesis.[37] The only pectic enzyme capable of catalyzing reaction of glycosyl transfer appears to be *Selenomonas ruminantum* ExoPG.

Due to the complexity of "hairy" regions of pectin, some accessory enzymes, listed in Table 10.6, take part in their degradation.[38] Certain *Aspergillus* exopolygalacturonases release galacturonic acid not only from polygalacturonic acid but also from sugar beet pectin and xylogalacturonan. They also produce β-D-xylopyranosylo-(1,3)-D-galacturonate from the latter polymer. Deacetylation of rhamnogalacturonan (RG) by respective acetylesterase (RGAE) is a prerequisite for its further digestion by endohydrolase (RHG) and lyase (RGL), and two exo-acting enzymes such as RG-RH and RG-GH, which attack oligosaccharides liberated by RHG from the nonreducing end.

Preparations of pectic enzymes applied for the fruit and vegetable processing produced by microorganisms, mainly filamentous fungi, used to contain a mixture of pectinesterase and depolymerases. Currently, GMO-derived pectinesterases, free from pectin depolymerases, have been tapped to improve the rheologic quality of ketchup and to provide firmness of fruit and vegetable slices introduced to yoghurts and other dairy products.[39] Preparations of accessory enzymes responsible for the degradation of hairy regions of pectin have been applied for their cleavage from smooth regions. This operation facilitates production of juices with stable and uniform haze of partially depolymerized pectin, whereas the colloid formed by native pectin shows a tendency to precipitate on storing. Undoubtedly, microbial pectinolytic enzymes are among the most significant industrial biocatalysts, applicable not only in fruit and vegetable processing but also in textile and pulp industries.

TABLE 10.6

Pectin- and Pectic-Acid-Degrading Enzymes

Common Name	EC Number	Systematic Name	Application
Protopectinase — PPase			Solubilization of pectin from protopectin
Pectinesterase	EC 3.1.1.11	Polymethylgalacturon ate esterase	Removal of methanol residues
Pectin acetylesterase			Removal of acetic acid residues
Endopolygalacturonase (PG)	EC 3.2.1.15	Poly(1,4-α-D-galacturonide) glycanohydrolase	Random hydrolysis of α-1,4 bonds (adjacent to demethylated carboxylate groups) in "smooth" pectin regions
Exopolygalacturonases (ExoPG)	EC 3.2.1.67 EC 3.2.1.82	Poly (1,4-α-D-galacturonide) galacturonohydrolas e, Poly(1,4-α-D-galactosiduronate digalacturonohydrol ase	Hydrolysis of α-1,4 bonds in pectic acids, from nonreducing ends; mono- and digalacturonate-producing, respectively
Pectate lyase	EC 4.2.2.2	Poly(1,4-α-D-galacturonide) lyase	Random cleavage of α-1,4 glycosidic bonds in pectic acids (via β-elimination)
Pectin lyase	EC 4.2.2.10	Poly(methoxy-α-D-galacturonide) lyase	Random cleavage of α-1,4 glycosidic bonds in pectin (via β-elimination)
Exo-pectin lyase	EC 4.2.2.9	Poly(1,4-α-D-galacturonide) exo-lyase	Cleavage of α-1,4 glycosidic bonds (via β-elimination) from nonreducing ends
RG Acetylesterase (RGAE)			Deacetylation of rhamno-galacturonan (RG)
RG Endohydrolase (RHG)			Random hydrolysis of RG to oligosaccharides
RG Rhamnohydrolase (RG-RH)			Exohydrolase, releasing rhamnose molecules from nonreducing ends of oligomers produced by RHG
RG Galacturonohydrolase (RG-GH)			Exohydrolase, releasing galacturonic acid from nonreducing ends of oligomers produced by RHG

-- *continued*

TABLE 10.6 (continued)
Pectin- and Pectic-Acid-Degrading Enzymes

Common Name	EC Number	Systematic Name	Application
α-rhamnosidase	EC 3.2.1.40	α-L-Rhamnoside rhamnohydrolase	Release of rhamnose molecules from α-L-rhamnosides
Endo-xylo-galacturonase			Digestion of xylose-substituted polygalacturonate fragments
Endoarabinase	EC 3.2.1.99	1,5-α-L-arabinan 1,5-α-L-Arabinohydrolase	Random hydrolysis of 1,5-α-L-arabinan chains
α-L-arabinofuranosidase	EC 3.2.1.55	α-L-Arabino-furanoside arabinofuranohydrolase	Hydrolysis of α-1,5-L-arabinan chains in exo manner
Endogalactanase	EC 3.2.1.89	Arabinogalactan 4-β-D-galactanohydrolase	Hydrolysis of galactan side chains
Exogalactanases			Hydrolysis of galactan side chains (galactose- or galactobiose-producing)
β-Galactosidase	EC 3.2.1.23	β-D-Galactoside galactohydrolase	Releasing of galactose molecules from β-galactosides

10.4.4 XYLAN

Hemicelluloses produce several problems on processing of fruits and vegetables and production of animal feed and paper. Xylan is the major polymer among them. It contains a backbone of β-1,4-linked D-xylose residues, substituted by numerous side groups such as L-arabinose; D-galactose; and residues of acetyl, feruoyl, *p*-coumaroyl, and glucuronic acids.

The main chain of xylan is attacked by xylanases that catalyze hydrolysis of the internal β-1,4-glycosidic bonds, thus decreasing the degree of polymerization and facilitating further digestion to xylose by exo-acting β-xylosidases.[40] Endoxylanases are retaining hydrolases, classified for their structural differences into Families 10 and 11 of GHs. Producing smaller oligosaccharides and acting both on *p*-nitrophenyl-xylobiose and *p*-nitrophenyl-cellobiose, Family 10 xylanases have higher molecular weights than these of Family 11, usually close to 30 kDa. The latter enzymes are more specific for xylan.

Enzymes of both families possess three to five xylopyranose binding sites, neighboring with the catalytic site. The aromatic side groups of Tyr, and not those of Trp, play the basic role in binding of the xylopyranose units.

Among all known bacterial and archaeal xylanases, the most heat stable are enzymes from *Pyrococcus furiosus*, *Pyrodictum abysii*, *Sulfolobus solfataricus* MT-4, and *Thermotoga* strains, with optimum activities at 105, 110, 90, and 100°C, respectively.[30]

Produced by fungi and some bacteria, exo-acting β-xylosidases release D-xylose from short oligosaccharides and xylobiose.[2] Because the latter disaccharide is an inhibitor of endoxylanases, the exo-acting hydrolases improve the effectiveness of xylan digestion. The most heat-stable and thermoactive (at 50 to 100°C) β-xylosidases are synthesized by *Thermotoga* species, together with α-L-arabinofuranosidases, which are active against branched arabinoxylans, arabinose-substituted xylooligosaccharides, and *p*-nitrophenyl-α-L-arabinofuranoside.[30]

10.4.5 FRUCTANS

Table 10.7 presents the enzymes involved in the degradation of inulin and levan, which are relatively short plant reserve fructans (see Chapter 13). Inulin is hydrolyzed to fructose by inulinase and fructan β-fructosidase. The latter enzyme is active against inulin, levan, and sucrose, whereas inulinase is also active against sucrose, similar to invertase. Levan is also hydrolyzed by levanase, but this enzyme is not

TABLE 10.7
Fructan-Converting Enzymes

Enzyme	EC Number	Systematic Name	Application
Inulinase	EC 3.2.1.7	2,1-β-D-Fructan fructanohydrolase	Hydrolysis of inulin to fructose; active against sucrose; some inulinases yield also fructooligosaccharides with DP of 2–7
Fructan β-fructosidase	EC 3.2.1.80	β-D-Fructan fructanohydrolase	Hydrolysis of inulin, levan, and sucrose
Levanase	EC 3.2.1.65	2,6-β-D-Fructan fructohydrolase	Hydrolysis of levan to fructose
Inulinase II	EC 2.4.1.93	Inulin D-fructosyl-D-fructosyltransferase (1,2':2,3'-dianhydride forming)	Inulin conversion to di-D-fructose 1,2':2,3' dianhydride (DFA III)
Inulinase III	EC 2.4.1.200	Inulin D-fructosyl-D-fructosyltransferase (1,2':2',1-dianhydride-forming)	Inulin conversion to 1,2':2',1-dianhydride

commercially available, in contrast to thermostable inulinase preparations usually applied for inulin or sucrose hydrolysis in immobilized form.[41] Apart from free fructose, *A. ficum* inulinase also produces fructooligosaccharides with a degree of polymerization from 2 to 7.

Another inuline-derived product is di-D-fructose 1,2′:2,3′ dianhydride (DFA III),[2] which can be used as a low-calorie and less-cariogenic sucrose replacer in the human diet. This compound is obtained through inulin treatment with the so-called inulinase II. The latter is produced as an extracellular enzyme by certain *Arthrobacter* species. Production of this enzyme by recombinant *E. coli* strains is promising for its commercialization. The second type of inulin-derived anhydride can be prepared by the conversion of inulin with a 1,2′:2′,1-dianhydride-forming transferase.[2]

REFERENCES

1. Rodriguez, E. and McDaniel, R., Combinatorial biosynthesis of antimicrobials and other natural products, *Curr. Opin. Microbiol.*, 4, 526, 2001.
2. http://afmb.cnrs-mrs.fr/ ~pedro/CAZY/gtf.html
3. Barton, S., Bullock, C., and Weir, D., The effect of ultrasound on the activities of some glycosidase enzymes of industrial importance, *Enz. Microb. Technol.*, 18, 190, 1996.
4. Saxena, I.M., Brown, R.M. Jr., and Dandekar, T., Structure-function characterization of cellulose synthase: relationship to other glycosyltransferases, *Phytochemistry*, 57, 1135, 2001.
5. Kandra, L. et al., Chemoenzymatic synthesis of 2-chloro-4-nitrophenyl β-maltoheptaoside acceptor-products using glycogen phosphorylase b, *Carbohydr. Res.*, 333, 129, 2001.
6. Endo, T. and Koizumi, S., Large-scale production of oligosaccharides using engineered bacteria, *Curr. Opin. Struct. Biol.*, 10, 536, 2000.
7. Watt, G.M., Lowden, P.A.S., and Flitsch, S.L., Enzyme-catalyzed formation of glycosidic linkages, *Curr. Opin. Struct. Biol.*, 7, 652, 1997.
8. Monsan, P. et al., Homopolysaccharides from lactic acid bacteria, *Int. Dairy J.*, 11, 675, 2001.
9. Bourne, Y. and Henrissat, B., Glycoside hydrolases and glycosyltransferases: families and functional modules, *Curr. Opin. Struc. Biol.*, 11, 593, 2001.
10. Withers, S.G., Mechanisms of glycosyl transferases and hydrolases, *Carbohydr. Polym.*, 44, 325, 2001.
11. MacGregor, E.A., Janecek, S., and Svensson, B., Relationship of sequence and structure to specificity in the α-amylase family of enzymes, *Biochim. Biophys. Acta*, 1546, 1, 2001.
12. Henrissat, B. and Davies, G., Structural and sequence-based classification of glycoside hydrolases, *Curr. Opin. Struc. Biol.*, 7, 637, 1997.
13. Williams, S.J. and Withers, S.G., Glycosyl fluorides in enzymatic reactions, *Carbohydr. Res.*, 327, 27, 2000.
14. Schulein, M., Protein engineering of cellulases, *Biochim. Biophys. Acta*, 1543, 239, 2000.
15. Silvestra, M.M. and Jonas, R., The biotechnological production of sorbitol, *Appl. Microbiol. Biotechnol.*, 59, 400, 2002.

16. Leitner, C. et al., Enzymatic redox isomerization of 1,6-disaccharides by pyranose oxidase and NADH-dependent aldose reductase, *J. Mol. Cat. B: Enz.*, 11, 407, 2001.

17. Wymer, N. and Toone, E.J., Enzyme-catalyzed synthesis of carbohydrates, *Curr. Opin. Chem. Biol.*, 4, 110, 2000.

18. Schurmann, M., Schurmann, M., and Sprenger, G.A. Fructose 6-phosphate aldolase and 1-deoxy-D-xylulose 5-phosphate synthase from *Escherichia coli* as tools in enzymatic synthesis of 1-deoxysugars, *J. Mol. Cat. B: Enz.*, 19/20, 247, 2002.

19. Voragen, A.G., Technological aspects of functional food-related carbohydrates, *Food Sci. Technol.*, 9, 328, 1998.

20. Yun, J. W., Fructooligosacharides: occurence, preparation, and application, *Enz. Microbiol. Technol.*, 19, 107, 1996.

21. Tanriseven, A. and Gokmen, F., Novel method for production of a mixture containing fructooligosaccharides and isomaltooligosaccharides, *Biotechnol. Tech.*, 13, 207, 1999.

22. Alcade, M. et al., Immobilization of native and dextran-free dextransucrases from *Leuconostoc mesenteroides* NRRL B-512F for the synthesis of glucooligosaccharides, *Biotechnol. Tech.*, 13, 749, 1999.

23. Chen, C.-S., Hsu, C.-K., and Chiang, B.-H., Optimization of the enzymic process for manufacturing low-lactose milk containing oligosaccharides, *Proc. Biochem.*, 38, 801, 2002.

24. Bonnin, E., Vigouroux, J., and Thilbault, J.F., Kinetic parameters of hydrolysis and transglycosylation catalyzed by an exo-β-(1,4)-galactanase, *Enz. Microb. Technol.*, 20, 516, 1997.

25. Giacomini, C. et al., Enzymatic synthesis of galactosyl-xylose by *Aspergillus oryzae* β-galactosidase, *J. Mol. Cat. B: Enz.*, 19/20, 159, 2002.

26. Hansson, T., and Adlercreutz, P., Enzymatic synthesis of hexyl glycosides from lactose at low water activity and high temperature using hyperthermostable β-glycosidases. *Biocat. Biotransform.*, 20, 167, 2002.

27. Shabalin, K.A. et al., Enzymatic properties of α-galactosidase from *Trichoderma reesei* in the hydrolysis of galactooligosaccharides, *Enz. Microb. Technol.*, 30, 231, 2002.

28. Nielsen, J.E. and Borchert, T.V., Protein engineering of bacterial α-amylases, *Biochim. Biophys. Acta*, 1543, 253, 2000.

29. Norman, B.E. et al., The development of a new heat-stable α-amylase for calcium-free starch α-amylase, Novo Nordisk Information Handout A 6505, 1997.

30. Sunna, A. et al., Glycosyl hydrolases from hyperthermophiles, *Extremophiles*, 1, 2, 1997.

31. van der Maarel, M.J.E.C. et al., Properties and applications of starch-converting enzymes of the α-amylase family, *J. Biotechnol.*, 94, 137, 2002.

32. Degn, P. et al., Two-step enzymatic synthesis of maltooligosaccharide esters, *Carbohydr. Res.*, 329, 57, 2000.

33. Di Lernia, I. et al., Trehalose production at high temperature exploiting an immobilized cell bioreactor, *Extremophiles*, 6, 341, 2002.

34. Gilkes, N.R. et al., Attack of carboxymethylcellulose at opposite ends by two cellobiohydrolases from *Cellulomonas fimi*, *J. Biotechnol.*, 57, 83, 1997.

35. Sukwattanasinitt, M. et al., Utilization of commercial non-chitinase enzymes from fungi for preparation of 2-acetamido-2-deoxy-D-glucose from chitin, *Carbohydr. Res.*, 337, 133, 2002.

36. Ferreyra, O.A. et al. Influence of trace elements on enzyme production: protopecti-
    nases expression by a *Geotrichum klebahnii* strain, *Enz. Microb. Technol.*, 31, 498,
    2002.
37. Biely, P. et al., Inversion of configuration during hydrolysis of α-1,4-galacturonidic
    linkage by three *Aspergillus* polygalacturonases, *FEBS Lett.*, 382, 249, 1996.
38. De Vries, R.P. and Visser, J., *Aspergillus* enzymes involved in degradation of plant
    cell wall polysaccharides, *Microbiol. Mol. Biol. Rev.*, 65, 497, 2001.
39. Heldt-Hansen, H.P. et al., Application of tailor-made pectinases, in *Progress in Bio-
    technology, 14. Pectins and Pectinases*, Visser, J. and Voragen, A.G.J., Eds., Elsevier
    Science, Amsterdam, 1996, p. 463.
40. Jeffries, T.W., Biochemistry and genetics of microbial xylanases, *Curr. Opin. Bio-
    technol.*, 7, 337, 1996.
41. Ettalibi, M. and Baratti, J.C., Sucrose hydrolysis by thermostable immobilized inuli-
    nases from *Aspergillus ficuum*, *Enz. Microb. Technol.*, 28, 596, 2001.

11 Role of Saccharides in Texturization and Functional Properties of Foodstuffs

Vivian M.F. Lai and Cheng-yi Lii

CONTENTS

11.1 Introduction 159
11.2 General Functions of Saccharides 160
    11.2.1 Low-Molecular-Weight Saccharides 160
    11.2.2 Polysaccharides 162
11.3 Functions of Polysaccharide Blends in Food 163
    11.3.1 Starch-Containing Products 164
        11.3.1.1 High-Moisture Starch Products 164
        11.3.1.2 Low-Moisture Starch Products 165
    11.3.2 Gels, Drinks, Dairy, and Emulsified Products 165
        11.3.2.1 Food Gels and Drinks 166
        11.3.2.2 Dairy Products 169
        11.3.2.3 Whipped and Emulsified Products 169
        11.3.2.4 Low- or Reduced-Calorie Desserts, Dressings,
                Emulsions, and Spreads 169
    11.3.3 Processed Meat Products 172
    11.3.4 Other Products 172
References 175

11.1 INTRODUCTION

Saccharides found in food play an essential role in determining the texture and structure of foodstuffs. Their function depends principally on their molecular weight.[1] Low-molecular-weight saccharides, such as mono- and oligosaccharides, generally interact strongly with water and food components (e.g., polysaccharides and proteins), have excellent properties of moisture and flavor retention, and reduce extensive aggregation between biopolymers. In other words, they produce plasticization, have antistaling effects, and protect food biopolymers. Polysaccharide gums provide a wide variety of functionality in foodstuffs, including gelling, thickening, suspending, adhering, binding, bulking, inhibiting crystallization, clarifying, flocculating, emulsifying, stabilizing, whipping, and coating.[2–4] The particular functions of polysaccharide blends that are frequently used in real food systems are very attractive. They can be partially explained in terms of molecular interactions between polysaccharides.[5] This chapter focuses on the key functions of saccharides and polysaccharide blends in controlling the texture of food systems.

11.2 GENERAL FUNCTIONS OF SACCHARIDES

11.2.1 LOW-MOLECULAR-WEIGHT SACCHARIDES

Low-molecular-weight saccharides that help control food texture can be roughly categorized into three groups: sugars (including mono- and disaccharides and syrups), oligosaccharides, and sugar derivatives.

Sugars and sugar alcohols serve mainly as humectants, plasticizers, and antistaling agents, and as protective or cryoprotective agents of food components during freezing, storage, and dehydration (Table 11.1).[1] In comparison with other sugars, fructose and some sugar alcohols (e.g., sorbitol) are superior food humectants because of their high hygroscopicities.[1] They are widely applied in low-moisture food products such as fruit preserves, peanut butter,[8] or chewing gum.[9–11] In dough and other bakery or starch-containing products, sugars and their alcohols (e.g., sorbitol, mannitol, xylitol, and lactitol) also produce plasticizing or antistaling

TABLE 11.1
Key Functions of Low-Molecular-Weight Saccharides in Foodstuffs

Function	Food Example	Saccharide Example	References
Humectant	Fruit preserves, peanut butter, chewing gums	Fructose, sugar alcohols	6–11
Plasticizing or antistaling effect	Bakery products, films	Sugars, sugar alcohols	1,12–15
Shiny coating	Biscuits	Syrups, MD	17
Protecting agent	Dried and frozen products	Trehalose, sugar alcohols	1,18–20
Bulking agent	Low-calorie bakery products	MD, PD	16, 21, 22
Flavor encapsulation	Drinks, emulsions	CD, MD	16, 23, 25
Emulsifier, softener	Bakery products, coffee drinks	Sucrose esters	26, 27
Fat substitute	Low-fat dressings, fried products	Sucrose polyesters	28

effect,[1,12,15] enhancing texture, structure, and storage stability. Coating with a mixture containing syrups and maltodextrins (MD), starch hydrolyzates with a dextrose equivalent (DE)[16] of less than 20 can improve the crispness and shiny surface of low-fat snacks such as pretzels.[17] MD with DE greater than 6 are reported to exert antiplasticization effects on food biopolymers.[1] In frozen dough and meat products, low-molecular-weight saccharides provide good protective effect against freezing by reducing freezing point of water components and the amount of free water species.[1] Sugar alcohols, especially sorbitol, are most popular in this category. Trehalose is an excellent humectant and cryoprotectant and is expected to become an attractive alternative ingredient in processed foods.[18]

Compared with sugars, the higher-molecular-weight oligosaccharides frequently used include MD, polydextroses (PD) and cyclodextrins (CD). The first two act mainly as bulking agents, replacing sucrose in low-calorie bakery products.[16,21–23] The use of PD rather than MD may improve processing, bulking, and textural characteristics of baked foods, however, with the accompanying effect of deterioration of taste proportionally to the amount of the added agents and the increasing baking temperature.[24]

CD with superior encapsulation ability to flavors and, generally, hydrophobic compounds,[25] can be used alone in oil-in-water systems or together with MD when dry mixes are prepared. Fatty acid esters of sugars are good emulsifiers or fat substitutes, depending on the degree of esterification on the sugar molecule. Frequently used in food, fatty acid esters of sucrose are digestible when their degree of esterifiaction is less than or equal to 3,[26,27] and those with a degree of esterification of 4 to 8 are nondigestible.[28]

The functions of sugars and sugar alcohols in most foods containing biopolymers can be explained in terms of molecular interactions among water, sugar, and biopolymer.[1,13,14] Sugar alcohols closely resemble the molecular weight and hydration properties of saccharides.[14] The antifreezing, antistaling, and plasticization effects (e.g., reducing starch retrogradation in bakery goods) decrease with an increase in their molecular weight.[1,12,14] However, because of their stoichiometry, fructose and trehalose are better at moisture retention than other saccharides with the same molecular weight. Despite their antistaling effects at DE greater than 6,[1] commercial MD (DE 1 to 20)[16] potentially promote starch retrogradation. The higher-molecular-weight MD are more effective.[14] On the other hand, low-molecular-weight saccharides under sufficiently high concentration show antiplasticizing effects on starch gelatinization. They retard starch swelling and elevate the melting temperature of microcrystallites in a starch granule.[13] The effectiveness of antiplasticization is linearly related to the dynamic hydration number of saccharides, which varies with the starch variety.[13] Similar effects are also reported on elevating the melting temperatures of many polysaccharide gels. Sugars at sufficiently high concentrations are able to synergistically increase the gel strength or elasticity of some polysaccharide systems, for example, agar,[29,30] pectins,[31,32] and locust bean gum (LBG).[33,34] The incorporation of sugars and MD helps to prevent lumping of dry polysaccharide mixes.[31,34]

11.2.2 POLYSACCHARIDES

The function of polysaccharides in foodstuffs generally originates from chain entanglement, formation of junction zones between the ordered polysaccharide chains, and subsequent aggregation.[3,4] These effects are essentially governed by molecular properties (e.g., molecular weight, degree of branching, chain rigidity, and functional groups) of polysaccharides[35] and such factors as the concentration and variety of the polysaccharides and the presence of biopolymers (e.g., proteins), sugars, or salts.

Table 11.2 lists key functions of polysaccharides. Generally, native and modified starches provide the widest variety of functions among food polysaccharides due to variations in their amylose content, molecular and granular structure, and physicochemical state. Therefore, they are not mentioned again in the following descriptions. Typical gelling agents from agar and alginate to carrageenans (CAR, especially κ- and ι-type) and pectins exhibit gel characteristics from a hard and brittle texture with high syneresis potential to a soft, elastic, and cohesive texture with insignificant syneresis.[2] Alginate and carrageenans can also be used as thickening, suspending, and stabilizing agents in diverse foodstuffs, by adjusting their concentrations and selecting proper cations in the systems.[31,38] Most of the commonly used thickening agents — guar gum (GG), locust bean gum (LBG), CAR, carboxymethyl cellulose (CMC), and xanthan gum (XG) — are also good at suspension, and stabilization and crystallization inhibition, partially due to their branched structures (for GG, LBG, and XG),[33] extended rigid chains (for CMC and XG),[35] or polyelectrolyte nature (for CAR, CMC, and XG). Gum arabic, starch (often used to reduce cost), and starch hydrolyzates (used as encapsulation aids)[23] are the best choices for emulsification and flavor encapsulation.[34] Polysaccharides with a cellulose backbone [i.e., β-(1→4)-D-glucan, found in XG, CMC, methyl cellulose (MC), hydroxypropylmethyl cellulose (HPMC), and microcrystalline cellulose (MCC)] usually have higher resistances to shear, temperature, and drastic pH change than do other polysaccharides.[42,44] They are therefore suitable for preparation of foods such as emulsions, salad dressings, and sauces. Because of their hydrophobicity, these nonanionic cellulose derivatives exhibit anticaking and bulking properties. In various foods, they act as fat mimetics. They are also binding agents in tablets.[41] Bulking agents and fat mimetics can be roughly grouped into water-soluble dietary fiber (psyllium, inulins, and β-glucans from oat bran) and insoluble dietary fiber [MCC, hemicelluloses, tamarind seed xyloglucan (TSX), wheat β-glucans, and partially resistant starch and konjac flour]. Their key functions are revealed when they are dispersed or they form microparticulate gels.[56,57,59,60] In fabricated or low-calorie products, they provide a fat-like texture and creamy appearance. Among the anionic polysaccharides that have considerable potential to interact with proteins, CMC, pectin, alginate and propylene glycol alginate (PGA) can be used to coat the surface of fresh meat products. CAR and alginate are superior in binding processed meat products. Biodelivery hydrogels can be easily prepared from alginate or chitosan solutions because of their calcium-ion-mediated[36,37] or alkaline-pH-induced[66] gelation, respectively.

TABLE 11.2
Key Functions of Polysaccharides in Foodstuffs

Function	Food Example	Typical Saccharide	References
Gelling agent	Confectionery, desserts	Agar, alginate, CAR, pectin, high-amylose starch	2, 3, 14, 29, 36, 40
Thickening agent	Dairy drinks, juices	GG, LBG, CMC, CAR, XG, modified starch	33, 34, 41, 43
Suspending agent	Fruit purée, jams, jellies, dressings, fillings	XG, CMC, pectin	31, 32, 41, 42, 44
Stabilizer, crystallization inhibitor	Emulsions, milks, frozen products	GG, LBG, CAR, PGA, modified starch, CMC	4, 33, 36–38, 41, 45–47
Emulsifier	Emulsions	Gum arabic, starch and cellulose derivative	41, 42, 45, 46
Flavor encapsulation	Chewing gums	Gum arabic, starch hydrolyzate, MD	9, 34, 48
Anticaking agent	Confectionery, bakery products	MCC	3
Bulking agent and dietary fiber	Bakery products, ready-to-eat cereals, drinks	Hemicelluloses, TSX, psyllium, konjac flour, cereal β-glucan	22, 49, 54
Fat mimetic	Low-fat or calorie products	MCC, resistant starch, TSX, inulin, konjac flour, hemicelluloses	55–60
Film coating	Fresh products	CMC, PGA, pectin, chitosan, modified starches	61, 62
Binding agent	Processed meat products	Alginate, CAR, modified starch	36, 37, 63, 65
	Tablets	MC, HPMC, MCC	42, 57, 59, 62
Biodelivery system	Hydrogels	Alginate, chitosan	36, 62, 66–68

11.3 FUNCTIONS OF POLYSACCHARIDE BLENDS IN FOOD

In natural food products, polysaccharide blends are frequently used to reduce cost, increase processing benefits, obtain desired texture and storage stability, or to design novel products. Interpreting the concomitant functions of polysaccharide mixtures is more complicated. In the following paragraphs, additional functions of polysaccharides in natural food from a second minor polysaccharide component are emphasized.

11.3.1 STARCH-CONTAINING PRODUCTS

11.3.1.1 High-Moisture Starch Products

Table 11.3 shows that blending various starches in native, pregelatinized, or modified state are commercial approaches in preparing high-moisture starch systems such as pastes, gels, and mixes. Waxy starches are beneficial in quick thickening. They exhibit a low hot-paste viscosity (η), which facilitates heat penetration.[47] Nonwaxy

TABLE 11.3
Polysaccharide Blends in Starch Pastes, Gels, and Mixes

System	Polysaccharide	Function	References
Starch paste	Fill-η starch (mainly waxy maize) + η-building starch (cross-linked starch)	Thickening quickly, no residual η, increasing heat penetration and process resistance	47
Starch paste and gel	Waxy and Indica rice starches	Adjusting pasting viscosity and elastic module	69,73
Dry mix	Pregelatinized starch + hydroxypropylated starch	Easy to "cook-up," increasing freeze–thaw stabilization	47
Dry mix	"Cook-up" starch + MD + gum tragacanth + sugars	Increasing hydration and η development, decreasing lumping	34
Batter mix	Corn flour + mildly oxidized starch	Excellent adherence	46
Starch-based gum and jelly	Thin-boiling starch + gelatin	Giving soft, medium-long texture, medium clarity	46
Starch-based gum and jelly	Oxidized starch +gum arabic	Giving hard, long texture, high clarity	46
Starch-based jam and filling	Cross-link and acetylated waxy maize starch and pectin	Increasing bakery and freeze–thaw stabilities, giving desired texture	46
Instant pudding, pie filling, and cream	Roll-dried maize starch, modified starches + Na-alginate and CAR	Thickening, stabilizing, gelling	46
Ready-to-serve dessert	Starch + CAR	Giving desired texture	46
Jellies, pudding	Starch or κ-CAR + MCC	Gelling, creating new texture, decreasing syneresis	59

starches, especially those with an amylose content over 25%, and cross-linked starches contribute to the paste viscosity with high η and gels with high elasticity after cooling. This is a result of restricted swollen granules and retrogradation in soluble matrices.[69] In dry mixes and instant pudding, pregelatinized, "cook-up," and roll-dried starches provide benefits of easy cooking, high hydration, and rapid thickenings. Hydroxypropylated or acetylated waxy maize starches generally help products have high freeze–thaw stability.[47] Products with oxidized starches generally have an adhesive and a soft to hard texture, depending on the extent of oxidization (and degradation) of the starch used.[46] In starch-based confectionery, gum tragacanth,[34] gum arabic, pectin, sodium alginate,[46] CAR,[46,59] MCC,[59] and gelatin[46] are incorporated to modify starch texture, increase storage stability, reduce syneresis, or create fat-like smooth texture (for MCC). Generally, high-moisture starch systems consist of soluble matrices and dispersed granules. Changes in elastic modulus resulting from addition of starch compositions to natural starch products may be interpreted as involving the isostrain or isostress Takayanagi's polymer blending rule.[70] The isostrain system is a composite containing a strong continuous matrix dispersed with weak particles. Its overall elasticity equals the weighted average of the individual elasticities. The isostress composite with weak matrices and strong fillers exhibits the overall compliance equivalent to the weighted average of individual compliances. It has been found that the gelatinized starch blend systems without shear disintegration or before retrogradation follow the isostress model.[71] Heat-induced gels from whey protein isolate–starch mixtures show an elastic modulus, with a maximum when the weight fraction of starch is 0.2 to 0.3.[72]

11.3.1.2 Low-Moisture Starch Products

For low-moisture starch products (Table 11.4), mixing native starches with different amylose content can produce the desired texture and structure for dough and noodles. Formulations with a higher amylose content tend to increase hardness, elasticity, and integrity of the products, and reduce the oil absorption of fried foods.[47,48] This is mainly due to amylose retrogradation. Pregelatinized starch cross-linked starch blends are chosen for snacks made by extrusion, baking, or frying.[47] In addition to conditioning starch formulation, alginate, or alginate blends with different glucuronic acid compositions and molecular weights, gum karaya, CAR, CMC, or GG generally retard staling and increase freeze–thaw stability.[33,34,37–75,76] CMC and GG are especially good for preparation of high-fiber breads,[75,76] whereas HPMC can improve the eating quality of gluten-free breads based on rice flour and potato starch,[76] possibly relating to its thermoreversible gelation property.[41,42]

11.3.2 Gels, Drinks, Dairy, and Emulsified Products

Table 11.5 to Table 11.8 list the wide applications of polysaccharide blends in confectionery gels, drinks, emulsions, whipped, and dairy goods. Benefits of using polysaccharide blends in these products include controlling organoleptic texture; reducing cost due to their synergy in gel strength or elasticity; increasing the suspension stability on preparation and storage; improving effectiveness of flavor

TABLE 11.4
Polysaccharide Blends in Processed Starch Products

System	Polysaccharide	Function	References
Dough	Starch mixtures with various amylase contents	Conditioning dough, decreasing oil absorption	47
Instant noodles	Mungbean starch + potato starch	Giving desired texture and structure	74
Snack	Pregelatinized starch + cross-linked starch	Giving desired properties during extrusion, baking, and frying	47
Bakery product	Alginate blend + pregelatinized starch	Increasing heat resistance, baking fast, giving freeze–thaw stability	37
Bakery product	Gum karaya + alginate or CAR	Retarding staling	34
Gluten-free bread	CMC + HPMC in rice flour + potato starch	Modifying structure	75, 76
Hih-fiber bread	CMC + GG in wheat flour and bran	Providing acceptable texture	76
Deep-frozen ready meal	Starch adipate + GG	Increasing freeze–thaw resistance	33

encapsulation; creating novel fabricated low-calorie foodstuffs; and facilitating process. The following paragraphs mainly focus on the additional functions of commonly found beneficial combinations besides those of the polysaccharides mentioned in Table 11.2.

11.3.2.1 Food Gels and Drinks

For food gels and drinks (Table 11.5) and dairy products (Table 11.6), red algal polysaccharides (agar, κ-, ι and λ-CAR, and furcellaran) are most commonly used as the main components for controlling end products with diverse rheological and textural characteristics, depending on the concentration and variety of the polymer used. Table 11.5 indicates that they are usually applied with the other red algal polysaccharides to control gel textures of dessert gels,[38–40,77] acidified milk yogurt,[82] or pasteurized milks and cold prepared custard.[40,82] High-gelling-ability agars can gel synergistically with a small amount of LBG or sodium alginate (about 10 to 20% on an agar basis) in sweet potato sweets[30] and canned mitsumame gel.[29] Synergistic gels formed with gelling CAR (especially κ-type) and galactomannans (mainly LBG)[5,38–40] are also used in dessert gels, either with or without fruit juices,[39,40] and milk shakes.[33] Synergistic gels with CAR and XG or konjac flour are suitable to prepare dessert gels or chocolate milks with high solid contents, due to

TABLE 11.5
Polysaccharide Blends in Food Gels and Drinks

System	Polysaccharide	Function	References
Dessert gel, sweet potato, sweet	Agar + LBG	Synergy, increasing gel strength and elasticity, decreasing rigidity, brittleness and syneresis	30
Water dessert gel	κ- + ι-CAR or + konjac flour	Gelling, decreasing syneresis	39, 77
Dessert gel	κ- + ι-CAR + LBG, κ-CAR + LBG	Gelling, gelatin-like texture, synergy	39, 40
Dessert gel, jam, jelly	Na-alginate + HM-pectin, pH < 4.0	Synergy, gelling	37
Fruit-juice-based product	CMC and/or XG, PGA	Thickening, suspension	42
Cake glaze and flan gel, dessert gel	κ- + ι-CAR and/or +LBG; κ-CAR + furcellaran	Gelling, synergy, elastic, cohesive, and gelatin-like mouth feel	38, 40
Sherbet (sweet frozen fruit dessert)	CMC + pectin, LBG or gelatin	Gelling, stabilizing, desired texture	42
Canned mitsumame gel	Agar + Na-alginate	Gelling, synergy	41
Instant mix for drink, soup, and dessert	XG + CMC or GG	Increasing stabilizing, thickening, suspension	44
Fruit pieces, confectionery gel	Konjac flour + CAR or XG	Synergy, brittle–elastic texture, decreasing cost	39
Confectionery	Gum arabic + agar, modified starch, or gelatin	Increasing stabilizing effect, controlling flavor, thickening, gelling	34
Low-NaCl food and beverage	Gelatinized starch + MD + gum arabic	Carrier, encapsulating edible ammonium salts	78
Ca-enriched drink (milk, soy milk)	Colloidal MCC + low strength agar	Ca stabilizer	79
Milk shake, chocolate milk	Galactomannans + XG or + CAR	Increasing creamier mouth feel, foam stabilizing, gelling	33, 80
Low-η chocolate milk	κ-CAR + MCC	Stabilizing, thickening, increasing thermal stability of κ-car–κ-casein complex	81
Acidified milk yogurt	κ- + ι-CAR	Body, fruit suspension	82

TABLE 11.6
Polysaccharide Blends in Some Dairy Products

System	Polysaccharide	Function	References
Gelled dairy dessert	Starch + hydrocolloids (mainly CAR)	Giving desired texture, decreasing syneresis	42, 47
UHTST-processed milk dessert	Starch + CAR	Less pudding-like, increasing fluidity	46
Canned ready-to-eat milk pudding	Starch + ι-CAR	Decreasing η, increasing heat transfer, controlling degree of set, increasing flavor release, decreasing syneresis	77
Pasteurized and sterilized chocolate milks, cold prepared custard	κ- + λ- and/or ι-CAR	Suspending, bodying, thickening, gelling, stabilizing emulsion	40, 82
Cottage and creamed cheese product	κ-CAR +LBG	Stabilizing, increasing curd formation and shape retention	77, 83
Fermented milk product (fresh cheese)	CMC, GG, LBG, gelatin	Stabilizing on shear	33
Cream cheese and spread	κ-CAR + LBG	Gelling, moisture binding	40
Pourable, spreadable food	Coprecipitates of GG or LBG and glucomannan	Increasing gelling, thickening, pourable, spreadable texture	84
Milk-based milk protein system	CAR + GG, CMC, or LBG, CAR + starch	Hard packing and soft serving, gelling, bodying	39
Hard milk ices and water ice	CMC + CAR, GG or LBG, and corn syrup blends	Stabilizing, increasing desired texture and body, decreasing whey separation	42
Dairy product (flan and instant mousse)	κ-CAR + inulin	Greatly increasing stabilizing effect, creamier mouth feel	60

additional stabilization effects.[39,77] However, CAR are not usually considered for products with high acidity (pH < 4.0) or fruit pieces because of their lability to acid.[38] Instead, blends with sodium alginate, highly methylated pectin,[37] CMC, XG, and PGA[42,44] are better as thickening and stabilizing agents, due to their resistances to high acidity, high viscosity, or thixotropic characteristics (for XG). Synergistic effects between two of these polysaccharides[5] are also expected for blend gels. Gum arabic blends with starch components are suitable to stabilize products with encapsulated flavor[34] or ammonium salts.[78] Mixtures of colloidal MCC and low-strength agar, both prone to form dispersed microparticulates, are calcium stabilizers in calcium-enriched drinks.[79] Commonly, blending gelling and nongelling polysaccharide gums makes gel and meat products with increased gel strength, elasticity, and cohesiveness, and with decreased hardness, brittleness, and syneresis.

11.3.2.2 Dairy Products

Table 11.6 shows that in most of dairy products, CAR (especially κ-CAR) are able to interact synergistically with milk components (especially κ-casein), and are therefore excellent polysaccharide partners in controlling textures of these products. The second polysaccharides generally include starches and galactomannans (especially LBG). The additional benefits of using their blends come from increasing fluidity, heat transfer, and flavor release, and reducing syneresis for milk desserts and ice cream;[42,46,47,77] stabilizing and increasing curd formation and moisture retention for cheese products;[40,47,83] and giving the property of hard packing and soft serving to milk protein systems.[39] Besides, using CMC instead of CAR with galactomannans can improve the shear stability of fresh cheese products.[33] The use of coprecipitates of galactomannans and glucomannans can enhance the pourable and spreadable texture of foods.[84] The incorporation of inulin notably improves the stabilizing effect on κ-Car–κ-casein complexes and provides creamier mouth feel in dairy products.[60]

11.3.2.3 Whipped and Emulsified Products

In whipped and emulsified foodstuffs (Table 11.7), the key function of the polysaccharide blends used is generally to thicken aqueous matrices, and lead to improved body texture and excellent stabilization in emulsion on preparation and storage.[87] Accordingly, blends of two CAR, galactomannans, and XG are applied due to their synergistic stabilization effects.[33,40,77,85] Starches and XG or GG are also used for some cases of salad dressings[44] and ketchup,[33] with the benefits of high viscosity and suspension, and low retrogradation. Blends with two or more cellulose derivatives (CMC, MC, and HPMC), extrudate gums (gum arabic, gum tragacanth, and gum karaya) and PGA can provide additional functions in emulsification, stabilization, pourable property, and preparation tolerance.[34,42,86]

11.3.2.4 Low- or Reduced-Calorie Desserts, Dressings, Emulsions, and Spreads

The general functions of polysaccharide blends in this food category are to act as bulking agents, stabilizing agents, and fat-mimetics (Table 11.8). Generally, MCC, resistant starch,[59] cereal β-glucans,[33,50] inulins,[60] and pulverized curdlan gel[94] are important. The polysaccharide blends developed are mainly colloidal MCC aggregates or particulates blended with one or two of GG[57,88,91] konjac flour,[93] CMC, CAR, XG,[57] starch components,[57,92] and microparticulated proteins. These formulations show excellent bulking and stabilizing effects and produce smooth fat-like or oil-like and spreadable textures as well as creamy appearance to the end products. Modified

TABLE 11.7
Polysaccharide Blends in Whipped and Emulsified Products

System	Polysaccharide	Function	References
Aerosol-propelled cream topping	κ-CAR + LBG	Stabilizing	33, 77
Whipped cream and mousse	XG + LBG	Increasing stabilizing, synergy	44
Ice cream	Gum karaya + LBG	Stabilizing	34
Low η-mix (ice cream, milk ice)	CMC + CAR	Stabilizing, thickening, decreasing whey separation and syneresis	42
Ice cream mix	κ-CAR + (GG, LBG, XG, or Na-alginate)	Stabilizing	40
Vinaigrette-style salad dressing	ι-CAR + XG	Thickening, stabilizing	40
Salad dressing	Starch + XG and/or depolymerized GG	High η, suspension, stabilizing, decreasing retrogradation	44
Sauce or dressing	(CMC, MC, HPMC) + (GG, LBG, XG)	Thickening, stabilizing, desired texture	42
Sauce or dressing	Gum tragacanth + PGA, gum arabic or cellulose derivatives; gum tragacanth + XG and GG; gum karaya + gum arabic	Emulsifying, thickening, stabilizing	34
Ketchup and dressing	Starch, GG or GG + XG	Thickening, decreasing syneresis	33
O/W emulsion, mayonnaise	CAR + XG	Stabilizing, decreasing oil imitation, body, cling	85
Sterilized soup and sauce	GG + XG	Thickening	33
Dry mix with emulsion	XG + CMC, GG, tragacanth, karaya, or PGA	Pourable and stable viscosity, increasing preparation tolerance	86

starches coupling with gellan gum,[90] galactomannans–XG blends,[33] TSX,[51] or cellulose[95] are also reported to give low-calorie products with margarine-like spreadable texture and desired mouth feel. Commonly, the fat-mimetic properties of these polysaccharide blends arise from the presence of gel particles with a controlled size of 0.1 to 100 μm.[93,96] Other additional functions such as coating chocolate, control setting stability,[88] and absorbing lipophilic or hydrophilic materials[93] are found in the use of blends with galactomannans or konjac glucomannans.

TABLE 11.8
Polysaccharide Blends in Low- or Reduced-Calorie Desserts, Dressings, Emulsions, and Spreads

System	Polysaccharide	Function	References
Reduced-fat dispersion	MCC aggregate, GG or LBG-konjac mixture	Chocolate coating, inclusions, setting stability	88
Low-calorie O/W emulsion	Low η-MCC/CMC	Bulking, thixotropic	57
Reduced-fat peanut butter	Modified starch + MD	Fat replacers	89
Low-fat dairy spread	MD or gelatin + inulin	Increasing spreadability, fat-like texture, stabilizing	60
No-fat gellan gum spread	Non-gelling starch + gellan gum	Margarine-like texture, mouth feel and spreadability	90
Coconut milks and dessert	Cereal β-glucans + soy flour	Bulking, fat-like texture	50
Reduced-fat frozen dessert, dry mix, instant food	Medium η-colloidal MCC + CAR–GG or + MD–XG	Bulking, stabilizing, fat-like texture	57
Reduced-fat processed cheese	MCC–GG aggregates + CAR	Bulking, stabilizing, fat-like texture	57
Frozen dessert, salad dressing	Colloidal MCC–GG particulates	Fat-like consistency, creamy mouth feel, increasing body and opacity	91
Low-fat mayonnaise and salad dressing	MCC–GG particulate + XG + modified waxy maize starch	Bulking, full-fat body, mouth feel, and creaminess	92
Dry mix no-oil creamy Italian dressing	Cereal β-glucans + XG + GG	Bulking, oil-like texture	50
Low- or reduced-calorie mayonnaise	Starch + galactomannans + XG	Desired texture, stabilizing	33
Fat-reduced dressing and mayonnaise product	Tamarind seed xyloglucan (TSX) + XG, starches or dextrins	Bulking, fat-like texture	51
Reduced-fat food	Colloidal MCC/CMC	Bulking, stabilizing	57
Low-fat food	MCC + glucomannan	Bulking, absorbing lipophilic and/or hydrophilic materials	93

11.3.3 PROCESSED MEAT PRODUCTS

Low-fat formulations for processed meat products, especially frankfurter sausages, bologna, and surimi gels have been intensively developed. Fat reduction generally results in reduced water holding capacity; visual density and sensory preference; and increased cohesiveness, gumminess, chewiness, and cooking loss or purge loss for processed meat products.[63,100] Starch-based fat replacers are often used at preferably ~5 wt% to compensate for most of these shortages. CAR[101] and konjac gels[56] are also reported to have excellent functions in low-fat meat products. The incorporation of agar or CAR can increase moisture retention, cooking yield, slicing property, and mouth feel of aspic products (canned meat preserves).[30,40] Commonly, these fat replacers possess some good water-binding ability, gelation, stabilization, or protein-binding property. However, use of a pure polysaccharide is often unable to satisfy all requirements for developing a desired low-fat meat product. The fat contents of the meat products (sausages or frankfurters) with acceptable texture can only be reduced to about 7 to 11%.[56,63] Fat replacement by using a single polysaccharide is more limited in emulsified pork products such as kuang-wan.[101] Accordingly, polysaccharide blends are attractive as in the case of low-calorie foodstuffs.

Table 11.9 displays that blends of κ- and ι-CAR or LBG in meat broth and fish cans or jars create a gelling texture with benefits of suspending seasonings and preserving flavors.[77,82] Similar blend formulations[40,44] and XG-galactomannan mixes[44] are also used for canned pet foods because of the thickening, suspending, gelling, and stabilizing effects. In pastry filling, fat-free frankfurter sausage, surimi products,[39] or sardine mince gels,[64] blends of konjac flour or starch with a minor amount of CAR or XG can create desired elastic and cohesive textures that are cost effective, due to the polysaccharides synergy involved[5] or the use of starch. Furthermore, additional functions such as soluble dietary fiber and fat replacers can be found in low-fat comminuted meat products incorporated with blends of cereal hydrolyzates and CAR or XG–LBG mixture,[97] and in low-fat pork sausages with blends of CAR/whey protein isolate (WPI) performed gels and starch.[65] Generally, use of triple polysaccharide components can produce a sausage product with an acceptable texture and the fat content reduced to less than 3%.[65,97]

11.3.4 OTHER PRODUCTS

Polysaccharide blends are also widely applied in structured and encapsulated products as well as those that require to be coated with edible films. Table 11.10 lists some examples. Generally, agars with galactomannans and curdlan that exhibit thermoreversible gelation are major components for preparation of multiplayer sweetened confectionery,[62] due to their excellent gelling abilities[29,30] and thermally controllable gelation behavior,[103] respectively. The resultant gel layers have high adhesiveness and identical texture. Because of its excellent gelling and texturizing ability, agar can also be applied for gelled texturized fruits.[62] On the other hand, restructured blackcurrant, onion ring, and pimiento olive filling can

TABLE 11.9
Polysaccharide Blends in Processed Meat Products

System	Polysaccharide	Function	References
Meat broth, fish can or jar	κ + ι-CAR, κ-CAR + LBG	Gelling, suspending seasonings, preserving flavors	77, 82
Pastry filling, fat-free frankfurter sausage	Konjac flour + CAR or XG	With synergy, gives brittle–elastic texture, decreasing cost	39
Low-fat comminuted meat product	Cereal hydrolyzates + hydrocolloids (e.g., CAR, XG-LBG)	Soluble dietary fibers, fat replacers	97
Low-fat pork sausage	CAR/WPC performed gels + starch	Increasing fatty texture for final product (<3%)	65
Low-fat bologna model system	Konjac blends (KF + κ-CAR and/or starch)	Increasing moisture, cooking yield and texture	98, 99
Surimi product	Konjac flour + CAR or XG	With synergy, giving brittle–elastic texture, decreasing cost	39
Sardine mince gel	Starch + ι-CAR	Increasing elasticity, cohesiveness, folding scores	64
Pet food	XG + GG or LBG	Thickening, stabilizing, increasing smooth texture	44
Canned pet food	ι-CAR + GG; κ-CAR + LBG or GG	Thickening, suspending, gelling, stabilizing fat	40, 44

be produced by using alginate as a texturizing agent and cross-linked potato starch–CMC, wheat flour, or GG as partners.[36,37] Selection of salts (NaCl or $CaCl_2$) as well as ionic strength are important for alginate texturization.[36,37] Incorporation of chitosan produces calcium-induced alginate gel beads that can control nicotinic acid release and absorb bile acids.[67] For biodelivery adhesive systems,[62] herbal medicate capsules,[68] and other structured or extruded products,[42] cellulose derivatives play a key role in carrying, protecting, encapsulating, structuring, bulking, and coating functions. In low-fat or fat-free snacks[17] and flavor or oil-encapsulated powders,[23] mixtures of modified starch, starch hydrolyzate (including MD), or their hydrogenated products are often employed with great cost effectiveness. The edible films formed by agar–starch blends are used to protect gels and powdered medicines.[29,102]

TABLE 11.10
Polysaccharide Blends in Some Structured, Encapsulated, and Coated Systems

System	Polysaccharide	Function	References
Multilayer sweetened confectionery	Agar, starch, curdlan, agar + galactomannan, XG, CAR, konjac mannan	Gelling, adhesiveness	62
Gelled texturized fruit	Agar + LBG + fruit pulps	Gelling, texturizing	62
Restructured blackcurrant	Outer high-M alginate and inner –cross-linked potato starch + CMC	Outer and inner layer with similar viscosities	36
Restructured onion ring	Alginate + wheat flour + NaCl	Texturizing, increasing adhesiveness, retaining shape	37
Restructured pimiento olive filling	Alginate + GG + CaCl$_2$	Texturizing	37
Gel beads with nicotinic acid	Ca-induced alginate gel + chitosan	With controlled releasing ability and absorbing bile acids	67
Biodelivery adhesive systems	(1) HPC, HPMC, gum karaya, PGA 400; (2) EC, CMC, HPC; (3) agar	Carrying and protecting bioactive materials	62
Herbal medicate capsules	HPMC, other cellulose derivatives + gelatin	Encapsulation	68
Structured, extruded, coated products	CMC, MC, HPMC	Structuring, bulking, coating	42
Coated low-fat and fat-free snacks (pretzels)	Modified starch + MD + corn syrup solids	Increasing texture, taste, flavor, color, shiny surface	17
Encapsulation shell materials	Hydrogenated starch hydrolyzate + MD	Encapsulation	23
Oblate edible paper for Yokan (agar hard gel), jelly, powdered medicines	Agar + starch	Warping and protecting	29, 102

REFERENCES

1. Slade, L. and Levine, H., Mono- and disaccharides: selected physicochemical and functional aspects, in *Carbohydrates in Food*, Eliasson, A.-C., Ed., Marcel Dekker, New York, 1996, p. 41.
2. Glicksman, M., Food applications of gums, in *Food Carbohydrates*, Lineback, D.R. and Inglett, G.E., Eds., AVI Publishing, Connecticut, 1982, p. 270.
3. Williams, P.A. and Phillips G.O., Introduction to food hydrocolloids, in *Handbook of Hydrocolloids,* Phillips, G.O. and Williams, P. A., Eds., Woodhead Publishing, Cambridge, 2000, p. 1.
4. Doublier, J.-L. and Cuvelier, G., Gums and hydrocolloids: functional aspects, in *Carbohydrates in Food*, Eliasson, A.-C., Ed., Marcel Dekker, New York, 1996, p. 283.
5. Morris, E.R., Mixed polymer gels, in *Food Gels*, Harris, P., Ed., Elsevier Applied Science, London, 1990, p. 291.
6. Bollenback, G.N., Sucrose and health, in *Food Carbohydrates,* Lineback, D.R. and Inglett, G.E., Eds., AVI Publishing, Connecticut, 1982, p. 62.
7. Hanover, L.M., Crystalline fructose: production, properties, and applications, in *Starch Hydrolysis Products*, Schenck, F.W. and Hebeda, R.E., Eds., VCH Publishers, New York, 1992, p. 201.
8. Rudan, B.J. et al., U.S. Patent 5 366 754, 1995.
9. Cherukuri, S.R., Vink, W., and Friello, D.R., U.S. Patent 4 271 199, 1981.
10. Cherukuri, S.R et al., U.S. Patent 4 271 198, 1981.
11. Christen, F. and Kracher, F. U.S. Patent 4 804 544,1989.
12. Slade, L. and Levine, H., Non-equilibrium melting of native granular starch. Part I: Temperature location of the glass transition associated with gelatinization of A-type cereal starches, *Carbohydr. Polym.*, 8, 183, 1988.
13. Lii, C.-y. et al., Influences of polyols on thermal and dynamic viscoelastic properties of rice starches during gelatinization, *Starch/Stärke*, 49, 346, 1997.
14. Lii, C.-y., Lai, M.-F., and Liu, K.-F., Factors influencing the retrogradation of two rice starches in low-molecular-weight saccharide solutions, *J. Cereal Sci.*, 28, 175, 1998.
15. Cesàro, A. and Sussich, F., Plasticization: the softening of materials, in *Bread Staling*, Chinachoti, P. and Vodovotz, Y., Eds., CRC Press, Boca Raton, 2001, p. 19.
16. Alexander, R.J., Maltodextrins: production, properties, and applications, in *Starch Hydrolysis Products*, Schenck, F.W. and Hebeda, R.E., Eds., VCH Publishers, New York, 1992, p. 233.
17. Lanner, D.A. et al., U.S. Patent 6 352 732, 2002.
18. Levine, H. and Slade, L., An alternative view of trehalose functionality in drying and stabilization of biological materials, *BioPharm*, 5(4), 36, 1992.
19. Emodi, A., Polyols: chemistry and application, in *Food Carbohydrates*, Lineback, D.R. and Inglett, G.E., Eds., AVI Publishing, Connecticut, 1982, p. 49.
20. Voirol, F.A., Xylitol, its properties and applications, in *Sugar: Science and Technology*, Birch, G.G. and Parker, K.J., Eds., Applied Science Publishers, London, 1979, p. 325.
21. Craig, S.A.S. et al., Polydextrose as soluble fiber and complex carbohydrate, in *Complex Carbohydrates in Food*, Cho, S.S., Prosky, L., and Dreher, M., Eds., Marcel Dekker, New York, 1999, p. 229.
22. Cho, S.S. and Prosky, L., Application of complex carbohydrates to food product fat mimetics, in *Complex Carbohydrates in Food*, Cho, S.S., Prosky, L., and Dreher, M., Eds, Marcel Dekker, New York, 1999, p. 411.

23. Subramaniam, A., U.S. Patent 5 506 353, 1997.
24. Huang, R.-M., Yeh, S.-C., and Lii, C.-y., Investigation on the optimizing formulation of low-sugar chiffon cake, *Taiwanese J. Agric. Chem. Food Sci.*, 38, 445, 2000.
25. Hedges, A.R., Cyclodextrin: production, properties, and applications, in *Starch Hydrolysis Products*, Schenck, F.W. and Hebeda, R.E., Eds., VCH Publishers, New York, 1992, p. 319.
26. Yin, Y., Walker, C.E., and Deffenbaugh, L.B., Emulsification properties of sugar esters, in *Carbohydrate Polyesters as Fat Substitutes*, Akoh, C.C. and Swanson, B.G., Eds., Marcel Dekker, New York, 1994, p. 111.
27. Pomeranz, Y., Sucrose esters in baked products, in *Carbohydrate Polyesters as Fat Substitutes*, Akoh, C.C. and Swanson, B.G., Eds., Marcel Dekker, New York, 1994, p. 137.
28. Glueck, C.J., Sucrose polyester, cholesterol, and lipoprotein metabolism, in *Carbohydrate Polyesters as Fat Substitutes*, Akoh, C.C. and Swanson, B.G., Eds., Marcel Dekker, New York, 1994, p. 169.
29. Matsuhashi, T., Agar, in *Food Gels*, Harris, P., Ed., Elsevier Applied Science, London, 1990, p. 1.
30. Armisén, R., Agar, in *Thickening and Gelling Agents for Food*, 2nd ed., Imeson, A., Ed., Blackie Academic & Professional, London, 1997, p. 1.
31. Rolin, C. and de Vries, J., Pectin, in *Food Gels*, Harris, P., Ed., Elsevier Applied Science, London, 1990, p. 401.
32. May, C.D., Pectins, in *Thickening and Gelling Agents for Food*, 2nd ed., Imeson, A., Ed., Blackie Academic & Professional, London,1997, p. 230.
33. Fox, J.E., Seed gums, in *Thickening and Gelling Agents for Food*, 2nd ed., Imeson, A., Ed., Blackie Academic & Professional, London, 1997, p. 262.
34. Wareing, M.V., Exudate gums, in *Thickening and Gelling Agents for Food*, 2nd ed., Imeson, A., Ed., Blackie Academic & Professional, London, 1997, p. 86.
35. Launay, B., Doublier, J. L., and Cuvelier, G., Flow properties of aqueous solutions and dispersions of polysaccharides, in *Functional Properties of Food Macromolecules*, Mitchell, J.R. and Ledward, D.A., Eds, Elsevier Applied Science, London, 1986, p. 1.
36. Sime, W.J., Alginates, in *Food Gels*, Harris, P., Ed., Elsevier Applied Science, London, 1990, p. 53.
37. Onsøyen, E., Alginates, in *Thickening and Gelling Agents for Food*, 2nd ed., Imeson, A., Ed., Blackie Academic & Professional, London, 1997, p. 22.
38. Stanley, N.F., Carrageenans, in *Food Gels*, Harris, P., Ed., Elsevier Applied Science, London, 1990, p. 79.
39. Thomas, W.R., Carrageenan, in *Thickening and Gelling Agents for Food*, 2nd ed., Imeson, A., Ed., Blackie Academic & Professional, London, 1997, p. 45.
40. Imeson. A. Carrageenan, in *Handbook of Hydrocolloids*, Phillips G.O. and Williams, P.A., Eds., Woodhead Publishing, Cambridge, 2000, p. 87.
41. Coffey, D.G., Bell, D.A., and Henderson, A., Cellulose and cellulose derivatives, in *Food Polysaccharides and Their Applications*, Stephen, M., Ed., Marcel Dekker, New York, 1995, p. 123.
42. Zecher, D. and Gerrish, T., Cellulose derivatives, in *Thickening and Gelling Agents for Food*, 2nd ed., Imeson, A., Ed., Blackie Academic & Professional, London, 1997, p. 60.
43. Seidel, W.C., Stahl, H.D., and Orozovich, G.E. U.S. Patent 4 303 451.
44. Urlacher, B. and Noble, O., Xanthan gum, in *Thickening and Gelling Agents for Food*, 2nd ed., Imeson, A., Ed., Blackie Academic & Professional, London, 1997, p. 284.

45. Wurzburg, O.B., Modified starches, in *Food Polysaccharides and Their Applications*, Stephen, M. Ed, Marcel Dekker, New York, 1995, p. 67.
46. Rapaille, A. and Vanhemelrijck, J., Modified starches, in *Thickening and Gelling Agents for Food*, 2nd ed., Imeson, A., Ed., Blackie Academic & Professional, London, 1997, p. 199.
47. Murphy, P., Starch, in *Handbook of Hydrocolloids*, Phillips G.O. and Williams, P.A., Eds., Woodhead Publishing, Cambridge, 2000, p. 41.
48. Blanchard, P.H. and Katz, F.R., Starch hydrolysates, in *Food Polysaccharides and Their Applications*, Stephen, M., Ed., Marcel Dekker, New York, 1995, p. 99.
49. Thomas, W.R., Konjac gum, in *Thickening and Gelling Agents for Food*, 2nd ed., Imeson, A., Ed., Blackie Academic & Professional, London, 1997, p. 169.
50. Morgan, K., Cereal β-glucans, in *Handbook of Hydrocolloids*, Phillips G.O. and Williams, P.A., Eds., Woodhead Publishing, Cambridge, 2000, p. 287.
51. Nishnari, K., Xyloglucan, in *Handbook of Hydrocolloids*, Phillips G.O. and Williams, P.A., Eds., Woodhead Publishing, Cambridge, 2000, p. 247.
52. Dreher, M., Food sources and uses of dietary fiber, in *Complex Carbohydrates in Food*, Cho, S.S., Prosky, L., and Dreher, M., Eds., Marcel Dekker, New York, 1999, p. 327.
53. Cho, S.S. and Bussey, M., Estimation of psyllium content in ready-to-eat cereals, in *Complex Carbohydrates in Food*, Cho, S.S., Prosky, L., and Dreher, M., Eds., Marcel Dekker, New York, 1999, p. 317.
54. Kincaid, J.G. and Talbot, M.W., U.S. Patent 5 382 443, 1996.
55. Roberfroid, M., Dietary fiber, inulin and oligofructose: a review comparing their physiological effects. *Crit. Rev. Food Sci. Nutr.*, 33, 103, 1993.
56. Osburn, W.N. and Keeton, J.T., Konjac flour gel as fat substitute in low-fat prerigor fresh pork sausage, *J. Food Sci.*, 59, 484, 1994.
57. Imeson, A.P. and Humphreys, W., Microcrystalline cellulose. In *Thickening and Gelling Agents for Food*, 2nd ed., Imeson, A., Ed., Blackie Academic & Professional, London, 1997, p. 180.
58. Würsch, P., Production of resistant starch, in *Complex Carbohydrates in Food*, Cho, S.S., Prosky, L., and Dreher, M., Eds., Marcel Dekker, New York, 1999, p. 385.
59. Iijima, H. and Takeo, K., Microcrystalline Cellulose: an overview, in *Handbook of Hydrocolloids*, Phillips G.O. and Williams, P.A., Eds., Woodhead Publishing, Cambridge, 2000, p. 331.
60. Sensus Operations CV, Frutafit®-inulin, in *Handbook of Hydrocolloids*, Phillips G.O. and Williams, P.A., Eds., Woodhead Publishing, Cambridge, 2000, p. 397.
61. Krochta, J.M., Baldwin, E.A., Nisperos-Carriedo, M.O., *Edible Coatings and Films to Improve Food Quality*, Technomic Publishing, Lancaster, Basel, 1994.
62. Nussinovitch, A., Gums for coatings and adhesives, in *Handbook of Hydrocolloids*, Phillips G.O. and Williams, P.A., Eds., Woodhead Publishing, Cambridge, 2000, p. 347.
63. Carballo, J. et al., Morphology and texture of bologna as related to content of fat, starch and egg white, *J. Food Sci.* 61, 652, 1996.
64. Gómez-Guillén, M.C. and Montero, P., Addition of hydrocolloids and non-muscle proteins to sardine (*Sardina pilchardus*) mince gels, *Food Chem.*, 56, 421, 1996.
65. Lyons, P.H. et al., The influence of added whey protein/carrageenan gels and tapioca starch on the textural properties of low fat pork sausages, *Meat Sci.*, 51, 43, 1999.
66. Winterowd, J.G. and Sandford, P.A., Chitin and chitosan, in *Food Polysaccharides and Their Applications*, Stephen, M., Ed., Marcel Dekker, New York, 1995, p. 441.

67. Murata, Y. et al., Preparation of alginate gel beads containing chitosan nicotinic acid salt and the functions, *Eur. J. Pharm. Biopharm.*, 48, 40, 1999.
68. Wang, X., U.S. Patent 6 238 696, 2001.
69. Chen, J.-J., Lai, V.M.-F., and Lii, C.-y., Effects of compositional and granular properties on the pasting viscosity of rice starch blends, *Starch/Stärke*, 55, 2003, in press.
70. Takayanagi, M., Harima, H., and Iwata, Y., Viscoelastic behavior of polymer blends and its comparison with model experiments, *Mem. Fac. Eng. Kyushu Univ.*, 23, 1, 1963.
71. Chen, J.-J., Lai, V.M.-F., and Lii, C.-y., Rheological properties of rice starch blends interpreted with polymer blending law, personal communication, 2003.
72. Aguilera, J.M. and Rojas, E., Rheological, thermal and microstructure properties of whey protein-cassava starch gels, *J. Food Sci.*, 61, 962, 1996.
73. Lii, C.-y., Lai, M.-F., and Tsai, M.-L., Studies on starch gelatinization and retrogradation with dynamic rheometry: the influence of starch granular structure and composition. *Food Technol. Qual. (Poland)*, 2, 27, 1996.
74. Toh, T.S., Europe Patent 0 738 473 A2, 1997.
75. Ylimaki, G. et al., Application of response surface methodology to the development of rice flour yeast breads: objective measurements, *J. Food Sci.*, 53, 1800, 1988.
76. Foda, Y.H. et al., Special bread for body weight control, *Ann. Agric. Sci.*, 32, 397, 1987.
77. Guiseley, K.B., Stanley, N.F., and Whitehous, P.A., Carrageenan, in *Handbook of Water-Soluble Gums and Resins*, Davidson, R.L.. Ed., McGraw-Hill, New York, 1980, p. 5-1.
78. Lee, E.C. and Tandy, J.S., U.S. Patent 5 370 882, 1995.
79. Kaji, N. et al., Europe Patent 0 715 812 A2, 1997.
80. Lawrence, J., Coutant, A.F., and Swayhoover, F., U.S. Patent 5 387 427, 1996.
81. Kawagoe, S., Cocoa drink, Japanese Patent 6 169737.
82. Davidson, R.L., *Handbook of Water-Soluble Gums*, McGraw-Hill, New York, 1980.
83. Anon., Cottage cheese dressing, in *Application Bulletin E-27*, Marine Colloids Division, FMC Corporation, 1988.
84. Modliszewski, J.J. and Ballard, A.D, U.S. Patent 5 498 436, 1996.
85. Anon., Imitation mayonnaise, in *Application Bulletin C-61*, Marine Colloids Division, FMC Corporation, 1988.
86. Cassanelli, R.R. et al., U.S. Patent 4 278 692, 1981.
87. Stanley, D.W., Goff, H.D., and Smith A.K., Texture–structure relationship in foamed dairy emulsions, *Food Res. Int.*, 29, 1, 1996.
88. Hartigan, S.E., Izzo, M.T., Stahl, C.A., and Tuazon, M.T., U.S. Patent 5 709896, 1998.
89. Franklin, K.K., U.S. Patent 5 302 409, 1995.
90. Chalupa, W.F. and Sanderson, G.R. U.S. Patent 5 534 286, 1997.
91. Novagel™ Cellulose Gel Fat Replacer, Food Ingredients Division, FMC Corporation, Philadelphia, 1992.
92. Bernadini, D.L. and Harkenbus, E.M., Europe Patent 0 477 827, 1991.
93. McGinley, E.J. and Tuason, D.C. U.S. Patent 5 462 761, 1996.
94. Okura, Y., Tawada, T., and Nakao, Y. U.S. Patent 5 360 624, 1995
95. Tang, P.S., WO Patent 96/11587 A1, 1997.
96. Hoefler, A.C., Sleap, J.A., and Trudso, J.E., U.S. Patent 5 324 531, 1995.
97. Jenkins, R.K. and Wild, J.L., U.S. Patent 5 294 457, 1995.
98. Chin, K.B. et al., Functional, textural and microstructural properties of low-fat bologna (model system) with a konjac blend, *J. Food Sci.*, 63, 801, 1998.

99. Chin, K.B. et al., Low-fat bologna in a model system with varying types and levels of konjac blends. *J. Food Sci.*, 63, 808, 1998.
100. Khalil, A.H., Quality characteristics of low-fat beef patties formulated with modified corn starch and water, *Food Chem.*, 68, 61, 2002.
101. Hsu, S.Y. and Chung, H.-Y., Comparisons of 13 edible gum-hydrate fat substitutes for low fat kung-wan (an emulsified meatball), *J. Food Eng.*, 40, 279, 1999.
102. Meer, W., Agar, in *Handbook of Water-Soluble Gums and Resins*, Davidson, R.L., Ed., McGraw-Hill, New York, 1980, p. 5-1.
103. Nishinari, K. and Zhang, H., Curdlan, in *Handbook of Hydrocolloids,* Phillips G.O. and Williams, P.A., Eds. Woodhead Publishing, Cambridge, UK, 2000, p. 269.

12 Pectic Polysaccharides

Alistair J. MacDougall and Stephen G. Ring

CONTENTS

12.1 Introduction 181
12.2 Pectin Structure 182
12.3 Extraction of Pectins 186
12.4 Pectin Modification 187
12.5 Pectin Conformations... 188
12.6 Pectin Networks 189
12.7 Concluding Remarks... 193
References 193

12.1 INTRODUCTION

In this chapter, we examine the relationship among the structure of pectic polysaccharides, their physical chemistry, and functional behavior relevant to food use. Pectins are found in the primary cell wall of higher plants[1–3] where they have a range of functions, including contributing to the mechanical properties of the cell wall and its hydration characteristics and acting as a barrier to the invasion of pathogens. Pectin structure shows variation with botanical origin, cell wall type, and cell wall development. Extracted pectins find use in the food industry as gelling, thickening, and film-forming agents. Commercially useful pectins come from a relatively limited range of materials such as apple pulp or citrus peel. It is inevitable that the extraction procedure modifies the polysaccharide to some extent, and it may be further modified, after extraction, by chemical or enzymatic means. Studies on pectin may have a number of motivations, including trying to explain its role in the plant cell wall or understanding its behavior as a food ingredient. Cell wall studies often focus on the complexity of pectin structure and its variation with plant type and cell development. Food-related studies are often more concerned with the functional behavior of extracted polysaccharides of somewhat simpler structure. In this chapter, we draw on literature from both areas, as both are relevant to understanding the relationship between pectin structure and its functionality. The functionality that forms the focus of this chapter is the contribution of pectic polysaccharides to mechanical properties of materials.

A difficulty with establishing structure–function relationships for pectic polysaccharides is the potential complexity of structure. To put the structural information

0-8493-1486-0/04/$0.00+$1.50
© 2004 by CRC Press LLC

into some sort of perspective, it is perhaps worthwhile initially to consider the structure from a somewhat more abstract point of view. A major component of pectic polysaccharides is the α-D-galacturonosyl unit and its methyl ester. The intrinsic pK of the uronic acid[4] is about 3, and as pH increases the uronic acid becomes charged and the pectin changes from a neutral polymer to a polyelectrolyte. This change can have an enormous effect on behavior. At low charge levels, the repulsion between charges expands the polymer coil in aqueous solution. If the pectin is present in a network, associated monovalent counterions can increase osmotic pressure and network hydration through a Donnan effect.[5] If charged divalent or polyvalent counterions are present, there is also the possibility for charge–charge interactions, leading to aggregated or network structures. At higher charge densities, even monovalent counterions can "condense" on the polymer backbone, leading to collapse of the polymer coil.[6,7] From a functional point of view, it is therefore very important to characterize the charge on the pectin molecule, and the way that charge is distributed. Another important aspect is branching of the pectin molecule. Pectin can contain neutral sugar side-chains, and the main backbone may be branched. Long chain branching of a polymer can have important effects on the rheology of concentrated systems, as the entanglement of the branches and the difficulty of separating these branched entanglements tends to increase the elasticity of the concentrated viscoelastic polymer solution.

Pectins also form network structures such as gels. In a gel, pectin chains are interconnected to form a three-dimensional network. The connectedness can be provided by secondary interactions at junction zones or perhaps by covalent cross-linking. In the synthetic polymer area, there is an abundant and established literature[5,8] on the effect of cross-linking on the stiffness of networks. In the theory of ideal rubber-like elasticity, the mechanism of energy storage is entropic. Deformation of the network restricts the number of conformations that the network chain can access, and, as a result, there is a restorative force. Real networks must also have energetic contributions to gel elasticity. However, it is still worthwhile to estimate the cross-link density that might be required to give a certain elasticity. This would seem to be particularly important for cell wall chemists who are searching for covalent cross-links in the pectin network of the plant cell wall. How sensitive should the methodology be? For a 2% w/w pectin gel formed by physical association with a modulus of 2000 N m^{-2}, it was estimated that there was on average one cross-link for every 130 or so residues.[9] Therefore, a relatively small number of cross-links are needed to account for the observed elasticity of a pectin gel. In the more concentrated cell wall network, a similar number of cross-links could provide a substantial contribution to the stiffness of the cell wall network. At this level of cross-linking, relatively sensitive chemical methods would be required to identify and characterize the chemical nature of the cross-link.

12.2 PECTIN STRUCTURE

Models of the primary cell wall of dicotyledons show cellulose microfibrils, coated with a hemicellulosic xyloglucan, dispersed in a pectic polysaccharide network.[1] In addition, a pectic polysaccharide network forms the middle lamella. For structural studies, pectic polysaccharides, or fragments of them, need to be extracted from the plant cell wall and middle lamella. The pectin network must be disrupted to enable extraction. This may involve extraction with calcium-chelating agents, dilute alkali, or dilute acid. Alternatively, pectic polysaccharide fragments can be released through the use of degrading enzymes. The extracted materials may then be subjected to detailed chemical analysis. The extracted materials will be polydisperse, having a range of molecular weights and sizes, and polymolecular, having a diversity of chemical structures. In most cases, pectins are extracted from plant materials with cells of different maturity and tissue type. This heterogeneity of cell type might be expected to introduce further heterogeneity into the polysaccharide population. Even within the cell wall from a single cell, there is a heterogeneity in pectin structure as assessed by antibody probes that recognize different structural motifs.[10] A further relevant aspect is how the cell wall material is produced prior to pectin extraction. Its method of isolation may cause chemical or enzymatic modifications of pectins in the cell wall. A common procedure for chemical studies is to use an alcohol insoluble residue (AIR), which precipitates cytoplasmic polymers onto the cell wall preparation. The cell wall preparation is then frequently dried. These treatments encourage chain association of the pectic polysaccharides and can render extraction more difficult. Consideration should also be given to the use of methodologies that solubilize cytoplasmic proteins with the purified cell wall material being maintained in a hydrated state.[9] Pectin chemistry is relatively complex, and the reader is referred to a number of excellent recent review articles which summarize the complexity of pectin structure and its variability.[2,3] It is not the intention in this chapter to cover all the possible polysaccharide structures that can be classified as pectic polysaccharides, but rather to focus on a few representative structures.

Both chemical and enzymatic procedures for isolation of pectic polysaccharides from cell wall materials involve some inevitable degradation. Although this may be relatively mild for some chemical procedures, it is still potentially relevant. At this stage we focus on the structural characterization of the extracted elements, and then go on to consider the impact of extraction procedures on the isolated pectins.

The major sugar residue found in pectins is D-galacturonic acid, a fraction of which is esterified with methanol. Other sugars commonly present include L-rhamnose, L-arabinose, D-galactose, and, to a lesser extent, D-xylose. A typical representation of pectin structure[2,11–13] consists of linear regions of $(1\rightarrow4)$-linked α-D-galacturonosyl residues, some of which are methyl esterified, interspersed with highly branched regions of a rhamnogalacturonan (RG-I). In some plant species such as soy and pea, the backbone of $(1\rightarrow4)$-linked α-D-galacturonosyl residues may contain single-unit side chains of β-D-xylopyranosyl residues linked to O-3 of the galacturonic acid.[14,15] The backbone α-D-galacturonosyl residues may also be partially esterified with acetic acid on C-2 or C-3, or both.

The rhamnogalacturonan region[2,3,11,12] consists of a backbone of alternating $(1\rightarrow2)$-linked α-L-rhamnopyranosyl and $(1\rightarrow4)$-linked α-D-galacturonosyl residues. A fraction of the rhamnosyl residues contain side chains linked to O3 or O4, or both. The side chains linked to rhamnose may vary in length from 1 to ~50 units and consist of short chains of D-galactopyranosyl and L-arabinofuranosyl residues, or a mixture of both. A range of linkages are found in these side chains, including

(1→4)- and (1→4,6)-linked D-galactose; (1→3)- and (1→3,6)-linked D-galactose; and (1→5)-, (1→3)-, and (1→3,5)-linked L-arabinose. In some models of pectin structure, the arabinose- and galactose-containing side chains are quite distinct. More recently,[16] it has been demonstrated that the (1→4)-linked galactopyranosyl-containing side chains of some pectins may be terminated by arabinopyranosyl residues and may contain (1→5)-linked arabinofuranosyl residues within the chain. In some plant families such as the Chenopodiaceae, of which the most investigated is the sugar beet (*Beta vulgaris*), these neutral sugar side-chains may also contain some ferulic and coumaric acids.

Another pectin fragment with a more conserved structure is also a rhamnogalacturonan (RG-II).[17–19] This consists of a backbone of at least seven (1→4)-linked α-D-galacturonosyl residues which carries four structurally different oligosaccharide side-chains. These side chains contain a number of uncommon sugars such as apiose and aceric acid. RG-II may be cross-linked as a 1:2 borate–diol ester formed between C2-OH and C3-OH of the apiosyl residues in each monomeric subunit. This demonstration of a chemical cross-link that exists in the primary cell wall of all higher plants is important in the understanding of cell wall growth. This is supported by recent observations on *Arabidopsis thaliana* mutants, where reduced cross-linking is associated with dwarfism.[17] It has been postulated that in the plant cell wall, the galacturonan, RG-I, and RG-II are covalently linked together,[3] and recent research has supported this suggestion, although clearly further research is needed. The proportion of various structural elements varies with cell wall type and plant species. In the apple cell wall, for example,[2] pectic polysaccharides comprise about 40% w/w of the cell wall material, of which ~10% consists of the RG-II fragment, ~40% the galacturonan (xylogalacturonan) backbone, ~4% the RG-I backbone.

This discussion was concerned with the chemical characterization of pectin structure. Recently, other methods such as atomic force microscopy (AFM) have been used to examine pectin structure.[20,21] In this case, the presence of long chain branches was observed. For pectin extracted from unripe tomato with a chelating agent, the mean backbone length was 350 nm, corresponding to about 800 monomer residues. About 30% of the individual polymers showed the presence of long chain branches, with an average branch length of 150 nm, or about 350 monomer residues. The branching observed by AFM did not correlate with the observed neutral sugar content. Therefore, it was proposed that the observed branching was of the backbone galacturonic acid segments. This level of branching of the backbone would be very difficult to detect by chemical techniques.

In summary, pectic polysaccharides are very complex polysaccharides. It seems premature at this stage to attempt to produce a single representative model of their structure. Various representations of structure are available, and the reader is referred to the relevant articles for a more detailed discussion.[2,3] There is broad agreement that the main structural elements are a homogalacturonan (or xylogalacturonan), an RG- I, and an RG-II.

The isolated materials find industrial use, and it is important to consider the impact of the isolation procedures on the characteristics of isolated materials. Cell wall studies have often used sequential procedures of increasing chemical severity to isolate the polysaccharides in an attempt to understand their role in plant cell wall organization. Procedures for the production of pectins for commercial use have to maximize the yield of useful polymer, and these procedures may also introduce chemical degradation. Various chemical degradations can be envisaged. At low pH, there is the possibility of acid hydrolysis of relatively labile glycosyl linkages such as arabinofuranosyl residues. At higher pH, there is the possibility of deesterification both of the methyl ester of galacturonic acid and potential ester linkages that may be involved in cross-linking the pectin network. At pH values in the region of neutrality and above, a β-eliminative degradation, involving the methyl ester of galacturonosyl residues, occurs, which results in chain cleavage. Another potential degradation that may be important for the isolation of pectic polysaccharides is degradation of the borate ester cross-link involving RG-II.[18]

An important structural characteristic of both cell wall and isolated pectins is the way that charge is distributed along the galacturonic acid backbone. The charge may be distributed randomly or it could occur in blocks of a certain average length. This distinction is not necessarily clear cut. For example, even if a low-methoxyl pectin was randomly esterified, there would be blocks of galacturonosyl residues present. Recently, methodologies have been developed that attempt to measure charge distribution. In one approach,[13,22–24] an endopolygalacturonase is used to degrade pectin and the composition of the degradation products determined with high performance anion exchange chromatography. The endopolygalacturonase requires a sufficient number of contiguous uronic acid residues to permit enzymolysis. The approach enables determination of the percentage of nonesterified residues liberated as mono-, di-, and trimers, and characterization of the amount of oligomers containing a mixture of esterified and free uronic acid. From this information and the known degree of methyl esterification, the sequence similarity of charge distribution in different pectins can be compared and the characteristics of the charge distribution discerned, that is, the tendency to a random distribution or a blockwise distribution. Direct examination of the residue sequence in pectins has been carried out by ^{1}H NMR[25] to identify the occurrence of specific diads, triads, and tetrads within the overall pattern of esterification. Spectra from pectins deesterified with alkali and pectins deesterified enzymatically have been compared and clear differences can be seen in the relative abundance of particular sequences, notably the unesterified tetrad. This approach is to some extent complimentary to the polygalacturonase degradation method discussed previously as it highlights changes in the relative abundance of short-residue sequences. However, neither method is able to satisfactorily provide an insight into the relative abundance of the more extended sequences of unesterified residues that have such a strong effect on the overall physical behavior of the pectin molecule.

Alternative approaches use chemical degradation procedures. Mort[26] reduced the methyl-esterified galacturonic acid residues of pectin to galactose, which were then selectively removed by hydrolysis with anhydrous hydrogen fluoride to produce unesterified regions of galacturonic acid. Because of the use of highly corrosive hydrogen fluoride, this procedure has not been widely used. In another approach, Needs[27] converted the methyl ester of uronic acid into hydroxamic acids. Following treatment with a carbodiimide, the product undergoes a Lossen rearrangement to an isocyanate. Following its hydrolysis, ring opening, and subsequent reduction, there is chain cleavage at esterified uronic acid residues to form galacturonic acid oligomers terminated by an arabinitol residue. Subsequent analysis of the galacturonic acid containing fragments by high performance anion exchange chromatography directly gives the distribution of unesterified uronic acid sequences. Coupling the chromatography to electrospray-ionization mass spectroscopy provides confirmatory evidence on structure of the oligomers produced. Obtaining this structural information is a key step that should enable the refinement of structure–function relationships which are sensitive to polymer charge and its distribution.

12.3 EXTRACTION OF PECTINS

Procedures for the extraction of pectic polysaccharides from cell wall materials have developed in a semiempirical way, without necessarily a very detailed understanding of the underlying physical and chemical mechanisms involved in the extraction procedure. For example, suppose that it has been possible to isolate a cell wall material with a minimum of physical and chemical modification. One common cross-link involved in the pectin network of the plant cell wall involves calcium counterions. If a powerful chelating was available, then this might potentially disrupt the cross-link and enable the solubilization of pectic polysaccharides. One such powerful Ca^{2+} chelating agent is cyclohexane diamine tetraacetic acid (CDTA). Treatment of parenchymatous tissues, such as tomato pericarp, with an aqueous solution of CDTA at neutral pH causes cell separation, and it is possible to solubilize a pectin fraction from a cell wall material at room temperature.[9] CDTA extracts only a fraction of the pectin when used at neutral pH. It becomes a much more effective calcium-chelating agent as pH is increased, but it is surprising that it has rarely been used as a pectin extractant at higher pHs. Although CDTA is popular as a pectin extractant, it is worth remembering that its effect on the plant cell wall is unlikely to be limited to the disruption of calcium-mediated ionic cross-linking. Gel-forming interactions between pectin and basic peptides have been shown to be disrupted by CDTA.[28] Even at neutral pH and room temperature, CDTA might cause some degradation of the borate cross-link of RG-II.

In cell wall studies, increasing concentrations of alkali are then used to extract a series of pectin fractions and hemicellulosic polysaccharides. The key linkages or associations disrupted in this alkali treatment are not known, but it has been suggested that ester linkages are included.

For commercial extractions of pectins, acid conditions are typically used from pH 1 to 3 and at elevated temperature of 50 to 90°C.[2] These acidic conditions inevitably result in some hydrolysis of glycosyl linkages, including the hydrolysis of the rhamnosyl linkage in the RG-I backbone and the more labile arabinofuransyl linkages, galacturonosyl sequences being relatively resistant to hydrolysis. It has been shown that the borate ester of RG-II is hydrolyzed under relatively mild acidic conditions, being completely hydrolyzed within 30 min at room temperature at pH 1.[18] A further consequence of the acidic extraction conditions is the suppression of charge on the uronic acid and the solubilization of the associated calcium counterions. Therefore, the acidic conditions used disrupt a range of linkages and associations. Unfortunately, pectin solubility is reduced at acidic pH, which is why presumably elevated temperatures are required to increase the effectiveness of water as a solvent for the polysaccharide.

Commercial acid-extracted citrus pectins typically have a uronic acid content of 85–90%, a neutral sugar content of less than 10%, and a degree of esterification of 55–80%.[2] In comparison, careful purification of cell wall material and extraction with CDTA yields pectins with 80% uronic acid, 15% neutral sugars, and a degree of methyl esterification of 68%.[29] Acid extraction appears to cause some degradation of the pectin, but the overall chemical composition is broadly similar to that of the native material. In contrast, in a recent study of pectin extraction from sugar beet pulp,[30] acid-extracted pectins had a uronic acid content of 30 to 52%, depending on the conditions of pH and temperature used. Decreasing pH led to the extraction of pectins richer in RG-I. The difference observed between acid-extracted citrus and sugar beet pectins clearly reflects differences in pectin composition of the starting material rather than variability in method of extraction.

12.4 PECTIN MODIFICATION

The ester content of pectin may be modified through the use of acid or alkali to deesterify the pectin. If ammonia is used as the alkali, some of the uronic acid residues become amides, and an amidated pectin is formed.[2] Chemical deesterification occurs in a random way. Pectin may also be deesterified with pectin methyl esterases. Fungal enzymes have a random attack pattern, whereas plant pectin methyl esterases tend to result in a block distribution of deesterified uronic acid residues.[4,31] This categorization may require some modification in the future, as in higher plants there is a whole range of pectin methyl esterases whose mode of action has not been fully characterized. The conditions of deesterification are also important. For an apple pectin methyl esterase,[31] a blockwise distribution of uronic acid residues was generated. At pH 7, only a fraction of the chains in the pectin polysaccharide preparation were modified in this way. At pH 4.5, shorter blocks were produced in the entire chain population. One of the reasons for the interest in different methods of deesterification is that pectins produced should behave differently in terms of their interactions with counterions. It is suggested that blockiness should favor network formation, whereas more randomly distributed charge might have a more pronounced effect on the affinity of the polysaccharide for water. The combination of methodologies for the production of different patterns of esterification and those for characterizing the pattern of charge distribution in pectins should help provide structure–function relationships for patterns of charge distribution.

The availability of different pectin methyl esterases and enzymes for selective cleavage of neutral sugar pectin side chains makes it possible to produce a range of pectins with different functional behavior. For example,[32] it was found that removal of side chains decreased the viscosity of semidilute aqueous solutions of pectins in the absence of calcium, whereas the use of pectin methyl esterase enhance gel properties.

12.5   PECTIN CONFORMATIONS

From a physicochemical point of view, it is of interest to know something of both the solution and solid-state conformations of the pectic polysaccharides. As pectin is potentially such a complex macromolecule, it is not possible to deal exhaustively with all the possible conformations of the main backbone of the pectin molecule and that of the side chains. The reader is referred to an excellent review on the modeling of conformations of the pectic polysaccharides.[33] In this chapter we focus on the conformation of the backbone of pectic polysaccharides.

From the point of view of solution behavior, it is of interest to know something of the relationship between molecular weight and molecular size and how it is affected by solution conditions such as pH and ionic strength. For the rheological behavior of polymer solutions, an important parameter is the concentration at which the polymer chains in solution become entangled.[34] On increasing concentration, the progressive entanglement of polymer coils changes rheological behavior. The dependence of viscosity on concentration becomes more marked, and the temporary cross-links formed through entanglement contribute to the transient elasticity of the now viscoelastic polymer solution. In addition, if the solution conditions are changed such that parts of a polymer chain can form more permanent intermolecular associations, there is the possibility of network formation and formation of cross-linked elastic gels and films. What is the molecular size in solution of the pectin molecule? Indications can be gained from examining the dilute solution viscosity as a function of concentration to obtain an intrinsic viscosity on extrapolation to zero concentration. The intrinsic viscosities of pectins in solution[9,35] can approach 1000 ml g^{-1}, and even for pectins extracted under acidic conditions, intrinsic viscosities in the region of 400 ml g^{-1} are found.[30] These are very high values given the molecular weight of the pectin. If we consider a neutral linear polysaccharide such as the starch polysaccharide amylose,[36] then intrinsic viscosities of 100 ml g^{-1} are observed for materials with a molecular weight of 8×10^5 (4500 residues per molecule on average). Typical molecular weights for pectin are much lower than this, in the region of 1×10^5 (500 residues per molecule on average). High values of intrinsic viscosity for a given molecular weight, compared with those of amylose, indicate that the pectin molecule has an extended conformation in solution. Interpretation of the relationship between pectin structure and solution viscosity is, however, complicated by the fact that pectin is not a linear molecule but carries a proportion of neutral sugar side-chains, which varies with source of the pectin. In general, branching may be expected to lower molecular size in solution. The complexity of structure means that no simple relationship between molecular size (in solution) and molecular weight exists for pectins. As an indication of the variation that can be encountered,[30] fractions of sugar beet pectin with an intrinsic viscosity of 454 and 293 ml g^{-1} had molecular weights of 1.28×10^5 and 3.77×10^5, respectively; that is, the lower-molecular-weight fraction had the highest intrinsic viscosity. Compared to the neutral polysaccharide amylose, for a given molecular weight, the intrinsic viscosities obtained are very high, indicating that the pectin chain in solution has an expanded or extended conformation. Molecular modeling[33] studies predict that the galacturonic acid backbone of pectin has an extended chain confirmation with a persistence length (a measure of the flexibility of the polymer chain) of about 10 nm or 20 monomer units. For amylose, the corresponding figure is about 5 monomer units, and clearly shows that pectin has a more extended conformation. The introduction of methyl esterification in the galacturonic acid model had little effect on chain extension. Similarly, insertion of L-rhamnosyl residues in the chain has a relatively small effect on predicted chain dimensions.

In the solid state, a number of chain conformations are possible. To investigate experimentally these conformations, it is necessary to crystallize the galacturonic acid chain. Analysis of fiber diffraction data on Ca^{2+} and Na^+ pectate gels yield models of threefold right-handed helices with repeating lengths of 1.3 nm.[37,38] Powder diffraction data of highly crystalline Na^+ and K^+ oligogalacturonates have also been recently obtained, which were prepared by crystallization from a relatively dilute solution (0.5% w/w).[39] Interestingly, by this approach, only poorly crystalline forms of Ca^{2+} oligogalacturonate were obtained. These observations suggest the following. First, in pectin solutions with a sufficient concentration of oligogalacturonate sequences in the pectin backbone, their interaction with monovalent counterions can lead to chain association in microcrystallites. Second, Ca^{2+} has such a strong interaction with the oligogalacturonate sequence that it acts as a very effective precipitant. Oligomers and polymers tend to produce highly crystalline materials at relatively modest quench below the crystalline melting temperature, T_m. As the Ca^{2+} oligogalacturonate is poorly crystalline, this probably indicates that the depth of quench is very large, and as a consequence there is a large amount of amorphous material in a kinetically arrested form. On the basis of other experimental approaches,[40] including circular dichroism, it was proposed that in calcium pectate gels the galacturonate backbone forms a twofold helix having a repeat of 0.870 nm. This conformation is the well-known "egg-box," which is frequently encountered in schematic representations of pectin gel structure. Molecular modeling studies indicate that the various helical forms are almost equally favoured.[33] It is somewhat unusual that the proposed structure for the solid-state conformation of oligogalacturonate sequences in a gel is different from that obtained in the crystalline solid.

12.6 PECTIN NETWORKS

A characteristic of pectins is their ability to form network structures. Depending on the fraction of methyl ester, pectins are classified as either high methoxyl (>50%) or low methoxyl (<50%). A classical textbook view is that high-methoxyl pectins gel at acid pH (less than 3.5) in the presence of sugar. Low-methoxyl pectins, on the other hand, gel at higher pH in the presence of some divalent counterions, of which the most relevant is Ca^{2+}. Gels are produced when the polymer chains interact to form a continuous three-dimensional polymer network within which the solvent water is held. If the interactions are sufficiently permanent, a free-standing elastic solid is formed. To understand these different pectin gels, it is necessary to find out more about the factors that affect chain association. Some association is needed to form the network; however, it might be expected that more extensive association leads to the formation of a collapsed structure or precipitate.

For high-methoxyl pectins at low pH, the pectin molecule becomes essentially neutral. The addition of sucrose promotes gel formation. What physicochemical factors might be involved in promoting the necessary chain associations? One possible starting point is to examine predictive relationships describing the composition dependence of polymer melting.[5] The classical description of the compositional dependence of polymer melting in the presence of a diluent is given by:

$$\frac{1}{T_m} = \frac{1}{T_m^0} + (R / \Delta H_u) (V_u / V_1) [v_1 - \chi v_1^2] \tag{12.1}$$

where T_m^0 is the melting temperature of the pure polymer; V_u and V_1 are the molar volumes of polymer repeating unit and diluent, respectively; and v_1 is the diluent volume fraction. The focus on molar volumes comes from the use of a lattice model to represent the entropy of mixing. ΔH_u is the enthalpy of fusion per repeating unit and χ is the Flory–Huggins interaction parameter characterizing the interaction energy per solvent molecule.

This relationship predicts that the smaller the diluent size relative to that of the polymer repeating unit and the stronger the favorable interaction between the diluent and polymer, the stronger the depression in T_m. The relatively small molecular size of water is predicted to lead to a strong depression in T_m. In the case of the high-methoxyl pectin in solution, marked chain association does not occur in the absence of sucrose. The replacement of some of the water by sucrose produces a ternary system, the single interaction parameter χ is replaced by three interaction parameters describing the interactions between water and polymer, water and low-molecular-weight solute (sucrose), and polymer and low-molecular-weight solute. Although imbalance in the interactions between the various components can promote phase separation in the polymer solutions, in carbohydrate systems these effects are likely to be small. The strongest effect would appear to come from replacing water, a relatively small molecule, with sucrose a very much larger one. The model would predict that this replacement would elevate the T_m of the pectin chain association (whatever that is) until eventually the association is sufficiently stable to form large clusters of pectin chains and eventually kinetic arrest through the formation of the gel network.[41] It would be interesting to investigate the effect of molecular size of the added solute on the observed gelation behavior.

For low-methoxyl pectin gels, the interaction with counterions has an important effect on the physicochemical properties of the gel. Of the counterions, the interaction with Ca^{2+} has received the most attention, although interactions with monovalent and polyvalent counterions are also potentially relevant. Different effects may be produced. For example, the counterion may lead to cross-linking of the pectic polysaccharide network; on the other hand, counterions within the gel can generate an osmotic pressure that tends to swell and expand the polymer network. What then is the physicochemical basis for these effects?

One approach is to measure the affinity of the polymer chain for the counterion[42-44] and describe that affinity by calculation of a stability constant K.[42] For Ca^{2+} ions interacting with a pectin, the equilibrium can be quantitatively described by

$$K = [Ca^{2+}2COO^-]/[Ca^{2+}][2COO^-] \qquad (12.2)$$

where, following convention and the observed stoichiometry of binding, one Ca^{2+} ion is considered to bind to a fragment of pectin containing two carboxyl functions. This is inevitably somewhat of a simplification. It is unlikely that the uronic acid sites on the chain are independent, and the presence of uronic acid sequences of different length will lead to a spread of affinities. In addition, anticooperative or cooperative behavior may be observed depending on ionic strength. Nevertheless, the approach has much to recommend it. The change in stability constant with both degree of methyl esterification and galacturonate chain length has been determined.[42] The stability constant increases with decrease in degree of methyl esterification. For a pectin with a degree of methyl esterification of 65%, the interpolated value of log K is ~2.4. This constant describes the binding in solution, which can involve both intra- and intermolecular associations. For a tomato pectin[45] with a carbohydrate composition of galactose 10%, arabinose 3%, rhamnose 0.7%, galacturonic acid 71% w/w, and degree of methyl esterification 68%, the Ca^{2+}-binding behavior in aqueous solution is comparable to that of other pectins (log K ~ 2.7) with a similar degree of methyl esterification. As well as interacting with a single chain, the interaction with Ca^{2+} can cause interchain association and the formation of network structures. The binding of Ca^{2+} in a tomato pectin network was determined by following the dissolution of a pectin gel in water, and log K ~ 3.9 was found. This is more than an order of magnitude greater than the comparable solution case, indicating that a cross-link of sufficient permanence to create an elastic gel has a higher affinity for Ca^{2+} than the chain in solution. In addition, the observation that a high-methoxyl pectin, degree of methyl esterification 68%, can form a gel in the presence of Ca^{2+} counterions indicates that the assertion that only low-methoxyl pectins gel in the presence of Ca^{2+} is too much of a simplification. Current research is also examining the effect of charge distribution on the Ca^{2+}-binding behavior, with clear differences in the interaction between pectin and the counterion being observed for pectins chemically deesterified (random distribution of charge) and pectins deesterified with a plant pectin methyl esterase (blockwise distribution of charge).[4] The interaction of pectins with Ca^{2+} can lead to network formation; if this interaction is very extensive, then collapse of the polymer network becomes a possibility. This can be observed as a slow, time-dependent shrinkage of the polymer gel, with the separation of a watery layer at the gel surface, a process known as syneresis. On the other hand, one of the characteristics of some pectin gels is to swell and take up water.

The physicochemical basis of hydration of polymer networks has been the subject of continuing study.[5] For neutral polymer networks, hydration can be characterized in terms of a single parameter describing the affinity of polymer and solvent (the term χ in Equation 12.1). At high polymer concentrations, this contribution to network hydration can be large. However, available literature on the hydration behavior of synthetic polymers[46-48] suggests that in the plant cell wall and pectin gel a more important potential contribution to hydration comes from the polyelectrolyte characteristic of the pectin network. For polyelectrolyte gels, the requirement for electrical neutrality leads to an excess of counterion within the gel compared with that in the external medium.[5] This excess generates an osmotic pressure difference between the gel and external medium, which increases with decrease in ionic strength. The excess osmotic pressure leads to the swelling of the gel until a balance is achieved between the osmotic pressure that drives swelling and the restorative force arising from the deformation of the cross-linked network. At intermediate salt concentrations, an estimate of the contribution to osmotic pressure, π, due to a polyelectrolyte may be obtained from

$$\pi \approx \frac{RTc^2}{A(c + 4Ac_s)} \tag{12.3}$$

for univalent electrolytes, where c and c_s are the molar concentrations of polymer segment and salt, respectively, and A is the number of monomers between effective charges. The greater the charge on the polymer and the lower the ionic strength, the greater the osmotic pressure generated. However, at high charge densities, the phenomenon of counterion condensation[6,7] can reduce the counterion fraction that can contribute to this Donnan effect. In pectin networks where the cross-links are formed by specific ionic complexation, a further proportion of the ionizable residues on the polyelectrolyte is excluded from contributing to the generation of the hydration force. Where pectin networks are cross-linked by interaction with Ca^{2+} ions, the galacturonate sequences can play a dual role. They can contribute to swelling when ionized, but when involved in Ca^{2+}-mediated cross-linking they no longer contribute, but, instead, play a role in resisting network expansion. It can be speculated that regions having a blockwise distribution of charge will tend to be involved in network cross-linking whereas other regions where the charges are more separated will play more of a role in generating a swelling pressure.

The preceding discussion was concerned with the interaction of pectins with small inorganic counterions. There is also the possibility of cross-linking through interaction with organic counterions. These can include basic, charged peptides[28] such as poly-L-lysine or basic oligosaccharides such as chitosan. More generally, charge interactions between pectins and food proteins affect their use as texturizing and stabilizing agents.

As well as the use of pectins as network-forming agents for the production of food gels, their use in more concentrated systems should also be considered. The plant cell wall is after all a very much more concentrated system than the pectin gel, with estimates of concentration in the region of 30% w/w being typical.[29,49] Unfortunately, much of the discussion on the structure of the pectin gel network in the plant cell wall has not really considered the effect of this tenfold change in polymer concentration. Schematic representations of the cell wall network structure still focus solely on the role of Ca^{2+} in ionic cross-linking. However, at the very much higher concentration of pectin found in the plant cell wall, other ionic interactions, involving K^+ and Mg^{2+}, could potentially be sufficiently strong to lead to cross-linking of the network. Similarly, in considering the use of pectins to produce edible film coatings, these ionic interactions should also be considered as potentially contributing to cross-linking of the polymer film.

12.7 CONCLUDING REMARKS

From a physicochemical and materials perspective, pectin is potentially a very useful macromolecule. Being a polyelectrolyte, its behavior is very sensitive to ionic effects; also, its charge and its distribution along the pectin backbone can be varied in a rather defined way through the use of chemical and enzymic methods. The sensitivity of material properties to ionic environment is potentially important in its role as a structural component of the plant cell wall. Simply by changing the ionic environment of the apoplast it might be possible to induce large changes in material properties of the cell wall. From a physicochemical point of view, this would seem rather attractive compared with the requirement to export an enzyme into the plant cell wall (which is a very dense network and not that porous to macromolecules) to modify a covalent cross-link. The ability to alter charge distribution opens up the possibility of using pectins more widely. For example, it is possible to envisage the fabrication of pectin films that maintain their integrity under one set of environmental conditions of pH and ionic strength yet swell markedly under another set of conditions. Such devices may well find use for the encapsulation and then release of active species.

REFERENCES

1. Carpita, N.C. and Gibeaut, D.M., Structural models of primary cell walls in flowering plants: consistency of molecular structure with the physical properties of the walls during growth, *Plant J.,* 3, 1, 1993.
2. Seymour, G. B. and Knox, J. P., *Pectins and their Manipulation*, CRC Press, Boca Raton, 2002.
3. Ridley, B.L., O'Neill, M.A., and Mohnen, D.A., Pectins: structure, biosynthesis, and oligogalacturonide-related signaling, *Phytochemistry,* 57, 929, 2001.
4. Ralet, M.C. et al., Enzymatically and chemically de-esterified lime pectins: characterisation, polyelectrolyte behaviour and calcium binding properties, *Carbohydr. Res.,* 336, 117, 2001.
5. Flory, P.J. *Principles of Polymer Chemistry*, Cornell University Press, Cornell, 1953.
6. Manning, G.S. and Ray, J., Counterion condensation revisited, *J. Biomol. Struct. Dyn.,* 16, 461, 1998.
7. Manning, G.S., The critical onset of counterion condensation: a survey of its experimental and theoretical basis, *Ber. Buns.-Ges.-Phys. Chem. Chem. Phys.,* 100, 909, 1996.
8. Treloar, L.R.G., *The Physics of Rubber Elasticity*, Oxford University Press, Oxford, 1958.
9. MacDougall, A.J. et al., Calcium gelation of pectic polysaccharides isolated from unripe tomato fruit, *Carbohydr. Res.,* 293, 235, 1996.
10. Willats, W.G.T. et al., Modulation of the degree and pattern of methyl-esterification of pectic homogalacturonan in plant cell walls: implications for pectin methyl esterase action, matrix properties, and cell adhesion, *J. Biol. Chem.,* 276, 19404, 2001.
11. Schols, H.A., Posthumus, M.A., and Voragen, A.G.J., Structural features of hairy regions of pectins isolated from apple juice produced by the liquefaction process, *Carbohydr. Res.,* 206, 117, 1990.

12. Schols, H.A. et al., Different populations of pectic hairy regions occur in apple cell walls, *Carbohydr. Res.*, 275, 343, 1995.

13. Daas, P.J.H., Voragen, A.G.J., and Schols, H.A., Study of the methyl ester distribution in pectin with endo-polygalacturonase and high-performance size-exclusion chromatography, *Biopolymers*, 58, 195, 2001.

14. Le Goff, A. et al., Extraction, purification and chemical characterisation of xylogalacturonans from pea hulls, *Carbohydr. Polym.*, 45, 325, 2001.

15. Huisman, M.M.H. et al., The CDTA-soluble pectic substances from soybean meal are composed of rhamnogalacturonan and xylogalacturonan but not homogalacturonan, *Biopolymers*, 58, 279, 2001.

16. Huisman, M.M.H. et al., The occurrence of internal (1→5)-linked arabinofuranose and arabinopyranose residues in arabinogalactan side chains from soybean pectic substances, *Carbohydr. Res.*, 330, 103, 2001.

17. O'Neill, M.A. et al., Requirement of borate cross-linking of cell wall rhamnogalacturonan II for *Arabidopsis* growth, *Science*, 294, 846, 2001.

18. ONeill, M.A. et al., Rhamnogalacturonan-II, a pectic polysaccharide in the walls of growing plant cell, forms a dimer that is covalently cross-linked by a borate ester: in vitro conditions for the formation and hydrolysis of the dimer, *J. Biol. Chem.*, 272, 3869, 1997.

19. ONeill, M.A. et al., Rhamnogalacturonan-II, a pectic polysaccharide in the walls of growing plant cell, forms a dimer that is covalently cross-linked by a borate ester: in vitro conditions for the formation and hydrolysis of the dimer, *J. Biol. Chem.*, 271, 22923, 1996.

20. Round, A.N. et al., Unexpected branching in pectin observed by atomic force microscopy, *Carbohydr. Res.*, 303, 251, 1997.

21. Round, A.N. et al., Investigating the nature of branching in pectin by atomic force microscopy and carbohydrate analysis, *Carbohydr. Res.*, 331, 337, 2001.

22. Daas, P.J.H. et al., Nonesterified galacturonic acid sequence homology of pectins, *Biopolymers* 58, 1, 2001.

23. Daas, P.J.H., Voragen, A.G.J., and Schols, H.A., Investigation of the galacturonic acid distribution of pectin with enzymes. Part 2: Characterization of non-esterified galacturonic acid sequences in pectin with endopolygalacturonase, *Carbohydr. Res.*, 326, 120, 2000.

24. Daas, P.J.H. et al., Investigation of the non-esterified galacturonic acid distribution in pectin with endopolygalacturonase, *Carbohydr. Res.*, 318, 135, 1999.

25. Andersen, A., Larsen, B., and Grasdalen, H., Sequential structure by H-1 NMR as a direct assay for pectinesterase activity, *Carbohydr. Res.*, 273, 93, 1995.

26. Mort, A.J., Qiu, F., and Maness, N.O., Determination of the pattern of methyl esterification in pectin. Distribution of contiguous nonesterified residues, *Carbohydr. Res.*, 247, 21, 1993.

27. Needs, P.W. et al., Specific degradation of pectins via a carbodiimide-mediated Lossen rearrangement of methyl esterified galacturonic acid residues, *Carbohydr. Res.*, 333, 47, 2001.

28. MacDougall, A.J. et al., The effect of peptide behaviour–pectin interactions on the gelation of a plant cell wall pectin, *Carbohydr. Res.*, 335, 115, 2001.

29. Ryden, P. et al., Hydration of pectic polysaccharides, *Biopolymers*, 54, 398, 2000.

30. Levigne, S., Ralet, M.C., and Thibault, J.F., Characterisation of pectins extracted from fresh sugar beet under different conditions using an experimental design, *Carbohydr. Polym.*, 49, 145, 2002.

31. Denes, J.M. et al., Different action patterns for apple pectin methylesterase at pH 7.0 and 4.5, *Carbohydr. Res.,* 327, 385, 2000.
32. Schmelter, T. et al., Enzymatic modifications of pectins and the impact on their rheological properties, *Carbohydr. Polym.,* 47,99, 2002.
33. Perez, S., Mazeau, K., and du Penhoat, C.H., The three-dimensional structures of the pectic polysaccharides, *Plant Physiol. Biochem.,* 38, 37, 2000.
34. Ferry, J.D., *Viscoelastic Properties of Polymers,* John Wiley & Sons, New York, 1980.
35. Renard, C.M.G.C. and Thibault, J.-F, Structure and properties of apple and sugar-beet pectins extracted by chelating agents, *Carbohydr. Res.,* 244, 99, 1993.
36. Ring, S.G., Ianson, K.J., and Morris, V.J., Static and dynamic light scattering studies of amylose solutions, *Macromolecules,* 18, 182, 1985.
37. Walkinshaw, M.D. and Arnott, S., Conformations and interactions of pectins. I. X-ray diffraction analyses of sodium pectate in neutral and acidified forms, *J. Mol. Biol.,* 153, 1055, 1981.
38. Walkinshaw, M.D. and Arnott, S., Conformations and interactions of pectins. II. Models for junction zones in pectinic acid and calcium pectate gels, *J. Mol. Biol.,* 153, 1075, 1981.
39. Rigby, N.M. et al., Observations on the crystallization of oligogalacturonates, *Carbohydr. Res.,* 328, 235, 2000.
40. Grant, G.T. et al., Biological interactions between polysaccharides and divalent cations: the egg box model, *FEBS Lett.,* 32, 195, 1973.
41. Bulone, D. et al., Role of sucrose in pectin gelation: static and dynamic light scattering experiments, *Macromolecules,* 35, 8147, 2002.
42. Kohn, R., Ion binding on polyuronates: alginate and pectin, *Carbohydr. Res.,* 42, 371, 1975.
43. Garnier, C., Axelos, M.A.V., and Thibault, J.-F., Selectivity and cooperativity in the binding of calcium ions by pectins, *Carbohydr. Res.,* 240, 219, 1993.
44. Thibault, J.F. and Rinaudo, M., Interactions of mono and divalent counterions with alkali and enzyme deesterified pectins in salt-free solutions, *Biopolymers,* 24, 2131, 1985.
45. Tibbits, C.W., MacDougall, A.J., and Ring, S.G., Calcium binding and swelling behaviour of a high methoxyl pectin gel, *Carbohydr. Res.,* 310, 101, 1998.
46. Skouri, R. et al., Swelling and elastic properties of polyelectrolyte gels, *Macromolecules,* 28, 197, 1995.
47. Rubinstein, M. et al., Elastic modulus and equilibrium swelling of polyelectrolyte gels, *Macromolecules,* 29, 398, 1996.
48. Patel, S.K. et al. Elastic modulus and equilibrium swelling of poly(dimethylsiloxane) networks, *Macromolecules,* 25, 5241, 1992.
49. MacDougall, A.J. et al., Chlorodeoxy derivatives: swelling behaviour of the tomato cell wall network, *Biomacromolecules,* 2, 2, 2001.

13 Fructans: Occurrence and Application in Food

Werner Praznik, Ewa Cieślik,
and Anton Huber

CONTENTS

13.1 Introduction 197
13.2 Occurrence and Analytical Approach to Fructan Crops 198
13.3 Fructans in Crops: Biosynthesis and Properties 200
13.4 Fructan Crops 204
    13.4.1   Composites 204
    13.4.2   Chicory (Cichorium intybus L.) 204
    13.4.3   Jerusalem Artichoke (Helianthus tuberosus L.) 206
    13.4.4   Globe Artichoke (Cynara Scolymus L.) 209
    13.4.5   Monocotyledons 209
    13.4.6   Liliaceae 210
    13.4.7   Allium L. Species 211
    13.4.8   Poaceae (Cereals, Grasses) 211
    13.4.9   Asparagaceae 212
    13.4.10 Amaryllidaceae 212
    13.4.11 Agavaceae 212
References 213

13.1 INTRODUCTION

Worldwide, fructans are important components in human nutrition, with slight differences due to geographic and cultural peculiarities. Sprouts of chicory; tubers of Jerusalem artichoke; globes of artichoke, onion, leek, garlic; and sprouts of asparagus are used as vegetables in different ways of preparation. Onion, garlic, and chives are used as spices or taste-providing components in many kinds of food. All the listed cultivars contain fructans in different amounts as reserve carbohydrate. Another source of fructans are caryopses of grasses and flours of wheat, rye, oat, and barley, which contain between 1 and 7% fructans, depending on the degree of fine milling and applied technological processing. As the human intestinal tract lacks enzymes that depolymerize fructans, their metabolic processing is rather low, and thus they may be attributed as dietary fiber materials. Due to this fact, food manufacturers are showing an increased interest in fructans, which has resulted in increasing numbers of quality profiles being developed and tested, with particular focus on functional food. A professional qualification of these profiles needs comprehensive information about potential sources and details about biosynthesis pathways and molecular characteristics.

13.2 OCCURRENCE AND ANALYTICAL APPROACH TO FRUCTAN CROPS

Approximately 15% of all crop cultivars store fructans as water-soluble polysaccharides in vacuoles of their cells. Besides the primary feature of stored fructans being an energy resource, many crop varieties show pronounced adaptability on environmental stress, such as long drought resistance in case of lack of water, frost tolerance for long periods, and survival during significant oxygen deficiency in root sections.[1,2] The major reason for this adaptability is the easy and fast mobilization of fructans: their immediate polymerization or depolymerization on varying conditions, which provides a flexible and efficient osmoregulatory surviving strategy.[3] Crops with pronounced frost tolerance, for instance, such as *Helianthus tuberosus*, form low-molecular-weight carbohydrates from polymeric fructans rather fast when temperatures go below 0°C to favor the formation of tiny ice crystals, and, by that, avoid cell damages. The most important crop varieties that take advantage of fructans are dicotyledons such as Asteracea; this is, Compositacea (composites), Campanulacea (bellflower), and Boraginacea. However, a much higher number of fructan-containing species are found in monocotyledon varieties such as Amaryllidaceae, Liliaceae, Agavaceae, Haemodoraceae, Iridacea, and the family of Poaceae (cereals, grasses).

The aqueous soluble fraction of fructan crops dominantly contains glucose, fructose, sucrose, and low- and high-molecular-weight fructans. Such profiles are quantified by enzymatic assays (glucose, fructose, and sucrose) before and after enzymatically or acid-catalyzed total hydrolysis[3-5] (Table 13.1). Nonpolar materials in leaves and stems must be eliminated by hexane/acetone extraction before fructan analysis. To guarantee quantitative results, in particular for fructans of the inulin type, hot-water extractions at 80°C are required. Carbohydrate profiles are most efficiently quantified by chromatographic techniques. Mono-, di-, and oligosaccharides with degree of polymerization (DP) values up to 15 can be investigated by thin layer chromatography (TLC). Complementary anion-exchange chromatography combined with amperometric detection (HPAEC-PAD, Dionex) provides a qualitative column-chromatographic approach to separate mono-, di- and oligosaccharides with DPs up to ca. 50 as long reference materials are available. Quantification may be achieved for low-DP components, but not for the midrange- and high-DP components.[6-8] Another method for investigating mono-, di-, and trisaccharides are anionic exchange columns with Ca and Pb as counter ions. However, they are of low stability if extracts are not precleaned properly.

Fructans may be separated by reversed-phase (rpHPLC) or amine-modified high-performance liquid chromatography ($_{NH2}$HPLC) techniques combined with quantitative mass detection by vaporization light scattering detection (ELSD).[9,10]

TABLE 13.1
Analytical Strategy for Fructan-Containing Crop Materials[a]

Fructan Containing Crop Material
tubers / roots / leaves / stem: Jerusalem artichoke, Chicory

1. Smashing / suspending in water Separation from fibers by sedimentation Freeze-drying Storage conditions	1.1 Pretreatment	

	1.2 Homogenized
	dry material
		1.2.1 Identification /
		  analysis:
		solid body / surface
		  spectroscopy

	1.3 Nonpolar extraction
	(defatting)
		1.3.1 Dissolution in
		  hexane / acetone:
		lipids, phenolic
		  compounds,
		  chlorophyll;
		elimination of rest
		  moisture
		(in particular for
		  leaves and stem)

2. Identification / quantification / analysis Dissolution in aqueous media, 80°C: mono- / di- / oligo-/ polysaccharides peptides / proteins, salts	2.1 Polar extraction	
	2.2 Insoluble residues	2.2.1 Hemicellulose, cellulose, lignin

Aqueous Dissolved Components

1. Acidic / enzymatic Hydrolysis	1.1 Enzymatic analysis of Glu + Fru + Suc	1.1.1 Analytical fractionation: size-exclusion chromatography

2. Enzymatic +
  chemical
Glc + Fru -
  analysis
	2.1 Fructan	2.1.1 Molecular weight distribution,
		degree of polymerization distribution

3. Quantification
  of
Glc + Fru +
  fructans
	3.1 Quantification of
	Glc + Fru + fructans
		3.1.1 Nondestructive molecular
		  analysis:
		dimension, conformation, interactive
		  properties

[a] In the procedure, follow order of numbers

In particular, for midrange- and high-DP fructans, calibrated size-exclusion chromatography (SEC) with mass detection by refractive index provides quantitative information about molar mass and DP distribution.[11,12]

Application of preparative and semipreparative soft-gel systems such as Biogel P-2 or P-4 provides fructan fractions and pools for subsequent detailed structural analyses such as reductive methylation, acetylation, and GC/MS or GC/FID detection. The resulting 1,5-D-anhydrosorbits and 2,5,-D-anhydromannits provide quantitative information about the kind of glycosidic linkages and enable computation of molecules with mean structural conformations which typically are checked by simultaneously accomplished NMR spectra.[13–15]

13.3 FRUCTANS IN CROPS: BIOSYNTHESIS AND PROPERTIES

Biosynthesis of fructans in crops differs significantly from pathways in bacteria or fungi. The only key enzyme for the formation of high-DP fructans — exopolysaccharides (EPS) — starting from sucrose in bacteria, such as in *Pseudomonas* species, is levansucrase (EC 2.4.1.10). These readily aqueous-soluble EPSs are nonreducing levans,[16–18] (2→6)-linked β-D-fructans with branching option at O-1 and molecular weights up to 10^6 (g/mol).

Fructosyltransferases of fungi *Aureobasidium pullulans* and *Aspergillus niger* form fructooligosaccharides with DPs of 3 to 6 at high initial sucrose concentrations; however, the final mixture still contains sucrose and fructosyl-transfer-equivalent concentrations of glucose.[19,20]

The key compound in the formation of fructans is the disaccharide sucrose (Figure 13.1). Polymerization, and thus the formation of low-molecular-weight fructans (fructooligosaccharides, FOSs) and high-molecular-weight fructans runs via transfer of fructofuranosyl units from a donor sucrose to an acceptor sucrose or to the already formed acceptor fructan.

In contrast to glucans, fructans contain no reducing terminal hemiacetal but have a single heteromonomer glucopyranosyl residue. Those amounts of fructans without terminal glucose are the result of internal rearrangements or depolymerization reactions in fructan metabolism, which are amplified by cell-specific propagation processes.

For optimization of fructan metabolism, biosynthesis of fructan in crops is most probably a two-step process (Figure 13.2). In the initial steps, polymerization runs without a precursor, catalyzed by sucrose:sucrose 1-fructosyltransferase (1-SST: EC 2.3.1.99), with donor sucrose and acceptor sucrose forming trisaccharide 1-kestose. The remaining glycosyl residue immediately gets processed in the cytosolic pathway by sucrose synthase. If necessary, the glucosyl residue may even be transferred to the starch metabolism pathway or to pathways for the formation of cell wall structurizing materials (Figure 13.1).

The second step is the formation of fructans, starting from kestose and catalyzed by several crop-specific fructosyltransferases (fructan:fructan fructosyltransferases,

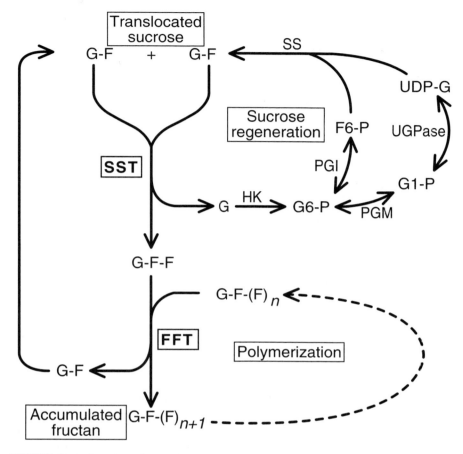

FIGURE 13.1 Combined fructan biosynthesis and regenaration of glucose in fructan plants. SST: sucrose:sucrose fructosyltransferase, FFT: fructan:fructan fructosyltransferases, HK: hexokinase, PGM: phosphoglucomutase, UGPase: UDP-glucosepyrophosphorylase, PGI: phosphoglucoisomerase, SS: sucrosesynthase.

FFT: EC 2.4.1.100), with different, more or less uniform, but variety-specific gly-cosidic linkage patterns.

Inulin-type $\beta(2{\to}1)$-linked fructans are formed by 1-FFT, starting from trisac-charide 1-kestose by a step-by-step transfer of β-D-fructofuranosyl residues from sucrose. Levan-type $\beta(2{\to}6)$-linked fructans are formed by 6-FFT, starting from 6-kestose by a step-by-step transfer of β-D-fructofuranosyl residues from sucrose. Mixed-type fructans are formed by the combined activities of 1-FFT and 6-FFT, and contain both $\beta(2{\to}1)$- and $\beta(2{\to}6)$-linked fructofuranosyl residues.

Additionally, a β-D-fructofuranosyl is often introduced to these fructans at O-6 of the terminal glucosyl residue by 6^G-FFT, which apparently indicates neo-kestose as the starter triose. The high variability of branching and elongation of primary

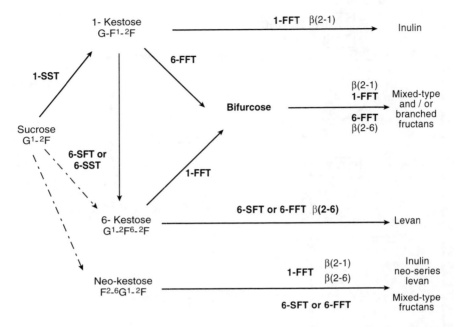

FIGURE 13.2 Biosynthesis of kestoses and fructans in crops. 1-SST: sucrose:sucrose fructosyltransferase; 1-FFT, 6-FFT: fructan:fructan fructosyltransferases; 6-SFT: sucrose:fructan fructosyltransferase.

chains and branches generates rather complex molecules, which contribute significantly to the broad-band functional properties of these fructans.[21]

Three categories need to be distinguished in fructan metabolism of crops: synthesis/two-step-mechanism polymerization, utilization/depolymerization, and mobilization to respond on environmental stress. For instance, in tubers of *Helianthus tuberosus* (Jerusalem artichoke), high SST activities are observed at high photosynthetic activities and fast transformation of initial carbohydrates to fructans. SST activity decreases and finally vanishes in the course of the vegetation period with increasing amounts of stored reserve fructans, whereas FFT activities stay even in the final vegetation states. FFT activity in tubers increases in stress conditions such as frost and causes depolymerization to support osmoregulatory mechanisms. At the state of expelling after the winter frost period, fructosyl exohydrolase (FEH) activity appears additionally and increases to mobilize depolymerized fructosyl residues in the cytosolic pathway forming sucrose by transfer to a glycosyl residue. Obviously, fructan-forming and transforming enzymes are closely correlated with sucrose metabolism.[22]

Fructans are obtained from crop raw materials by hot-water extraction and different processing drying steps as a white powder. Depending on their content of mono- and disaccharides, low-molecular-weight fructans (fructooligosacchaides) are more or less sweet in taste, whereas high-molecular-weight fructans are more or less of neutral taste.

Solubility of fructans in aqueous media is governed by their molecular structure and variations in temperature of the solvent. The more symmetric the fructan, the harder is its solubility; the more branched, mixed-type the fructan, the easier and better is the solubility. Also, higher the temperature of the solvent, better the solubility of the fructan. Highly symmetric inulin tends to form crystalline structures, and elevated temperatures between 50 and 80°C are needed to obtain a solution. However, once dissolved, such solutions are stable even at room temperature.

Figure 13.3 presents the degradation of aqueous solution of high-molecular-weight inulin from chicory (10 mg/ml) at 80°C. Stability of the molecular composition after 3 days was very high, but after 7 days the molecular weight distribution decreased in low-molecular-weight components and the fructose portion increased about fivefold. Nevertheless, even at room temperatures, inulin swells in an aqueous

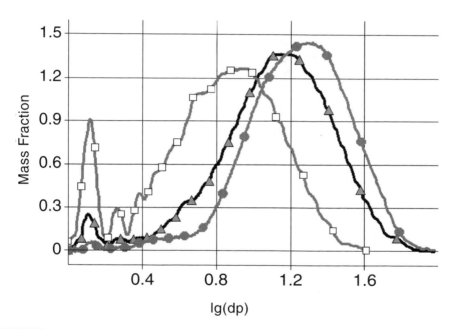

FIGURE 13.3 Distribution (m_dpD_d) of high-molecular-weight fractions of chicory inulin according to their degree of polymerization (DP) determined calibrated size-exclusion chromatography (SEC at 80°C). ——●——: initial inulin ($t = 0$ days), mean values of DP from m_dpD_d are $DP_w = 21$, $DP_n = 14$ (including mono- and disaccharides); ——▲——: 3 days 80°C, mean values of DP from m_dpD_d are $DP_w = 16$, $DP_n = 9$ (including mono- and disaccharides); ——■——: 7 days at 80°C, mean values of DP from m_tdpD_d are $DP_w = 8$, $DP_n = 5$ (including mono- and disaccharides). SEC column system: precolumn (Fractogel HW40; 10 ×100 mm) + Superose 12 (10 × 300 mm) + 2× Fractogel HW40 (10 × 300 mm); eluent: 0.05 mol/l NaCl; flowrate: 0.6 ml/min; injection volume of sample: 0.1 ml of 0.1% carbohydrate; mass detection by DRI; internal standard for flow control: 2 mg NaCl/ml; SEC calibration: dextran 20, dextran 1 (Pharmacosmos, DK) and sucrose; data processing with CODAwin32 and CPCwin32 (a.h group, Graz, Austria).

environment and forms as a 30 to 40% suspension a hysteresis in rheological experiments which corresponds to rheopexplastic materials.[23]

The texture of rather stable inulin suspensions is very similar to that of fats and is neutral in taste (see Chapter 11). Consequently, inulin suspensions are considered as fat replacers in food; on the other hand, there are interesting interaction properties between inulin and fats. Water-binding capacity of inulin is ca. 1 g/g. Inulin as an additive stabilizes emulsions and suspensions in different food systems, for instance, in dairy products. Additionally, texture and mouth-feel of food is enhanced by inulin by soft gel formation. Due to its neutral taste and basically good solubility in aqueous media, inulin has become of increasing interest from the pharmaceutical viewpoint as a filler binder and carrier for tablets and capsules. An important argument in favor of fructans in nutrition and medical applications is their noncarcinogenicity, as fructans are not metabolized by *Stryptococcus mutans*. Fructans contribute only negligibly to the glucose level, due to which fructan-containing goods are perfect for diabetics. FOSs are used as sweeteners to substitute sucrose, being perfectly soluble with only a minor tendency toward crystallization (see Chapter 24). Hydrolysis of FOSs at short-time temperature treatment is rather low; however, on longer periods such as in backing processes, hydrolysis rate goes up to 50%. Hydrolysis of inulin at identical conditions is 35% only. Stability of FOSs against combined impact of fruit acid, temperature, and sunlight, such as in fruit juices, is rather low. FOSs in industrial fruit juices are significantly hydrolyzed to form 20 to 30% fructose. Compared to FOSs, inulin is much more resistant to pH and temperature variations. At a pH of 3.5 and temperature of 70°C, degradation to fructose is <5% after 30 min and <10% after 60 min.

13.4 FRUCTAN CROPS

13.4.1 Composites

Industrial utilization of fructans within the last decades has focused on two varieties of composite crops: chicory (*Cichorium intybus* L.) and Jerusalem artichoke (*Helianthus tuberosus* L.). Both crops contain fructans of the inulin type (Figure 13.4); however, quantity and composition of carbohydrates and distribution of fructan polymers significantly depends on crop species and period of harvesting.

13.4.2 Chicory (*Cichorium intybus* L.)

Chicory is a biannual crop. The root body forms in the first year, and expelling of roots and flowering takes place in the second year. Seeds for breeding of chicory are relatively small and need to be prepared by pilling for optimum chicory fields and annual yields between 40 and 60 t. Chicory beets contain 18–20% soluble carbohydrates with a fructan content of 14–18%. Besides roots, even sprouts of chicory (chicoree) are used in several regions, in particular in Belgium and The Netherlands, as salad and vegetable of light bitter taste, which is caused by the taste-dominating bitter compound intybine (lactucoricin) in roots and sprouts of chicory.[24]

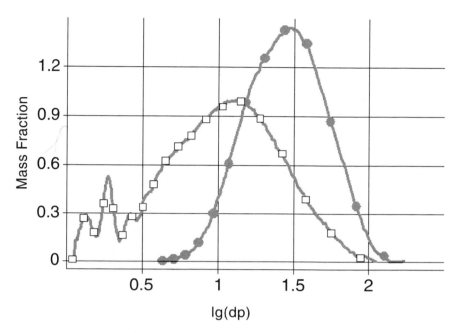

FIGURE 13.4 Degree of polymerization distribution (m_dpD_d) for inulin from chicory by calibrated SEC (for details see Figure 13.3). ——■——: Raftilin®ST (commercial product), mean values of DP from m_dpD_d are $DP_w = 15$, $DP_n = 7$ (including mono- and disaccharides); ——●——: Raftilin® HP (commercial product), mean values of DP from m_dpD_d are $DP_w = 34$, $DP_n = 24$ (including mono- and disaccharides).

Chicory is used in several applications. Combination of bitter compounds and fructans from sliced chicory roots with additional ingredients such as malt, barley, and fig in a roasting process provides a "healthy" and compliant coffee surrogate. Hydrolytic depolymerization of inulin to fructose even provides basic material for manufacturing nutrition products for diabetics.

Nondigestibility of inulin in the human intestinal tract makes this material a tool to test renal clearance. In the last decade, the fact of indigestibility became increasingly interesting from the point of view of soluble dietary fibers. As consumer request for functional and healthy food increases exponentially, tailored fructooligo-saccharides and inulin perfectly match the request of application profiles of prebiotic goods in various market segments (Figure 13.4). Domination of large-scale fructan production from chicory is supported by the relatively simple cultivation of the crops, easy harvesting of the roots, and utilization of existing sugar beet processing technology. The most crucial point in isolation and purification of inulin is the quantitative elimination of salts and bitter compounds. Finally obtained distribution of fructan polymers may be controlled by chicory cultivar and period of harvesting and may be determined prior to harvesting by chromatography.[25,26]

Isolation of inulin-type fructans from chicory by hot-water extraction from sliced chicory roots runs similar to the processing of sugar beets. Their current annual production is ca. 100,000 t and is on the rise. If specific distribution profiles of inulin polymers are required for certain products, membrane filtration and chromatography might be appropriate ways to cut sections of interest from total distribution profiles.

Another way of obtaining distribution profiles can be partial hydrolysis with endoinulinases, which form fructose oligomers from inulin. Hydrolysis of inulin, however, results in an increase of reducing molecules in a mixture of $\beta(2\rightarrow1)$-fructofuranosyl residues with and without terminal glucosyl residues. These reducing oligomers of fructans have high affinity to secondary amino groups, amino acids, peptides, and proteins forming Maillard products.

Inulin from chicory is available on the Western European market as Frutafit® manufactured by Sensus, The Netherlands; Raftiline® series by Orafti, Belgium; and Fibroline® series by Cosucra, Belgium. A standard preparation of Raftiline®, for instance, contains ca. 92% inulin and high-molecular-weight compounds with an average DP of 25 and no mono- or disaccharides. Raftiline® is characterized by neutral taste and has a wide range of applications in the food industry, in particular for manufacturing dietetic food. Raftiline® as a supplement improves the performance of low-fat products such as pasteurized fresh cheese, prebiotic advantages in general and improved calcium absorption in particular.

FOSs are endoinulinase-depolymerized oligomers from chicory inulin and are available under the trade names Frutalose™ (Sensus), Raftilose® (Orafti), and Fibrolose® (Cosucra). They all contain ca. 95% FOSs and are applied for similar purposes. FOSs with a DP of 3 to 5 are achieved by transfructosylation of β-fructosidase from *Aspergillus niger* on sucrose Actilight® (Beghin Meiji, Eridania Béghin-Say, France) and Neosugar® (Meiji Seika Kaisha Ltd., Japan). The approval of FOSs in Japan prompted the establishment of an acceptable daily intake of ca. 0.8 g/kg body weight/day.

13.4.3 JERUSALEM ARTICHOKE (*HELIANTHUS TUBEROSUS* L.)

Another source of inulin and FOSs are a wide variety of Jerusalem artichoke cultivars, which can be cultivated as early, middle, or late species with individual, however, specific, carbohydrate composition profiles. Appropriate selection of varieties and time of harvesting enables a continuous harvesting period starting from September till the appearance of the first frost. Because they are frost resistant, Jerusalem artichoke tubers can stay on the field over wintertime, which enables another harvesting in spring. However, carbohydrate compositions of spring tubers differ significantly from those of autumn tubers: a shift from high- to low-molecular-weight components with a ca. fivefold increase in sucrose concentration and ca. 35% decrease in FOS and inulin concentration[27–29] (Figure 13.5, Table 13.2).

In some areas of the U.S. (California, Washington, Minnesota), Canada, and Europe (Belgium, France, Germany, Poland, Ukraine), Jerusalem artichoke tubers are known as sunchoke or topinambur and are popular vegetables. Their taste is rather similar to artichoke hearts of globe artichoke, delightfully sweet — in particular those made from spring-harvested tubers with a high sucrose content — with a nutty or easy, earthy flavor. Tubers are consumed crude or boiled, backed, or fried. In particular, products made from autumn tubers are perfect basic material for diabetics as they are low in sucrose.

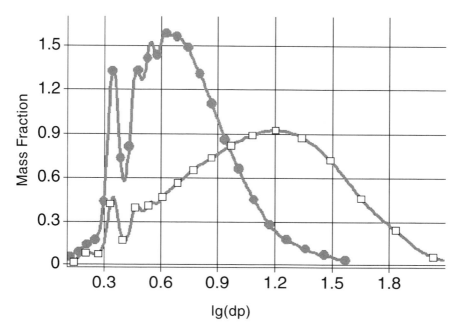

FIGURE 13.5 Degree of polymerization distribution (m_dpD_d) of inulin from Jerusalem artichoke by calibrated SEC (for details see Figure 13.3). ──■──: autumn harvest, mean values of DP from m_dpD_d are $DP_w = 19$, $DP_n = 9$ (including mono- and disaccharides); ──●──: spring harvest, mean values of DP from m_dpD_d are $DP_w = 6$, $DP_n = 4$ (including mono- and disaccharides).

Compared to chicory, tubers of Jerusalem artichoke contain no bitter compounds and can be processed without preprocessing. Additionally, FOSs of Jerusalem artichoke tubers are nonreducing with terminal a glucosyl residue as they are not made by partial hydrolysis but are of native origin. Because of this, these FOSs form no Maillard products and do not become brownish during low-temperature processing. However, enzymatically catalyzed oxidation is often observed in juice production if phenoloxidases are not deactivated. Powders or dried and milled slices of Jerusalem artichoke tubers can be utilized for manufacturing bread and bakery products.[28]

According to current studies,[29] the yield of Jerusalem artichoke is rather high: 40 to 80 t/ha, with a dry matter content between 20 and 28% if there is appropriate choice of crop species, soil profile, and climatic conditions. Utilizable total carbohydrates typically range from 18 to 20%, with ca. 90% fructans with a DP over 3. Due to the potential high yields of Jerusalem artichoke, production of high-fructose syrups and processing for biofuel are promising future perspectives.[29] The German food industry utilizes Jerusalem artichoke tubers for manufacturing a high-fructose syrup containing ca. 52% FOSs characterized by a low energy value (550 kJ/100 g) and a low-calorie powder (flour).[30–34] Powders from native Jerusalem artichoke tubers used in bakery products (wheat/rye bread) show a significantly different tendency for hydrolysis: 25% degradation to fructose of powders from autumn harvesting

TABLE 13.2
Water-Soluble Carbohydrates of Fructan Crops

Crop	Content in Fresh Matter (g/100 g)							Mean DP of Fructan
	DM	Glu	Fru	Suc	Fructan			
					DP 3–10	DP ›10	Total	
Compositae								
Chicory root (autumn harvest)	25	0.02	0.2	1.2	6.5	9.7	16.2	17
Jerusalem artichoke tuber (autumn harvest)	24	0.02	0.2	1.6	7.5	7.6	15.1	12
Jerusalem artichoke tuber (spring harvest)	18	0.03	0.2	3.2	7.2	2.5	9.7	6
Globe artichoke leaves	12.0	0.2	1.1	0.7	1.5	5.3	6.8	40
Globe artichoke hearts	20.7	0.2	2.0	1.5	1.2	10.5	11.7	42
Globe artichoke globe	18.2	0.2	1.5	1.1	1.3	7.7	9.0	41
Liliaceae								
Onion 1	10.9	0.9	0.85	1.1	2.0	1.4	3.4	5
Onion 2	13.5	0.8	1.9	2.1	4.6	0.7	5.3	4
Leek	12.0	1.1	1.0	0.5	3.0	3.6	6.6	n.d.
Chives	11.5	0.4	0.2	0.5	n.d.	n.d.	6.5	n.d.
Garlic	32.5	0.1	0.2	1.9	0.2	18.9	19.1	32
Poaceae (Cereals)								
Wheat	89.1	0.2	0.05	0.8	n.d.	n.d.	2.2	n.d.
Barley	89.4	0.1	0.05	0.5	n.d.	n.d.	2.7	n.d.
Rye	88.9	0.5	0.1	1.3	n.d.	n.d.	4.8	n.d.
Oat	87.1	0.15	0.03	1.0	n.d.	n.d.	0.4	n.d.
Asparagaceae								
Asparagus	6.5	0.3	0.4	0.1	1.0	0.8	1.8	n.d.
Amaryllidaceae								
Banana	25.7	1.2	0.1	4.3	n.d.	n.d.	1.4	n.d.
Agavaceae								
Agave leaves	19.0	0.2	0.3	0.8	4.9	2.4	7.3	5
Agave trunks	31.5	—	—	0.2	9.9	4.4	14.3	9

Note: DM, dry matter; Glu, glucose; Fru, fructose; Suc, sucrose; DP, degree of polymerization; n.d., not determined.

(DP_w = 22, DP_n = 10) and 41% degradation to fructose of powders from spring harvesting (DP_w = 17, DP_n = 8). Obviously, the tendency for hydrolysis depends strongly on the initial degree of polymerization: higher the DP, lower the tendency for degradation.

13.4.4 GLOBE ARTICHOKE (*CYNARA SCOLYMUS* L.)

Another source of fructans is the composite vegetable globe artichoke, which was originally from Ethiopia and entered South European countries via Egypt. Currently, globe artichoke is cultivated in France, Italy, Romania, Bulgaria, and the U.S. (California). The crop grows to a height of 60 to 200 cm and forms pronounced sprouts, which are utilized as vegetables. To be used in many different ways, the sprouts typically are boiled.

Extracts of globe artichoke roots and leaves are known to have been used for therapeutic purposes for over 400 years. The effect of such extracts is caused by phenolcarbon acids (cynarin, caffeic acid, chlorogenic acid), leteolines (luteolin, cynarosid, scolymosid), and the bitter compound sesquinterpenic lacton (cynaropikrin).

Dry matter content of artichoke globes ranges from 15 to 18%, with significant differences between leaves (12%) and hearts (20–22%). Content of the basic carbohydrates glucose, fructose, and sucrose in dry matter is ca. 20%, and the content of fructans ranges from 50 to 60%. Composition of inulin and FOSs (low-DP fructans) strongly depends on the status of the shoots. Table 13.2 gives the composition of aqueous soluble carbohydrates of an adult artichoke globe containing ca. 70% inulin with a DP between 60 and 120, which is nearly tenfold compared to that of inulin from other composites.

13.4.5 MONOCOTYLEDONS

Many monocot crops utilize fructans as an energy resource and a stress modulator. For human nutrition, the most important cultivars are Liliaflorae families such as Liliacae, Asparaguacae, and Agavaceae. Additionally, components of Liliaceae (onion, garlic) are used in prophylactic and therapeutic medication and, due to their biological background, are increasingly requested and well accepted by patients and consumers.

As inulin- and levan-type fructans, mixed-type fructans in monocotyledons are nonreducing molecules but dominantly contain a mix of $\beta(2\rightarrow1)$- and $\beta(2\rightarrow6)$ linkages (Figure 13.6). The first such fructan, sinistrin, was isolated from red squill, and analogous mixed-type fructans have now been found in garlic, onion, leek, chives, and asparagus.[31,32]

A minor amount of fructose in *Allium*, has been identified as the inulin type. The highly branched structure of these molecules favors a compact spherical conformation and a rather limited crystallization. Consequently, sinistrin is perfectly soluble in aqueous media and stays stable for long periods.

Mixed-type fructans additionally contain polymers with neo-kestose as the basic triose. Fructofuranosyl residues are transferred to the glucosyl residue of the initial

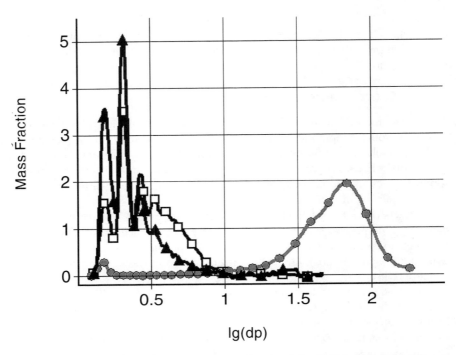

FIGURE 13.6 Degree of polymerization distribution (m_dpD_d) of fructan from *Allium* L. species by calibrated SEC (for details see Figure13.3). ——●——: garlic fructan (DP 12, mixed type) fresh from autumn harvest, mean values of DP from m_dpD_d are $DP_w = 61$, $DP_n = 26$ (including mono- and disaccharides); ——■——: onion "Centurio" fructan fresh from autumn harvest, mean values of DP from m_dpD_d are $DP_w = 3.1$, $DP_n = 2.6$ (including mono- and disaccharides); ——▲——: autumn harvest after 2 months of cool storage, mean values of DP from m_dpD_d are $DP_w = 2.6$, $DP_n = 2.1$ (including mono- and disaccharides).

acceptor sucrose, and polymerization continues either exclusively on this terminal or on both terminal fructosyl residues. Conformational differences between inulin and sinistrin have been proved by small-angle x-ray scattering (SAXS), NMR, and anionic-exchange chromatography with amperometric detection (Dionex). There exists a high variability in the composition of fructans of monocotyledons and the actually found structural fructan spectrum strongly depends on cultivars, vegetative status of the crops, climate, and environmental conditions.

13.4.6 LILIACEAE

Members of the Liliaceae, for example, onion, leek, garlic and chives, are used worldwide as vegetables, spices, and tasty components. Most of the well-known health-preserving and even therapeutical effects of onion and garlic are most probably due to their sulfur-containing compounds. However, the aqueous-soluble low- and high-molecular-weight fructans, which are present in onion and garlic in high percentages, are also responsible for the positive effects such as stimulating the intestinal tract and supporting consumers' well feeling. Carbohydrate profiles and the spectrums of low- and high-molecular-weight fructans strongly depend on cultivar, climate, environmental conditions, and harvesting period.

13.4.7 *ALLIUM* L. SPECIES

Only few of the more than 700 *Allium* species are utilized in human nutrition. These include *A. cepa* L. (onion, shallot), *A. ampeloprasum* L. var. *prooum* (leek), *A.schoenoprasum* L. (chives), and *A. sativum* L. (garlic).

Dry matter contents of onion and shallot bulbs (thickened leaves for storage reserve substances) are 9–14%, with 60–80% aqueous soluble carbohydrates. About half the soluble carbohydrates are FOSs and fructans, the latter having DPs between 3 and 10. Once again, there are significant differences in the contents of fructose, glucose, sucrose, and fructans, depending on cultivars, periods of harvesting, and storage conditions. For instance, concentrations of mono- and disaccharides can be increased two- or threefold, with a simultaneous decrease or absence of high-DP fructans, if storage temperature is changed from room temperature to 2°C (Figure 13.6). Table 13.2 gives the fructan composition of a commercially available freshly harvested onion with significant amounts of components with DP of 5 or lower.

Dry matter contents of chives and leek are similar to those of onion and shallot, between 10 and 14%, with aqueous soluble carbohydrates between 60 and 80%. However, fructan content is significantly higher (80%), and the content of high-molecular-weight fructans with DPs over 10 is ca. threefold that of onion.

Garlic cultivars differ significantly from onions with respect to their carbohydrate composition. Dry matter content of freshly harvested garlic bulbs ranges from 32 to 45%, with slight variations depending on cultivar and storage conditions. Fructan content is ca. 22 to 24% of the fresh weight, ca. sixfold that of onion. Fructan content in dry matter ranges from 58 to 68%, with only a minor content (1%) of low-molecular-weight fructans. Fructose and glucose contents are extremely low (below 1%), and sucrose content ranges from 4 to 7%. Table 13.2 lists the carbohydrate profile of commercially available garlic.

13.4.8 POACEAE (CEREALS, GRASSES)

Members of the family Poaceae, in particular, seeds of the cereals wheat (*Triticum aestivum* L.), rye (*Secale* cereale L.), oat (*Avena sativa* L.), and barley (*Hordeum vulgare* L.), play an important role in human nutrition (Table 13.2). In the growth period of roots, leaves, and stems, grasses utilize fructans as storage metabolites, which are later found as costorage polysaccharides with glucans in the seeds (caryopses). In early development stages, the caryopses contain a dry matter fructan content of 2 to 4%. In adult seeds of wheat grains, this content is reduced to ca. 1.3 to 2.5%. After milling, the fructan content in flour ranges from 1.0 to 1.3%, and in the bran is ca. 3%.

Adult rye grains contain 4–6% fructans, with 2.5–4% in the flour and 7% in the shorts after milling.[34] Barley grains contain 0.5 to 1.5% and oat grains ca. 0.5%

fructans. Historically, levans of grasses are often named phleins, after *Phleum prat-ense* L., from where fructans were first isolated.

Grain fructans are of the levan type (see Chapter 1) with $\beta(2\rightarrow6)$-linked fructofuranosyl residues and a terminal glucosyl residue. Their polymerization starts primarily from 1-kestose; however, additionally, a significant percentage of 6-kestose is found. Wheat and rye fructans dominantly (50 to 70%) have DPs over 10. As high-molecular-weight fructans become increasingly branched (mixed-type fructans), the DPs increase.

13.4.9 ASPARAGACEAE

Sprouts of Asperagaceae species, in particular *Asparagus officinalis* L. (asparagus), are popular vegetables worldwide. Dry matter contents range from 40 to 50%, depending on cultivar and cultivation conditions, with 10–15% being glucose and fructose and 1–2% sucrose. Fructan contents are similar to those of onion (25 to 30%), with a rather homogeneous distribution of low- and high-molecular-weight fructans (Table 13.2). Fructans are of the mixed type, and include the neo-kestose series similar to that in Liliaceae.

13.4.10 AMARYLLIDACEAE

The most important species from the Amaryllidaceae family is banana, in particular flour banana (*Musacea paradiiaca* L.) and fruit banana (*Musacea sapentium* L.), which are used as vegetable and fruit. Besides starch, glucose, fructose, and sucrose, fruit bananas contain more than 5% fructans, which corresponds to ca. 1.4% fresh weight at a dry matter content of ca. 26%.

13.4.11 AGAVACEAE

The family Agavaceae grows in Mediterranean and hot climates and needs ferrous and calcareous soil for optimum development of plant organelles. More than 200 species are known worldwide, with ca. 120 located in Mexico. A few varieties are used to grow raw material for food and for fermentation for tequila production. Tequila is a liquor from fermented agaves dominantly produced in Mexico. Today, ca. 120 tequila brands of different quality levels are offered by 17 producers. The preferred raw material is the Agavaceae species *Agave tequila* Weber L. var. *azul*.

Blue agave is cultivated in specific plantations, with up to 3000 vegetative plant seedlings/ha, and it is harvested 6 to 8 years before blooming, which guarantees maximum carbohydrate content. Trunk weight is ca. 240 t/ha.

Dry matter contents of the trunk range from 30 to 32% if harvested after a dry period. Total carbohydrate content is 22–24%, with fructan contents of ca. 20–23%. The amount of FOSs (3 < DP > 10) is 8–10%, and is identical with the amount of midrange-DP fructans (11 < DP > 30), and 4–8% of fructans have DPs between 20 and 50. Agave fructans are of the mixed type with $\beta(2\rightarrow1)$- and $\beta(2\rightarrow6)$ linkages and a basic backbone of β-D-$(2\rightarrow1)$-linked fructofuranosyl residues. Short branches having two or three fructofuranosyl residues are β-D-$(2\rightarrow6)$ linked. The terminal segment either is a α-D-glucopyranosyl-residue or a $\beta(2\rightarrow6)$ linked to a glucose fructofuranosyl residue of the neo-kestose type. Different agave species show different DP distribution profiles, with DPs even up to 100. However, in general, the mixed-type fructans are rather similar to those of Liliaceae (such as fructans of garlic).

For tequila production, 150 to 200 t of fructans are hydrolyzed per day to fructose by continuous boiling for 48 h at 2 to 3 bar. The temperature-treated agaves are milled and separated in cellulose and fructose partitions. Total fermentation of fructose and FOSs is achieved slowly for over 50 h by specific *Saccharomyces* species and yields 6 to 8% alcohol. Purification by distillation provides a 55% concentrate of neutral taste. The typical phenolic tequila taste and slight brownish colour is achieved on 2 to 3 years of storage in 50- to 100-l oak-wood tanks. Finally, blending of different batches and an additional period of storage achieve tailored performance profiles.

Due to the negligible content of monosaccharides (contents of glucose and fructose less than 0.2%) and sucrose (0.5 to 1.5%), agave is a perfect raw material to obtain highly pure fructan. Even hydrolyzed agave fructan is a valuable source for agave juice, consisting of up to 70% fructose, minerals (Ca and Mg), and vitamins dispersed in ca. 25% water with high sweetening power and is suited for healthy, dietary, and diabetic foods such as energizers and cereal bars. Organic agave juice is a new commercial application for agaves: a high-fructose (70% of totally contained carbohydrates) juice with high sweetener potential as fructose is ca. 1.8 sweeter than sucrose. Additionally, vital components such as Ca, Mg, and different vitamins make this juice a healthy and dietary-valuable basic material for dietary foods, energizers, and cereal bars.

REFERENCES

1. Albrecht, G., Kammerer, S., Praznik, W., and Wiedenroth, E.M., Fructan content of wheat seedlings (*Triticum aestivum* L.) under hypoxia and following re-aeration, *New Phytol.* 123, 471, 1993.
2. Neefs, V., Leuridan, S., Van Stallen, N., De Meulemeester, M., and De Proft, M.P., Frost sensitiveness of chicory roots (*Cichorium intybus* L.), *Sci. Hortic.*, 86, 185, 2000.
3. Suzuki, M. and Chatterton, N.J., Eds., *Science and Technology of Fructans*, CRC Press, Boca Raton, 1993.
4. Praznik, W., Baumgartner, S., and Huber, A., Molecular weight characterization of inulin: application of enzymatic and chromatographic methods, in *Proceedings of the Sixth Seminar on Inulin*, Fuchs, A., Schittenhelm, S., and Frese, L., Eds., CRF, The Hague, 1997.
5. Prosky, L. and Hoebregs, H., Methods to determine food inulin and oligofructose, *J. Nutr.*, *129 (Suppl. S)*, 1418, 1999.
6. Ernst, M.K., Chatterton, N.J., Harrison, P.A., and Matitschka, G., Characterization of fructan oligomers from species of the genus *Allium* L., *J. Plant Physiol.*, 153, 53, 1998.
7. Shiomi, N., Onodera, S., Chatterton, N.J., and Harrison, P.A., Separation of fructooligosaccharide isomers by anion-exchange chromatography, *Agric. Biol. Chem.*, 55, 1427, 1991.

8. Timmermans, J.W., Van Leeuwen, M.B., Tournois, H., De Wit, D., and Vliegenthart, J.F.G., Quantitative analysis of the molecular weight distribution of inulin by means of anion exchange HPLC with pulsed amperometric detection, in *Proceedings of the Fourth Seminar on Inulin,* Fuchs, A., Ed., NRLO, The Hague, 1994 (NRLO report 94/4).

9. Heinze, B. and Praznik, W., Separation and purification of inulin oligomers and polymers by reversed-phase high-performance liquid chromatography., *J. Appl. Pol. Sci.: Appl. Pol. Symp.,* 48, 207, 1991.

10. Shaw, P.E. and Wilson, C.W., Separation of sorbitol and mannoheptulose from fructose, glucose and sucrose on reversed-phase and amine-modified HPLC columns, *J. Chromatogr. Sci.,* 20, 209, 1982.

11. Praznik, W., Beck, R.H.F., and Nitsch, E., Determination of fructan oligomers of degree of polymerization 2-30 by high-performance liquid chromatography, *J. Chromatogr.,* 303, 417, 1984.

12. Beck, R.H.F. and Praznik, W., Molecular characterization of fructans by high-performance gel chromatography, *J. Chromatogr.,* 369, 208, 1986.

13. Carpita, N.C., Housley, T.L., and Hendrix, J.E., New features of plant-fructan structure revealed by methylation analysis and carbon-13 NMR spectroscopy, *Carbohydr. Res.,* 217, 127, 1991.

14. Mischnick, P., Determination of the constituent distribution of O-(2-hydroxy-propyl) inulin by methylation analysis, *Starch/Stärke,* 50, 33, 1998.

15. Spies, T., Strukturuntersuchungen an Fruktanen, Ph.D thesis, Universität für Bodenkultur, Wien, 1991.

16. Gross, M. and Rudolph, K., Studies on the extracellular polysaccharides (EPS) produced in vitro by *Pseudomonas phaseolicola* III. Kinetics of levan and alginate formation in batch culture and demonstration of levansucrase activity in crude EPS, *J. Phytopathol.,* 119, 289, 1987.

17. Jang, K.H., Song, K.B., Kim, J.S., Kim, C.H., Chung, B.H., and Rhee, S.K., Production of levan using recombinant levansucrase immobilized on hydroxyapatite, *Bioproc. Eng.,* 23, 89, 2000.

18. Fett, W.F., Osman, S.F., Wijey, Ch., and Singh, S., *Expolysacharides of rRNA Group I Pseudomonads in Carbohydrates and Carbohydrate Polymers,* Yalpani, M., Ed., ATL Press, Science Publishers, Shrewsbury, MA, 1993.

19. Yun, J.W., Jung, K.H., Jeon, Y.J., and Lee, J.H., Continuous production of fructooligosaccharides from sucrose by immobilized cells of *Aureobasidium pullulans,* *J. Microbiol. Biotechnol.,* 2, 98, 1992.

20. Park, Y.K. and Almeida, M.M., Production of fructooligosaccharides from sucrose by a transfructosylase from *Aspergillus niger, World J. Microbiol. Biotechnol.,* 7, 331, 1991.

21. Van den Ende, W., Michiels A., De Roover, J., and Van Laere, A., Fructan biosynthetic and breakdown enzymes in dicots evolved from different invertases: expression of fructan genes throughout chicory development, *Sci. World J.,* 2, 273, 2002.

22. John, P., Fructan, in *Biosynthesis of the Major Crop Products,* John Wiley & Sons, Chichester, 1992, chap. 4.

23. Berghofer, E., Cramer, A., Schmidt, U., and Veigl, M., Pilot-scale production of inulin from chicory roots and its use in foodstuffs, in *Inulin and Inulin-Containing Crops, Studies in Plant Science, Vol. 3,* Fuchs, A., Ed., Elsevier Science, Amsterdam, 1993, pp. 77–84.

24. Van Beek, T.A., Maas, P., King, B.M., Leclercq, E., Voragen, A.G.J., and DeGroot, A., Bitter sesquiterpene lactones from chicory roots, *J. Agric. Food Chem.*, 38, 1035, 1990.

25. Dersch, G., Kammerer, S., and Praznik, W., Harvest dates and varietal effects on yield, concentration and composition of carbohydrates in chicory roots (*Cichorium intybus*), in *Abstracts of the Second European Symposium on Industrial Crops and Products, Pisa*, 22–24 November 1993.

26. Baert, J. and Van Waes, C., Effects of cultivar and harvest date on chicory root yield and inulin composition, in *Proceedings of the Seventh Seminar on Inulin*, Fuchs, A. and Van Laere, A., Eds., Leuven, Belgium, 22–23 January 1998.

27. Praznik, W. and Beck, R.H.F., Inulin composition during growth of tubers of *Helianthus tuberosus*, *Agric. Biol. Chem.*, 51, 1593, 1987.

28. Praznik, W., Cieslik, E., and Filipiak-Florkiewicz, A., Soluble dietary fibres in Jerusalem artichoke powders: composition and application in bread. *Nahrung/Food*, 46, 151, 2002.

29. Beck, R.H.F. and Praznik, W., Inulinhaltige Pflanzen als Rohstoffquelle. Biochemische und pflanzenphysiologische Aspekte, *Staerke*, 38, 391, 1986.

30. Barta J., Jerusalem artichoke as a multipurpose raw material for food products of high fructose or inulin content, in *Inulin and Inulin-Containing Crops*, Elsevier Science, Amsterdam, 1993, p. 323.

31. Eigner, W.-D., Abuja, P., Beck, R.H.F., and Praznik, W., Physicochemical characterization of inulin and sinistrin, *Carbohydr. Res.*, 180, 87, 1988.

32. Spies, T., Praznik, W., Hofinger, A., Altmann, F., Nitsch, E., and Wutka, R., The structure of the fructan sinistrin from *Urginea maritima*, *Carbohydr. Res.*, 235, 221, 1992.

33. Baumgartner, S., Dax, T.G., Praznik, W., and Falk, H., Characterisation of the high-molecular weight fructan isolated from garlic (*Allium sativum* L.), *Carbohydr. Res.*, 328, 177, 2000.

34. Fretzdorff, B., Kuhlmann, T., and Betsche, T., Fructan contents in wheat and rye grains and in milling fractions, poster presented at the 5th Karlsruhe Nutrition Congress, Karlsruhe, Germany, 22–24 October 2000.

14 Structure–Property Relationships in Chitosan

Aslak Einbu and Kjell M. Vårum

CONTENTS

14.1 Introduction 217
14.2 Composition and Molecular Mass 218
    14.2.1 Chemical Structure 218
    14.2.2 Sequence 220
    14.2.3 Molecular Mass 220
    14.2.4 Molecular Mass Distribution 221
14.3 Physical Properties 222
    14.3.1 Ion Binding 222
    14.3.2 Solubility and Charge Density 222
14.4 Chemical Stability 223
14.5 Enzymatic Degradation 225
14.6 Technical Properties 226
    14.6.1 Film-Forming Properties 226
    14.6.2 Gelling Properties 226
    14.6.3 Antimicrobial Properties 226
References 227

14.1 INTRODUCTION

Chitin, occurring as a structural polysaccharide in the outer skeleton of animals belonging to the phylum Arthropoda (animals with an outer skeleton) and a component of the cell walls of certain fungi and algae, is quite abundant. In contrast, chitosans are much less abundant in nature than chitin and have so far been found only in the cell walls of certain fungi. Chitin is the raw material for all commercial production of chitosans, with an estimated annual production of 2000 t.[1] Chitin is also used for a substantial production of glucosamine, with an estimated annual production of 4000 t.[1]

Our intention in this chapter is to give a state-of-the-art review of the structure–function relationships of chitosans as polysaccharides, and not to discuss how a certain physical property is utilized in a food product. Derivatives of chitin and chitosan are not covered, and those interested in this part are referred to a recent review article.[2] Functional properties such as water uptake, solubility parameters, stability, ion-binding properties, film forming, and antimicrobial properties are examples relevant for the use of chitin and chitosans in food applications. In addition, the application of chitosans in food for human consumption involve regulatory approval, although chitin and chitosan can generally be recognized as safe based on their traditional uses in different national food products; for example, in unpeeled shrimp products and the traditional Norwegian "old cheese" ("gamalost"), the yeast *Mucor mucedo*, which contains chitosan in the cell wall, is used in the fermentation process. Chitosan is currently used only in Asia only in human food applications. However, a number of patents describe the use of chitosan in food applications, some of which are listed in Table 14.1.

TABLE 14.1
Food Applications of Chitosan in Patent Literature

Diet sauces
Adsorption of triglycerides/cholesterol
Alcoholic beverages
Food preservation
Solution for cooking rice
Health-care enzymatic chitosan food
Structure-forming agent
CO_2-permeable packing films
Chitosan particle drink
Chitosan-stabilized peanut butter

Source: From Sandford, P.A., *Advances in Chitin Sciences,* Vårum, K.M., Domard, A., and Smidsrød, O., Eds., NTNU Trondheim, Trondheim, 6, 35, 2003. With permission.

14.2 COMPOSITION AND MOLECULAR MASS

14.2.1 CHEMICAL STRUCTURE

Chitin is a linear polymer of $(1\rightarrow4)$-linked 2-acetamido-2-deoxy-β-D-glucopyranose (GlcNAc; A-unit), which is insoluble in aqueous solvents. Chitin shares many structural similarities with cellulose, such as conformation of the monomers and diequatorial glycosidic linkages. Chitosans may be considered as a family of linear binary copolymers of $(1\rightarrow4)$-linked A-units and 2-amino-2-deoxy-β-D-glucopyranose (GlcN; D-unit) (Figure 14.1). The term *chitosan* does not refer to a uniquely defined compound; it merely refers to polysaccharides having different composition of A- and D-units. It has been proposed to define chitin and chitosan based on their solubility in aqueous acetic acid, that is, chitosan as soluble and chitin as insoluble

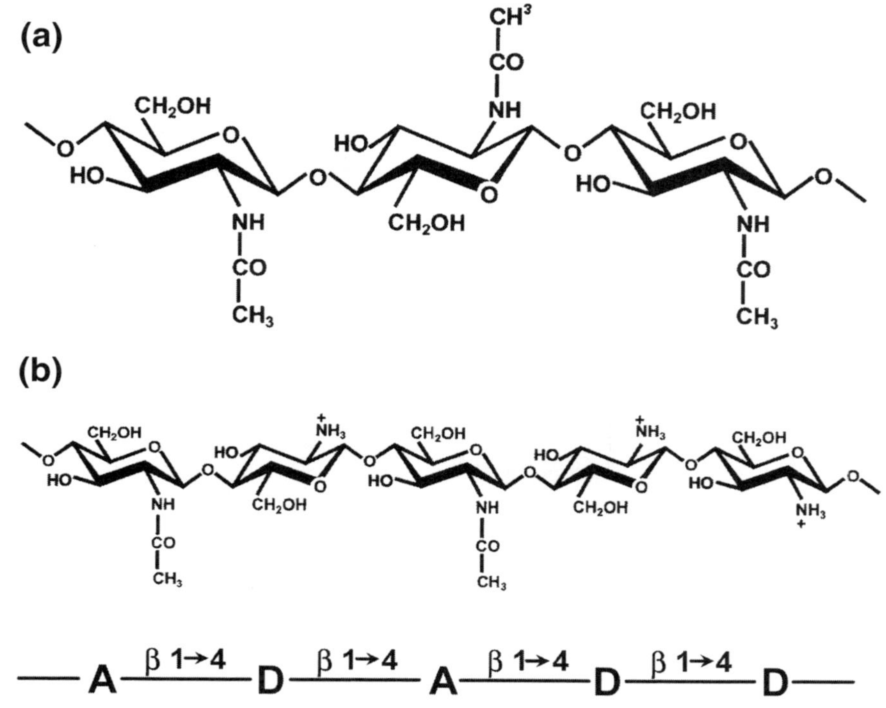

(a)

(b)

$$— \mathbf{A} \xrightarrow{\beta\ 1\to4} \mathbf{D} \xrightarrow{\beta\ 1\to4} \mathbf{A} \xrightarrow{\beta\ 1\to4} \mathbf{D} \xrightarrow{\beta\ 1\to4} \mathbf{D} —$$

FIGURE 14.1 Schematic representation of the chemical structure of chitin (a) and chitosan (b).

in 0.1 M acetic acid.[3] Commercially produced chitosans are normally prepared by alkaline deacetylation of chitin.

Several methods have been proposed and used for determining the degree of N-acetylation of chitosans: IR, UV, gel permeation chromatography, colloid titration, elemental analysis, dye adsorption, and acid–base titration methods.[3] These methods have repeatedly been reported to be only partially quantitative and relatively cumbersome. An enzymatic method based on complete hydrolysis of chitosan for determining the fraction of N-acetylated units, F_A, has been proposed,[4] in which quantification of GlcNAc and GlcN was done either colorimetrically or by HPLC analysis. Also, a similar method based on acid hydrolysis of the N-acetyl groups and detection of acetic acid by HPLC analysis and UV spectrometry has been reported.[5] This method is advantageous in that both chitosan and chitin (soluble and insoluble material) can be analyzed. Proton NMR spectroscopy is a convenient and accurate method for determining the chemical composition of chitosans,[6,7] and a typical proton NMR spectrum used to determine F_A is shown in Figure 14.2. NMR measurements of aqueous solutions of chitosans are, however, limited to samples that are soluble in the solvent, which limits the analysis to chitosans with F_A values lower than ca. 0.7 in aqueous solutions.

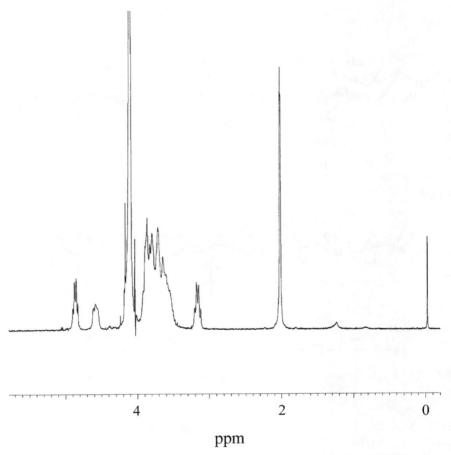

FIGURE 14.2 Typical ^{1}H-NMR spectrum (300 MHz) of solutions of a depolymerized chitosan with F_A 0.43 (10 mg/ml) in D_2O at pH 3 and 90°C. The resonance at 0 ppm is the internal reference TSP [sodium-(trimethylsilyl)-propionate-d_4], whereas the resonances at 4.8 and 4.6 ppm are H-1 of D-units and A-units, respectively. For further assignments of the resonances in the spectrum, see Vårum, K.M. et al., *Carbohydr. Res.*, 211, 17, 1991.

14.2.2 SEQUENCE

High-field ^{1}H and ^{13}C NMR spectroscopy has been used to determine sequences of the monomers in chitosans. However, it should be noted that high-resolution spectra are required to obtain information on sequence parameters of chitosans.[7,8] The units in the chitosan chain have been found to be randomly distributed in water-soluble, partially N-acetylated chitosans prepared from chitin by alkaline de-N-acetylation.[7,8]

14.2.3 MOLECULAR MASS

As polysaccharides in general, chitosans are polydisperse with respect to molecular weight. Because of this, the "molecular weight" of a chitosan is an average over the whole distribution of molecular weights. The molecular weight heterogeneity of polysaccharides can be described by several types of average molecular weight. The two most common methods in use for averaging are the number-average, M_n (which weighs the polymer molecules according to the *number* of molecules having a specific molecular weight) and the weight-average, M_w (which weighs the polymer molecules according to the *weight* of molecules having a specific molecular weight). In a population of molecules where N_i is the number of molecules and w_i the weight of molecules having a specific molecular weight M_i, these two averages are defined as:

$$M_n = \frac{\sum_i N_i M_i}{\sum_i N_i} \tag{14.1}$$

$$M_w = \frac{\sum_i w_i M_i}{\sum_i w_i} = \frac{\sum_i N_i M_i^2}{\sum_i N_i M_i} \tag{14.2}$$

Static light scattering has been used to determine M_w of chitosan.[9,10] However, osmometry is also a convenient method to determine M_n for chitosans.[11] The latter method is much less influenced by the presence of aggregates.[11,12]

14.2.4 MOLECULAR MASS DISTRIBUTION

The fraction M_w/M_n is called the polydispersity index (PI). In a polydisperse molecule population, the relation $M_w > M_n$ is always valid, whereas in a monodisperse molecule population $M_w = M_n$. Processes occurring during the production process or in the raw material prior to extraction may affect molecular weight distribution. Because of these processes, molecular weight distribution of chitin in its native form may differ from the distribution observed in the extracted chitin solutions. For a randomly degraded polymer, $M_w = 2M_n$. A PI of less than 2.0 may suggest that some fractionation has occurred during the production process. Certain procedures can also cause loss of the high-molecular-weight fraction or the low-molecular-weight tails of the distribution. A PI of more than 2.0 indicates a wider distribution and may suggest nonrandom degradation or the presence of aggregates.

By combining classical GPC (gel permeation chromatography) with a light-scattering detector, the molecular weight distribution can be determined for a sample. It is important that the GPC column separates the molecules over the entire molecular weight distribution, which may be a problem for high-molecular-weight samples. Ottøy et al.[13] have reported on analytical and semipreparative GPC of chitosans. Reversible interactions between chitosans and different column packings were found to occur, which strongly influenced the relationship between elution volume and molecular weight. Thus, care should be taken in the analysis of molecular weight distribution of chitosans by GPC systems calibrated with, for example, pullulan standards.

14.3 PHYSICAL PROPERTIES

14.3.1 ION BINDING

Chitosan form complexes with metal ions, particularly transition metal and post-transition ions.[3] In relation to food applications of chitosans, including the well-documented application of chitosans as a cholesterol-lowering agent[14,15] and the much more controversial use of chitosans as a weight-reducing agent,[16] knowledge on the (selective) binding of essential metal ions to chitosans is important. Most studies of ion binding to chitosan have been aimed at determining whether chitosan binds to a given ion, whereas only a few studies have involved determining the selectivity of binding of different ions to chitosans. However, recently Rhazi et al.[17] determined the selectivity of mixtures of the ions Cu > Hg > Zn > Cd > Ni > Co = Ca, using potentiometric and spectrometric methods. Vold et al.[18] reported the selectivity of different chitosans in binary mixtures of Cu^{2+}, Zn^{2+}, Cd^{2+}, and Ni^{2+}, showing that chitosan could bind Cu in large excess of the other metal ions.

14.3.2 SOLUBILITY AND CHARGE DENSITY

Solubilty at acidic pH and insolubility at basic pH is a characteristic property of commercial chitosans. Generally, three essential parameters determine the solubility of chitosans in water. The pH is the most obvious parameter, which is linked to the charge of the D-units. The ionic strength of the solvent is also an important parameter (salting-out effects). Furthermore, the content of ions in the solvent that specifically interact with chitosans (e.g., Cu and multivalent negative ions such as molybdate) can also limit the solubility of chitosans.

All chitosans are soluble at pHs below 6, but solubility decreases as pH increases. At a pH of 6 to 8, commercial chitosans precipitate on increase of pH. However, solubility increases with increasing F_A, as shown in Figure 14.3.[19] Chitosans of medium molecular weight with an F_A ca. 0.5 can be regarded as neutral soluble. Solubility differences can have a profound effect on the accessibility of chitosans to enzymes and on biological effects in the physiological pH range.

The charge density of chitosan, that is, the degree of protonation of amino groups, is determined by the chemical composition of the chitosan (F_A), the molecular weight, and external variables such as pH and ionic strength. Dissociation constants (pK_a) for chitosan range from 6.2 to 7, depending on the type of chitosan and conditions of measurement.[20–24] The titration behavior of chitosans has been studied by three different methods: colloid titration, ^1H NMR spectroscopy, and electrophoretic light scattering (ELS). The chemical shift of the H-2 resonance of the D-units relative to the internal standard TSP can be monitored as a function of increasing pH in proton NMR studies, as shown in Figure 14.4. This chemical shift is directly influenced by the charge of the nearby nitrogen atom. In experiments as shown in Figure 14.4, chitosans with F_As of 0.01 and 0.13 precipitate on increase of pH to approximately 6.5, making the titration curves incomplete. The chitosan with an F_A of 0.49 is soluble over the entire pH interval studied and a complete titration curve is recorded. It is apparent from Figure 14.4 that all chitosans show the same titration behavior. The same type of logistic regression as given previously

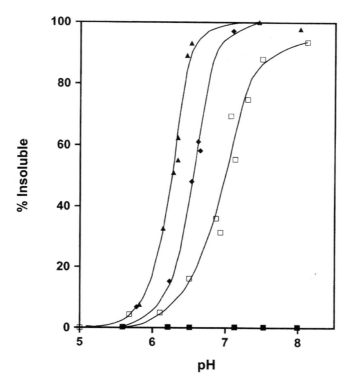

FIGURE 14.3 Solubility versus pH curves of chitosans. ▲: $F_A = 0.01$, ◆: $F_A = 0.17$, ⬚: F_A = 0.37, ■: $F_A = 0.60$. (From Vårum, K.M., Ottøy, M.H., and Smidsrød, *Carbohydr. Polym.*, 25, 65, 1994. With permission.)

made it possible to extrapolate beyond the measured points and determine the pK_a values. All three chitosans had the same pK_a values of 6.7, only slightly higher than those derived from the ELS study.

14.4 CHEMICAL STABILITY

For many applications of chitosans, including food applications, it is important to know the factors that determine and limit the stability of chitosans both in the solid state and in solution, and the chemical reactions responsible for the degradation. The glycosidic linkages are susceptible to both acid and alkaline degradation and oxidation by free radicals. However, because most applications of commercial chitosans are limited by its solubility at acidic pH, mostly the stability at low pH has been studied.

The acid-catalyzed degradation rates of chitosans have been found to be dependent on F_A (Figure 14.5), and the initial degradation rate constant was found to increase in direct proportion to F_A.[25] Acid hydrolysis was found to be highly specific to cleavage of A–A and A–D glycosidic linkages, which were found to be hydrolyzed three orders of magnitude higher than D–D and D–A linkages. This is probably due

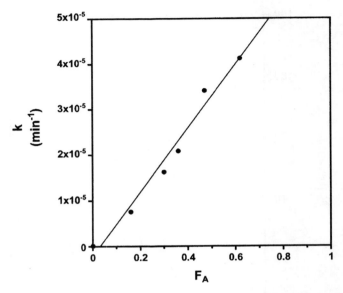

FIGURE 14.4 Titration curves of chitosans monitored by ^{1}H NMR. The chemical shifts differences between H-2 of D-units and the internal reference sodium-(trimethylsilyl)-propionate-d$_4$ (TSP) as a function of pH for chitosan. ●: $F_A = 0.01$, _: $F_A = 0.13$, and ▲: $F_A = 0.49$. Solid lines indicate logistic regression. (From Strand, S.P. et al., *Biomacromolecules,* 2, 1310, 2001. With permission.)

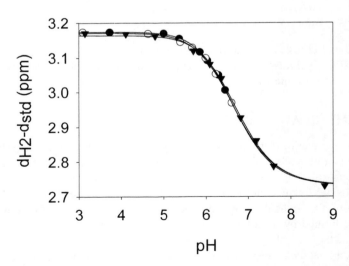

FIGURE 14.5 Degradation rate constants (k) as a function of F_A of chitosans, determined by the viscosity assay at a chitosan concentration of 1.5 mg/ml in 0.4 *M* HCl at 60°C. (From Vårum, K.M., Ottøy, M.H., and Smidsrød, O., Acid hydrolysis of chitosans, *Carbohydr. Polym.*, 46, 89, 2001. With permission).

to the combination of the decrease in the rate when a positively charged amino group is present close to the glycosidic linkage to be hydrolyzed, and the substrate-assisted mechanism by which a glycosidic linkage following an A-unit is cleaved.

Chitosan solutions can be sterilized by autoclaving (typically 120°C for 20 min), which may lead to depolymerization, depending on the pH and the chemical composition of the chitosan. We have found that autoclaving of chitosan solutions (F_As of 0.01, 0.35, and 0.60) at pH 4.5 at 120°C for 20 min does not reduce the intrinsic viscosity of the chitosan with an F_A of 0.01, whereas the intrinsic viscosities of the other two chitosans are only moderately reduced, from 760 to 550 ml/g for the chitosan with an F_A of 0.35 and from 820 to 600 ml/g for the chitosan with an F_A of 0.60.[26] It seems therefore that autoclaving can be used to sterilize chitosan solutions without severe depolymerization.

The stability of chitosan hydrochloride powders with different chemical compositions has been studied at 60, 80, 105, and 120°C.[27] It was concluded that the dominant degradation mechanism in the solid state of these chitosan salts is acid hydrolysis.

14.5 ENZYMATIC DEGRADATION

Chitosans may be enzymatically degraded by hydrolases widely distributed in nature. Lysozyme can also, in addition to its natural substrate (the glycosidic linkage of certain bacterial cell walls peptidoglycans), hydrolyze chitin and chitosans.[28,29]

Two different approaches have been used to investigate degradation rates and specificity of lysozyme-catalyzed degradation of chitosans. Viscometric studies, in which chitosans with widely varying F_As and known Bernoullian distribution of monomers have been used,[30,31] offer an easy and convenient experimental technique to study initial degradation rates, which avoids the transglycosylation reaction. Table 14.2 gives initial lysozyme degradation rates of chitosans with widely varying F_A

TABLE 14.2
Relative Lysozyme Degradation Rates of Chitosans with Different F_As

F_A	Relative Rate of Degradation
0.04	0.033
0.12	1
0.17	2
0.27	12
0.42	44
0.47	53
0.51	125
0.53	169
0.59	280

values, showing a very pronounced increase in the rate with increasing F_A. NMR studies of enzymatic degradation of chitosans[32,33] give more direct information on the specificities with respect to cleavage of the four differerent glycosidic linkages (A–A, A–D, D–A, and D–D) as determined by the identity of the new reducing and nonreducing ends, and in some cases also the variation in the identity of the nearest neighbours to the new reducing and nonreducing ends. From both viscometric and enzymatic studies of lysozyme (using both hen egg white and human lysozyme) degradation of chitosans, it was concluded that minimum four A-units had to be contained in the active site of lysozyme in order to obtain maximum degradation rates, explaining the dependence of lysozyme degradation rates of chitosans on F_As given in Table 14.2. It was also found that chitosans with a low degree of N-acetylation may bind specifically to lysozyme[34,35] without cleaving the polysaccharide, which was recently demonstrated by using immobilized lysozyme to fractionate chitosan chains containing A-units from fully de-N-acetylated chains.[36]

14.6 TECHNICAL PROPERTIES

14.6.1 FILM-FORMING PROPERTIES

Chitin and especially chitosan have been extensively studied for applications as films or membranes. The use of such films to improve the quality of foods has been examined[37] and is especially interesting due to the antimicrobial action of chitosan.[38] The preparation of a composite film of cellulose and an almost fully de-N-acetylated and highly viscous chitosan has been reported.[39] A Schiff's base is formed between the carbonyl groups on the cellulose and the amino groups on chitosan, resulting in a film in which the chitosan does not dissolve in water and has good wet tensile strengths. Uragami has reported the preparation and characteristics of different chitosan membranes.[40]

14.6.2 GELLING PROPERTIES

The preparation of chitin gels has been reported.[41–43] The preparation and characterization of a few hydrogels of chitosan have also been reported, such as thermoreversible chitosan-oxalate gels.[44,45] So far, no simple ionic and nontoxic cross-linking agent has been found that gives reproducible chitosan gels at low concentrations, such as calcium ions for gelling of alginates. However, aqueous chitosan gels cross-linked with molybdate polyoxy-anions have been reported, resulting in transparent, thermoirreversible gels that are able to swell several times their original size in aqueous solutions, depending on the ionic strength.[46]

Different chitosan gels made with covalent cross-linking have been reported, with cross-linking with glutaraldehyde being the most widely applied.[47–49] Recently, an enzymatic gelling system with chitosan has been reported.[50,51]

14.6.3 ANTIMICROBIAL PROPERTIES

Although the mechanism(s) involved in the antimicrobial activity of chitosans has not been explained, a number of studies support that the polycation chitosan does actually prevent the growth of certain microorganisms.[52-55] Such studies have again led to suggestions that chitosan can be used as a food preservative, and the reader is referred to a recent review on this subject.[38]

REFERENCES

1. Sandford, P.A., Commercial sources of chitin and chitosan and their utilization, in *Advances in Chitin Sciences*, Vårum, K.M., Domard, A., and Smidsrød, O., Eds., NTNU Trondheim, Trondheim, 6, 35, 2003.
2. Kurita, K., Controlled functionalization of the polysaccharide chitin, *Prog. Polym. Sci.* 26, 1921, 2001.
3. Roberts, G.A.F., *Chitin Chemistry*, Macmillan, Houndmills, UK, 1992.
4. Nanjo, F., Katsumi, R. and Sakai, K., Enzymatic method for determination of the degree of deacetylation of chitosans, *Anal. Biochem.*, 193, 164, 1991.
5. Niola, F. et al., A rapid method for the determination of the degree of *N*-acetylation of chitin-chitosan samples by acid hydrolysis and HPLC, *Carbohydr. Res.*, 238, 1, 1993.
6. Hirano, S. and Yamaguchi, R., *N*-Acetylchitosan gel: a polyhydrate of chitin, *Biopolymers*, 15, 1685, 1.
7. Vårum, K.M. et al., Determination of the degree of *N*-acetylation and the distribution of acetyl groups in partially *N*-deacetylated chitins (chitosans) by high-field NMR spectroscopy. Part I: High-field NMR spectroscopy of partially *N*-deacetylated chitins (chitosans). *Carbohydr. Res.*, 211, 17, 1991.
8. Vårum, K.M. et al., ¹³C-NMR studies of acetylation sequences in partially *N*-deacetylated chitins (chitosans). Part II: High-field NMR spectroscopy of partially *N*-deacetylated chitins (chitosans). *Carbohydr. Res.*, 217, 19, 1991.
9. Terbojevich, M. et al., Chitosan: chain rigidity and mesophase formation, *Carbohydr. Res.*, 209, 251, 1991.
10. Terbojevich, M. et al., Chitosans from *Euphasia superba*. Part 1: Solution properties, *Carbohydr. Polym.*, 18, 35, 1992.
11. Anthonsen, M.W. et al., Aggregates in acidic solutions of chitosans detected by static laser light scattering, *Carbohydr. Polym.*, 25, 13, 1994.
12. Amiji, M.M., Pyrene fluorescence study of chitosan self-association in aqueous solutions, *Carbohydr. Polym.*, 26, 211, 1995.
13. Ottøy, M.H. et al., Preparative and analytical size-exclusion chromatography of chitosans, *Carbohyd. Polym.*, 31, 253, 1996.
14. Koide, S.S., Chitin-chitosan: properties, benefits and risks, *Nutr. Res.*, 18, 1091, 1998.
15. Ormrod, D.J., Holmes, C.C., and Miller, T.E., Dietary chitosan inhibits hypercholesterolaemia and atherogenesis in the apolipoprotein E: deficient mouse model at atherosclerosis, *Atherosclerosis*, 138, 329, 1998.
16. Ernst, E. and Pittler, M.H., Chitosan as a treatment for body weight reduction? A meta-analysis, *Perfusion*, 11, 461, 1998.
17. Rhazi, M. et al., Contribution to the study of the complexation of copper by chitosan and oligomers, *Polymer*, 43, 1267, 2002.
18. Vold, I.M.N,, Vårum, K.M., Guibal, E., and Smidsrød, O., Binding of ions to chitosan: selectivity studies, submitted, *Carbohydrate Polymers*, 2003.

19. Vårum, K.M., Ottøy, M.H., and Smidsrød, O., Water-solubility of partially N-acetylated chitosans as a function of pH: effect of chemical composition and depolymerization, *Carbohydr. Polym.*, 25, 65, 1994.

20. Domard, A., pH and CD measurements on a fully deacetylated chitosan: application to copper(II)-polymer interactions, *Int. J. Biol. Macromol.*, 9, 98, 1987.

21. Rinaudo, M. and Domard, A., Solution properties of chitosan, in *Chitin and Chitosan,* Skjåk-Bræk G., Anthonsen T., and Sandford P., Eds., Elsevier Applied Science, London, 1989, 71.

22. Anthonsen, M.W. and Smidsrød, O., Hydrogen ion titration of chitosan with varying degrees of N-acetylation by monitoring induced ¹H-NMR chemical shifts, *Carbohydr. Polym.*, 26, 303, 1995.

23. Sorlier, P. et al., Relation between the degree of acetylation and the electrostatic properties of chitin and chitosan, *Biomacromolecules,* 2, 765, 2001.

24. Strand, S.P. et al., Electrophoretic light scattering studies of chitosans with different degrees of N-acetylation, *Biomacromolecules,* 2, 1310, 2001.

25. Vårum, K.M., Ottøy, M.H., and Smidsrød, O., Acid hydrolysis of chitosans, *Carbohydr. Polym.*, 46, 89, 2001.

26. Vårum, K.M., unpublished results, 1996.

27. Holme, H.K. et al., Thermal depolymerization of chitosan chloride, *Carbohydr. Polym.*, 46, 287, 2001.

28. Berger, L.R. and Weiser, R.S., The glucosaminidase activity of egg-white lysozyme, *Biochim. Biophys. Acta,* 26, 517, 1957.

29. Amano, K.I. and Ito, E., The action of lysozyme on partially deacetylated chitin, *Eur. J. Biochem.*, 85, 97, 1978.

30. Nordtveit, R. J., Vårum, K.M., and Smidsrød, O., Degradation of fully water-soluble, partially N-acetylated chitosans with lysozyme, *Carbohydr. Polym.*, 23, 253, 1994.

31. Nordtveit, R.J., Vårum, K.M., and Smidsrød, O., Degradation of partially N-acetylated chitosans with hen egg white and human lysozyme, *Carbohydr. Polym.*, 29, 163, 1996.

32. Stokke, B.T. et al., Sequence specificities for lysozyme depolymerization of partially N-acetylated chitosans. *Can. J. Chem.*, 73, 1972, 1995.

33. Vårum, K.M. et al., Determination of enzymatic hydrolysis specificity of partially N-acetylated chitosans. *Biochim. Biophys. Acta,* 1291, 5, 1996.

34. Kristiansen, A., Vårum, K.M., and Grasdalen, H. Competitive binding of highly de-N-acetylated chitosans and N-N'-diacetylchitobiose to lysozyme from chicken egg white studied by ¹H NMR spectroscopy. *Carbohydr. Res.*, 289, 143, 1996.

35. Kristiansen, A., Vårum, K.M., and Grasdalen, H., Interactions between highly de-N-acetylated chitosans and lysozyme from chicken egg white studied by ¹H-NMR spectroscopy, *Eur. J. Biochem.*, 251, 335, 1998.

36. Sasaki, C., Kristiansen, A., Fukamizo, T., and Vårum, K.M., Fractionation of chitosan, submitted, *Biomacromolecules*, 2003.

37. Labuza, T.P. and Breene, W.M., Applications of 'active packaging' for improvement of shelf-life and nutritional quality of fresh and extended shelf-life foods, *J. Food Proc. Preserv.*, 13, 1, 1989.

38. Roller, S, The antimicrobial action of chitosan: laboratory curiosity or novel food preservative? in *Advances in Chitin Sciences,* Vårum, K.M., Domard, A., and Smidsrød, O., Eds., NTNU Trondheim, Trondheim, 6, 43, 2003.

39. Hosokawa, J. et al., Reaction between chitosan and cellulose on biodegradable composite film formation, *Ind. Eng. Chem. Res.*, 30, 788, 1991.

40. Uragami, T. Preparation and characteristics of chitosan membranes. In *Chitin Handbook*, Muzzarelli, R.A.A. and Peter, M.G., Eds., 1997, 451.

41. Hirano, S., Kondo, Y., and Ohe, Y., Chitosan gel: a novel polysaccharide gel. *Polymer*, 16, 622, 1975.
42. Hirano, S., Tsuchida, H., and Nagao, N., *N*-acetylation in chitosan and the rate of its enzymic hydrolysis, *Biomaterials*, 10, 574, 1989.
43. Vachoud, L., Zydowicz, N., and Domard, A., Formation and characterisation of a physical chitin gel, *Carbohydr. Res.*, 302, 169, 1997.
44. Hayes, E.R. and Davies, D.H., Characterization of chitosan. I. Thermoreversible chitosan gels, in *Proceedings of the 1st International Conference on Chitin and Chitosan*, 193, 1978.
45. Hirano, H. et al., A chitosan oxalate gel: its conversion to an *N*-acetylchitosan gel via a chitosan gel, *Carbohydr. Res.*, 201, 145, 1990.
46. Draget, K.I. et al., Chitosan cross-linked with Mo(VI) polyoxyanions: a new gelling system, *Biomaterials*, 13, 635, 1992.
47. Hirano, S. et al., Chitosan-aldehyde gel: a novel polysaccharide gel produced from chitosan and aldehydes. *Agric. Biol. Chem.*, 41, 1547, 1977.
48. Roberts, G.A.F. and Taylor, K.E., Chitosan gels. 3. The formation of gels by reaction of chitosan with glutaraldehyde. *Macromol. Chem.*, 190, 951, 1989.
49. Draget, K.I., Associating phenomena in highly acetylated chitosan gels, *Polym. Gels Net.*, 4, 143, 1996.
50. Kumar, G. et al., *In situ* chitosan gelation using the enzyme tyrosinase, *Polymer*, 41, 2157, 2000.
51. Chen, T. et al., Enzymatic coupling of natural products to chitosan to create functional derivatives, in *Advances in Chitin Sciences,* Vårum, K.M., Domard, A., and Smidsrød, O., Eds., NTNU Trondheim, Trondheim, 6, 63, 2003.
52. Wang, G.-H., Inhibition and inactivation of five species of foodborn pathogens by chitosan, *J. Food Prot.,* 55, 916, 1992.
53. Simpson, B.K. et al., Utilization of chitosan for preservation of raw shrimp (*Pandalus borealis*), *Food Biotechnol.,* 11, 25, 1997.
54. Sudharshan, N.R., Hoover, D.G. and Knorr, D., Antibacterial action of chitosan, *Food Biotechnol.*, 6, 257, 1992.
55. El Ghaouth, A. et al., Antifungal activity of chitosan on post-harvest pathogens: induction of morphological and cytological alterations in *Rhizopus stolonifer, Mycol. Res.*, 96, 769, 1992.

15 Plant and Algal Gums and Mucilages

Lawrence Ramsden

CONTENTS

15.1 Introduction 232
    15.1.1 Structure of Plant and Algal Gums 234
    15.1.2 Gum Rheology 234
15.2 Wall Polysaccharides 234
    15.2.1 β-Glucans 236
    15.2.2 Arabinoxylans 237
    15.2.3 Arabinogalactans (Larch) 237
    15.2.4 Pectins (Rhamnogalacturonans) 238
15.3 Soluble Mucilage Polysaccharides 238
    15.3.1 Yellow Mustard Seed Gum 238
15.4 Reserve Polysaccharides 239
    15.4.1 Legume Seed Galactomannans 239
        15.4.1.1 Locust Bean Gum...239
        15.4.1.2 Guar Gum 240
        15.4.1.3 Tara Gum 240
        15.4.1.4 Cassia Gum 241
        15.4.1.5 Fenugreek Gum 241
    15.4.2 Glucomannans 241
    15.4.3 Flaxseed Gum 241
    15.4.4 Psyllium Gum 242
    15.4.5 Tamarind Gum 242
    15.4.6 Quince Seed Gum 242
15.5 Exudates 243
    15.5.1 Arabinogalactan Proteoglycans 243
        15.5.1.1 Gum Arabic 243
        15.5.1.2 Karaya Gum 244
        15.5.1.3 Gum Tragacanth 244
        15.5.1.4 Gum Ghatti 245
        15.5.1.5 Okra Gum 245

0-8493-1486-0/04/$0.00+$1.50
© 2004 by CRC Press LLC

　　　　　　15.5.1.6　Marigold Flower Gum 245
　　　　　　15.5.1.7　Other Arabinogalactan Protein Gums 245
15.6　Algal Gums 246
　　　15.6.1　Alginates 246
　　　15.6.2　Agars 247
　　　15.6.3　Carrageenans 247
　　　15.6.4　Furcellarans 249
15.7　Closing Remarks 249
References..249

15.1　INTRODUCTION

Plant gums are generally taken to include all plant polysaccharides other than starch that can be hydrated and have a commercial application. The most abundant plant polysaccharide, cellulose, is in a slightly ambiguous position, as naturally occurring, it is clearly insoluble and would appear difficult to consider as a gum. However, although microcrystalline cellulose and soluble modified cellulose are important commercial gums, these are outside the scope of this chapter, which is concerned solely with naturally occurring gums. A polysaccharide can be regarded as mucilage when a high degree of solubility is accompanied by easy hydration and a reduced viscosity. There is also a difference in the origin of gums and mucilages. Mucilages are usually of secretory nature whereas the gums are either cell wall components or intended as reserve polysaccharides to provide energy store for the plant. These categories are not entirely exclusive because highly soluble wall components such as pectins can be readily liberated from the wall and function as mucilage when the wall is fully hydrated. Also, an important class of gums are produced as plant surface exudates, usually as a consequence of plant damage. Algal gums are derived from the wall mucilage of seaweeds.

This chapter focuses on the structure and functional properties of the gums and their value in food applications. Gums have been described in many different ways and are often categorized according to their plant source or their chemical composition. Because gums with related properties are often present in more than one plant species, gums are considered here by their functional source and chemical properties. Despite their importance to the food industry and high research interest, fully comprehensive reviews are comparatively rare, but detailed chemistry and structure of many gums have been covered in reviews.[1-5] Other works have focused on particular hydrocolloids, and references to these are given under the appropriate section.

The market for natural plant and algal gums for food use has been dominated for the past 50 years by the "big five:" pectins, galactomannans and gum arabic from plants and alginate and carrageenan from seaweeds. Although many other plant gums are available, including those known since antiquity and those introduced more recently, there has been no major change in this position. Recently, there have been greater incentives to look for new polysaccharides of natural origin, leading to the emergence of several new gum candidates as well as reemergence of some older gums that have yet to realize their full potential.

As regards the overall gum market, there has been a dramatic shift due to the introduction of new chemical derivatives of natural polysaccharides and of new products of microbial synthesis. The market for gums and food additives has been reviewed in 1987[6] and more recently in 2000.[7] There is a continuing trend for expansion of the market for new gums, but as the market overall has increased, there has been no drop in demand for traditional plant and algal products.

The major uses of plant and algal gums are as hydrocolloid texture modifiers and viscosity enhancers to control the rheology of manufactured food products. Some gums are also used as stabilizers or emulsifiers. For particular gums, references are made to rheological properties that are of importance for their food uses, and the significance of these properties is discussed in more detail in Chapter 11.

Polysaccharide hydrocolloids for the food industry are also available from another source since the development of the gums of microbial origin, notably xanthan and gellan. These gums are produced from microbial cultures and are discussed further in Chapter 10. Gums have provided food producers with new rheologies and have been rapidly adopted by manufacturers. Table 15.1 gives the relative contribution of gums from different sources to the hydrocolloid market.

Gums of both plant and algal origin are generally considered as indigestible and forming part of the dietary fiber. Most of the polysaccharide linkages found in gums are resistant to digestion as the human gut lacks the enzymes necessary to break these bonds. Gums are, however, susceptible to microbial degradation, and break-down of gums by intestinal microbes can release absorbable monosaccharides or other absorbable products of microbial metabolism. The relative contribution of such sources to human nutrition is a matter of current debate for various gums. In the case of gum arabic, which is included at relatively high levels in some confectionery, a significant contribution to energy intake can be presumed.[8]

TABLE 15.1
World Markets for Nonstarch Natural Food Hydrocolloids

Gum	% by Volume%	% by Value
Algal polysaccharides	36.9	32.7
Reserve polysaccharides	22.7	16.4
Plant exudates	17.0	13.6
Wall polysaccharides	14.2	22.7
Microbial polysaccharides	9.1	14.5
World total	100	100

Source: From Phillips, G.O. and Williams, P.A., Handbook of Hydrocolloids, CRC Press, Boca Raton, 2000 and Lillford, P.J., in Gums and Stabilisers for the Food Industry, Vol.10, Williams, P.A. and. Phillips, G.O., Eds., RSC Cambridge, 2000, p. 387.

15.1.1 STRUCTURE OF PLANT AND ALGAL GUMS

As natural products, plant gums display a wide range of structures, which can vary both between and within species. Composition and size of polysaccharide molecules can further fluctuate, depending on genotype of the source crop, environmental conditions during cultivation, and age at harvest. Precise structures are therefore not so useful, and typically a gum is described in terms of the percentage or ratio of its main monosaccharide components, which can carry a range of minor substituents. The impact such variation in structure might have must be determined experimentally for each batch of gum unless guaranteed by the supplier. Algal gums show a similar variation in structure but differ as a class from plant gums by the inclusion of some different residues that are not represented in plants and by the high levels of sulfation that can occur. Table 15.2 summarizes the principal monosaccharide components of major gums.

The functional properties of gums largely depend on the conformation the polysaccharide can adopt in solution, which, in turn, depends on the composition and linkages of the main polysaccharide chain and the presence of side groups.[9] The properties of polysaccharides in solution are dominated by hydrogen bonding, and the two major conformations adopted by polysaccharide chains are either a random coil or a helix stabilized by regular intrachain hydrogen bonds. In helix-forming polysaccharides, double helices can often form where hydrogen bonds occur between two chains.

15.1.2 GUM RHEOLOGY

The key property of plant and algal gums that has led to their widespread use in food is their ability to modify the rheology of aqueous systems at low levels of addition. Apart from pectins, plant gums do not readily form gels, but due to their large molecular size, when hydrated, they provide large increases in viscosity. Rheology is generally pseudoplastic, with a decrease in viscosity at high shear rates, though some gums show Newtonian behavior where viscosity remains constant. Neutral polysaccharides can show rheology that is pH-stable, but where ionizable groups are present an optimum pH for maximum viscosity is observed. Table 15.3 summarizes the viscosity for solutions of different hydrocolloids. However, all gums can show a wide range of viscosities, depending on the average chain length of the sample used, and the values given are only broadly indicative of the general range for the class of hydrocolloid. Algal gums are typically gel forming. Gel formation requires the ability for cross-links to form between chains without binding along the entire length of the chain. A number of books on the rheology of food polysaccharides provide good coverage of plant and algal gums.[10]

15.2 WALL POLYSACCHARIDES

A major site of polysaccharide synthesis in any plant cell is in the production of polysaccharides for the formation of the cell wall. The principal component of all plant cell walls is cellulose. Natural cellulose cannot function as a hydrocolloid due

TABLE 15.2

Major Monosaccharide Components of Different Plant and Algal Gums

Gum	Monosaccharides Present (In Order of Abundance)	Main Chain
Plant Cell Wall Polysaccharides		
β-Glucans (cereals)	Glucose	Glucan
Arabinoxylans	Xylose, arabinose	Xylan
Larch arabinogalactan	Galactose, arabinose	Galactan
Pectins	Galacturonic acid, rhamnose	Galacturonan
Plant Reserve Polysaccharides		
Yellow mustard seed	Glucose, galacturonic acid, mannose	Glucan
Guar gum	Mannose, galactose	Mannan
Locust bean gum	Mannose, galactose	Mannan
Tara gum	Mannose, galactose	Mannan
Cassia gum	Mannose, galactose	Mannan
Fenugreek gum	Mannose, galactose	Mannan
Glucomannan	Glucose, mannose	Glucomannan
Tamarind gum	Glucose, xylose, galactose	Glucan
Flaxseed gum	Glucose, xylose, galactose, rhamnose	Xylan
Okra gum	Galactose, rhamnose, galacturonic acid, arabinose	Galactan
Psyllium gum	Xylose, arabinose, rhamnose, galactose	Galactan
Quince seed gum	Galactose, arabinose, xylose	Galactan
Plant Exudates		
Gum Arabic	Galactose, arabinose, rhamnose, glucuronic acid,	Galactan
Karaya gum	Galactose, rhamnose	Galactan
Gum tragacanth	Galactose, fucose, xylose, arabinose	Galactan
Gum ghatti	Arabinose, galactose	Galactan
Algal Polysaccharides		
Alginate	Guluronic acid, mannuronic acid	Guluromannuronan
Agar	D-Galactose, 3,6-anhydro-L-galactose	Galactan
Carrageenan (kappa)	3,6-Anhydro-D-galactose, D-galactose-4-sulfate	Galactan
Furcelluran	Galactose, D-galactose 2, 4, or 6 sulfated, 3,6-anhydro-D-galactos	Galactan

TABLE 15.3
Indicative Viscosities of Gum Solutions

Gum	Conc.(%)	Apparent Viscosity (cps)	Reference
Arabic	1	50	2
Karaya	0.5	250	2
Gum ghatti	5	288	2
Tragacanth	1	3500	1
Arabinoxylan	1	3000	2
Larch arabinogalactan	10	2	2
Guar gum	1	4800	38
Locust bean gum	1	2000	32
Tara gum	1	3000	39
Glucomananan, konjac	1	30,000	49
Flaxseed gum	1	600	13
Psyllium gum	0.5	300	2
Tamarind gum	1	200	62
Quince seed gum	1	4000	2
Alginate	1	500	2
Alginate	2	5000	2
Agar	1.5	4[a]	98
Carrageenan kappa	1	300[a]	49
Furcelluran	1.5	80[a]	2

Note: Values for neutral pH, room temperature, and moderate shear (20 rpm).
[a] Values for hot viscosity at 60°C.

to the very large size and insolubility of the molecule. Cell walls also contain a large variety of other polysaccharides of different sizes, composition, and solubility, many of which can act as hydrocolloids. In the plant, these noncellulosic wall components provide essential structural support for the wall. The more water-soluble polysaccharides can be extracted from the wall and readily purified.

15.2.1 β-Glucans

β-glucans are a class of wall polysaccharides found in cereal grains, with a linear chain of glucose residues joined by β(1-3) and β(1-4) linkages. β-Glucans are typically obtained by alkaline hot-water extraction of oat and barley, which can remove about 50% of the glucans present. Higher yields can be obtained by more complicated processes for both oats[11] and barley.[12] β-glucans have a linear chain with stretches of (1-3) linkages involving three or four glucose residues interspersed with stretches of three or four (1-4)-linked glucose. β-glucans are generally cold-water soluble, though some insoluble fractions might be present. Solutions have moderate viscosity and show pseudoplastic behavior. Presence of some low-molecular-weight fractions in β-glucan samples can give rise to unusual rheology.[13]

β-Glucans at a 6% concentration can form gels after storage at low temperature. Junction zones can form from the parallel alignment of cellulose-like stretches of β(1-4)-linked glucoses. Regions of the chain containing β(1-3)-linked glucose will not bind to other chains as they have a tendency to form helical structures. The gels are thermoreversible, with melting points ca. 65°C.

β-Glucans were first observed as a problem in the brewing industry due to the viscosity of barley β-glucan, which prevented free flow of brewery products during fermentations. Subsequently, oat β-glucans were found to have beneficial properties in lowering serum cholesterol, with the potential for reducing the risk of heart disease.[14] Oat β-glucans have now been included in formulations of many health-related and dietary products.[15]

15.2.2 ARABINOXYLANS

Cell walls of some grasses contain arabinoxylans, and these are particularly abundant in the cereals wheat and rye. Arabinoxylans (sometimes called pentosans) differ in their solubility, with some being water soluble and others water insoluble. Water-soluble fractions are extracted from the cereal flour by stirring with water, followed by centrifugation to remove insoluble matter. Contamination by starch and protein can be reduced by incubation with amylases and proteases. Water-insoluble arabinoxylans require alkaline conditions for their extraction, which is improved by the addition of barium ions.

The basic structure of an arabinoxylan is a β(1-4)-linked chain of xylose units, which are substituted with varying degrees of arabinose linked (1-2) or (1-3) to the xylose.[16] The difference in solubility is related to the degree of substitution. Insoluble arabinoxylans have few arabinose side substituents, and the long regions of smooth xylan chains can associate with each other to produce large insoluble complexes. In water-soluble arabinoxylans, close association of the chains is prevented by the higher number of arabinose substituents.[17]

Arabinoxylans form highly viscous solutions that show pseudoplastic behavior, and can form weak gels at low shear. However, in presence of some oxidizing agents, a much stronger and more stable gel can be formed, which involves participation of ferulic acid in formation of cross-links between arabinoxylan chains.[18] Cross-linking of arabinoxylans occurs during baking and adds texture to bread. Natural levels of arabinoxylans in wheat flour are an important determinant of bread quality but so far have not found application as food additives. Potential applications can arise from the behavior observed with mixed gums present naturally in cereals. Wheat contains not only arabinoxylans but also β-glucans, and water-soluble extracts of wheat bran containing arabinoxylan and β-glucan in a ratio of 3:1 show gelling behavior at a low concentration (2%).[19]

15.2.3 ARABINOGALACTANS (LARCH)

Arabinogalactans are obtained from larch wood, where they are present in high concentration as a wall polysaccharide. The polysaccharide shows high water solubility and can be extracted from wood chips in warm water, low temperatures reducing contamination by wood phenolics. The gum is soluble in water up to 60% concentration and has a low viscosity with Newtonian behavior, which is largely stable to changes in pH and salts.[20] The gum has a molecular weight between 50,000 and 100,000, probably due to aggregate formation between smaller chains,[21] and the main components galactose and arabinose are present in a variable ratio between 8:1 and 23:1 as an α(1-3)-linked galactan backbone with (α-6)- linked arabinose and galactose substituents.[22] The gum has achieved only limited use in the food market since its introduction as a stabilizer in sweeteners. The advantage of the gum is that it can be produced as a by-product from the wood milling industry, but the main disadvantage compared with other gums is that it has to be added at high levels to exert an effect.

15.2.4 Pectins (Rhamnogalacturonans)

Pectins are a major class of wall-related plant polysaccharides that are used in food and are considered in more detail in Chapter 12. Pectins are charged, acidic polysaccharides with a variety of structures based on a rhamnogalacturonan. Unlike other plant gums, they can exhibit gelling properties which are of considerable value to the food industry. Here we discuss only those water-soluble rhamnogalacturonans that are not considered as traditional pectins in the food industry.

15.3 SOLUBLE MUCILAGE POLYSACCHARIDES

15.3.1 Yellow Mustard Seed Gum

Yellow mustard seed gum (YMG) is obtained from the seeds of yellow mustard (Sinapsis alba), which release a mucilage from the seeds when wet. Commercially, the gum can be prepared from ground seeds by cold-water extraction[23] and the seed bran can also be used as raw material for hot-water extraction,[24] which is economically advantageous as bran is a by-product of mustard processing. YMG is a mixture of several polysaccharides containing two major water-soluble components, one an acidic rhamnogalacturonan and the other β(1-4)-glucans with varying degrees of ether substitution.[25]

YMG exhibits some useful functional properties. Emulsifiant action has been reported, with crude mucilage showing an emulsion capacity similar to that of xanthan gum.[26] Dilute solutions of YMG show shear-thinning behavior,[27] a viscosity decrease with increase in temperature, reasonable stability with changing ionic concentration, and a minimum of shear thinning at pH 7. The rheology of YMG shows synergistic effects with galactomannans. Mixtures of LBG and YBG give large increases in viscosity, and YMG–LBG mixtures in the ratio 9:1 show gel-like properties.[28] Synergy is also observed with guar and fenugreek galactomannans, but the effect is strongest with LBG.

YMG can act as a useful complement to other hydrocolloids as a stabilizer and has been tested in a salad cream, giving equivalent performance to that by commercial xanthan-stabilized products.[29] YMG is not digested, acts as soluble dietary fiber, and has shown some beneficial effects of reducing glycemic index when added to the diet.[30]

15.4 RESERVE POLYSACCHARIDES

Seed-bearing plants deposit energy-containing reserves to support growth of the embryo within the seed. Seed reserves can be of protein, lipid, or polysaccharide character. Although the major reserve polysaccharide is indisputably starch, a number of other polysaccharides are used, particularly within the plant family Leguminoseae. Unlike starch, which is stored in amyloplasts in the cytoplasm, nonstarch reserve polysaccharides in seeds are deposited in the cell wall. Reserve gums are typically found in seeds of plants growing in arid regions, where the ability of the polysaccharide to bind water increases survival of germinating seeds in conditions of fluctuating water availability. Plants can also deposit energy reserves in underground storage organs such as roots or tubers, and some root-derived polysaccharides are used in the food industry.

15.4.1 LEGUME SEED GALACTOMANNANS

The most important plant gums by volume used in the food industry are the galactomannans derived from legume seeds. Four main gums used at present, though many other potential species might provide useful hydrocolloids as yet untested in the laboratory or the market. There is guar gum from the seeds of *Cyamopsis tetragonoloba*, locust bean gum (LBG) from seeds of *Ceratonia siliqua* (carob tree), tara gum from seeds of *Cesalpinia spinosa*, and cassia gum from seeds of Cassia obtusifolia. All seed galactomannans possess the same basic structure of a β(1-4)-linked mannan backbone with varying degrees of substitution with α(1-6)-linked galactose residues. The ratio of galactose to mannose is roughly constant for a given species and provides a useful diagnostic to identify gum samples. The distribution of galactose along the mannan chain in both LBG and guar gum is not uniform but tends to be grouped in blocks of high substitution (hairy regions), which are separated by intervening stretches with few galactose residues (smooth regions).

Preparation of the gums involves dehulling of seeds, crushing to remove the embryo, followed by milling of the endosperm to produce crude flour. The flour can be purified by dissolving in hot water, followed by filtration and precipitation with isopropanol to remove impurities.

15.4.1.1 Locust Bean Gum

In the Mediterranean and the Middle East, LBG has been used in traditional food preparation for hundreds of years as well as in other nonfood uses as described by classical authors. The main LBG-producing regions remain in the Mediterranean and Middle East.

LBG has galactose-to-mannose ratio of 1:4 and molecular weights around 300,000. The rheology shows pseudoplastic behavior and decreases with temperature.[31] Cold-water solubility of LBG is low and dispersions need to be heated to 85°C to achieve good dissolution, concentrations of up to 5% w/v being possible. Low solubility is due to the tendency of the linear mannan chains to strongly hydrogen bond to each other in unsubstituted regions of the chain, limiting opportunities for interaction with water molecules. Different grades of LBG are marketed on the basis of molecular size classes and can provide some differences in rheology.

Although LBG does not itself form gels, it can be used together with other hydrocolloids to provide gel formation. This synergistic gel formation can be observed with other nongelling polysaccharides such as xanthan gum as well as with the gelling algal polysaccharides, that is, the agars, carrageenans,[32] and furcelleran. This property was attributed to the ability of nonsubstituted regions of the linear mannan backbone to form hydrogen bonds with helical regions of other hydrocolloids and provide cross-linking.[33,34] More recent observations have indicated that in the interaction with xanthan gum, both polysaccharides adopt a helical conformation and the junction between xanthan and LBG may be formed by a mixed double helix in which chains from both polysaccharides participate.[35,36] Synergistic gel formation can also be observed with other galactomannans, depending on the frequency of galactose substitution on the mannan chain.

LBG is widely used as a thickener and stabilizer in many foods such as ice cream, cheese spreads, salad creams, processed meats products, and pie fillings. It can also be used to prepare a chocolate substitute, carob chocolate. A major use of LBG and guar is in ice cream, where they can act as stabilizers to prevent ice crystal growth at low temperatures. The presence of large crystals imparts a rough texture to ice cream, reducing its palatability.

15.4.1.2 Guar Gum

Guar gum has long been used as a food ingredient in India but was developed commercially in the aftermath of World War II, which interrupted traditional supplies of LBG. Subsequently, guar has been cultivated in the southern U.S., which is now a major source of the gum.

Guar is a linear $\beta(1\text{-}4)$ mannan with a higher proportion of galactose substituents than LBG has, having a galactose-to-mannose ratio of 1:2, and this is reflected in the easier dispersion of guar gum when compared with that of LBG. Guar gum can be dissolved at lower temperatures (typically at 20°C) as the extent of unsubstituted regions of the mannan chain is lower, limiting opportunities for interchain hydrogen bonding that would form aggregates and prevent hydration.

The rheology of guar gum is similar to that of LBG, pseudoplastic and decreasing with temperature, with good pH stability.[37] Gel formation is observed only when other polysaccharides are added, but the ability of guar gum to participate in synergistic actions is lower than that of LBG. Again, this can be attributed to the reduced extent of galactose-free regions of the mannan backbone that would be available for close hydrogen bonding with another polysaccharide. The uses of guar gum are similar to those listed for LBG, reflecting its original commercial introduction as a substitute for LBG.

15.4.1.3 Tara Gum

Tara gum has a galactose-to-mannose ratio of 1:3, intermediate between those of LBG and guar gum. Rheological properties are similar to those of LBG.[38] Although available commercially, tara gum does not appear to have specific applications, but it can be used as an alternative to other galactomannans.[39]

15.4.1.4 Cassia Gum

Cassia gum has an average galactose-to-mannose ratio of 1:5 and can be solubilized only after boiling, when a high viscosity solution can be obtained. The structure of the gum is rather variable, and fractions of different solubilities are obtained due to varying degrees of galactose substitution.[40]

15.4.1.5 Fenugreek Gum

Fenugreek (*Trigonella foenum-graecum*), another legume from the Mediterranean region, contains an endosperm galactomannan with a galactose-to-mannose ratio approaching 1:1.[41] Fenugreek gum is cold-water soluble, forming solutions with lower viscosity than the other galactomannans do. It has potential as an emulsifier and shows good ability to stabilize oil/water interfaces.[42,43]

15.4.2 GLUCOMANNANS

Glucomannans are one of the few gums in widespread use which are obtained from plant roots. Many plant roots can produce a variety of polysaccharides and mucilage, but few have other than local importance. The major commercial source of gluco-mannan is the tubers of *Amorphophallus konjac*, which are ground to flour. *Amorphophallus* is cultivated in east Asia and has been used for traditional foods in Japan and China. Other species of Amorphophallus and other aroids also contain gluco-mannans but are not exploited commercially. Other sources of glucomannans used commercially are tubers of orchid species found in the Middle East and used to produce Salep mannan.[44]

   Glucomannans principally consist of a linear chain of mixed residues of mannose and glucose linked $\beta(1\text{-}4)$, together with some side groups of glucose and mannose linked (1-3).[45] Glucomannans are soluble in cold water with some difficulty, and give solutions of high viscosity at low concentration.[46] Glucomannans can form gels at high pH[47] and can participate in synergistic interactions with helix-forming hydro-colloids such as xanthan, carrageenans, and agar,[48] as observed with the galacto-mannans. The main applications of glucomannans are as a thickener in sauces, a gel in confectionery products, or a fat substitute.[49]

15.4.3 FLAXSEED GUM

Flaxseed gum is obtained from the seeds of the flax plant (*Linum usitatissimum*), which is grown in temperate regions of the world. Flax is a traditional crop valued for the stem fibers, which can be woven to form cloth, and the oil, linseed oil, which can be expressed from the seeds. Flaxseeds readily hydrate to form mucilage that can easily be extracted by incubation in cold water.[50] A problem with flaxseed gum isolation has been contamination through extraction of water-soluble proteins, par-ticularly if higher temperatures are used.[51] Further investigation of extraction conditions have show that at appropriate dilutions, pH, and extraction time, high yields with low protein contamination can be obtained at temperatures around 90°C.[52]

15.4.4 PSYLLIUM GUM

Psyllium gum is obtained from seeds of various species of genus *Plantago*. The gum is present in the seed coat and can be extracted with boiling water, and yields can be increased under mild alkaline conditions.[53] The structure of psyllium gum is based on a xylan backbone substituted with arabinose and some uronic acids.[54] Solubility in water is not high and the gum swells to give a weak gel.[55] The gel shows a broad melting range around 80°C and is susceptible to syneresis on freeze–thawing but shows good stability over a range of ionic concentrations.

Psyllium gum is an indigestible soluble dietary fiber having the ability to lower plasma cholesterol levels.[56] It has traditionally been a component in laxatives, where its high swelling power and mucilaginous gel ease passage of gut contents.[57]

15.4.5 TAMARIND GUM

Tamarind gum is obtained from the seeds of the tamarind tree (*Tamarindus indica*), which is widely grown in India. The seeds are ground to flour, which can be dissolved in hot water to form a mucilaginous gel. The polysaccharide is a xyloglucan with a $\beta(1\text{-}4)$ glucan backbone carrying (1-6)-linked xylose and a few arabinose and galactose substituents with a galactose:xylose:glucose ratio of approximately 1:2:3.[58]

Tamarind gum forms a strong gel under acid conditions, which remains stable at alkaline pH. The gel strength increases in the presence of sucrose. Gels also form in the presence of ethanol with cross-linking occurring due to low solubility of some chain regions, allowing aggregates to form.[59] Tamarind gels have been traditionally commercially employed in India for a range of confectionery products. Potential applications for tamarind gum are similar to those for which pectins can be used,[60] and it is used as a food additive in Japan for a wide range of products.[61] Some other plants also produce seeds containing xyloglucans such as nasturtium seeds (*Tropaeolum majus*), and these can be developed as new gum sources.

15.4.6 QUINCE SEED GUM

Quince seed gum is obtained from seeds of the quince tree (*Cydonia oblonga*), which is mainly grown in the Middle East. Seeds can be directly extracted with hot or cold water and a mucilage is readily released, which is a mixture of cellulose microfibrils dispersed in a matrix composed principally of xylose and arabinose.[62] The gum is easily soluble in cold water up to 2% w/v. It forms highly viscous mucilage but is not gel forming. Dispersions show good stability with respect to pH, salt concentration, and temperature.

Quince seed gum is mainly used for cosmetics applications in Middle Eastern countries. A number of food uses have been suggested, such as a stabilizer in ice cream, but these have not been commercialized due to the high cost, poor availability, and variable quality of the gum.

15.5 EXUDATES

Plant exudates are a group of both polymeric and lower-molecular-weight substances that can be released from plant tissues as part of normal growth and development or as a response to environmental stress or damage. Many plant exudates play an important role in protecting the plant against the physical environment, herbivory, and infection. The major exudate gums of commerce are all collected as wound exudates, which in the plant can serve to seal gaps in a tissue by forming a protective barrier.

15.5.1 ARABINOGALACTAN PROTEOGLYCANS

Arabinogalactan proteoglycans (AGPs) are a widely distributed family of plant polysaccharides that have common structural features.[63] The backbone chain of the polysaccharide is a galactan to which are attached various substituents, with arabinose being the most abundant. The polysaccharide chains are then attached to a linker peptide, resulting in a large macromolecular complex with particular properties. AGP gums have a long history of usage in traditional industries and continue to maintain their position in the modern food sector.

15.5.1.1 Gum Arabic

The original source for gum arabic is the exudate of trees of *Acacia senegal*; however, some closely related Acacia tree species can also provide exudates with similar, but not identical, properties.[64] The main producing region is Africa, centered on Sudan, which provides the bulk of the world market, though difficulties in satisfying demand have led to the substitution of other sources and gums in recent years. The gum is collected from cuts made in the tree bark. The raw gum can be obtained in a crude granular form or as a fine powder after dissolving in water to remove impurities by filtration followed by spray drying.

The main components of gum arabic are arabinose and galactose in a 1:1 ratio as a $\beta(1\text{-}3)$ galactan chain with mainly (1-6)-linked arabinose side groups and lesser amounts of rhamnose and galacturonic acid.[65,66] A small amount of protein is associated with the gum, and this represents part of a proteoglycan structure with the arabinogalactan chains covalently bonded to hydroxyproline resides on the protein by links to arabinose residues.[67] The gum can adopt a very compact structure, with blocks of arabinogalactan attached to a central linker protein chain, the bottle brush or wattle blossom model originally proposed by Fincher for plant AGPs.[68] Naturally available gum arabic can be separated into various fractions that represent different polymers, with about 90% of the gum present as arabinogalactan with only a small amount of protein present, 10% as AGP, and a small, mainly protein, fraction.[69] Initially, the gum may all be present as the AGP form and subsequent degradation probably gives rise to the separate arabinogalactan and protein components.

The viscosity of gum arabic solutions is much lower than for other large polysaccharides due to the compact structure of AGPs. Pseudoplastic behavior is observed at high concentrations (above 25%)[70] and also at low concentration at low shear.[71] The AGP structure includes some charged uronic acid residues and is most stable at acid pH. Highest viscosity is observed around neutral pH, at which the structure is expanded by charge repulsion. The protein part of AGP plays an important role in the emulsifying capability of gum arabic,[72,73] allowing the molecule to stabilize oil–water interfaces and providing one of the main commercial applications of gum arabic.[74,75] However, much of the polysaccharide in commercial gum arabic is dissociated from the protein part and thus cannot act as an emulsifying agent. It has been shown that other acacia gums with higher nitrogen content can act as better emulsifiers.[76] Gum arabic can be used in functional foods as a soluble dietary fiber, with beneficial action in aiding digestive system function.[77]

15.5.1.2 Karaya Gum

Karaya gum is obtained from several *Sterculia* species, notably *Sterculia urens*, mainly in central India. Other sources are from sub-Saharan Africa and Pakistan. Gum is collected from cuts in the tree bark and is available commercially in a crude form after mechanical cleaning.

The gum contains a mixed chain of rhamnose and galacturonic acid with a total high proportion of uronic acid residues.[78] Some protein is associated with the polysaccharide, which may play the same role as the linker proteins in AGP proteoglycans. Karaya dissolves easily in cold water and solubility can be further increased by deacetylation.[79] The gum can form weak gels which disintegrate at low stress. The rheology appears to be determined by the particulate nature of the dispersion, and at high shear rates viscosity decreases as particles are broken down. At high concentrations (50%), solutions have marked adhesive properties, which give rise to some nonfood uses.

Karaya finds a number of applications as a stabilizer for sauces, in which the weak gels formed can maintain other components in suspension. It is also used in frozen ices and in bakery products for glazes.

15.5.1.3 Gum Tragacanth

Gum tragacanth is produced in the Middle East as an exudate from stems or roots of *Astragalus gummifer* and related species of shrubs. The composition of the gum produced varies with the time of collection, with a lower-viscosity flake grade produced later in the season and a higher-viscosity ribbon grade produced earlier.

The structure of gum tragacanth contains two main fractions, one of which is a neutral arabinogalactan complexed with protein. The other fraction is more acidic and less soluble, containing more galacturonic acid, xylose, and other residues.[80] Solubility of the gum is not high and requires prolonged stirring and some heat. Solutions are pseudoplastic and viscous and form weak gels at 4%. The acid stability is good, with optimum viscosity being developed at ca. pH 5.5.[81]

Gum tragacanth can act as an emulsifying agent for the stabilization of oil-in-water emulsions but is not as effective as gum arabic. It is used to enhance mouthfeel in dressings and sauces, as coating for bakery products, and as a binder in pie fillings and confectionery products. Tragacanth has the interesting property of being able to form a synergistic interaction with gum arabic, leading to a reduction in viscosity.

15.5.1.4   Gum Ghatti

Gum ghatti is obtained from the tree *Anogeissus latifolia* (Combretaceae) that grows in Indian forests. Various other species of *Combretaceae* also produce AGP exudates with properties comparable to those of gum arabic.[82] The gum is exuded from incisions in the bark of the tree and is collected as "tears," which are sorted by color. Processing involves grinding to produce a powder, which can be dissolved in water to produce a colloidal dispersion. The basic structure of the gum is an arabinogalactan with minor substituents of mannose, xylose, and glucuronic acid.

15.5.1.5   Okra Gum

Okra (*Hisbiscus esculentis*) is a plant which produces seed pods that when immature contain viscous mucilage. The mucilage can be obtained by aqueous extraction of dried and ground pod powder. The composition suggests an arabinogalactan proteoglycan with amino acids present and arabinose and galactose present in a 10:3 ratio together with appreciable quantities of rhamnose and galacturonic acid.[83] Okra gum is only a weak emulsifiant but can act as a thickener.[84]

15.5.1.6   Marigold Flower Gum

The gum is extracted from marigold seed petals by hot water and contains arabinose, galactose, and glucose in the ratio 15:3:7. Protein is stably associated with the polysaccharide, which shows emulsifying properties though weaker than those of gum arabic.[85]

15.5.1.7   Other Arabinogalactan Protein Gums

AGPs are produced by nearly all plants. There is a wide range of potential plants to provide new gum resources. However, most plants produce only small quantities of AGPs, and only large gum secretors can be commercially developed. Various species of trees that do produce gum exudates have often been exploited for traditional food or medicinal purposes. Some of these are now receiving attention for potential food applications, and a few are mentioned here.

Mesquite gum is produced in Central America from species of *Prosopis* trees. Some trees produce high levels of tannins, giving the gum a dark color. The composition, structure,[86] and rheology of the gum are similar to those of gum arabic, and the gum is expected to be an AGP though this is not yet been confirmed.[87] The gum is used locally in Mexico but is not approved for food use in the U.S., and therefore current applications are limited.[88]

Enterolobium gum is exuded from trees of *Enterolobium cyclocarpum* and shows properties similar to those of gum arabic but at a much lower concentration.[89] The gum is water soluble and produced in large quantities by injured trees.

Gums from *Spondias* species may be related arabinogalactans and show a range of viscosities.[90] Gum from *Portulaca oleracea* contains an arabinogalactan and has emulsifying properties.[91]

15.6 ALGAL GUMS

Marine algae or seaweeds are an excellent source of hydrocolloids due to the requirements for growth in marine environment. The gel-like hydrated and charged polysaccharides of the seaweed cell wall mucilage assist the algae to cope with the physical stresses of wave motion and the chemical stresses of the high salt concentration of sea water. The polysaccharides are also a key protection for seaweeds growing in the intertidal zone, where they are regularly exposed to the threat of desiccation at low-tide, prevented only by the presence of the water-binding polysaccharide gel around their tissues.

15.6.1 Alginates

Alginates are present in the walls of many brown seaweeds (Phaeophyta), which grow in oceans throughout the world. The main commercial resources are found on the coasts of temperate regions in both the Northern and Southern hemispheres. Important commercial species are selected for yield of alginate and ease of growth and collection, which is usually associated with the large size of the seaweed. The most important algal resources are *Laminaria digitata* and *L. hyberborea* from Northern Europe and Asia and Kelp (*Macrocystis pyrifera*) from the Pacific coast of North America. Other species harvested in Northern Europe and America are *Ascophyllum nodusum* and *Fucus serratus*.

Alginates are acidic polysaccharides, linear polymers of β-D-mannuronic acid and α-L-guluronic acid, which are (1-4) linked. The two residues can be found alternating in the chain but more commonly form a block copolymer structure with regions containing only guluronic acid and regions with only mannuronic acid. The final chain is then composed of stretches of pure guluronate or mannuronate with some mixed regions.[92]

Extraction of alginates from raw seaweed fronds commences with an acid wash to remove acid-soluble impurities and convert alginates present to the acid form. The alginate can then be extracted with alkali to give a soluble alginate salt, which can be separated from other material by filtration before precipitation with acid. Alginate can be neutralized with different bases to give the various soluble salts that are used in the food industry, principally sodium, potassium, and ammonium alginate. These salts are all readily cold-water soluble.

In solution, soluble alginates can act as effective thickeners to increase viscosity at low concentration. The viscosity decreases with increasing temperature and shows pseudoplastic behavior. The solutions are not stable at low pH, because below pH 4 alginate is insoluble and forms a precipitate or acid gel.[93] In the presence of Ca^{2+}, alginates can form a thermostable gel by binding the cations between two opposed blocks of guluronic acid. The guluronic acid regions of the chain adopt a buckled conformation that provides a negatively charged pocket that can accommodate a Ca^{2+} ion and allow it to bind to charges on another guluronic region on another chain.[94] Guluronate regions provide the junction zones whereas mannuronate regions and mixed regions of the chain provide the nonbinding regions.

Alginates are widely used as thickeners and stabilizers as well as for their gel-forming properties and are found in ice creams, dairy products such as milk shakes, soups, and sauces. They are also an important ingredient in beer manufacture to stabilize dissolved solids in commercial beer and prevent sedimentation. Alginate gels are popular in the dessert and bakery sectors for pie fillings and also for restructuring meat. Alginates can also be used to form films that can be used as edible packaging. An important market for alginates uses chemically modified alginate, mainly propylene glycol alginate PGA. PGA has greater acid solubility than native alginate and can therefore be used in a range of applications where native alginate would start to precipitate out.

15.6.2 AGARS

Agars are obtained from red seaweeds (Rhodophyta) mainly from the genus *Gelidium*, of which some 15 species are now used to produce agar from the cell wall mucilage. The main center of production has historically been Japan, and this bias continues to this day. Production involves aqueous extraction of seaweed in hot water, with the extract frozen and thawed to precipitate polysaccharides and remove water-soluble contaminants. Agarose is a purified form of agar with typically shorter chains and a lower molecular weight.

Agars are formed from linear chains of both L- and D-galactose residues linked alternately $\alpha(1\text{-}3)$ and $\beta(1\text{-}4)$ and can be sulfated at the 2, 4, or 6 positions. Some of the L-galactose residues are converted to 3,6-anhydrogalactose by the formation of a bridging bond across the ring between the C-3 and C-6. Formation of the 3,6-anhydrogalactose changes the conformation of the sugar ring and favors the formation of a helix by the polysaccharide chain.[95] Gelling can occur by formation of junctions between the helical regions of two agar chains that intertwine to form a double helix. Agar gels are thermoreversible and show significant hysteresis, a large difference between the melting (80–90°C) and gelling temperatures (35–45°C), which reflects the slow rate of double-helix formation and the strength of the association once it is formed.[96]

Agar gels are formed at low concentration and are hard and brittle. Generally, they are too hard for many food uses and are therefore more commonly used in nonfood applications. Food uses include confectionery and sweets, icings, and restructured meat pie fillings.

15.6.3 CARRAGEENANS

Carrageenans are a family of sulfated polymers with a linear chain of D-galactose linked alternately $\alpha(1\text{-}3)$ and $\beta(1\text{-}4)$. The sulfate groups are present at the 2, 4, or 6 positions. An important feature of carrageenan structure is that some of the galactose residues are converted to 3,6-anhydrogalactose by the formation of a bridging bond across the ring between the C-3 and C-6. This transformation can occur after chain synthesis catalyzed by an enzyme in the algae or it can be promoted during extraction by the use of alkaline conditions which result in the loss of some sulfate groups from galactose residues. Formation of 3,6-anhydrogalactose changes the conformation of the sugar ring and favors the formation of a helix by the polysaccharide chain.[97] Two helices are able to intertwine to form a double helix, which can act as a junction zone in gel formation.[98] Nonbonding regions are provided by stretches of the chain where sulfate substitution is still present and prevents helix formation.

Variations in the basic structures provide the distinctions between the three main types of carrageenan: (i) κ-carrageenan, with the chain containing nonsulfated 3,6-anhydrogalactose which is able to form helices that can participate in double-helix junctions to give strong gels. Increase in temperature causes the two chains to separate, so the gels are thermoreversible. Carrageenans are charged polymers and gel stability is influenced by the presence of cations, especially potassium. The small K$^+$ ion can act as a counterion to reduce charge repulsion within the helices between nearby sulfate groups, which permits the helices to pack more closely and strengthens the gel junction. No effect is observed with sodium due to the greater size of the ion. Calcium has some effect, possible by acting as a cross-link between two sulfate groups, which can give rise to difficulty in dissolving κ-carrageenan in hard water. (ii) ι-Carrageenan, with the chain containing 3,6-anhydrogalactose sulfated at C-2, providing elastic gels. Potassium ions have no effect on gelation, but gel strength increases in the presence of calcium. (iii) λ–Carrageenan, with the chain without 3,6-anhydrogalactose and containing instead a high percentage of 2-sulfated galactose. No gels are formed and there is little effect from the presence of salts.

Carrageenans are obtained from a variety of different red seaweeds (Rhodophyta) in different coastal regions. The original source of carrageenan was from *Chondrus crispus* (Irish moss) present in Northern European waters together with *Gigartina stellata*. A number of other *Gigartina* species are exploited off the coast of South America and North Africa. Asian tropical waters around the Philippines and Indonesia have become a major commercial source of carrageenans from *Eucheuma spinosum* and *Eucheuma cottonii*, which are cultivated on nets or rafts.

Carrageenans are obtained from the seaweeds by extraction in hot water, followed by precipitation with ethanol, drying, and milling to a powder. The solubility of different carrageenans is related to their gel-forming ability. κ-carrageenan is soluble in hot water (70°C) only, because it has a strong tendency to form gels, limiting hydrogen bonding with water. ι-Carrageenan has a higher degree of sulfation can more readily can dissolve in warm water (50°C), and λ-carrageenan dissolves in cold water.

Carrageenans find applications in a large number of food products. Historically, the most important use of carrageenans has been in milk-based products where the polysaccharide can stabilize the fat micelles in milk by binding to the milk protein casein and preventing close approach and coalescence of the micelles.[99] Carrageenans can also form gels with milk, and this provides much greater temperature stability to the milk[100] and enables use and processing of hot drinks as well as frozen products and milk jellies. ι-Carrageenan solutions have thixotropic properties and are a useful component of sauces and coatings. λ-Carrageenan is used as a nongelling thickener. The three carrageenans can be readily blended in varying proportions to achieve a desired rheology.

15.6.4 FURCELLARANS

Furcellarans are extracted from the red seaweed *Furcellaria fastigiata* in the coastal waters of Denmark and Canada. The cleaned seaweed is boiled in water, and the extract separated from residues by filtration. The extract is precipitated as a gel by addition of KCl. The gel is then frozen. After thawing in KCl, the purified material is ground and marketed as the potassium salt.

Furcelleran is soluble in hot water (70°C) and gives a pseudoplastic solution with moderate viscosity. On cooling, the solution forms a thermoreversible gel at around 35°C. Generally, the properties of furcelleran are intermediate between those of agar and κ-carrageenan in terms of gel strength and formation. Furcellaran can also form milk gels through interaction with milk casein. At lower pH, furcelleran gels become more brittle and their texture can be improved by interaction with LBG.

Furcellerans are mainly produced for the food market and are used in bakery as tart and flan fillings and coverings, in milk puddings, for jams and jellies, and in restructured meat products.

15.7 CLOSING REMARKS

The last century has seen a considerable expansion in the range and quantity of plant gums used in the food industry. The earliest gums used were natural products, which were extracted and purified to become food ingredients. Interest in the expanding sector of manufactured foods has already led to the exploitation of new sources, many of nonplant origin, and provided new attributes to support new product development. Several examples have been given of potential gum candidates not yet exploited commercially but with rheologies close to those already used in the food industry. The number of plant species used for gum production is only a small fraction of the potentially available number, and there is little doubt that there are many as yet undiscovered gum candidates present in nature. The metabolism of most plants and algae has the capacity for the synthesis of many polysaccharides related or unrelated to those already used in food production, but whether these will emerge as new gums depends on a number of factors. Yield of the gum, ease of extraction, and ease of cultivation are as important in determining commercial success as the basic rheological and functional properties. Finally, access to the food ingredients market depends on regulatory approval and the satisfaction of health requirements.

REFERENCES

1. Phillips, G.O. and Williams, P.A., *Handbook of Hydrocolloids*, CRC Press, Boca Raton, 2000.
2. Whistler, R.L. and BeMiller, J.N., Eds., *Industrial Gums,* 3rd ed., Academic Press, New York, 1993.
3. Davidson, R.L., *Handbook of Water Soluble Gums and Resins*, McGraw-Hill, New York, 1980.
4. Glicksman, M., Ed., *Food Hydrocolloids,* CRC Press, Boca Raton, 1986.
5. Imeson, A., *Thickening and Gelling Agents for Food*, 2nd ed., Blackie, London, 1999.

6. Robbins, S.R.J., A review of recent trends in selected markets for water soluble gums. Bulletin No. 2, *Overseas Development Natural Resources Institute*, 1987.
7. Lillford, P.J., Commercial requirements and interests: an update, in *Gums and Stabilisers for the Food Industry*, Vol. 10, Williams, P.A. and. Phillips, G.O., Eds., RSC, Cambridge, 2000, p. 387.
8. Anderson, D.M.W. and Eastwood, M.A., The safety of gum arabic as a food additive and its energy value as a food ingredient, *J. Human Nutr. Diet*, 2, 137, 1989.
9. Rees, D.A., *Polysaccharide Shapes*, Chapman & Hall, London, 1977.
10. Dickinson, E., *An Introduction to Food Hydrocolloids*, Oxford University Press, Oxford, 1992.
11. Wood, P.J. et al., Large scale preparation and properties of oat fractions enriched in (1-3)(1-4)beta-D-glucan concentrations, *Cereal Chem.*, 68, 48, 1991.
12. Bohm, N. and Kulicke, W.M., Rheological studies of barley (1-3)(1-4)-β-glucan in concentrated solution: investigation of the viscoelastic flow behaviour in the sol-state, *Carbohydr. Res.*, 315, 293, 1999.
13. Cui, W., *Polysaccharide Gums from Agricultural Products*, Technomic Publishing, Lancaster, PA, 2001, p. 134.
14. Wood, P.J., *Oat Bran*, American Association of Cereal Chemists, St. Paul, 1993.
15. Wood, P.J. and Beer, M., Oat β-glucan, in *Functional Foods*, G. Mazza, Ed., Technomic Publishing, Lancaster, PA, 1998, p. 1.
16. Izydorczyk, M.S. and Biliaderis, C.G., Studies on the structure of wheat endosperm arabinoxylans, *Carbohydr. Polym.*, 24, 61, 1994.
17. Andrewartha, K.A., Phillips, D.R., and Stone, B.A., Solution properties of wheat flour arabinoxylans and enzymatically modified xylans, *Carbohydr. Res.*, 77, 191, 1979.
18. Hoseney, R.C. and Faubion, J.M.A., A mechanism for the oxidative gelation of wheat flour water soluble pentosans, *Cereal Chem.*, 58, 421, 1981.
19. Cui, W., Wood, P.J., and Wang, Q., Gelling mechanisms of non-starch polysaccharide from pre-processed wheat bran fraction, in *Gums and Stabilisers for the Food Industry*, Vol. 10, Williams, P.A. and Phillips, G.O., Eds., RSC, Cambridge, 2000, p. 156.
20. Aspinall, G.O., Hemicelluloses, gums, and pectic substances, *Annu. Rev. Biochem.*, 31, 79, 1962.
21. Ponder, G.R. and Richards, G.N., Arabinogalactan from western larch. Part III., *Carbohydr. Polym.*, 34, 251, 1997.
22. Timell, T.E., Wood hemicelluloses. Part II, *Adv. Carbohydr. Chem.*, 20, 409, 1965.
23. Bailey, K. and Norris, F.W., The nature and composition of the mucilage from the seed of white mustard (*Brassica alba*), *Biochem. J.*, 26, 1609, 1932.
24. Weber F.E., Taillie, S.A., and Stauffer, K.R., Functional characteristics of mustard mucilage, *J. Food Sci.*, 39, 461, 1974.
25. Cui, W. et al., NMR characterization of a 4-O-methyl beta-D-glucuronic acid containing rhamnogalacturonan from yellow mustard (Sinapsis alba), *Carbohydr. Polym.*, 27, 117, 1996.
26. Cui, W., Eskin, M.N.A., and Biliaderis, C.G., Chemical and physical properties of yellow mustard (*Sinapsis alba*) mucilage, *Food Chem.*, 46, 169, 1993.
27. Cui, W., Eskin, M.N.A., and Biliaderis, C.G., Water-soluble yellow mustard (*Sinapsis alba*) polysaccharides: partial characterization, molecular size distribution and rheological properties, *Carbohydr. Polym.*, 20, 215, 1993.
28. Cui, W. et al., Synergistic interactions between yellow mustard polysaccharides and galactomannans, *Carbohydr. Polym.*, 27, 123, 1995.

29. Cui, W. and Eskin, N.A.M., Interaction between yellow mustard gum and locust bean gum: impact on a salad cream product, in *Gums and Stabilisers for the Food Industry,* Vol. 8, Phillips, G.O., Williams, P.A., and Wedlock, D.J., Eds., Oxford University Press, Oxford, 1996, p.161.

30. Jenkins, A.L. et al., Effect of mustard seed fibre on carbohydrate tolerance, *J. Clin. Nutr. Gastroenterol.*, 2, 81, 1987.

31. Doublier, J.L. and Launey, B. Rheology of galactomannan solutions, *J. Text. Stud.,* 12, 151, 1981.

32. Oakenfull, D., Naden, J., and Paterson, J., Solvent structure and the influence of anions on the gelation of κ-carrageenan and its synergistic interaction with LBG, in *Gums and Stabilisers for the Food Industry,* Vol. 10, Williams, P.A. and Phillips, G.O., Eds., RSC Cambridge, 2000, p. 221.

33. Dea, I.C.M., McKinnon, A.A., and Rees, D.A., Tertiary and quaternary structure in aqueous polysaccharide systems which model cell wall cohesion, *J. Mol. Biol.,* 68, 153, 1972.

34. Cairns, P.A., Miles, M.J., and Morris, V.R., Intermolecular binding of xanthan gum and carob gum, *Nature,* 322, 89, 1986.

35. Rinaudo, M. et al., Physical properties of xanthan, galactomannan and their mixtures in aqueous solution, in *Applications of Polymers in Foods,* Cheng, H.N., Cote, G.L., and Baianu, L.C., Eds., Wiley-VCH, Weinheim, 1999, p. 115.

36. Chandrasekaran, R,. X-ray and molecular modelling studies on the structure function correlations of polysaccharides, in *Applications of Polymers in Foods,* Cheng, H.N., Cote, G.L., and Baianu, L.C., Eds., Wiley-VCH, Weinheim, 1999, p. 17.

37. Robinson, G.R., Ross-Murphy, S.B., and Morris, E.R., Viscosity-molecular weight relationships, intrinsic chain flexibility and dynamic solution properties of guar, *Carbohydr. Res.,* 107, 17, 1982.

38. Wielinga, W.C., Galactomannans, in *Handbook of Hydrocolloids*, Phillips, G.O. and Williams, P.A., Eds., CRC Press, Boca Raton, 2000, p. 137.

39. Jud. B, and Lössl, U., Tara gum, a thickening agent with perspective, *Int. Zeit. Lebensm.Tech., Verfrahrenstechnik,* 37, 28, 1986.

40. Bayerlein, F., Technical applications of galactomannans, in *Plant Polymeric Carbo-hydrates*, Meuser, F., Manners, D.J., and Seibel, W., Eds., RSC, Cambridge, 1993, p. 191.

41. Reid, J.S. and Meier, H., Formation of a reserve galactomannan in the seeds of *Trigonella foenum graecum, Phytochemistry,* 9, 513, 1970.

42. Garti, N.M. and Reichman, D., Surface properties and emulsification activity of galactomannans. *Food Hydrocoll.,* 8, 155, 1994.

43. Huang, X., Kakuda, Y., and Cui,W., Hydrocolloids in emulsions: particle size distri-bution and interfacial activity. *Food Hydrocoll.,* 15, 533, 2001.

44. Buchala, A.J., Franz, G., and Meier, H., A glucomannan from the tubers of Orchis morio, *Phytochemistry,* 13, 458, 1974.

45. Kato, K.Y. and Matsuda, K., Isolation of oligosaccharides corresponding to the branching point of konjac mannan, *Agric. Biol. Chem.,* 37, 2045, 1973.

46. Kishida, N, Okimasu, S., and Kamata, T., Weight and intrinsic viscosity of konjac glucomannan, *Agric. Biol. Chem.,* 42, 1645, 1978.

47. Yoshimura, M. and Nishinari, K. Dynamic viscoelastic study on the gelation of konjac glucomannan with different molecular weights. *Food Hydrocoll.,* 13, 227, 1999.

48. Takigami, S., Konjac mannan, in *Handbook of Hydrocolloids*, Phillips, G.O. and Williams, P.A., Eds., CRC Press, Boca Raton, 2000, p. 413.

49. Tye, R.J., Konjac flour: properties and interactions with carrageenan and starch, forms heat stable gels suitable for use in fat substitutes, *Food Technol.*, 45, 82, 1990.
50. Susheelamma, N. and Salimath, P.V., Isolation and properties of flaxseed mucilage, *J. Food Sci.*, 24, 103, 1987.
51. Fedeniuk, R.W. and Biliaderis, C.G., Composition and physicochemical properties of linseed (Linum usitatissimum) mucilage, *J. Agric. Food Chem.*, 42, 240, 1994.
52. Cui, W. et al., Optimization of an aqueous extraction process for flaxseed gum by response surface methodology, *Food Sci. Techol.*, 27, 363, 1994.
53. Tomoda, M., Ishikawa, K., and Yokoi, M., Plant mucilages: isolation and characterization of a mucous polysaccharide, plantago mucilage-A, from the seeds of Plantago major var. asiatica, *Chem. Pharm. Bull.*, 29, 2877, 1981.
54. Tomoda, M. et al., Plant mucilages: the location of O-acetyl groups and the structural features of plantago mucilage-A, the mucous polysaccharide from the seeds of Plantago major var. asiatica, *Chem. Pharm. Bull.*, 32, 2182, 1984.
55. Haque, A. et al., Xanthan like weak gel rheology from dispersions of ispaghula seed husk, *Carbohydr. Polym.*, 22, 223, 1993.
56. Anderson, J.W. et al., Cholesterol lowering effects of psyllium hydrophilic mucilloid in the hamster, sites and possible mechanisms of action, *Arch. Intern. Med.*, 148, 292, 1988.
57. Bone, J.N. and Rising, L.W., An in vitro study of various commercially available bulk-type laxatives, *J. Am. Pharm. Assoc.*, 43,102, 1953.
58. Gidley, M.J. et al., Structure and solution properties of tamarind seed polysaccharide, *Carbohydr. Res.*, 214, 299, 1991.
59. Yamanaka, S. et al., Gelation of tamarind seed polysaccharide xyloglucan in the presence of ethanol, *Food Hydrocoll.*, 14, 125, 2000.
60. Marathe, R.M. et al., Gelling behaviour of polyose from tamarind kernel polysaccharide, *Food Hydrocoll.*, 16, 423, 2002.
61. Nishinari, K. Yamatoya, K., and Shirakawa, M., Xyloglucan, in *Handbook of Hydrocolloids*, Phillips, G.O. and Williams, P.A., Eds., CRC Press, Boca Raton, 2000, p. 247.
62. Morvay, J. and Szendrei, K., Cydonia seed polysaccharide, *Gyogyszereszet*, 11, 178, 1967.
63. Clarke, A.E., Anderson, R.L., and Stone, B.A., Form and function of arabinogalactans and arabino-galactan proteins, *Phytochemistry*, 18, 521, 1979.
64. Karamallah, K.A., Gum arabic: quality and quantity assured, in *Gums and Stabilisers for the Food Industry*, Vol. 10, Williams, P.A. and Phillips, G.O., Eds., RSC, Cambridge, 2000, p. 37.
65. Defaye, J. and Wong, E., Structural studies of gum arabic, the exudate polysaccharide from Acacia senegal, *Carbohydr. Res.* 150, 221, 1986.
66. Randall, R.C., Phillips, G.O., and Williams, P.A., Fractionation and characterisation of gum from Acacia senegal, *Food Hydrocoll.*, 3, 65, 1989.
67. Conolly, S, Fenyo, J.C., and Vandevelde, M.C., Effect of a proteinase on the macromolecular distribution of Acacia senegal gum, *Carbohydr. Res.*, 8, 23,1988.
68. Fincher, G.B., Stone, B.A., and Clarke, A.E., Arabinogalactan proteins: structure biosynthesis and function, Annu. Rev. *Plant Physiol.*, 34, 47, 1983.
69. Osman, M.E. et al., The molecular characterisation of the polysaccharide gum from Acacia senegal, *Carbohydr. Res.*, 246, 303, 1993.
70. Williams, P.A., Philiips, G.O., and Randall, R.C., Structure-function relationships of gum arabic, in *Gums and Stabilisers for the Food Industry,* Vol. 5, Phillips, G.O., Wedlock, D.J., and Williams, P. A., Eds., Oxford University Press, Oxford, 1990, p. 25.

71. Sanchez, C. et al., Structure and rheological properties of acacia gum dispersions, *Food Hydrocoll.,* 16, 257, 2002.
72. Randall, R.C., Phillips, G.O., and Williams, P.A., The role of the proteinaceous component on the emulsifying properties of gum arabic, *Food Hydrocoll.,* 2, 131, 1988.
73. Dickinson, E. et al., Surface activity and emulsifying behaviour of some acacia gums, *Food Hydrocoll.,* 2, 447, 1988.
74. Dickinson, E., Galazka, V.B., and Anderson, D.M.W., Emulsifying behaviour of gum arabic, Part I, *Carbohydr. Polym.,* 14, 373, 1991.
75. Snowden, M.J., Phillips, G.O., and Williams, P.A., Functional characteristics of gum arabic, *Food Hydrocoll.,* 1, 291, 1987.
76. Dickinson, E. et al., Surface activity and emulsifying behaviour of some acacia gums, *Food Hydrocoll.,* 2, 477, 1988.
77. Kravtchenko, T.P., The use of acacia gum as a source of soluble dietary fibre, in *Gums and Stabilisers for the Food Industry,* Vol. 9, Williams, P.A. and Phillips, G.O., Eds., RSC, Cambridge, 1998, p. 413.
78. Stephen, A.M. and Churms, S.C., Gums and mucilages, in *Food Polysaccharides and their Applications,* Stephen, A.M., Ed., Marcel Dekker, New York, 1995, p. 377.
79. Le Cerf, D., Irinei, F., and Muller, G., Solution properties of gum exudates from Sterculia urens, *Carbohydr. Polym.,* 13, 375, 1990.
80. Anderson, D.M.W. and Grant, D.A.D., Gum exudates from four Astragalus species, *Food Hydrocoll.,* 3, 217, 1989.
81. Stauffer, K.R., Gum tragacanth, in *Handbook of Water Soluble Gums and Resins,* Davidson, R.L., Ed., McGraw-Hill, New York, 1980, p. 111.
82. Anderson, D.M.W. and Bell, P.C., The composition of gum exudates from some Combretum species, the botanical nomclature and systematics of *Combretaceae, Carbohydr. Res.,* 57, 215, 1977.
83. El Amin, S., Mucilages of *Hibiscus esculentus* and *Corchorus oliforius*, *J. Chem. Soc.,* 828, 1956.
84. Ndjouenkeu, R. Akingbala, J.O., and Oguntimein, G.B., Emulsifying properties of three African food hydrocolloids, *Plant Foods Hum. Nutr.,* 51, 245, 1997.
85. Medina, A.L. and BeMiller, J.N., Marigold flower meal as a source of an emulsifying gum, in *New Crops,* Janick, J. and Simon, J.E., Eds., Wiley, New York, 1993, p. 389.
86. Churms, S.C., Merrifield, E.H., and Stephen, A.M., Smith degradation of some gum exudates from some prosopis species, *Carbohydr. Res.,* 90, 261, 1981.
87. Goycoolea, A.M. et al., Processing and functional behaviour of low-tannin mesquite gum, in *Gums and Stabilisers for the Food Industry,* Vol. 9, Williams, P.A. and Phillips, G.O., Eds., RSC, Cambridge, 1998, p. 305.
88. Anderson, D.M.W. and Wang, W., The characterisation of proteinaceous Prosopis gums which are not permitted food additives. *Food Hydrocoll.,* 3, 235, 1989.
89. Leon de Pinto, G. et al., Relevant structural features of the gum from *Enterolobium cyclocarpum,* in *Gums and Stabilisers for the Food Industry,* Vol. 10, Williams, P.A. and Phillips, G.O., Eds., Cambridge, 2000, p. 59.
90. Leon-de-Pinto,G. et al.,The composition of two spondias gum exudates, *Food Hydrocoll.,* 14, 259, 2000.
91. Garti, N., Slavin, Y., and Aserin, A., Surface and emulsification properties of a new gum from Portulaca oleracea L., *Food Hydrocoll.,* 13, 145, 1999.
92. Grasdalen, H., High field H-NMR spectroscopy of alginate: sequential structure and linkage conformation, *Carbohydr. Res.,* 118, 255, 1983.

93. Draget, K.I., Skjåk-Braek, G., and Smidsrød, O., Alginic acid gels: the effect of alginate chemical composition and molecular weight, *Carbohydr. Polym.*, 25, 31, 1994.

94. Grant, G. T. et al., Biological interactions between polysaccharides and divalent cations: the egg-box model, *FEBS Lett.*, 32, 195, 1973.

95. Rees, D.A., Shapely polysaccharides, *Biochem. J.*, 126, 257, 1972.

96. Armisen, R., Agar, in *Thickening and Gelling Agents for Food*, Imeson, A., Ed., Blackie, London, 1999, p.1.

97. Lawson, C.J. and Rees, D.A., An enzyme for the metabolic control of polysaccharide conformation and function, *Nature*, 227, 392, 1970.

98. Rees, D.A., Proposed gelation mechanism for carrageenan, *Adv. Carbohydr. Chem. Biochem.* 21, 267, 1969.

99. Dalgleish, D.G. and Morris, E.R., Interactions between carrageenans and casein micelles: electrophoretic and hydrodynamic properties of the particles. *Food Hydrocoll.*, 2, 311, 1988.

100. Tziboula, A. and Horne, D.S., Effect of heat treatment on κ-carrageenan gelation in milk, in *Gums and Stabilisers for the Food Industry*, Vol. 10, Williams, P.A. and Phillips, G.O., RSC, Cambridge, 2000, p. 211.

16 Carbohydrates of Animal Tissues

Tadeusz Kołczak

CONTENTS

16.1 Introduction 255
16.2 Glycogen 257
    16.2.1 Glycogenesis 258
    16.2.2 Glycogenolysis and Glycolysis 259
16.3 Glucosylaminoglycans 264
    16.3.1 Proteoglycans 265
    16.3.2 Metabolism of Glucosylaminoglycans 265
    16.3.3 Utilization of Glucosylaminoglycans 267
16.4 Glycoproteins 267
References..269

16.1 INTRODUCTION

Carbohydrates constitute quantitatively a much smaller component of the tissue in animals than they do in plants. They are present, however, in all animal tissues and tissue fluids as free compounds (D-glucose and glycogen), as components of nucleic acids, nucleosides, some proteins, and lipids.[1] For instance, D-glucose and its polymer glycogen, stored in liver and skeletal muscles, are the primary source of energy for organisms. Pentoses constitute cellular nucleic acids, which are involved in the transfer of genetic code and protein biosynthesis. Glycoproteins, with their oligosaccharide components, are essential components of cell membranes. Polymers of hexosoamines (proteoglycans) occur in connective tissues in which they function as intracellular cement substances or as lubricants in joint fluids. Conjugated glycolipid containing D-galactose is essential for the functioning of nerve tissue.

The quality of meat products depends more on carbohydrates, their metabolites, and compounds with carbohydrate components than on any other organic substances. The rate and extent of chemical processes in postmortem transformation of muscle to meat and their effect on its ultimate properties and quality are controlled, to a great extent, by reactions of carbohydrates. Sensory and inherent mechanical properties of meat depend on the quantitative and qualitative composition of glyc-osylaminoglycans of connective tissue. The extent of browning developed by thermal meat treatment is controlled, among others, by reducing sugars in meat. Apart from the Maillard reaction (Chapter 17), carmelization of endogenic sugars also contrib-utes to browning.

Most of the carbohydrates of animal tissues are present in the form of complex polysaccharides, and many of them are bonded to protein moieties. Figure 16.1 presents metabolism of carbohydrates in animal organisms in their life period.

Saccharides available from the decomposition of forage in the gastric system pass through the circulatory system to the tissues where they are transformed into glycogen, glucosylaminoglycans, and glycoproteins as well as glycolipids. Glycolip-ids usually contain monosaccharide residues bound to a phospholipid component.

Glucose is converted in tissues to glucose-6-phosphate (G-6-P). G-6-P is an intracellular product, which is hydrolyzed in the cells of liver, kidney, and intestines only and not in muscle. In living cells, G-6-P is used for glycogen synthesis or is entirely degraded to CO_2 by a process known as glycolysis which occurs in the cytoplasm, and in citric acid cycle which occurs in the mitochondria. Under anaer-obic conditions, pyruvate is decomposed to lactate, which is removed from tissue through the circulatory system to the liver, where it is resynthesized into glucose

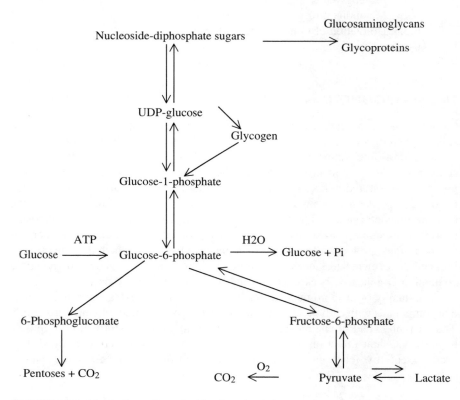

FIGURE 16.1 Metabolism of saccharides in animal tissues.

and glycogen, or to the heart, where it is metabolized to CO_2 and H_2O. The citric acid cycle is active under aerobic conditions only. When one molecule of glucose is split off glycogen and transformed through reactions under aerobic conditions, the net energy yield is 30 molecules of ATP, whereas under anaerobic conditions only 2 molecules of ATP are produced.

In the living muscle, ATP is the ultimate source of energy for the contractile process, the pumping of Ca^{2+} back into the sarcoplasmic reticulum during relaxation, and the maintenance of Na^+/K^+ gradients across the sarcolemma. Postmortem, as the oxygen supply is depleted, aerobic metabolism through citric acid cycle begins to fail. Less energy in the form of ATP is produced through anaerobic pathways. ATP is required for muscle to be maintained in the relaxed state. When ATP becomes limited, the muscle enters the onset phase of rigor mortis, loses its ability to relax, and transforms to meat.

16.2 GLYCOGEN

Molecular weights of glycogen (see Chapter 1) isolated from various tissues range from 103 to 10003 kDa. In the cytoplasm, glycogen forms 10- to 40-nm granules containing proteins and enzymes catalyzing synthesis and degradation of glycogen. In the period of animal growth and development, the number and the length of branches of glycogen are almost constant.

Glycogen isolated from various animal species, and even various breeds within a given species, shows significant differences. The structure of glycogen is a reflection of the steric specificity of the branching enzyme and may be affected by the physiological state of the animal.[2] For instance, a long-term tissue anoxia or stress influences the length of branches (usually at every tenth D-glucose unit) and trans-glycosidation in glycogen.

Peripheral chains of glycogen are longer than the chains in the interior of a particle.[3] Within the central part of the molecule, branching can occur at every fourth to fifth glucose unit, so the density of the interior of the molecule might be higher. The terminal glucose units are the nonreducing ends, which are abstracted in the process of phosphorylase-catalyzed phosphorylytic splitting of glycogen.

Even though glycogen is present in most body cells, it is most abundant in mammalian liver and comprises 2 to 8% of its wet weight. The normal glycogen content of mammalian muscle ranges from 0.5 to 1.0%, but because of the large mass of muscle, most of the glycogen is present in this tissue. Diet has a marked affect on glycogen stored in the liver, the muscle, and other tissues. Although the glycogen content of mammalian liver and muscle is relatively stable in the normal animal with free access to food, it is in constant flux, being degraded and synthesized continuously. Liver glycogen stores are much more sensitive to inanition than those of the muscle. Introducing fodder rich in glycogenic substance results in rapid return of tissue glycogen to normal levels. The presence of glycogen in all tissue is lowered by conditions, which increase energy demands such as exercise and exposure to stress. The extent of glycogen depletion is correlated to the degree and duration of such energy demands, and repletion is dependent on the length of recovery period. A long-term deficiency in glycogen caused by exercise, stress, metabolic acidosis, and anoxia can result in hyperglycemia, an excessive level of glucose in blood or in glucosuria, or both, and presence of glucose in urine.

Glycogenesis and glycogenolysis, that is, the formation and degradation of glycogen, are controlled by several hormones.[1] Insulin deficiency results in low level of tissue glycogen, caused by reduction of its synthesis from blood glucose. Administration of insulin produces increased level of muscle glycogen, though not necessarily of liver glycogen. Stress-induced enhanced level of epinephrine stimulates glycogenolysis, with greater effect in the muscles than in the liver. In contrast, stimulation of glycogenolysis by glucagon is greater in the liver than in the muscles. Glycogen synthesis from amino acids is also possible. This synthesis is stimulated by glucocorticoids and is greater in the liver than in the muscle. The hormones of the anterior pituitary gland also affect carbohydrate metabolism, although not directly. Growth hormone and prolactin produce a more direct affect on carbohydrate metabolism than do adrenocorticotropin (ACTH) or thyrotropic hormone (TSH).

16.2.1 GLYCOGENESIS

The process of glycogenesis in the muscles utilizes mainly glucose, which enters the cells by way of an active transport. In the liver, in the course of glyconeogenesis, glucose is delivered by diffusion and glycogenic substances such as lactate, amino acids, and glycerol are the substrates.[3] In each case, glucose is phosphorylated at 6-CH_2OH into G-6-P by means of a phosphate residue abstracted from ATP:

$$\text{Glucose} + \text{ATP} \rightarrow \text{G-6-P} + \text{ADP} \qquad (16.1)$$

The process is catalyzed by hexokinase, the enzyme activated by magnesium ions. In contrast to glucose, G-6-P does not pass through cellular membranes. Hydrolysis of G-6-P into free glucose and phosphoric acid (Pi) in liver is catalyzed by glucose-6-phosphatase:

$$\text{G-6-P} + \text{H}_2\text{O} \rightarrow \text{glucose} + \text{Pi} \qquad (16.2)$$

Because of the absence of glucose 6-phosphatase, G-6-P in muscles can be transformed into either glucose-1-phosphate (G-1-P) (glycogenesis) or fructose-6-phosphate (F-6-P) (glycolysis).

A metabolic transfer of the phosphate group results in the formation of G-1-P in liver and muscle cells. At the equilibrium state, the mixture of both monophosphates contains 95% G-6-P. An intermediate [glucose-1,6-bisphosphate (G-1,6-BP)], being a phosphoenzyme, participates in transformations of glucosophosphates:

$$\text{G-6-P} \leftrightarrow \text{G-1,6-BP} \leftrightarrow \text{G-1-P} \qquad (16.3)$$

Glucose for biosynthesis of glycogen is delivered by UDP-glucose available from G-1-P and UTP in the process catalyzed by UDP-glucose pyrophosphorylase:

$$\text{G-1-P} + \text{UTP} \leftrightarrow \text{UDP-glucose} + \text{pyrophosphate (PPi)} \qquad (16.4)$$

Pyrophosphate (PPi) hydrolyses into orthophosphate:

$$PPi + H_2O \rightarrow 2 \ Pi \qquad (16.5)$$

Activated glucose units are transferred on the terminal hydroxyl groups at C-4 of glycogen being attached thereto by means of α-1,4 glycosidic linkages. This process is catalyzed by glycogen synthase:

$$Glycogen + UDP\text{-}glucose \rightarrow Glycogen + 1 + UDP \qquad (16.6)$$

Glycogen synthase attaches these subsequent glucose moieties to the polysaccharide chains with at least four sugar units.

Synthesis of new glycogen requires initiation from glycogenin, a 37-kDa protein carrying an oligosaccharide moiety composed of glucose units linked with one another through α-1,4 glycosidic bonds.

Glycogen synthase catalyzes selectively the formation of α-1,4 glycosidic linkages. Branching takes place after formation of a sufficiently long, linear polymer. A branching enzyme, amylo-(1,4-1,6)-transglucosylase, splits the α-1,4 glycosidic bond and forms the α-1,6 glycosidic bond instead. This process produces a considerable number of nonreducing ends open for interaction with glycogen synthase and phosphorylase. Activity of glycogen synthase in liver depends on concentration of glucose in blood and the activity of that enzyme in muscles depends on the level of G-6-P. In either case, a high concentration of these factors is beneficial. Glycogen synthase is inactive when phosphorylase is active, and, vice versa, when glycogen synthase is active the phosphorylase is not. Extracellular factors such as epinephrine or glucagon and intracellular factors such as calcium ions, which stimulate phosphorylase activity, inhibit the synthesis of glycogen.

16.2.2 GLYCOGENOLYSIS AND GLYCOLYSIS

Degradation of glycogen into G-6-P and pyruvate under aerobic and anaerobic conditions and transformation of pyruvate into lactate taking place exclusively under anaerobic conditions is called glycogenolysis.[4] Decomposition of G-6-P to pyruvate under aerobic and anaerobic conditions and into lactate under anaerobic conditions exclusively is known as glycolysis.

Antemortem glycolysis reactions in animal tissues are reversible except in processes catalyzed by phosphofructokinase and pyruvate kinase. In the presence of oxygen, pyruvate decarboxylates and irreversibly turns into acetyl coenzyme A (acetyl-CoA). Acetyl-CoA is oxidized in the citric acid cycle into CO_2 and H_2O. In each case, electrons are accepted by NAD^+. Oxidized NAD^+ is recovered after electrons from NADH are transferred to the oxygen molecule.

After exsanguination, neither oxygen is transported to the tissues nor are metabolites removed from them. Postmortem complex of pyruvate dehydrogenase is inactive. Reduction of pyruvate into lactate takes place rather than its transformation into acetyl-CoA. Accumulation of lactate in tissues reduces their pH value. The pH of meat as well as the composition and concentration of glycolytic intermediates have a marked effect on its properties and quality as culinary and processing material.

Glycogenolysis and glycolysis take place in the cell cytoplasm. The processes within the citric acid cycle and electron transfer in the respiratory chain take place in mitochondria.

Glycogenolysis begins with the removal of glycosidic residues from the nonreducing ends of glycogen. In this phosphorylase-catalyzed reaction, G-1-P is formed. Phosphorylase splits only the α-1,4 glycosidic bonds, and the splitting stops when there are four glucose units between the position of branching and the terminal of the chain. The α-1,6 glycosidic bond is split by α-1,6-glucosidase. Transferase enzyme is also involved at this stage of glycogenolysis. This enzyme transfers a group of three glucose residues from one chain terminal to another. This transfer exposes the glucose residue, which is α-1,6-bound to the chain of glycogen. Thus, transferase and α-1,6-glucosidase change the structure of glycogen from branched to linear, making it susceptible to subsequent splitting with phosphorylase.

Phosphorylase exists in mutually transformable active (a) and inactive (b) forms. The form is controlled by several allosteric effectors, which either signalize the energy state of the cell or stimulate the cell by means of reversible phosphorolytic process governed by extracellular factors. The type and the action of effectors on phosphorylase in the skeletal muscles and in the liver are different.

In skeletal muscles, inactive (b) phosphorylase changes into active (a) phosphorylase with the involvement of phosphorylkinase in the presence of magnesium ions. High concentration of ATP and G-6-P inhibits the transformation. The latter is usually activated by a high concentration of AMP, calcium ions, and cyclic AMP.

High concentration of AMP indicates the cell's high demand for energy, as for example on a muscle contraction. An increase in the concentration of calcium ions in the sarcoplasm results from nervous stimuli, which depolarize the muscle fiber membrane. Action potential from a fiber membrane travels in a transverse tubular system to the sarcoplasmic reticulum, which releases the bound calcium ions to the sarcoplasm. Synthesis of cyclic AMP proceeds under the influence of adenyl cyclase residing in the sarcolemma. Adenyl cyclase is activated by such hormones as epinephrine, norepinephrine, and glucagon. Stress connected with animal stunning at slaughtering and exsanguination processes cause mass release of epinephrine from adrenal medulla. Concentration of epinephrine in blood on exsanguination increases by 200 times.[4] In consequence, the strong stimulation of phosphorylase activity in muscles during slaughter processes take place and the rate of postmortem glycolysis is increased.

In the liver, AMP does not activate the transformation of b-phosphorylase into its a-form; instead, phosphorylase reacts to concentration of glucose. Bonding to glucose deactivates the a-form of liver phosphorylase. Glycogen in liver decomposes in order to increase concentration of glucose in blood.

Phosphoglucomutase catalyzes the transformation of G-1-P into G-6-P, which subsequently in glycolysis is decomposed into pyruvate. Figure 16.2 shows the course of glycolysis.

G-6-P, the aldose, the main substrate in glycolysis, isomerizes into F-6-P, the ketose. This process, catalyzed by phosphoglucose isomerase is followed by *in vivo*

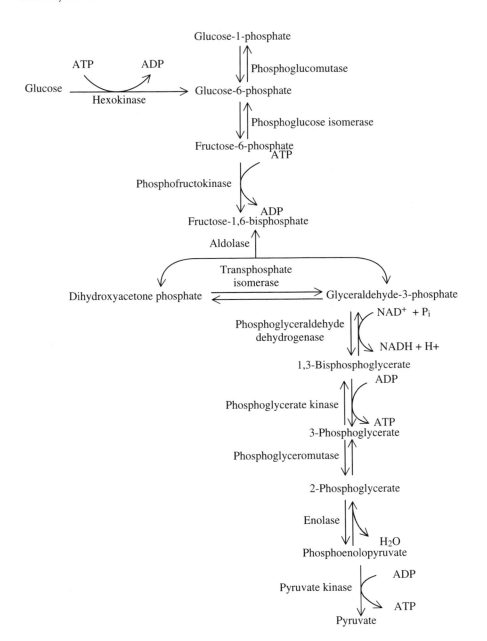

FIGURE 16.2 The reactions involved in glycolysis.

irreversible phosphorylation of F-6-P with ATP into fructose-1,6-bisphosphate (F-1,6-BP). The course and the rate of glycolysis depend to a great extent on the activity of that enzyme. The activity of phosphofructokinase is controlled by the level of all ATP, its metabolites, and calcium ions. Activity of the enzyme increases at low level of ATP and high level of AMP and calcium ions. Thus, an acceleration of glycolysis takes place on increased energy demand of cells and as an effect of nervous stimuli.

In the next step of glycolysis, F-1,6-BP is catalytically cleaved by aldolase into glyceraldehyde-3-phosphate and dihydroxyacetone phosphate. Up to this point, glycolysis does not produce energy; on the contrary, it consumes two ATP molecules. The energy is liberated in subsequent steps. Glyceraldehyde-3-phosphate directly enters further stages of glycolysis, and isomeric dihydroxyacetone phosphate can reversibly transform into glyceraldehyde-3-phosphate. *In vivo*, one F-1,6-BP molecule produces two molecules of glyceraldehyde-3-phosphate.

The redox reaction of glyceraldehyde-3-phosphate into 1,3-bisphosphoglycerate produces acetylophosphate, the mixed anhydride of phosphoric and acetic acids. The anhydride has a high potential of phosphate group transfer. The subsequent transfer of the phosphate group from acetylophosphate (1,3-bisphosphoglycerate) to ADP produces 3-phosphoglycerate and ATP.

In the last phase of glycolysis, 3-phosphoglycerate is intramolecularly rearranged into pyruvate and the second ATP molecule. ATP allosterically inhibits glycolysis at low energy demand of cells.

Glucokinase, which is located in liver, phosphorylates excessive glucose. Glucokinase produces G-6-P for synthesis of glycogen. Because the K_m of glucokinase is high, the production of glucose on its low level in blood is put into priority. The sequence of reactions leading from G-6-P to pyruvate is generally common for all organisms and types of cells but the fate of produced pyruvate is different.

In animal tissues, pyruvate is transformed into lactate under anaerobic conditions. This reduction with NADH from the oxidation of glyceraldehyde-3-phosphate is catalyzed by lactate dehydrogenase. Recovery of NAD$^+$ in this process provides continuous glycolysis. A portion of energy liberated on glycolysis under anaerobic conditions, although not high, is responsible for a postmortem increase in temperature of muscles.

Under aerobic conditions, pyruvate undergoes irreversible oxidative decarboxylation to acetyl-CoA:

$$\text{Pyruvate} + \text{NAD}^+ + \text{CoA} \rightarrow \text{acetyl-CoA} + \text{CO}_2 + \text{NADH} \qquad (16.7)$$

Postmortem glycogen degradation in muscles proceeds at a high rate until the resistance and capacitance of cell and plasma membranes (sarcolemma, sarcoplasmic reticulum, mitochondria, and lysosomes) are eliminated.[4] When pH from resting antemortem muscles decreases from approximately 7.4 to below 6.5, there is a free diffusion of ions through formerly semipermeable plasma membranes. In this state, the muscle loses its ability to contract, and there is a free diffusion of ions; this results in the equalization of pH throughout the tissue. From this point onward, glycolysis continues at a reduced rate until either glycogen deposits are completely depleted or the pH is low enough to completely inhibit the glycolytic enzymes. This occurs at a pH slightly below 5.4. Even though at this level of pH ample glycogen might still be present, glycogen breakdown ceases.

The ultimate pH of meat after rigor mortis may range between 7.0 and 4.8. Usually, pH of meat of standard quality ranges between 5.3 and 5.8 for pork, veal, beef, and lamb, and between 5.8 and 6.0 for poultry.[5]

The rate of postmortem glycolysis in skeletal muscles depends on the temperature at which the carcass is cooled. Also, composition of muscle fiber type; species, breed, and sex of animals; resistance to stress; and processing of the carcass after slaughter, as, for instance, electrical stimulation, storage temperature, prerigor cutting and curing, influence glycolysis.

The rate of glycolysis is higher in white muscle fibers than in the red, because the activity of muscle ATPases in white muscle fibers is higher than it is in red muscle fibers.

Muscle stimulation associated with the activity of the animal prior to its slaughter, electrical stunning at the time of slaughter, and electrical stimulation of freshly slaughtered carcass result in intensified release of calcium ions from the sarcoplasmic reticulum and mitochondria and increase the rate of postmortem glycolysis.

The rate of postmortem glycolysis in anatomically the same muscles but from another specimen can be entirely different. It is important in determining the ultimate properties and quality of meat (Figure 16.3).

A normal pH decline pattern in porcine muscle is represented by a gradual decrease from approximately pH 7.4 in living muscle to a pH of about 5.6 to 5.7 within 6 to 8 h postmortem, and then to ultimate pH (reached at approximately 24 h postmortem) in range of 5.3 to 5.7. In some animals, the pH drops only slightly during the first hour after slaughter and remains stable at a relatively high level, giving an ultimate pH in the range of 6.5 to 6.8. In other animals, muscle pH drops rapidly to around 5.4 to 5.5 during the first hour after exsanguination. Meat from these animals ultimately develops a pH in the range of 5.3 to 5.6.

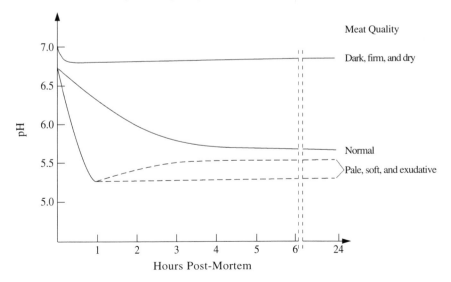

FIGURE 16.3 Effect of postmortem pH decline on the meat quality. (Modified from Briskey, E.J. et al., *J. Agric. Food Chem.*, 14, 201, 1966.)

Accumulation of muscle lactate early in the postmortem period may adversely affect meat quality.[6] Development of low pH (acidic) in the muscle before natural body heat and the heat of continuing metabolism (glycolysis) have been dissipated through carcass chilling causes denaturation of muscle proteins. Denaturation of proteins causes loss of protein solubility, water- and protein-binding capacity, and muscle color. Muscles that have very rapid and extensive pH decline will be pale, soft, and exudative (PSE). On the other hand, muscles that maintain a high pH during muscle to meat conversion may be dark in color, firm in texture, and dry on the exposed cut surface (DFD), because naturally occurring water is tightly bound to proteins. These changes are undesirable, whether the muscle is going to be used as fresh meat or subjected to further processing.

16.3 GLUCOSYLAMINOGLYCANS

Thus far, only eight glucosylaminoglycans (GAGs), mucopolysaccharides, have been isolated from animal tissues.[7] They are chondroitin, chondroitin sulfates A (Ch-4-S) [4→O-β-(Glc-UA)-1→3-O-β-D-(2-NHAc-Gal$p$-4-SO$_3$H)-1→] and C (Ch-6-S) [→4-O-β-D-GlcUA)-(1→3)-O-β-D-(2-NHAc-Gal$p$-6-SO$_3$H)-1→], hyaluronic acid (HA), heparin (H), heparan sulfate (HS), dermatan sulfate (DS) [4→O-α-D-IdoUA-(1→3)- β-D-(2-NHAc-Gal$p$-4-SO$_3$H)-1→], and keratosulfate (KS) [→3-O-β-D-Gal$p$-(1→4)-β-D-(2-NHAc-Gal$p$-6-SO$_3$H)-1→]. Apart from typical chondroitin monosulfates, chondroitin disulfates carrying the sulfate group in the 2-position of D-glucuronic acid (GlcUA) and in the 6-position of N-acetyl-D-galactosamine (chondroitin sulfate D), and in the 4- and 6-position of N-acetyl-D-galactosylamine (chondroitin sulfate E) have also been isolated.

In H and HS, either D-glucuronic acid or L-iduronic acid constitute the acidic function of these GAGs and D-glucosamine is the moiety carrying the amino group. Monosulfated polymers rich in D-glucuronic acid are accounted for HSs, and disulfated polymers rich in L-iduronic acid (IdoUA) belong to Hs. In Hs, over 80% of the D-glucosylamine moieties is N-sulfated. Degree of sulfation of D-glucosylamine in HS is significantly lower. Variation in sulfation is also observed among DS isolated from various tissues.

Number of disaccharide mers of GAGs varies between 25 and 25,000 such units. The highest (103 kDa) and the lowest (8 to 12 kDa) molecular weight was noted for HA and H, respectively. Molecular weights of Ch-4-S, Ch-6-S, and DS are similar and range from tens to several hundred kilodaltons.

Connective tissues contain various GAGs. However, in eye lens and synovia, solely HA is found. Cartilage tissue is the richest in GAGs, mainly in Ch-4-S and Ch-6-S. Dry matter cartilages of nose partition and trachea contain 10 to 12% glycosaminoglycans. Dry matter of eye cornea and skin contain 1% and 0.3–0.6% GAGs, respectively. KS and chondroitin sulfates contribute in 50 and 20% of the total eye cornea GAGs, respectively. The contribution of Ch-4-S and Ch-6-6 in cattle trachea cartilage reaches 56 and 38%, respectively. No DS is found in animal cartilages; it occurs mainly in connective tissue of skin, tendons, and heart valves.

The type and amount of GAGs are specific not only for a given type of connective tissue but also for the animal species. Rabbit skin contains mainly DS and considerable amount of Ch-6-S, whereas porcine skin has only a trace of Ch-6-S. Rat skin contains considerable level of H, which is absent in the human, porcine, and rabbit skin.

The level and composition of GAGs in tissues change with the growth and development of animals. The level of GAGs in the muscle epimysium of cattle decreases with age. Major changes are observed in the level of HA. In porcine skin, the contents of DS and KS increase with age. The content of GAGs depends also on sex, because their synthesis, especially the synthesis of HA, is stimulated by estrogens.

During postmortem ageing, the content of GAGs in meat decreases in cold storage. It is likely that the their degradation is caused by endogenic enzymes liberated from lysosomes and the bacterial enzymes chondroitinase and hyaluronidase.

GAGs are soluble in water and aqueous salt solutions. Differential solubility of GAGs in alcohol is used in their separation. Viscosity of GAG solutions depends on the molecular weight of the solute. Solutions of HA and H are the most and the least viscous, respectively. Aqueous GAG solutions are chiral. Specific rotation of the aqueous H solution is $+50°C$. This parameter for solutions of other GAGs is negative and varies between $-16°$ (Ch-6-S) and $-80°C$ (HA).

16.3.1 PROTEOGLYCANS

Except HS, GAGs are covalently bound to proteins, forming complexes called proteoglycans (PGs). PGs form macromolecules having a protein backbone to which chains of GAGs are covalently attached. In case of Ch-4-S and Ch-6-S, a Gal–Gal–Xyl trisaccharide participates in the bonding of those polysaccharides with the hydroxyl group of either serine or threonine fragment of the protein.[7]

Content and proportions of GAGs in various PGs of connective tissue are variable. Connective tissue contains also different PGs. PGs of epimysium and primysium of muscles contains mainly Ch-4-S, Ch-6-S, and DS. PGs complexes of muscle endomysium carryprimarily HS.[8] In tendons, low-molecular-weight PGs (50 to 600 kDa) are found.

Number of GAGs chains varies between 1 and 100. Each PG contains either one or two different GAGs.[9] Molecular weight of these complexes ranges between 50 kDa to several thousand kilodaltons. Among polysaccharide–protein complexes, PGs of cartilages (Figure 16.4) are best characterized. The dominating Ch-4-S and Ch-6-S are accompanied by a small amount of KS. Considerable number of negatively charged groups causes repulsion of GAG chains electrostatically bound to amino acid moieties of muscle collagen. A rigid cross-linked structure is formed. PGs form bridges connecting collagen fibers.

16.3.2 METABOLISM OF GLUCOSYLAMINOGLYCANS

PGs have a varying protein backbone. Synthesis of GAG starts from the attachment of a saccharide to a specific amino acid, usually either serine or threonine in the protein backbone. Chiefly, D-xylose is the first attached saccharide. Then, subsequent molecules join the terminal of the chain.

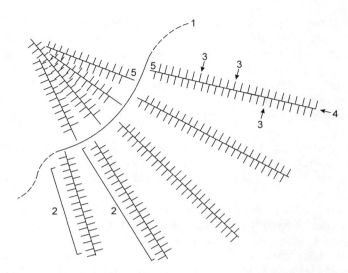

FIGURE 16.4 Structure of proteoglycans complex of cartilage. 1, hyaluronic acid; 2, proteoglycan; 3, chondroitin sulfates; 4, protein core; 5, binding protein. (Adapted from Briskey, E.J. et al., *J. Agric. Food Chem.*, 14, 201, 1966.)

Modifications such as sulfation, epimerization, and deacetylation occur after polymerization is completed. Apart from differentiation in the number and length of chains connected to the protein backbone, these modifications contribute to the structure and type of PG.

Decomposition of GAGs in connective tissue is catalyzed by galactosidases, glucuronidases, glycosidases, and hyaluronidase. Hyaluronidase decomposes substances cementing particular fibers and cells. This process facilitates exchange of matter in the extracellular space. Microorganisms producing hyaluronidase damage specific immunobarriers responsible for spreading of pathogens such as toxins, viruses, and bacteria throughout tissues. Primarily, hyaluronidase depolymerizes large HA molecules into simpler nonreducing polysaccharides. *N*-Acetylglucosamine and glucuronic acid are liberated in the subsequent phase.

Hyaluronidase also catalyzes the degradation of other GAGs. Both microbial hyaluronidase and hyaluronidase from male gonads catalyze hydrolysis of HA. Ch-4-S and Ch-6-S are decomposed solely by hyaluronidase of male gonads, and none of these enzymes can decompose DS and KS.

GAGs in animal organisms are distributed in the connective tissue. Together with other components of the ground substance of the tissue, they fill the space between collagen, elastin, and reticulin fibers. Sulfated GAGs play a key role in building structure and mechanical properties of connective tissue. HA which forms viscous, colloidal aqueous solutions that fill the extracellular space is a lubricant in joints, filler in eye lens, and shock absorber in disks of vertebral column.

Heparin is a blood anticoagulant, inhibiting formation of thrombin from prothrombin. Thrombin is responsible for transformation of fibrinogen into fibrin.

GAGs also influence functional properties of collagen. For instance, the presence of 30% chondroitin sulfates shifts the shrinkage temperature of collagen by 10°C. The type and number of associated PGs is essential for resistance of collagen to enzymatic degradation, which results in an increase in collagen solubility. An increase in beef tenderness on cold ageing is accompanied by increase in the activity of glucuoronidase. Structure of PGs and their ability of holding water is critical for the resistance of cartilage to deformation on stretching and compression.

Transformations of GAGs in animal organisms are influenced by the pituitary gland, thyroid, and testes hormones. Growth hormones induce the synthesis of GAGs, increase the number of mast cells in connective tissue, and stimulate the activity of hyaluronidase. Testosterone and estrogens stimulate synthesis of HA and chondroitin sulfates. Thyroxine and triiodothyronine, as well as glucocorticoids, inhibit GAG metabolism.

16.3.3 UTILIZATION OF GLUCOSYLAMINOGLYCANS

Ability of GAGs to hold water and salts and their interactions with protein can be useful in modifications of rheology of foodstuffs and in cosmetics. Effect of acidic GAGs on the viscosity and emulsifying properties can be applied in stabilization of meat batter.

GAGs can be helpful in reconstitution of collagen *in vitro*. In the presence of GAGs (DS, HA) in aqueous solution, collagen fibrils are formed even from fractions of soluble collagen that normally do not exhibit any tendency for reconstitution.

The pharmaceutical industry utilizes HA sodium salt in manufacturing medicine for ophtalmological surgery of cataract, cornea transplantation, therapy of glaucoma, and damage of bulbus oculi. HA is injected to joints in case of arthritis.

16.4 GLYCOPROTEINS

Most plasma proteins, particularly blood-group specific substances, many milk and egg proteins, mucins, connective tissue components, some hormones, and many enzymes are glycoproteins (GPs). GPs are found in all extracellular biological fluids. Several integral proteins of cells and intracellular plasma membranes contain an oligosaccharide component. The saccharide portion is 5 to 10% of the mass of GPs. Polysaccharide chains with GPs have up to 15 monosaccharide units.

GPs contain various sugar residues linked to one another by various glycosidic bonds. Usually, the carbohydrate component of GPs, an oligosaccharide, is attached to the oxygen atom of the side chain of either serine or threonine, and optionally forms an *N*-glycosidic linkage with the nitrogen atom of the side chain of aspartic acid residues.

GP oligosacharides linked through the *N*-glycosidic bond contain a common pentasaccharide core composed of three molecules of mannose (Man) and two *N*-acetyloglucosylamine (GlcNAc) molecules. Subsequent saccharides are variously bound to this core, which is responsible for a wide variety of GPs oligosacharides. In oligosaccharides rich in Man, the Man residues are attached to the oligosaccharide core. In complex oligosaccharides, various combinations of GlcNAc, galactose (Gal), sialic acid (Sia), and fucose (Fuc) residues are attached to the oligosaccharide core. Structurally, Sia (*N*-acetylneuraminic acid) is unique in that it originates from aldol condensation of pyruvic acid and 2-acetoamido-2-deoxy-D-mannose, the only known occurrence of this 2-amino sugar in the animal body.

Carbohydrate components in GPs control several bioprocesses. For instance, terminal carbohydrate residues can prompt liver cells to order protein intake from blood, and asialylglycoprotein is the surface receptor. Sialylases, enzymes that occur on the surface of blood vessels, remove sialylglycoproteins being secreted. Protruding Gal residues of GPs shortened in this manner are recognized by asialylglycoprotein receptors in the liver cell membranes. The complex of asialylglycoprotein with its receptor is then introduced into the cell interior by a process called endocytosis. In this manner, GPs pass from the blood to the liver cells.

Figure 16.5 presents the composition and distribution of dystrophin-associated glycoprotein complex (DGC) in skeletal muscle. DGC forms a critical link between intracellular cytoskeleton and extracellular matrix. The components of DGC include several proteins on the cytoplasmic side (among others dystrophin, the syntrophins and dystrobrevin) and two transmembrane glycoprotein complexes (the dystroglycans and sarcoglycans). DGC is composed of α-dystroglycan, a large extracellular laminin-binding glycoprotein, and β-dystroglycan, a transmembrane glycoprotein. The sarcoglycan complex is composed of four transmembrane glycoproteins: α-, β-, γ-, and δ-glycoproteins.

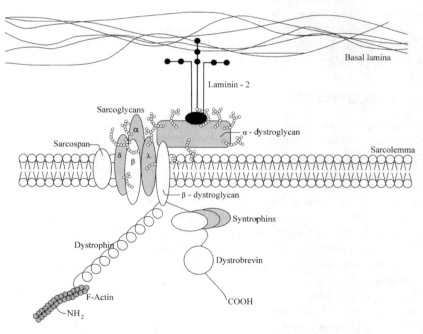

FIGURE 16.5 Schematic drawing of the dystrophin-associated complex in skeletal muscle. (Provided by J.T. den Dunnen, Leiden Muscular Dystrophy pages, http://www.dmd. nl.DGC.html. With permission.)

GPs play important role in the interactions between sperm and ovum during fertilization. The mammalian mature ovum is surrounded within an extracellular sheath, the zona pellucida (ZP). The receptor on the sperm surface recognizes oligosaccharides bound to ZP3, which is a GP of the ovum sheath. Oligosaccharides of this GP have Gal on its nonreducing end. Interaction of sperm receptors with ZP oligosaccharides results in the release of the sperm enzymes proteases and hyaluronidases. These enzymes dissolve ZP, providing penetration of the spermatozoon into the ovum.

REFERENCES

1. Merkel, R.A., Carbohydrates, in *The Science of Meat and Meat Products,* Price J.F. and Schweigert B.S, Eds., W.H. Freeman, San Francisco, 1971, p. 145.
2. Lawrie, R., *Meat Science,* Woodheat Publishing, London, 1998.
3. Stryer, L., *Biochemistry* (in Polish), Wydawnictwo Naukowe PWN, Warszawa, 2000.
4. Kołczak, T., *Biological Basis of Meat Technology* (in Polish), Cracow Agricultural University Editor, 1983.
5. Aberle, E.D. et al., *Principles of Meat Science*, Kendall/Hunt Publishing,Dubuque, IA, 2001.
6. Briskey, E.J. et al., Biochemical aspects of post-mortem changes in porcine muscle. *J. Agric. Food Chem.*, 14, 201, 1966.
7. Sadowska, M. and Łagocka, J., Mucopolysaccharides: physicochemical properties, isolation and applications (in Polish). *Żywność, Technologia, Jakość.* 2(11), 23, 1997.
8. Palka, K, Changes in the microstructure and texture of bovine muscles during post-mortem ageing and heating (in Polish). *Zeszyty Naukowe AR w Krakowie*, Ser. Rozprawy, 270, 2000.
9. Bailey, A.J. and Light, N.D., *Connective Tissue in Meat and Meat Products*, Elsevier Applied Science, London, 1989.

17 Cyclodextrins

József Szejtli

CONTENTS

17.1 Introduction 271
17.2 Cyclodextrins...272
    17.2.1 Chemistry and Physicochemical Properties 272
    17.2.3 Metabolism and Toxicology 277
    17.2.4 Cycylodextrin Derivatives 278
17.3 Cyclodextrin Inclusion Complexes 280
17.4 Applications of Cyclodextrins in Foods 282
    17.4.1 Cyclodextrin Complexes in Food: Flavors and Vitamins 282
    17.4.2 Cyclodextrins as Physical Property Modifiers 283
    17.4.3 Cyclodextrins as Taste Modifiers 284
    17.4.4 Selective Complexation and Sequestration 284
    17.4.5 Further Examples of Applications of Cyclodextrins to Food 285
17.5 Examples of Application of Cyclodextrins in Various Industrial Products
    and Technologies 285
    17.5.1 Pharmaceuticals 285
    17.5.2 Cosmetics and Toiletry 286
    17.5.3 Biotechnology 287
    17.5.4 Textiles 287
    17.5.5 Analytical Separation and Diagnostics 287
    17.5.6 Miscellaneous Industrial Applications 288
References 289

17.1 INTRODUCTION

The history of cyclodextrins (CDs) is divided into three stages. In the first period from 1891 (discovery by A. Villiers) until middle of the last century, the structure and fundamental properties of CDs were recognized. F. Schardinger isolated pure α- and β-CDs, and described their fundamental properties, and K. Freudenberg elucidated their structure. In the second stage, up to the 1970s, mainly applications and the inclusion complex forming properties of CDs were studied. D. French et al.[1] and F. Cramer et al.[2] began to work intensively on the enzymatic production of CDs, fractionating them and characterizing their chemical and physical properties. French[1] discovered that there are even larger CDs, whereas Cramer's group focused mainly on the inclusion complex forming properties of CDs. The first fundamental review on CDs was published in 1957 by French.[1] It was followed by monographs in 1965 by Thoma and Stewart[3] and in 1968 by Caesar.[4] French's otherwise excellent monograph contained a severe misinformation on the toxicity of CDs. The third stage began at the end of 1970s, with the industrial production of CDs and their steadily increasing utilization in numerous products and technologies.

The way Villiers discovered CDs suggested that they occur by chance or on purpose in "microbe-treated" starch-containing foods. The diversity of activities of different microbial amylases on starch implies a possibility of the formation of minute amount of CDs as an unusual by-product of amylolytic processes. Nishimura et al.[5] had reported in 1956 the formation of cyclic α-1,4-glucan with the involvement of bacterial α-amylases. In 2002, Harangi et al.[6] detected CDs in commercial beer samples and in bread crust. It is assumed that thermal treatment of dough yielded small amounts of branched CDs in probably a similar manner.[7] As beer and bread have been consumed by humans in considerable amounts since ancient times, these compounds cannot be toxic per se.

Adequate toxicological studies have proved that the toxicity attributed to CDs originated from complexed impurities, and there is no danger from an inadequate form of administration or extreme dosing. Because doubts on the inherent toxicity of CDs did not inhibit their widespread utilization anymore, the number of CD-related publications rose substantially. By the end of 2003, the number of CD-related publications will reach 26,000. About 7% of them are dedicated to the utilization of CDs in food, cosmetic, and toiletry products.[8]

17.2 CYCLODEXTRINS

17.2.1 CHEMISTRY AND PHYSICOCHEMICAL PROPERTIES[9–11]

CDs comprise a family of three well-known industrially produced major, and several rare minor, cyclic oligosaccharides. The three major CDs are crystalline, homogeneous, nonhygroscopic substances, which are torus-like macrorings built from glucopyranose units. α-CD (Schardinger's α-dextrin, cyclomaltohexaose, cyclohexaglucan, cyclohexamylose, αCD, ACD, C6A), β-CD (Schardinger's β-dextrin, cyclomaltoheptaose, cycloheptaglucan, cycloheptaamylose, β-CD, BCD, C7A), and γ-CD (Schardinger's γ-dextrin, cyclomaltooctaose, cyclooctaglucan, cyclooctaamylose, γ-CD, GCD, C8A) comprise six, seven, and eight glucopyranose units, respectively (Figure 17.1). Table 17.1 summarizes the most important characteristics of CDs.

As a consequence of the $^4C_1$ conformation of the glucopyranose units, all secondary hydroxyl groups are situated on one of the two edges of the ring, whereas all primary hydroxyl groups are placed on the other edge. The ring forms a conical cylinder, which is frequently characterized as a doughnut or wreath-shaped truncated cone. The cavity is lined with hydrogen atoms and glycosidic oxygen bridges. Orbitals of the nonbonding electron pairs of the glycosidic oxygen bridges are directed toward the cavity center, producing a high electron density therein and lending to the cavity a Lewis-base character.

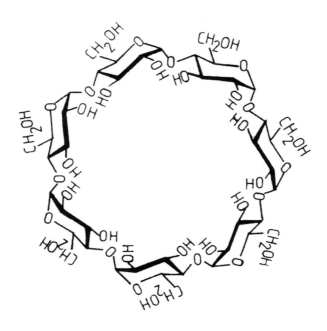

CAVITY VOLUME:

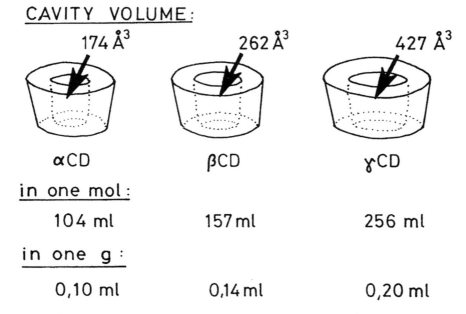

174 Å³   262 Å³   427 Å³

αCD   βCD   γCD

in one mol:

104 ml   157 ml   256 ml

in one g:

0,10 ml   0,14 ml   0,20 ml

FIGURE 17.1 Structure of β-CD and approximate dimensions of α-, β-, and γ-CDs.

The C-2-OH group of one glucopyranoside unit can form a hydrogen bond with the C-3-OH group of the adjacent glucopyranose unit. In the β-CD molecule, a complete secondary belt is formed by these hydrogen bonds. Therefore, β-CD is a

TABLE 17.1
Characteristics of α-, β- and γ-CD

Characteristic	α	β	γ
No. of glucose units	6	7	8
Molecular weight	972	1135	1297
Solubility in water (g/100 ml) at room temp.	14.5	1.85	23.2
$[\alpha]_D$ at 25°C	150 ± 0.5	162.5 ± 0.5	177.4 ± 0.5
Cavity diameter (Å)	4.7–5.3	6.0–6.5	7.5–8.3
Height of torus (Å)	7.9 ± 0.1	7.9 ± 0.1	7.9 ± 0.1
Diameter of outer periphery (Å)	14.6 ± 0.4	15.4 ± 0.4	17.5 ± 0.4
Approx. volume of cavity (Å³)	174	262	427
Approx. cavity volume in 1 mol CD (ml)	104	157	256
Approx. cavity volume in 1 g CD (ml)	0.10	0.14	0.20
Crystal forms (from water)	Hexagonal plates	Monoclinic parallelograms	Quadratic prisms
Crystal water (wt%)	10.2	13.2–14.5	8.13–17.7
Diffusion constants at 40°C	3.443	3.224	3.000
Hydrolysis by A. oryzae α-amylase	Negligible	Slow	Rapid
pK (by potentiometry) at 25°C	12.332	12.202	12.081
Partial molar volumes in solution (ml/mol)	611.4	703.8	801.2
Adiabatic compressibility in aqueous solutions (ml/mol/ bar ×10⁴)	7.2	0.4	−5.0

fairly rigid structure. The intramolecular hydrogen bond formation explains the lowest water solubility of β-CD among all CDs.

The hydrogen-bond belt in the α-CD molecule cannot be completed because one glucopyranose unit takes a distorted position. Consequently, only four hydrogen bonds are formed instead of the six required hydrogen bonds. γ-CD is the non-coplanar, more flexible molecule and is therefore most soluble among all the CDs.

Figure 17.2 demonstrates characteristic structural features of CDs. On the side with the secondary hydroxyl groups, the cavity is wider than on the other side carrying the freely rotating primary hydroxyl groups.

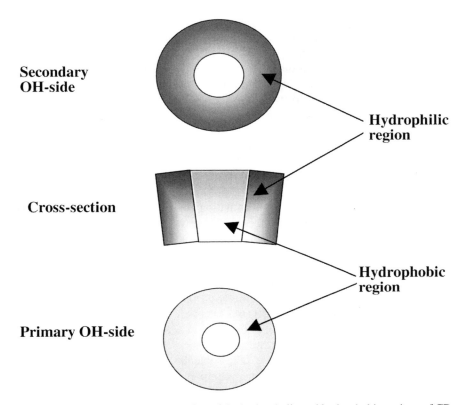

Secondary OH-side

Cross-section

Primary OH-side

Hydrophilic region

Hydrophobic region

FIGURE 17.2 Schematic representation of the hydrophylic and hydrophobic regions of CDs. The darker regions represent the more hydrophilic and the lighter regions the more hydrophobic regions.

For a long time, only three parent major CDs (α-, β-, and γ-CD) were known and well characterized. In the 1990s, a series of larger CDs was isolated. For example, the nine-membered δ-CD was chromatographically isolated from the commercially available CD-conversion mixture. It is more readily water soluble than β-CD, but less water soluble than α- and γ-CD. In acids it is the least stable among the CDs known; the rate of acid-catalyzed hydrolysis increases in the order α-CD < β-CD < γ-CD < δ-CD.

Larger CDs have irregular cylindrical structures. They are like collapsed cylinders, and their real cavity is even smaller than that in γ-CD. The driving force of the complex formation, the substitution of high-enthalpy water molecules in the CD cavity, is weaker in larger CDs. Therefore, their utilization as inclusion complexing agents will probably remain restricted.

17.2.2 PRODUCTION[10,11]

The cyclodextrin glucosyl transferase enzyme (CGT-ase) is produced by a large number of microorganisms, for example, *Bacillus macerans, Klebsiella oxytoca,*

Bacillus circulans, and *Alkalophylic bacillus* No. 38-2. Genetic engineering has provided more active enzymes, most of which will probably be used in the future for industrial CD production.

In the first step of CD production, starch is liquefied at elevated temperature. It is hydrolyzed to an optimum degree in order to reduce the viscosity of such fairly concentrated (around 30% dry weight) starch solution. The prehydrolyzed starch has to be free of glucose and low-molecular-weight oligosaccharides, which strongly reduce the yield of CDs. After cooling this solution to optimum temperature, CGT-ase enzyme is added. In the so-called nonsolvent technology, α-, β-, and γ-CDs have to be separated from the complicated partially hydrolyzed mixture. No complex-forming solvent is used in this method. Acyclic dextrins are removed by digestion with admixed glucoamylase. Glucoamylase does not digest CDs. This method applies only to β-CD, and this limitation, together with low yield (not higher than 20%), is the serious disadvantage of the method.

In solvent technology, an appropriate complex-forming agent is added to the conversion mixture (Figure 17.3). If toluene is added to this system, the toluene/β-CD complex formed is separated immediately, and the conversion is shifted toward β-CD formation. If *n*-decanol is added to the conversion mixture α-CD is mainly produced, whereas with cyclohexadecenol γ-CD is the main product. Various other complex-forming agents can be used, the selection depending mainly on the efficiency of the removal of the solvents from the crystalline final product. However, price, toxicity, and explosiveness are also factors. The insoluble complexes are filtered from the conversion mixture. The solvents are removed by distillation or extraction from the filtered and washed complex after suspending in water. The aqueous solution obtained after removing the complexing solvent is treated with activated carbon, and filtered. CDs are then separated from this solution by crystallization and filtration. The homogenicity of the industrially produced CDs exceeds 99%.

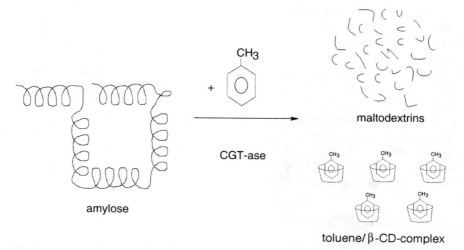

FIGURE 17.3 Production of β-CD.

17.2.3 METABOLISM AND TOXICOLOGY[11-13]

The fate of CDs in the mammal organisms is crucial for applications of CDs in the pharmaceutical and food industries. CDs orally administered with pharmaceuticals or food are either free CDs or CD inclusion complexes containing a drug, flavor, or other guest substance. Under physiological conditions, the inclusion complexes undergo instant dissociation in the gastrointestinal tract. Therefore, only the absorption of free CDs deserves attention. Summarizing the available data and experimental observations, the following conclusions can be drawn. CDs are relatively large molecules with strongly hydrophilic outer surfaces. They are true carriers bringing hydrophobic guests into solution, keeping them in solution, and transporting them to the lipophilic cell membrane. After delivery of the guest to the cell, because the cell has higher affinity to the guest than to the CD, the latter remains in the aqueous phase. Only an insignificant amount of orally administered CD is absorbed from the intestinal tract in the intact form. Eventual CD-elicited toxic phenomena might result from the solubilizing effect of CDs facilitating migration of insoluble, nonabsorbable toxic compounds such as fused-ring aromatic hydrocarbons and pesticides to and throughout the organism.

In contrast to starch which is metabolized in the small intestine, the preponderant part of orally administered CD is metabolized by the colon microflora. The maximum intensity of starch metabolism is achieved within the first two hours. Similar to starch, acyclic maltodextrins, maltose, and glucose, the primary metabolites, are absorbed, metabolized, and finally excreted in the form of CO_2 and H_2O. The apogee in metabolism of CDs is observed within the sixth to eighth hour after their uptake. The metabolism of α-CD is the slowest and that of γ-CD the fastest. Chemically modified CDs may be resistant to enzymatic degradation. Thus, methylated and hydroxypropylated CDs and perhaps all other more or less substituted CD derivatives can be absorbed from the intestinal tract to a quite considerable extent and enter circulation.

On parenteral administration, all CDs can interact with the components of the cell membranes, but the extent of this interaction may be entirely different. This cell-damaging effect can be illustrated by the CD concentration-dependent hemolysis of erythrocytes and decrease in the luminescence of *Escherichia coli* suspensions. More hydrophobic but still well-soluble CD derivatives such as dimethyl β-CD exerted the strongest effect. The cell-damaging effect from CD is caused by their complexation of cell wall components such as cholesterol and phospholipids. Therefore, among all CDs, β-CD, and particularly its more hydrophobic derivatives such as dimethyl β-CD, show the strongest cell-membrane damaging effect. γ-CD and its derivatives are less toxic, as due to the diameter of its cavity, affinity to cholesterol and phospholipids is lower. The immediate toxic consequence can be avoided when CDs are administered at a rate that maintains their actual concentration below the hemolytic threshold concentration (slow infusion, absorption from the intestinal tract, or through the skin). However, the toxic effect to kidneys cannot be eliminated. Parenteral administration of β-CD is impossible because it forms the most stable cholesterol/CD complex, which accumulates in the kidneys and destroys cells. Apparently, γ-CD is free from this noxious property. γ-CD is well soluble and does not form stable, readily crystallizing complexes with any vital component of the circulation. The cavity of γ-CD is too large and that of α-CD is too small to incorporate cholesterol.

Hydroxypropyl β-CD (HPβ-CD) does not form insoluble kidney-damaging cholesterol complex crystals. Therefore, its fairly large doses can be injected without any irreversible consequences, but the long-term treatment results in unwanted formation of soluble cholesterol/HPβ-CD complex in circulation. In the kidneys, this less stable complex dissociates, HPβ-CD is excreted, but cholesterol deposits in the kidneys. Only CDs that have a low affinity for cholesterol can be administered parenterally. CDs stable in aqueous solution, capable of good solubilization of drug, and rapidly decomposing after injection into the circulation are superior. Unfortunately, there is no such derivative yet.

17.2.4 CYCYLODEXTRIN DERIVATIVES[11,14,15]

For several reasons, such as price, availability, approval status, and cavity dimensions, β-CD is the most widely used and represents at least 95% of all produced and consumed CDs. It is used for many purposes; however, its low aqueous solubility and the low solubility of most of its complexes limit its common application. Fortunately, either chemical or enzymatic modifications strongly improve the solubility of all CDs, and concentration of available aqueous β-CD solutions can be increased from 18 g/l to over 500 g/l. Every glucopyranose unit of CD has three free hydroxyl groups, which differ in their functions and reactivity. The relative reactivities of the C-2 and C-3 secondary and the C-6 primary hydroxyl groups depend on reaction conditions such as pH, temperature, and reagents. Tewnty-one hydroxyl groups of β-CD can be modified by either chemical or enzymatic substitution of their hydrogen atom with a variety of substituting groups such as alkyl, hydroxyalkyl, carboxyalkyl, aminoalkyl, thioalkyl, tosyl, glucosyl, maltosyl, acyl, and other groups. In such a manner, several thousand of ethers, esters, anhydrodeoxy-, acidic, basic, and neutral derivatives can be prepared (Figure 17.4). The derivatization might (1) improve the solubility of CD (and its complexes); (2) improve the fitting, or the association, or both, between the CD and its guest with concomitant stabilization of the guest, reducing its reactivity or mobility, or both; (3) attach specific, for instance, catalyzing groups to the binding site (e.g., in enzyme modeling); or (4) form insoluble, immobilized CD-containing structures or polymers for chromatography.

Among several thousand CD derivatives described in a remarkable number of scientific papers and patents, only a dozen can be considered suitable for industrial scale manufacture and utilization. Complicated multistep reactions; the expensive, toxic, and environment-polluting reagents required; and chromatographic purification of the products limit relevant studies to the laboratory scale.

Among those industrially produced, standardized, and available in tons, the heterogeneous, amorphous, highly water-soluble methylated β-CDs and 2-HPβ-CDs are the most important products. Their heterogeneity inhibits crystallization, which is an important advantage in designing liquid drug formulations. However, derivatives that cannot form crystalline cholesterol complexes are much more important,

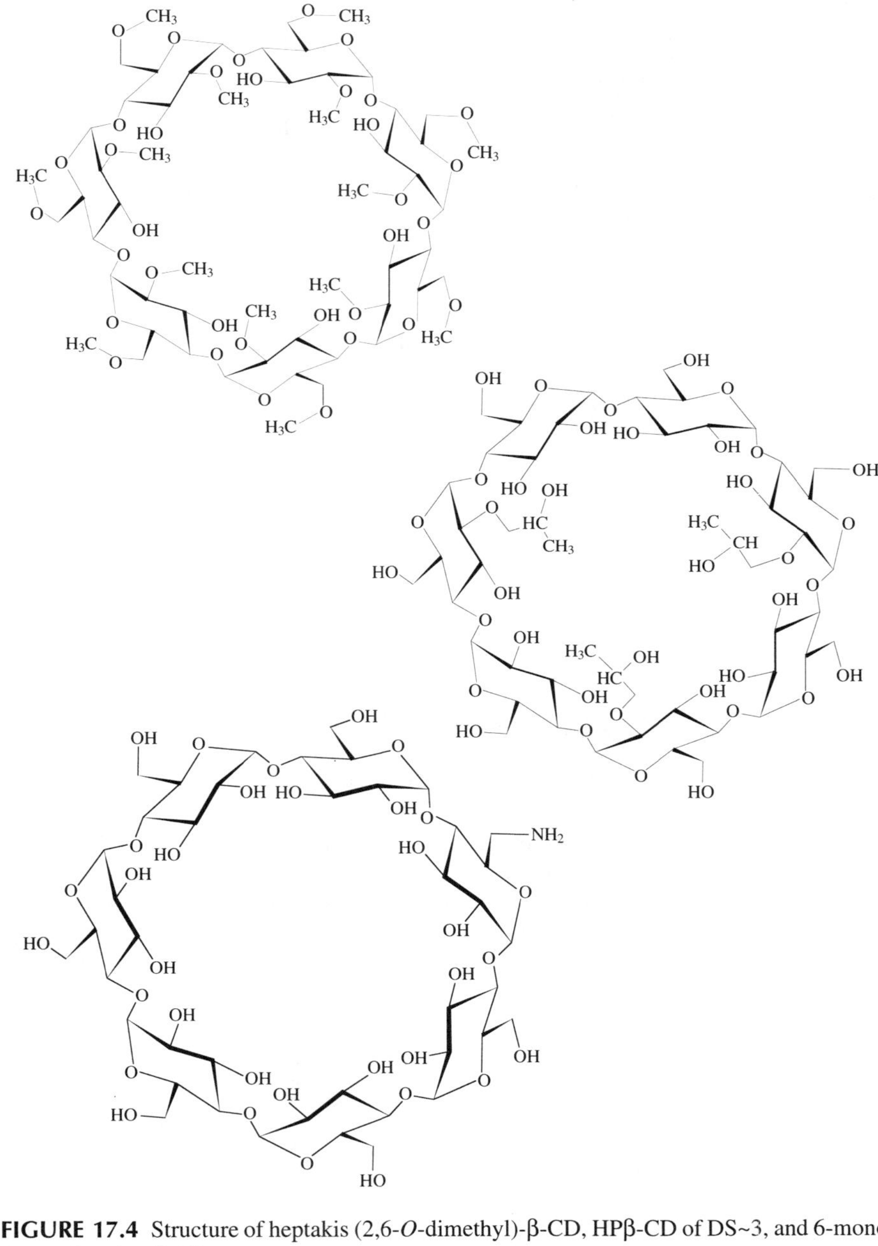

FIGURE 17.4 Structure of heptakis (2,6-*O*-dimethyl)-β-CD, HPβ-CD of DS~3, and 6-mono-deoxy-6-amino-β-CD.

because parent β-CD has a particularly high affinity for cholesterol. Drug formulations containing HPβ-CD and HPγ-CD are already approved and marketed in several countries.

Methylated β-CD is more hydrophobic than the parent β-CD. Therefore, it forms a more stable but still soluble cholesterol complex. Its high affinity to cholesterol provides extraction of cholesterol from blood cell membranes, resulting in hemolysis at ca. 1 mg/ml concentration.

One of the methylated β-CDs, heptakis (2,6-di-*O*-methyl-β-CD), called DIMEB, is a crystalline product. It is extremely soluble in cold water but insoluble in hot water; therefore, its purification and isolation of its complexes is technically very simple. To date, it is the most superior solubilizer among CDs. It is available at an isomeric purity above 95%, but for injecting drug formulations and widespread industrial applications the cheaper randomly methylated β-CD (called RAMEB) is produced and marketed.

In reaction of β-CD with starch in the presence of pullulanase, one or two either maltosyl or glycosyl groups may be attached with the α-1,6 glycosidic linkage to the primary side of the CD ring. The so-called branched CD (mono- or dimaltosyl or mono- or diglucosyl CD) is very well soluble in water, heterogeneous, and a noncrystallizing substance, produced and used in stable flavor powders for the food industry.

The heptakis(sulfobutyl)β-CD is very well soluble in water. Even at extremely high doses, it exhibits no toxic side effects. It can be used as a chiral separating agent in capillary zone electrophoresis, but intensive research focuses on its application as a parenteral drug carrier and component of aqueous injecting solutions of poorly soluble drugs.

17.3 CYCLODEXTRIN INCLUSION COMPLEXES[9–14]

In an aqueous solution, the slightly nonpolar CD cavity is occupied by water molecules, which are energetically nonfavored guests because of polar–nonpolar interactions. Therefore, water molecules can be readily substituted by other guests that are less polar than water (Figure 17.5). One, two, or three CD molecules (hosts) contain one or more entrapped guests. Most frequently, the host-to-guest ratio is 1:1. This is the essence of molecular encapsulation. However, commonly 2:1, 1:2, 2:2, or even more complicated associations and higher-order equilibria coexist.

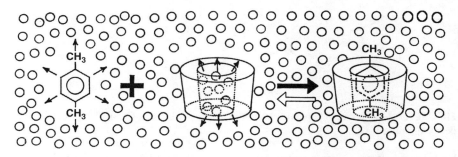

FIGURE 17.5 Formation of the inclusion complex. Small circles represent water molecules, which are repulsed both by the hydrophobic (potential guest) *p*-xylene and the hydrophobic cavity of the truncated CD cylinder.

Inclusion complexes can be isolated in stable crystalline form. On their dissolution, the equilibrium established between dissociated and associated species (Equation 17.1)

$$CD + D \rightleftharpoons CD.D \qquad (17.1)$$

can be expressed with the stability constant K_a defined in Equation 17.2:

$$K_{1:1} = \frac{[CD.D]}{[CD][D]} \qquad (17.2)$$

There are several essential consequences of the interaction between a poorly soluble
guest and a CD in aqueous solution.

1. The concentration of the guest in solution increases significantly, whereas
   concentration of dissolved CD decreases. The latter statement is not uni-
   versally true, because either ionizing guests or hydrogen bond containing
   compounds (e.g., phenols) might facilitate the solubility of CD.
2. Spectral properties of the guest are modified. For example, NMR chemical
   shifts of the anisotropically shielded atoms are modified. Achiral guests
   introduced into the cavity of the chiral CD become optically active,
   showing strong induced Cotton effects in the circular dichroism spectra.
   Sometimes, the UV spectral bands are shifted by several nanometers.
   Fluorescence is very strongly improved when fluorescing molecules are
   transferred from the aqueous milieu into a nonpolar environment.
3. Reactivity of included molecules is modified. Frequently, the reactivity
   decreases, i.e., the guest is stabilized, but in many cases CD behaves as
   an artificial enzyme, accelerating various reactions and modifying the
   reaction pathway.
4. Diffusion and volatility of the included guest strongly decrease.
5. On complexation, hydrophobic guests become hydrophilic, and therefore
   their chromatographic mobility also changes.
6. In the solid state, complexed substances are molecularly dispersed in the
   CD matrix, forming either microcrystalline or amorphous powder, even
   when guests are gaseous. Guests are effectively protected from any type
   of reaction, except reactions with the CD hydroxyl groups, and catalyzed
   by these groups. The complexes are hydrophilic, readily wettable, and
   rapidly soluble. If the formation of the CD inclusion complexes in aqueous
   solution can be detected, for example, by NMR spectroscopy, circular
   dichroism, or through a catalytic effect, this does not necessarily mean
   that a well-defined crystalline inclusion complex can be isolated. Repulsion between included water molecules and the nonpolar CD cavity on
   the one hand and between the bulk water and the nonpolar guest on the
   other are the main components of the driving force of the inclusion. The
   second type of repulsion is absent in the crystalline (dry) state. For this

sake, frequently the complex formation in solution cannot be convincingly proven. However, the isolated product is certainly a very fine dispersion of CD and the guest.

17.4 APPLICATIONS OF CYCLODEXTRINS IN FOODS[10,11]

The vast majority of practical applications of CDs are based on their inclusion complex forming capacity.

17.4.1 CYCLODEXTRIN COMPLEXES IN FOOD: FLAVORS AND VITAMINS

Flavor substances are generally volatile, easily deteriorating substances. Many volatile compounds, for instance, terpenoids and phenylpropan derivatives, form stable complexes with CDs. When dry, such complexes remain stable at room temperature for long time, without any further protection (Table 17.2). Such powder flavors are approved and produced in several countries. For example, the lemon-peel oil/β-CD complex, mixed with pulverized sugar, is used in pastries, spice-flavor mixtures complexed with CDs are used in canned meat and sausages, and the peppermint oil/β-CD complex adds flavor to chewing gum. In Germany, the garlic oil/β-CD

TABLE 17.2
Change of the Flavor Content (%) of β-CD Complexes on Storage Under Normal Conditions

β-CD Complex	1976	1978	1980	1984	1987	1992
Anethol/CD	9.7	9.2	9.4	8.8	8.0	—
Anis oil/CD	9.9	10.1	8.9	8.0	7.8	—
Bayleaf/CD	9.7	9.7	10.0	8.9	7.9	—
Cinnamaldehyde /CD	9.2	7.8	7.1	6.0	6.3	—
Caraway/CD	10.5	10.3	10.5	10.0	10.2	9.0
Bergamot/CD	10.2	—	—	9.6	9.8	9.2
Garlic/CD	10.3	10.0	10.4	—	10.7	10.3
Marjoram/CD	9.0	—	9.3	9.6	8.1	8.0
Mustard/CD	11.0	12.1	11.6	—	12.0	12.2
Dill/CD	9.0	8.8	—	8.0	8.4	8.0
Thyme/CD	9.8	9.7	—	—	10.0	10.2
Onion/CD	10.6	—	10.0	10.8	10.4	10.6
Camphor/CD	12.6	12.0	—	11.5	10.0	10.2
Menthol/CD	10.6	—	9.8	—	10.7	10.5
Vanillin/CD	6.1	6.3	—	3.2	2.8	—
Benzaldehyde /CD	—	9.2	8.0	7.1	4.5	2.4
Eugenol/CD	8.0	—	5.2	5.0	3.1	2.0

complex is marketed as odorless dragées substituting garlic and a number of various unstable garlic preparations for reducing blood cholesterol level.

In several cases, even the most sophisticated analytical instruments cannot substitute the human taste and smell sensors. Organoleptic evaluation, comparison of identical food products of different origin, for example beers, requires dosing of extremely small (ppm levels) but exactly known amounts of flavor components, such as thiols, butyric acid, and hydrogen sulfide. The storage and precision of dosing of such volatile additives are significantly improved when they are complexed with CDs.

Only fat-soluble vitamins and their derivatives form stable, crystalline complexes with CDs. Interaction of CDs with water-soluble vitamins was observed with vitamins B_1 and B_6. The bran-like odor of vitamin B_1 in B_1-enriched rice and barley can be eliminated with β-CD. The vitamin D_3/β-CD complex is the most thoroughly studied. The water solubility of vitamin D_3 is extremely poor. A remarkable water solubility of the D_3/β-CD complex is important for the biological activity or absorption of the vitamin. The D_3 vitamin entirely decomposes at 80°C within 24 h but in its complex with β-CD after 43 days, 49% of the original activity of the vitamin is still retained. Oxygen uptake has been determined as a major indicator of the stability of the vitamin. Thus, at 37°C, 1mg D_3 vitamin consumes 140 μl of oxygen. The β-CD complex containing the same amount of vitamin traps only 15.5 μl oxygen. Menadione, vitamin K_3, is another extensively studied fat-soluble vitamin. A considerable improvement of its solubility by the complexation has been reported. In the dry state, it is released from the complex only when CD is thermally decomposed at about 300°C. Because complexed menadione does not react with amino acids, it can be used in fodder premixes.

Vitamin A, its alcohol, acetate, and palmitate were the first examples of vitamin stabilization with CDs. Nowadays, a vitamin used in various dermacosmetic preparations is successfully stabilized with appropriate CDs. Highly water-soluble vitamin C (ascorbic acid) cannot be directly complexed in CDs, but its more lipophilic derivatives such as stearate, palmitate, and laureate form stable complexes with β-CD.

17.4.2 CYCLODEXTRINS AS PHYSICAL PROPERTY MODIFIERS[11]

CDs can be used for making stable water-in-oil emulsions such as mayonnaise and salad dressing. Natural food-coloring components in tomato ketchup can be stabilized with 0.2% β-CD. Color of such ketchup is stable for 2 h at 100°C, whereas the color of a control samples does not survive such treatment. Addition of CDs to emulsified foodstuffs and cheese can increase water retention and the shelf life of products. CDs admixed to processed meat products improve their water retention and texture. Solid compositions with sugar and amino acids designed for food manufacture and pharmaceuticals undergo browning and caking. Addition of at least 40% of an oligo- or polysaccharide, such as dextrin, starch or CD, inhibits, to a great extent, these symptoms. Simultaneously, decrease in the water content to 3% and below is recommended. Addition of 20% β-CD to a powdered juice consisting of anhydrous glucose, sodium L-aspartate, DL-alanine, citric acid, and inorganic salts results in a product of excellent stability. After 30 days at 40°C, there is no apparent discoloration or caking. The control β-CD-free sample begin to cake on the second day of storage and turn brown on the fourth day. Gelation is an important functional property of starches and starch-containing products as flours. β-CD considerably modifies gelation of wheat flour. Admixture of 1.5% β-CD increases swelling power and solubility of wheat starch granules and leaching of the amylose fraction. The 1.5% β-CD admixture produces a fourfold increase in viscosity of starch pastes. Probably, β-CD decomposes the amylase–lipid complexes, because after flour defatting the effect of CD is strongly reduced.

17.4.3 CYCLODEXTRINS AS TASTE MODIFIERS

β-CD has a taste. Its taste threshold values are lower than that of sucrose: detection 0.039% c.f. 0.27% and recognition 0.11% c.f. 0.25 %. A 0.5% aqueous solution of β-CD is as sweet as sucrose and a 2.5% β-CD solution is as sweet as a 1.71% sucrose solution. The sweetness of β-CD should not be ignored in food processing. Moreover, the sweetness of β-CD and sucrose is additive.

The free-to-complexed guest ratio in an aqueous CD solution depends on several factors, the important being stability or dissociation constant of the complex, temperature, and concentrations of both components. In cold concentrated solutions the equilibrium is shifted toward complexation and crystallization, whereas in warm, diluted solutions included guest molecules are released from the complexes. Therefore, when flavoring complexes contact the taste buds in the mouth as warm, diluted aqueous solutions, elimination of tastes and odors is fairly unlikely. Successful efforts were claimed in elimination of the phenol-like, unpleasant bitter components from coffee and tea liberated at 90°C by addition of 0.1% β-CD to the hot drinks. Although these advantageous results could not be reproduced elsewhere, unpleasant tastes and odors can be covered by CD complexation at lower temperature and at higher CD concentration.

Milk casein hydrolyzate is a readily digestible protein source, but its bitter taste limits its utilization. The bitter taste can be eliminated with a 10% admixture of β-CD. This method opens a route to the utilization of proteins otherwise useless for alimentary purposes. Similarly, the bitter taste of ginseng extract has also been eliminated with CD. The specific odor of mutton and fish, and the nonacceptable odors of bone powder used as the calcium supplement in animal fodder, sodium caseinate solution, and other products can be suppressed by CD. Soybean products free of the grassy smell and stringent taste are obtained by admixing CDs. Soybean lecithin kneaded with β-CD forms an odorless powder that can be used in nutrition. The bitter taste of grapefruit and mandarin juices decreases substantially when 0.3% β-CD is added prior to a 10-min heat treatment at 95°C because bitter naringin and limonin are complexed in β-CD. A partial complexation of chlorogenic acid and polyphenol components of coffee with β-CD provides an improvement of the taste and retardation of precipitate formation; the effect is particularly useful in canned coffee.

17.4.4 SELECTIVE COMPLEXATION AND SEQUESTRATION

Extraction of roasted coffee (ground beans) is more efficient in the presence of β-CD. β-CD-containing instant coffee powder produced by spray-drying or freeze-drying of the aqueous coffee extracts offers better aroma. The caffeine/CD complex can be prepared, but its properties (association constant and solubility) in aqueous solution are not appropriate to remove the caffeine from coffee. Various insoluble, swelling CD polymers were much less efficient in removing caffeine from aqueous coffee extracts. The CD polymers extracted hardly 4 to 26% of caffeine whereas charcoal provided a 77% yield of entrapment. In removal of chlorogenic acid, some β-CD polymers were as useful as charcoal. A 70% yield of isolation was achieved by both methods. Addition of 0.005 to 1% β-CD to canned citrus products prevented precipitation of poorly soluble bitter hesperidin and naringin. Excessive amounts of these compounds can be selectively removed chromatographically by using columns packed with CD polymers.

A low level of cholesterol in low-cholesterol butter, a Belgian product, was achieved by complexation of cholesterol from the molten butter with β-CD. The complex is easy to remove from the final product. Over 90% of cholesterol can be removed in one step and the butter remains CD-free. Also, low-cholesterol dairy products, such as cheese, cream, or even low-cholesterol egg, are available by using this technology.

17.4.5 FURTHER EXAMPLES OF APPLICATIONS OF CYCLODEXTRINS TO FOOD

CDs incorporated into food-packaging plastic films very effectively reduce the loss of aroma substances. Various CD complexes can be utilized in foodstuffs as anti-septics and preservatives. Iodine/β-CD applied at a 0.1% concentration provides a two-month inhibition against putrefying in fish paste and frozen marine products stored at 20°C. No growth of *Aspergillus, Penicillium,* or *Trichoderma* strains was observed on the surface of vinylacetate polymer or polypropylene foil containing the iodine complex.

Fungicide/CD complexes incorporated into films, for example, packaging of fermented cheese, inhibit the development of the surface mold colonies and thereby significantly extend the shelf life of products.

Edible films containing CDs and sugars have been described. The ethanol/CD complex packed in an ethanol-permeable film is an efficient antimold antiseptic for packaged foods. The CD complex or benzoic acid is also used as a food preservative.

17.5 EXAMPLES OF APPLICATION OF CYCLODEXTRINS IN VARIOUS INDUSTRIAL PRODUCTS AND TECHNOLOGIES

17.5.1 PHARMACEUTICALS[8,10,13]

More than 6000 papers, patents, and conference abstracts (i.e., about 25% of the available CD literature) are dedicated to the potential pharmaceutical uses of CDs. The number of CD-containing marketed drugs is steadily increasing. The active ingredients in drug formulations are frequently poorly soluble or instable, or they possess unacceptable taste. In many cases, application of a properly selected CD or its derivate can eliminate the problems. Complexation of a drug molecule with CD might be applicable under the following circumstances: (1) a drug is poorly soluble and its bioavailability in the oral, dermal, pulmonary, mucosal, and other applications is poor; (2) because of the low dissolution rate, even in case of a complete absorption, the time to reach the effective blood level of the orally administered drug is too long; (3) because of low solubility, no aqueous eye drop or injecting solution or other liquid formulation can be prepared; (4) a drug is chemically instable because of its self-decomposition, polymerization or degradation by atmospheric oxygen, absorbed humidity, light, etc., and no marketable formulation with satisfactory shelf-life can be produced; (5) a drug is physically unstable because of its volatility or sublimation, hygroscopicity, and when the migration the originally homogeneous product turns heterogeneous; (6) a drug has unpleasant odor or bitter or irritating taste; (7) a drug is liquid, but its preferred pharmaceutical form is a tablet, powder, aqueous spray, etc.; (8) a dose of the lipid-like, hardly homogenizing drug is extremely low and, therefore, uniformity of the product is problematic; (9) a drug is incompatible with other components of the formulation; (10) a drug develops serious side effects such as throat, eye, skin, and stomach irritation; and (11) extreme high biological activity operations with the pure drug is hazardous.

17.5.2 Cosmetics and Toiletry[10,11]

The cosmetic industry is another field of rapidly increasing successful applications of CDs. For instance, the inclusion of retinol in HPβ-CD leads to a water-soluble product, sufficiently stable, of higher bioavailability and lower toxicity compared to free retinol. Inclusion of other vitamins in CDs unquestionably increases their stability and bioavailability. The main advantages of using CD complexes in functional dermocosmetics are similar to those in the pharmaceutical formulations: (1) increase of bioavailability of incorporated active components such as vitamins, hormones, glycolic acid, depigmenting agents, and others; (2) protection from oxidation (vitamin and provitamin A, hydroquinone, arbutin, and others); (3) reduction of toxicity and aggressiveness (glycolic acid, hydroquinone, essential oils, fragrances, etc.); (4) formation of water-soluble complexes even if the incorporated guest is soluble in lipids; (5) increase in stability of emulsion and gel products; (6) easy formulation, by adding the complex to the final formulation and dispersing it in the aqueous phase at 35 to 40°C; and (7) avoiding preservatives.

Sunscreen agent/CD complexes provide increased photostability of creamy formulations, decreased cytotoxicity, improved functional activity, reduced odor, and reduced fabric staining. Thus, the markedly increased photostability is associated with decreased phototoxicity and photosensitization risk coming from the generation of free radicals primed by the sun photochemical attack on the filter molecule itself.

CDs are suitable not only for skin treatment products but also for formulation of make-up cosmetics, extending their stability even up to a whole day after their application. Several perfume concentrates, such as rose oil, citral, and citronellal, form CD complexes for application in solid preparations such as detergent powder. The collateral irritant effect caused by a scent in a shampoo can be reduced through the use of CD complexes; a solution of the iodine/β-CD complex can be used as a body deodorant, in bath formulations, and as a mouth refresher; CD complexes with detergents reduce foaming.

17.5.3 BIOTECHNOLOGY[16]

The majority of biotechnology processes are based on enzyme-catalyzed transformation of a substrate in an aqueous medium. The complexation with CDs can solve some of the difficulties encountered, for instance, if (1) a substrate is hydrophobic or sparingly soluble in water, (2) either the enzyme or enzyme-producing microbial cells are sensitive to toxic effects of the substrate or to inhibitors which can even be the products of the transformation, (3) the substrate or the product is unstable under the conditions of the enzymatic transformation, and (4) isolation of the product from a heterogeneous system is difficult.

CDs cause no damage to microbial cells or the enzymes. They cause intensification of the conversion of hydrocortisone to prednisolone, improvement in the yield of fermentation of lankacidine and podophyllotoxin, stereoselective reduction of benzaldehyde to L-phenylacetyl carbinol, reduction in the vanillin toxicity to yeast, and inhibition of detoxifying microorganisms. In the presence of an appropriate CD derivative (e.g., 2,6-dimethyl-β-CD), lipid-like inhibitor substances are complexed, the propagation of *Bordatella pertussis* occurs, and intensification of the production of pertussis toxin by two orders increases by up to hundred times. These are only a few examples of the numerous examples illustrating the rapidly growing and promising range of applications of CDs. CDs and their fatty acid complexes can substitute mammalian serum in tissue cultures.

Until recently, *Leprae bacillus* (*Mycobacterium leprae*) was considered to be not cultivable *in vitro*. Palmitic and stearic acids, the most important energy sources for bacillus, cannot penetrate through the strongly hydrophobic membrane of *Mycobacterium*. Solubilization of fatty acids and fatty alcohols with dimethyl-β-CD provides cultivation of *Mycobacterium* either *in vitro* or in a synthetic medium. This discovery might be useful for screening of other drugs for microorganisms similarly difficult to cultivate.

Biological remediation of soil polluted with carcinogenic polyaromatic hydrocarbons is very slow because these hydrocarbons very strongly adhere to soil particles. They are not available for soil microorganisms. Randomly methylated β-CD mobilizes hydrocarbons in soil, making them available for microorganisms.

17.5.4 TEXTILES

Either sorption or chemical binding of CDs to textile fiber opens new ways to prepare textiles to slowly release perfume scent and insect repellents. Simultaneously, CDs can bind distasteful smelling components of perspiration by acting as deodorants.

17.5.5 ANALYTICAL SEPARATION AND DIAGNOSTICS[8,11]

CDs and their specific derivatives are used widely in chromatography. From 1990 to 2000, about 25 papers were published monthly on the successful application of

CDs in gas or liquid chromatography. In gas chromatography, CDs serve only as the stationary phase, whereas in HPLC CDs are used either dissolved in the mobile phase or bound to the surface of the stationary phase. CDs and their anionic, cationic or alkylated derivatives are the most frequently used chiral selectors in capillary zone electrophoresis.

In medical diagnostic kits, frequently encountered problems of limited solubility of essential components, stability, and storability of the kit can be solved with the application of CDs. Frequently, intensity of color reactions or fluorescence, that is, sensitivity of the method, can be significantly enhanced by admixture of CDs.

17.5.6 MISCELLANEOUS INDUSTRIAL APPLICATIONS[11]

Conservation of wood-made products, such as window frames, doors, and wooden building constructions, prone to attack of microorganisms commonly involves impregnating with fungicides. The latter are usually water-insoluble, requiring application of environmentally toxic organic solvents. Dissolution of fungicides in aqueous CD solutions eliminates use of organic solvents for impregnation and conservation. Emulsion-type coatings and paints contain emulsion polymer binders, which provide after drying a resistant, continuous protecting film on the coated surface. The applied emulsions have to contain mutually compatible components such as solvent, pigment, thickener, and binder. The rheological properties of the paint are determined by the thickener, which is usually a hydrophobic polymer, for instance, polyurethanes, polyacrylamides, or cellulose ethers. To avoid a concomitant increase in viscosity obstructing the formation of uniform coating on the surface, viscosity suppressors have to be added to the emulsion. CDs in such emulsions associate with the hydrophobic sites of the polymer, and being strongly hydrated, inhibit the association of the macromolecules, resulting in a strong reduction of viscosity.

CD complexes of thymol, eugenol, isobutylquinoline, and other compounds in poly(vinylcarbazole) provide a natural leather-scent-releasing material for automobile upholstery.

CDs in light-sensitive photographic gelatin layers improve the quality of silver halide-containing photographic materials, improving their relative sensitivity and resistance to fogging. Explosives such as nitroglycerine or isosorbide dinitrate when complexed can be safely handled. The pernitro/γ-CD complex of the extremely explosive nitramine can be used as a controlled missile propellant. Complexation with CDs can provide the selectivity of insecticides. For example, an insecticide toxic for herbivorous insects would be safe for honey-bees. Stability against sunlight-accelerated decomposition of pyrethroids can be improved, ensuring a longer-lasting insecticidal effect. Wettability or solubility of poorly soluble benzimidazole-type antifungicides can be improved, offering reduction of the dose and the pollution of the environment.

REFERENCES

1. French D., The Schardinger dextrins, *Adv.Carbohydr. Chem.*, 12, 189, 1957.
2. Cramer F., *Einschlussverbindungen*, Springer Verlag, Berlin, 1954.
3. Thoma, J.A and Stewart, L., Cycloamyloses, in *Starch, Chemistry and Technology,* Vol. 1, Whistler, R.L. and Paschall, E.F., Eds., Academic Press, New York, 1965, p. 209.
4. Caesar, G.V., The Schardinger dextrins, in *Starch and its Derivatives*, Radley, J.A, Ed., Chapman and Hall, London, 1986, p. 290.
5. Nishimura, T. et al., Cyclic alpha-1,4 glycan formation by bacterial alpha-amylases, *J. Ferm. Bioeng.*, 81, 265, 1996.
6. Harangi, J. Szente, L., Greiner, M., Vikmon, M., Jicsinsky, L., and Szejtli, J., Cyclodextrins long consumed by humans: found in beer and bread, *Abstract Book of the 10th International Symposium on Cycylodextrins*, Ann Arbor, Michigan, May 21–24, 2000, Szejtli, J., Ed., Wacker Biochemical Corporation, Adrien, MI, p. 116.
7. Ammeraal, R.N. et al., The production of branched cyclodextrins by pyrolysis of crystalline βCD, *Proceedings of the 6th International Symposium on Cyclodextrins*, Chicago, April 21, 1992, Hedges, A.R., Ed., Editions de Sante, Paris, 1992, p. 69.
8. *Cyclodextrin News* (a monthly abstracting newsletter since 1986, dedicated exclusively to the cyclodextrin literature, production, marketing related conferences, etc.), Cyclolab Ltd., Budapest.
9. Szejtli, J, *Cyclodextrins and their Inclusion Complexes*, Akadémiai Kiadó, Budapest, 1982, p. 296.
10. Szejtli, J., *Cyclodextrin Technology*, Kluwer, Dordrecht, 1988, p. 450.
11. Cyclodextrins, in *Comprehensive Supramolecular Chemistry*, Vol. 3, Szejtli, J. and Osa, T., Eds., Pergamon, Oxford, 1996.
12. Duchene, D., Ed., *Cyclodextrins and their Industrial Uses*, Editions de Sante, Paris, 1987.
13. Frömming, K.H. and Szejtli, J., *Cyclodextrins in Pharmacy*, Kluwer, Dordrecht, 1993.
14. A thematic issue of 14 papers dedicated to cyclodextrins, *Chem. Rev.*, 98, 1741–2076, 1998.
15. Easton, N.J. and Lincoln, S.F., *Modified Cyclodextrins, Scaffolds and Templates for Supramolecular Chemistry*, Imperial College Press, London, and World Scientific Publishing, Singapore, 1999, p. 293.
16. Duchene, D., Ed., *New Trends in Cyclodextrins and Derivatives*, Editions de Santé, Paris, 1991.

18 Chemistry of the Maillard Reaction in Foods

Tomas Davidek and Jiri Davidek

CONTENTS

18.1   Introduction 291
18.2   Formation of Glycosylamines and Rearrangement to Amino
        Deoxysugars 292
18.3   Other Reactions of Glycosylamines 293
18.4   Degradation of Amino Deoxysugars 295
18.5   Degradation of 3-Deoxy-2-Osuloses 296
18.6   Degradation of 1-Deoxy-2,3-Diuloses 299
18.7   Degradation of 1-Amino-1,4-Dideoxydiketoses 304
18.8   Strecker Degradation of Amino Acids 305
18.9   Melanoidins 307
18.10   Closing Remarks 310
References 310

18.1 INTRODUCTION

The term *Maillard reaction* is used for a complex network of reactions involving carbonyl compounds, namely reducing sugars, and amino compounds, for example amino acids or peptides. These reactions are also called nonenzymatic browning, as they are accompanied by the formation of brown pigments (melanoidins). However, other reactions, such as caramelization, can also cause nonenzymatic browning of foods. Thus, the Maillard reaction should be considered as a special case of nonenzymatic browning.

The reaction is named after the French chemist Louis-Camille Maillard, who first described flavor and colour development on heating reducing sugars with amino acids.[1] Since then, the Maillard reaction has become of fundamental interest to food chemists and technologists, and also to scientists working in medicine and nutrition.

In the last two decades, the Maillard reaction has been the subject of seven international symposia.[2–8] The literature dealing with all aspects of the Maillard reaction, including under physiological conditions, has been covered in many excellent review articles and books.[2–13] Therefore, only recent articles and important older papers are cited here.

The significance of this reaction in foods stems from the fact that it alters important food attributes such as color, flavor, taste, nutritional value, antioxidant properties, and texture. These changes can be both desirable (e.g., flavor and color development on roasting and baking) and undesirable (e.g., darkening of dehydrated foods, formation of off-flavors, or development of toxic compounds). Despite ninety years of effort, the Maillard reaction is not fully understood and our abilities to control this reaction during food processing and storage are only very limited.

The intention of this chapter is to summarize recent findings on mechanistic aspects of the Maillard reaction, with focus on degradation of the sugar moiety. With some exceptions, the chapter does not cover the interactions and recombinations of sugar fragmentation or amino acid degradation products.

18.2 FORMATION OF GLYCOSYLAMINES AND REARRANGEMENT TO AMINO DEOXYSUGARS

The first step in the amino-carbonyl reaction between reducing sugars and amino acids is the addition of a primary amino group of an amino acid to a carbonyl group of a sugar, followed by the elimination of water. The resulting imine (Schiff base) (18.1) undergoes cyclization to the corresponding *N*-glycosylamine (18.2) (Reaction

Reaction 18.1

18.1). Whereas glycosylamines derived from aromatic and heterocyclic amines show a certain stability, those derived from amino acids are immediately converted into further compounds. Aldosylamines undergo Amadori rearrangement, forming 1-amino-1-deoxy-2- ketoses, the so-called Amadori compounds (Reaction 18.2). Ketosylamines rearrange to 2-amino-2-deoxy-aldoses via Heyns rearrangement (Reaction 18.3). Both rearrangements start with protonation, causing opening of the hemiacetal ring of glycosylamine and developing of the immonium ion. Loss of a hydrogen proton leads via enaminol (18.3a,b) to aminodeoxyketose (18.4a) (from aldosylamine) or aminodeoxyaldose (18.4b) (from ketosylamine). Amadori compounds have been found in various heated and stored foods such as malt, soy sauce, roasted cocoa, and dehydrated vegetables.[14–17]

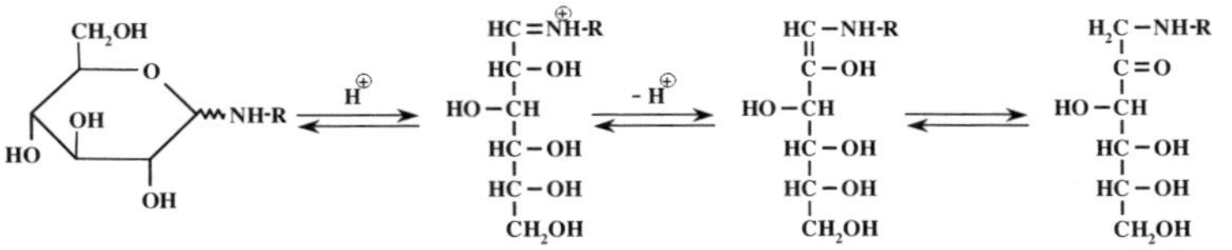

Reaction 18.2

18.2 18.3a 18.4a

Reaction 18.3

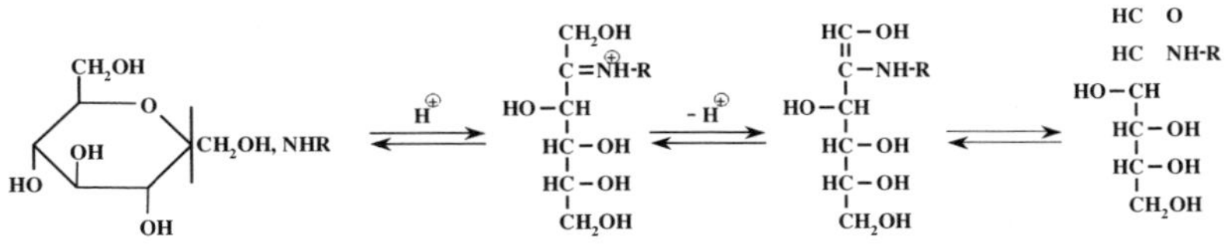

18.3b 18.4b

18.3 OTHER REACTIONS OF GLYCOSYLAMINES

Schiff bases 18.1 derived from sugars and amino acids may undergo transamination in which the amino acid is converted into the corresponding oxo-acid and the sugar into an amino alditol (18.5) (Reaction 18.4).[18] Alternatively, the Schiff base (18.1)

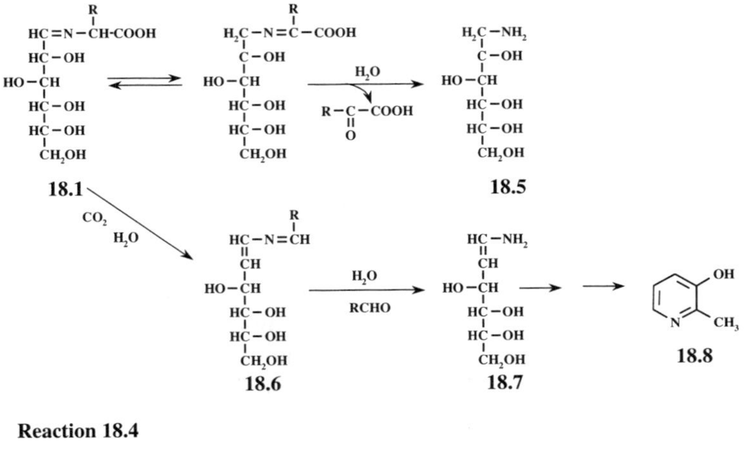

18.1 18.5

18.6 18.7 18.8

Reaction 18.4

can also be decarboxylated resulting in (18.6), which hydrolyses to form a Strecker aldehyde and unsaturated amino alditols (18.7).[19] The latter compound can be transformed to 3-hydroxy-2-methylpyridine (18.8).

Another mechanism that bypasses the Amadori rearrangement was reported by Namiki and coworkers.[20–22] This mechanism, active mainly under alkaline conditions, involves the cleavage of the sugar moiety of the Schiff base (18.1) (Reaction

18.5), leading to a C2 fragment (18.9), glycolaldehyde imine in the enol form, which

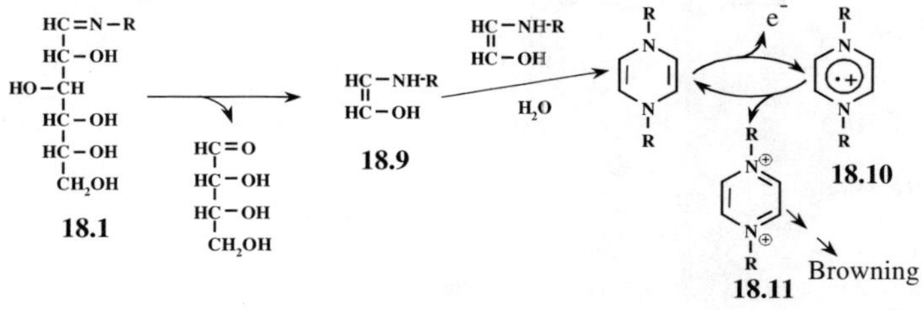

Reaction 18.5

on dimerization and oxidation forms a 1,4-dialkylpyrazinium radical cation (18.10). Disproportionation of the latter compound results in a 1,4-dialkylpyrazinium diqua-ternary salt (18.11), which has been shown to be a very reactive intermediate involved in color formation.[22,23]

Aldosylamines derived from primary amines can react with a second molecule of aldose and form dialdosylamines, which rearrange to diketosamines via Amadori rearrangement; e.g., di-D-glucosylglycine (18.12) yields di-D-fructosylglycine (18.13).

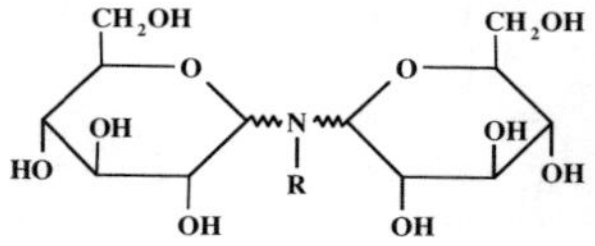

18.12

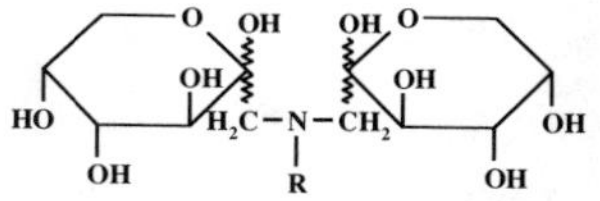

18.13

The reaction of sugars with multifunctional aromatic and heterocyclic amino acids can also bypass Amadori rearrangement and lead to the formation of various heterocyclic compounds. For example, thiazolidine derivatives (18.14) are formed

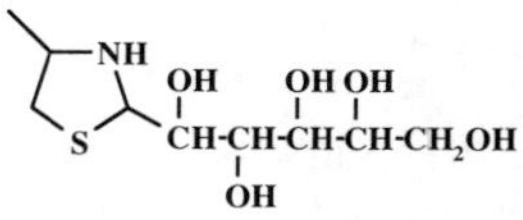

18.14

by reaction of aldoses with cysteine and glycotetrahydro-β-carboline-3-carboxylic acids (18.15) arise from the reaction of aldoses with tryptophan.[24,25]

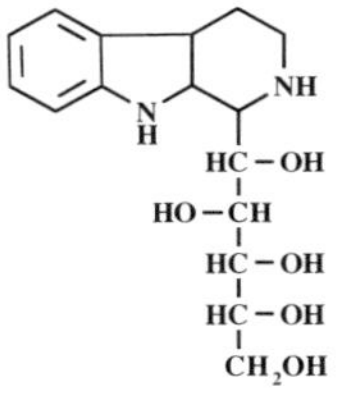

HC – OH
HO – CH
HC – OH
HC – OH
CH₂OH

18.15

18.4 DEGRADATION OF AMINO DEOXYSUGARS

Amadori compounds possess strong reducing properties. They are relatively stable in aqueous solutions but decompose slowly in acids and very rapidly in alkalis. An important characteristic of Amadori compounds is their tendency to easily undergo enolization. Further degradation sequences depend on whether 1,2-enaminol 18.3 or 2,3-endiol (18.16) is formed (Reaction 18.6). The former compound is transformed

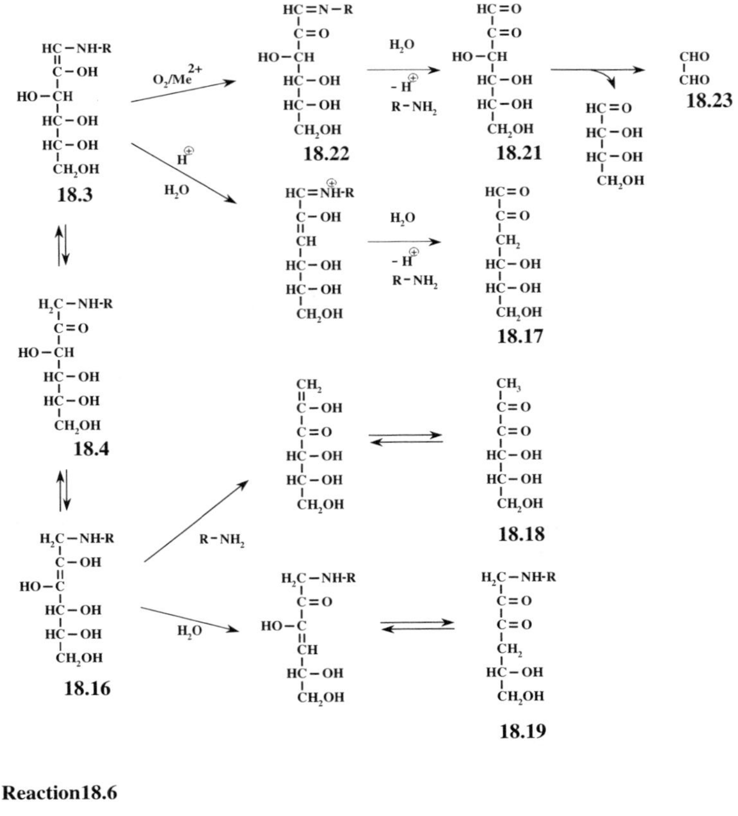

Reaction 18.6

via the so-called 1,2-enolization to 3-deoxy-2-osulose (18.17), whereas 2,3-endiol is transformed via 2,3-enolization to 1-deoxy-2,3-diulose (18.18). 3-Deoxy-2-osulose (18.17) can also be formed directly from glycosylamine without passing through the formation of the Amadori compound first.[26]

In parallel with 1-deoxy-2,3-diulose, another dicarbonyl compound, 1-amino-1,4-dideoxy-2,3-diulose (18.19), can be formed from 2,3-endiol (18.16) via elimination of the hydroxyl group at C-4 instead of the amino group at C-1.[12] The ratio of both dicarbonyls is highly dependent on the quality of the substituents on C-1 and C-4 as leaving groups, as shown by recent investigations. Whereas 1-amino-1,4-dideoxy-2,3-diulose is formed only in traces from monosaccharides, a significant amount of this compound is formed from disaccharides such as maltose or lactose.[27,28]

The formation of 1-amino-1,4-dideoxy-2,3-diulose (18.19) from oligosaccharides proceeds via the so-called peeling off mechanism. The reaction starts at the reducing end of the oligosaccharide. After 2,3-enolization of the Amadori compound, the cleavage via vinylogous β-elimination yields 1-amino-1,4-dideoxy-2,3-diulose and an oligosaccharide that contains one unit of monosaccharide less than the original oligosaccharide. The latter compound may react again in a second peeling off mechanism. By such a mechanism, for example, maltotriose (18.20) can be gradually degraded to maltose and glucose (Reaction 18.7).[29]

Reaction 18.7

In the presence of transition metals, the oxidation of 1,2-enaminol 18.3 leads to 2-osuloses glycosones (18.21), Reaction 18.6 that can yield glyoxal (18.23) by retro-aldol cleavage.[23,30] If the iminoketone (18.22) formed by oxidation of 1,2-enaminol (18.3) is not hydrolyzed, then Strecker aldehydes can be formed via decarboxylation see (18.8).[31]

Amadori compounds can also act as nucleophiles and react with another sugar molecule to form diketosyl derivatives (18.13). These derivatives are unstable compounds that easily decompose to 3-deoxy-2-osulose and the original Amadori compound.[32] Other degradation mechanisms of Amadori compounds involve their dehydration in the cyclic form and retro-aldol cleavage of further aminated Amadori compounds yielding diimine derivatives of methylglyoxal.[33,34]

18.5 DEGRADATION OF 3-DEOXY-2-OSULOSES

3-Deoxy-2-osuloses derived from pentoses and hexoses are relatively stable under neutral conditions, but their stability decreases under acidic and especially alkaline conditions.[35] Under alkaline conditions, 3-deoxy-2-hexosuloses, often called 3-deoxyhexosones, are easily transformed into so-called metasaccharinic acids (18.24).

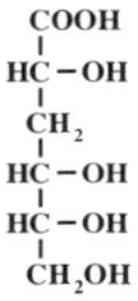

$$
\begin{array}{l}
COOH \\
| \\
HC-OH \\
| \\
CH_2 \\
| \\
HC-OH \\
| \\
HC-OH \\
| \\
CH_2OH
\end{array}
$$

18.24

Formation of formic acid has also been reported. On the other hand, degradation of 3-deoxypentosone and 3-deoxytetrosone in alkaline solutions leads to rapid browning; however, metasaccharinic acids are not detected.[35] Under acidic conditions, 5-hydroxymethyl-2-furaldehyde and 2-furaldehyde are formed as main products by dehydration and cyclization of 3-deoxyhexosones and 3-deoxypentosones, respectively. In the presence of larger amounts of primary amines, the formation of furan derivatives is suppressed in favor of pyrrole (18.25) or pyridinium derivatives (18.26).[36]

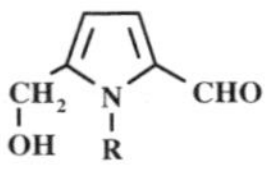

18.25

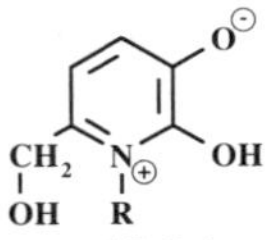

18.26

α-Dicarbonyl cleavage and the so-called β-carbonyl route are other mechanisms leading to the formation of furane and pyrrole derivatives from 3-deoxyosones. By the former mechanism, 3-deoxyhexosuloses are transformed into 2-deoxypentoses (18.27) that are further transformed into 2-hydroxymethyfurane (18.28) or 2-hydroxymethylpyrroles (18.29) (Reaction 18.8). The latter compounds have been shown to have outstanding polycondensation potential.[37]

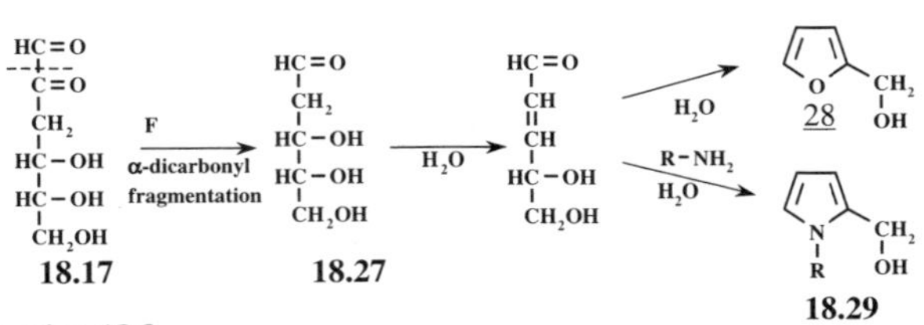

Reaction 18.8

The key intermediates in the formation of furane and pyrrole derivatives via the β-carbonyl route are 3-deoxy-2,4-diuloses and 1-amino-1,3-dideoxy-2,4-diuloses (18.30) (Reaction 18.9). The furane and pyrrole derivatives are formed by cycliza-

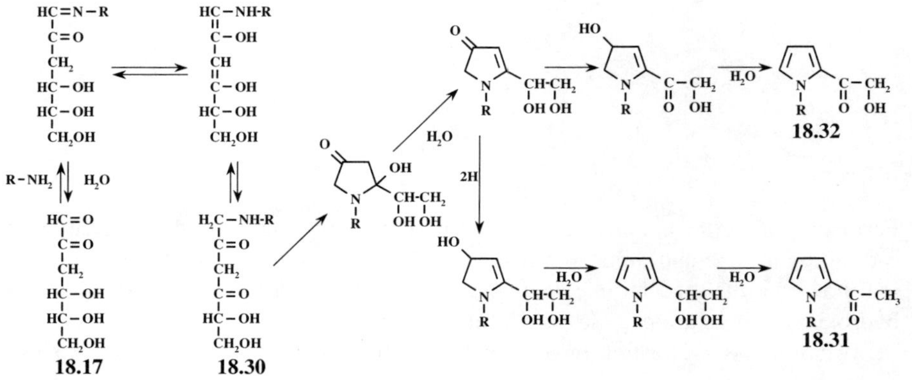

Reaction 18.9

tion, dehydration, and keto-enol tautomerization of these intermediates. The β-carbonyl route permits, for example, to explain the formation of two major products 2-acetylpyrroles (18.31) and (18.32) in the reaction of 4-aminobutyric acid with glucose.[38]

The reaction of 3-deoxyosuloses with Amadori compounds leads to polyhydroxyalkyl derivatives of 2-formylpyrroles.[39] In the presence of ammonia, the reaction yields polyhydroxyalkyl derivatives of pyrazines.[40]

3-Deoxyosuloses, especially those derived from pentoses, as well as certain their degradation products, for example, 2-furaldehyde, were identified as key intermediates in the formation of several low-molecular-weight colored compounds. Compounds (18.33) and (18.34) are cited as examples.[41–43]

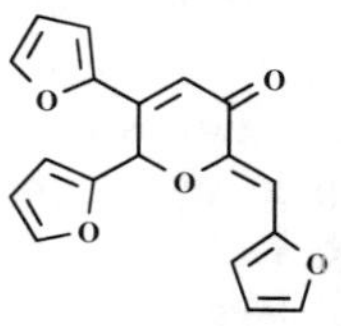

18.33

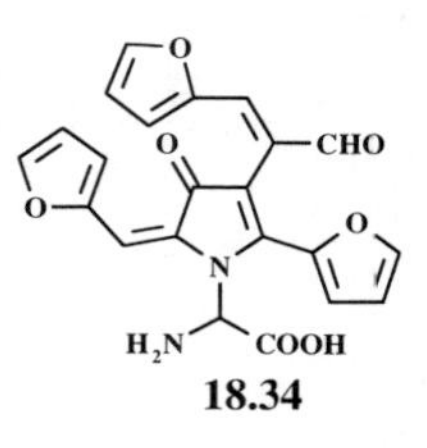

18.34

In the presence of certain amino acids, the degradation of 3-deoxyosones yields specific compounds not formed in other systems. For example, maltoxazine (18.35)

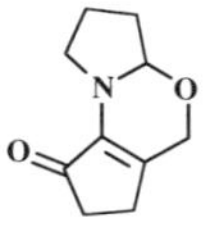

18.35

(an important component of malt or beer), tetrahydroindolizinone (18.36), and imidazolinone (18.37) are formed as proline, hydroxyproline, and arginine specific compounds, respectively.[44,45]

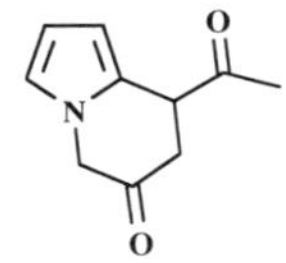

18.36

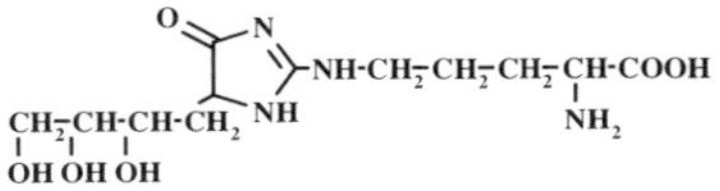

18.37

18.6   DEGRADATION OF 1-DEOXY-2,3-DIULOSES

1-Deoxy-2,3-diuloses are very reactive compounds that up to now were not isolated as such from foods or from model systems. Only recently, 1-deoxy-D-erythro-hexo-2,3-diulose was synthesized in a protected form that can be released under mild conditions and its structure present in aqueous solution was elucidated. [46]

Dehydration of 1-deoxy-2,3-diuloses gives rise to 3-furanones of type (18.38) (Reaction 18.10).[47,48] 4-Hydroxy-5-methyl-3(2H)-furanone (18.38b) and 4-hydroxy-

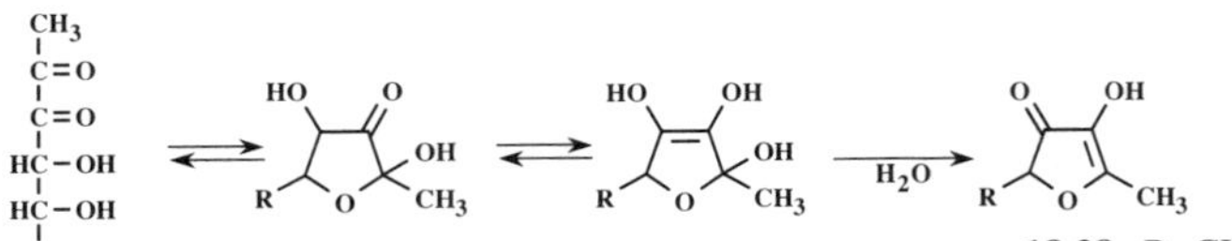

18.38a R=CH$_2$OH
18.38b R=H
18.38c R=CH$_3$

18.18

Reaction 18.10

2,5-dimethyl-3(2H)-furanone (18.38c) derived from pentoses and methylhexoses, respectively, are important aroma substances. In the presence of sulfur-containing compounds, these furanones may be transformed into their corresponding sulfur-containing analogs that are also flavor-active. (They often possess meaty notes.[49]) Besides aroma, furanones were shown to be important for color as they are precursors of many colored compounds.[41,50,51] For example, due to its highly reactive methylene group, furanone (18.38b) easily condenses with 2-formylpyrrole and forms compounds (18.39) and (18.40).[51]

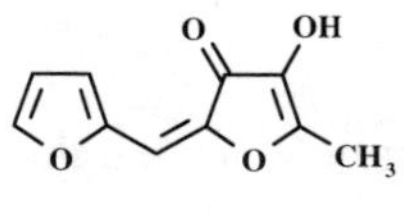

18.39

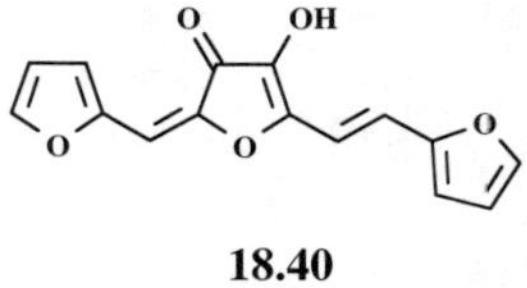

18.40

1-Deoxy-2,3-diuloses derived from hexoses can be easily converted to dihydro-pyranone (γ-pyranone) (18.41) (Reaction 18.11). Besides γ-pyranone (18.41), β-

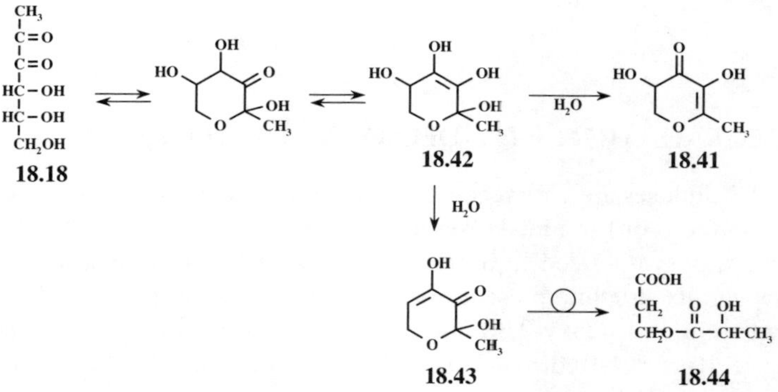

Reaction 18.11

pyranone (18.43) is formed by water elimination from endiol (18.42). This β-pyranone is not stable and rearranges to lactyl-β-propionic acid (18.44), which is finally hydrolyzed to corresponding acids.

Dihydropyranone (18.41), similarly to furanone (18.38b), easily condenses with carbonyl compounds. By this mechanism, colorant (18.45) is formed from dihydro-pyranone (18.41) and 2-formylpyrrole.[42] Reaction of dihydropyranone (18.41) with a primary amine yields aminoreductone (18.46).[52]

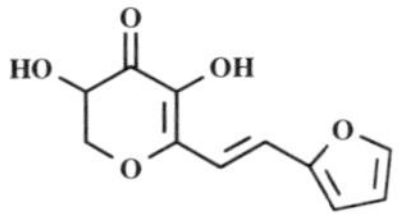

18.45

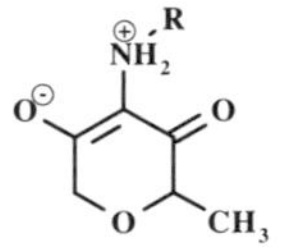

18.46

Dehydration of 1-deoxy-2,3-diuloses at position C-6 leads to acetylformoin (18.47a,b), another well-known degradation product of hexoses (Reaction 18.12).

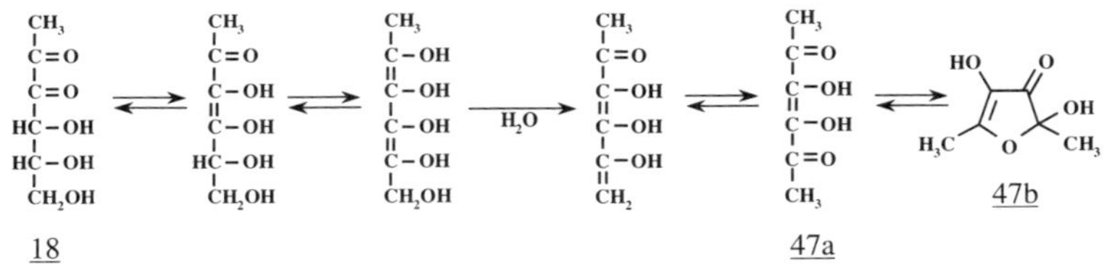

Figure 18.12

Acetylformoin is highly reactive and easily undergoes further reactions that lead to the formation of numerous compounds, some of them having flavor-, color- and taste-active properties. For example, reduction and dehydration of acetylformoin that leads to 3(2H)-furanone (18.38c) explains the formation of this important odorant from hexoses (known as furaneol®).[53] Reaction of glucose in the presence of proline can yield colorant (18.48a,b) (two forms in equilibrium), via condensation of acetyl-

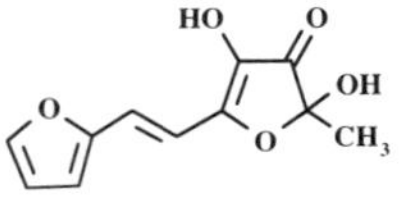

18.48a

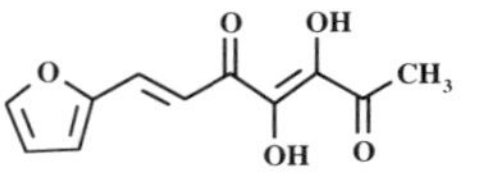

18.48b

formoin with 2-formylpyrrole. Another colorant, compound (18.49), is formed from acetylformoin and proline via bitter-tasting pyrrolidinohexose reductone (18.50) and bispyrrolidinohexose reductone (18.51).[54] In the presence of primary and secondary

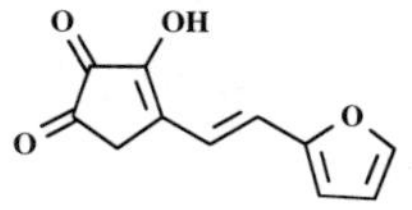

18.49

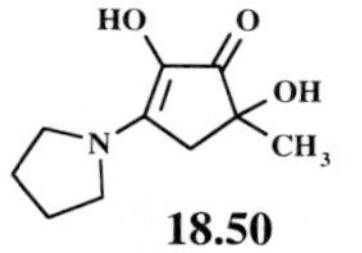

18.50

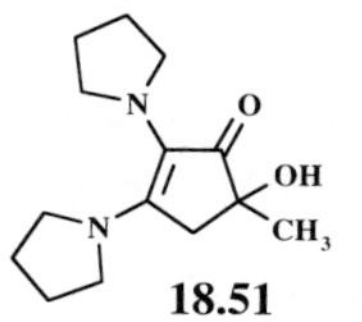

18.51

amino acids, acetylformoin reacts predominantly with primary amino acids, forming pyrrolinone-reductones of type (18.52). These compounds may further react with

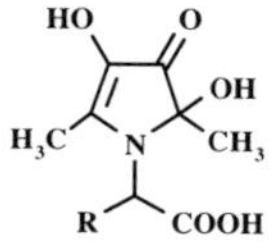

18.52

carbonyls such as 2-formylpyrrole and form colorants of type (18.53).[54]

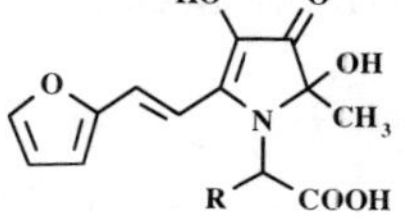

18.53

    Besides the above-mentioned reactions, 1-deoxy-2,3-diuloses easily breakdown to highly reactive fragments such as methylglyoxal, glyceraldehyde, diacetyl, and others, which are readily available for further reactions. Fragmentation of 1-deoxy-2,3-diuloses also seems responsible for the formation of high quantities of acetic acid in the early stages of the Maillard reaction.[55,56]

    In the presence of α-amino acids, 1-deoxy-2,3-diuloses may be transformed via Strecker degradation to 1,4-dideoxy-2,3-osuloses (18.54). These compounds are also formed as main α-dicarbonyl compounds during the peeling off degradation of oligosaccharides under almost water-free conditions.[29] 1,4-Dideoxy-2,3-osuloses are unstable and undergo further reactions.

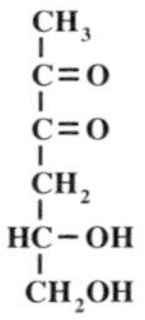

$$
\begin{array}{c}
CH_3 \\
| \\
C=O \\
| \\
C=O \\
| \\
CH_2 \\
| \\
HC-OH \\
| \\
CH_2OH \\
\mathbf{18.54}
\end{array}
$$

Degradation of 1-deoxy-2,3-diuloses derived from disaccharides with a 1,4 glycosidic bond partly differs from that of monosaccharides. Presence of this bond hinders the formation of γ-pyranone (18.41) from endiol (18.55). Instead, β-pyranone (18.56) and cyclopentenone (18.57) are formed as main products (Reaction 18.13).

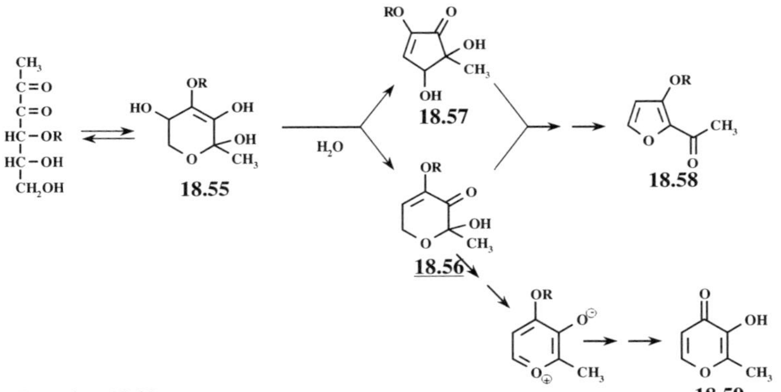

Reaction 18.13

On prolonged heating, these compounds are transformed into the more stable glycosylisomaltol (18.58).[57] β-Pyranone (18.56) can be also transformed via the pyrylium derivative to maltol (18.59). In the presence of primary amines, pyrrole derivatives (18.60) and pyridinium betaines (18.61) are formed. The latter compounds

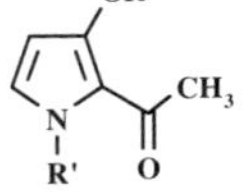

18.60

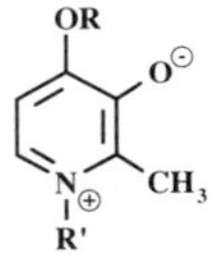

18.61

easily eliminate the sugar residue, yielding pyridones (18.62). In an excess of amine

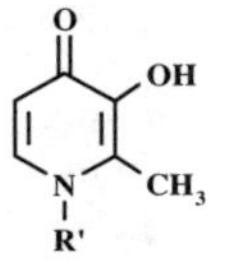

18.62

pyridone, imines (18.63) might be formed, which can participate in protein cross-

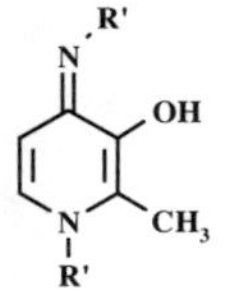

18.63

linking. Under slightly alkaline conditions, isomerization of pyridinium betaines (18.61) yields glycosylpyridones (18.64).[57] Another degradation product of 1-deoxy-

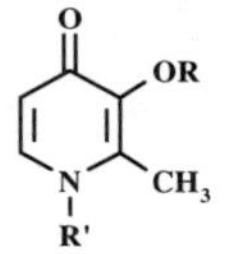

18.64

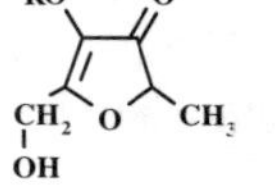

18.65

2,3-diuloses specific for disaccharide is 3(2*H*)-furanone (18.65).[58] This compound derived from lactose was recently identified in milk.[59]

18.7 DEGRADATION OF 1-AMINO-1,4-DIDEOXYDIKETOSES

The aminoreductones of Type 66 were identified only recently in reaction mixtures containing disaccharides and primary amines or amino acids.[28,60] These aminoreductones can be considered as the first stable tautomers of 1-amino-1,4-dideoxy-2,3-diuloses (18.19) and are also formed as derivatives of lysine side chains of proteins.[28] Due to their aminoreductone structure, these compounds undergo further reactions, including redox reactions.

The degradation of maltooligosaccharides in the presence of amino acids in almost water-free systems and in aqueous systems yields 1,4-dideoxy-2,3-diuloses (18.54) and 3-deoxypentosuloses (18.67), respectively, as main α-dicarbonyl compounds.[29] In both cases, 1-amino-1,4-dideoxy-2,3-diuloses (18.19) are considered as intermediates. The mechanism proposed for the formation of 3-deoxypentosuloses includes fragmentation of 1-amino-1,4-dideoxy-2,3-diulose via a retro-Claisen reaction, which leads to formamide or formic acid and a pentose intermediate (18.68) (Reaction 18.14). 3-Deoxypentosulose (18.67) is then formed by oxidation or H-abstraction of this pentose intermediate.[29]

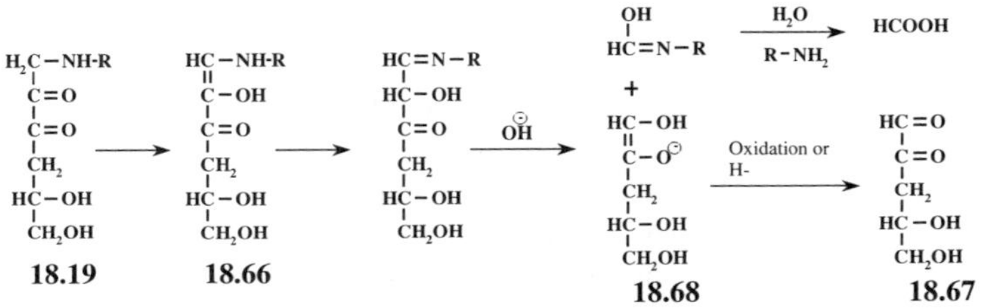

Reaction 18.14

Cyclization and dehydration of 1-amino-1,4-dideoxy-2,3-diuloses give rise to aminoacetylfuranes (18.69). Furosine (18.70) is the best-known representative of

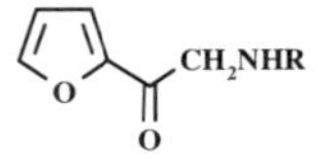

18.69

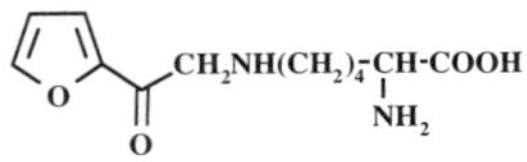

18.70

these compounds.[61] Aminoacetylfuranes easily undergo further reactions such as oxidation and condensation.

18.8 STRECKER DEGRADATION OF AMINO ACIDS

Strecker degradation of amino acids is an important reaction pathway that generates numerous flavor-active compounds. During this reaction (Reaction 18.15), α-dicarbonyl or vinylogous dicarbonyl compounds cause oxidative decarboxylation of α-amino acids to aldehydes with one less carbon atom (Strecker aldehyde).[62] Strecker aldehydes of certain amino acids, in particular methionine, phenylalanine, and leucine, have very low odor thresholds and contribute to the odor of thermally

$$
\begin{array}{c}
\text{R} \\
\text{H}_2\text{N}-\text{CH-COOH} \\
+ \\
\text{HC}=\text{O} \\
| \\
\text{C}=\text{O} \\
| \\
\text{R}_1
\end{array}
\xrightarrow{\text{H}_2\text{O}}
\begin{array}{c}
\text{R} \\
\text{HC}=\text{N}-\text{CH-COOH} \\
| \\
\text{C}=\text{O} \\
| \\
\text{R}_1
\end{array}
\xrightarrow{\text{CO}_2}
\begin{array}{c}
\text{R} \\
\text{HC}-\text{N}=\text{CH} \\
\| \quad \quad \\
\text{C}-\text{OH} \\
| \\
\text{R}_1
\end{array}
\xrightarrow{\text{H}_2\text{O}}
\begin{array}{c}
\text{R} \\
\text{HC}-\text{NH-CH-OH} \\
\| \quad \quad \\
\text{C}-\text{OH} \\
| \\
\text{R}_1 \\
\textbf{18.71}
\end{array}
\longrightarrow \text{RCHO}
$$

O₂/Me²⁺ / -H₂O₂

$$
\begin{array}{c}
\text{R} \\
\text{HC}=\text{N}-\text{CH-OH} \\
| \\
\text{C}=\text{O} \\
| \\
\text{R}_1
\end{array}
\rightleftarrows
\begin{array}{c}
\text{R} \\
\text{HC}-\text{N}=\text{C}-\text{OH} \\
\| \quad \quad \\
\text{C}-\text{OH} \\
| \\
\text{R}_1
\end{array}
\longrightarrow \text{RCOOH}
$$

H₂C−NH₂ | C=O | R₁

Reaction 18.15

18.72

processed foods.[63] During the reaction, α-dicarbonyls are transformed to α-aminocarbonyls, which might liberate ammonia or condense to numerous heterocyclic compounds such as pyrazines, pyrrols, and others.[64]

In the presence of oxygen, the corresponding acids are formed in parallel to Strecker aldehydes. The formation of acid is further increased in the presence of traces of transition metals. It has been demonstrated that the acids are not formed to a significant extent by the oxidation of aldehydes. The mechanism shown in Reaction 18.15 was proposed to explain their formation.[63] Up to the intermediate (18.71), the mechanism does not differ from that leading to Strecker aldehydes. This intermediate may either liberate a Strecker aldehyde or it may be oxidized to iminoketone (18.72), which on isomeration and hydrolysis yields the corresponding acid. Some "Strecker" acids are, similarly to Strecker aldehydes, important odorants of thermally processed foods.[63] Strecker aldehydes can be also formed directly from the Schiff base (18.1) (Reaction 18.4) or the Amadori compound (18.4) without passing through the formation of α-dicarbonyls. The mechanisms proposed for the formation from Amadori compounds include either oxidation of enaminol (18.3) (Reaction 18.16) or further amination of the Amadori

$$
\begin{array}{c}
\text{R}_1 \\
\text{H}_2\text{C}-\text{NH-CH-COOH} \\
| \\
\text{C}=\text{O} \\
| \\
\text{HO}-\text{CH} \\
| \\
\text{HC}-\text{OH} \\
| \\
\text{HC}-\text{OH} \\
| \\
\text{CH}_2\text{OH} \\
\textbf{18.4}
\end{array}
\rightleftarrows
\begin{array}{c}
\text{R}_1 \\
\text{HC}-\text{NH-CH-COOH} \\
\| \\
\text{C}-\text{OH} \\
| \\
\text{HO}-\text{CH} \\
| \\
\text{HC}-\text{OH} \\
| \\
\text{HC}-\text{OH} \\
| \\
\text{CH}_2\text{OH} \\
\textbf{18.3}
\end{array}
\xrightarrow[-\text{H}_2\text{O}_2/\text{Me}^{2+}]{\text{O}_2/\text{Me}^{2+}}
\begin{array}{c}
\text{R}_1 \\
\text{HC}=\text{N}-\text{CH-COOH} \\
| \\
\text{C}=\text{O} \\
| \\
\text{HO}-\text{CH} \\
| \\
\text{HC}-\text{OH} \\
| \\
\text{HC}-\text{OH} \\
| \\
\text{CH}_2\text{OH}
\end{array}
\rightleftarrows
$$

(pyranose ring) N−CH−COOH, R₁

H₂O / CO₂

(pyranose ring) N=CH R₁ → (pyranose ring) NH₂ + R₁CHO, H₂O

Reaction 18.16

compound (18.4) (Reaction 18.17).[19,31] An alternative pathway leading to Strecker

$$
\begin{array}{c}
\underset{|}{R_2} \\
H_2N-CH\text{-}COOH \\
+ \\
\underset{|}{R_1} \\
H_2C-NH\text{-}CH\text{-}COOH \\
\underset{|}{C=O} \\
HO-\underset{|}{CH} \\
H\underset{|}{C}-OH \\
H\underset{|}{C}-OH \\
CH_2OH
\end{array}
\quad \xrightarrow{H_2O} \quad
\begin{array}{c}
\underset{|}{R_1} \\
H_2C-NH\text{-}CH\text{-}COOH \\
C=N-\underset{|}{CH}\text{-}COOH \\
HO-\underset{|}{CH} \quad R_2 \\
H\underset{|}{C}-OH \\
H\underset{|}{C}-OH \\
CH_2OH
\end{array}
\quad \xrightarrow[H_2O]{CO_2} \quad
\begin{array}{c}
\underset{|}{R_1} \\
H_2C-NH\text{-}CH\text{-}COOH \\
\underset{||}{C}-N=CH \\
\underset{|}{CH} \quad R_2 \\
H\underset{|}{C}-OH \\
H\underset{|}{C}-OH \\
CH_2OH
\end{array}
\quad \xrightarrow{H_2O} \quad
\begin{array}{c}
\underset{|}{R_1} \\
H_2C-NH\text{-}CH\text{-}COOH \\
\underset{|}{C}=NH \\
\underset{|}{CH_2} \\
H\underset{|}{C}-OH \\
H\underset{|}{C}-OH \\
CH_2OH
\end{array}
+ R_2CHO
$$

18.4 $\updownarrow$

$$
\begin{array}{c}
\underset{|}{R_1} \\
HC=N-CH\text{-}COOH \\
H\underset{|}{C}-NH\text{-}CH\text{-}COOH \\
HO-\underset{|}{CH} \quad R_2 \\
H\underset{|}{C}-OH \\
H\underset{|}{C}-OH \\
CH_2OH
\end{array}
\quad \xrightarrow[H_2O]{CO_2} \quad
\begin{array}{c}
\underset{|}{R_1} \\
H\underset{||}{C}-N=CH \\
\underset{|}{CH} \\
HO-\underset{|}{CH} \\
H\underset{|}{C}-OH \\
H\underset{|}{C}-OH \\
CH_2OH
\end{array}
\quad \xrightarrow{H_2O} \quad
\begin{array}{c}
HC=NH \\
\underset{|}{CH_2} \\
HO-\underset{|}{CH} \\
H\underset{|}{C}-OH \\
H\underset{|}{C}-OH \\
CH_2OH
\end{array}
+ R_1CHO
$$

Reaction 18.17

aldehydes was described by Rizzi.[65] The reaction called reductive amination consists of the transamination of conjugated carbonyls in the presence of amino acids to unsaturated amines with concomitant formation of Strecker aldehydes. Other mechanisms yielding Strecker aldehydes propose decarbonylation followed by deamination of amino acids, or decarboxylation of imine formed from α-amino acid and α-keto acid.[18,66]

18.9 MELANOIDINS

The term *melanoidins* is used for high-molecular-weight brown polymers that arise at the final stage of the Maillard reaction. Although these polymers generally represent the principal reaction products of Maillard reaction in foods and despite extensive studies over the last two decades, the chemical nature of melanoidins is not yet clear. One theory is that the formation of high-molecular-weight colored products proceeds via polymerization of low-molecular-weight Maillard reaction intermediates. For example, Tressl and coworkers reported on the formation and characterization of a linear polymer (18.73) from N-methyl-2-

18.73 (X = O;

(hydroxymethyl)pyrrole as well as a branched polymer (18.74) from N-alkyl-2-formylpyrrole, N-alkyl-2-pyrrole, or 2-furaldehyde.[37,67] The corresponding pyrrole

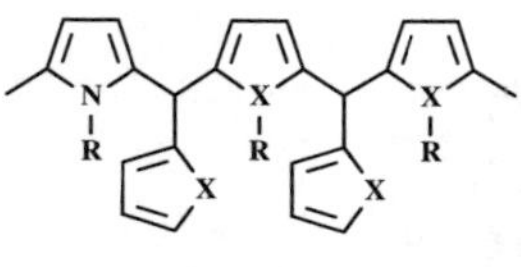

18.74 (X = O;

intermediates are formed from pentoses, hexoses, and disaccharides with 1,4 glyco-sidic bonds.

The condensation of early intermediates of the Maillard reaction, such as Amadori compounds or deoxyosones, is another mechanism that has been proposed to explain the formation of the melanoidin backbone.[68,69] Recently, it has been demonstrated that under water-free conditions, significant amounts of disaccharides or oligosaccharides with intact glycosidic bonds can be incorporated into melanoidins and cause branching of the melanoidin skeleton. Further side chains may also be formed by transglycosylation reactions. Further branching is possible via addition of amino compounds, such as amino acids or Amadori compounds, to carbonyl groups and α,β-unsaturated endiol structures of the melanoidin skeleton. The proposed melanoidin structures are shown in Reaction 18.18.[70]

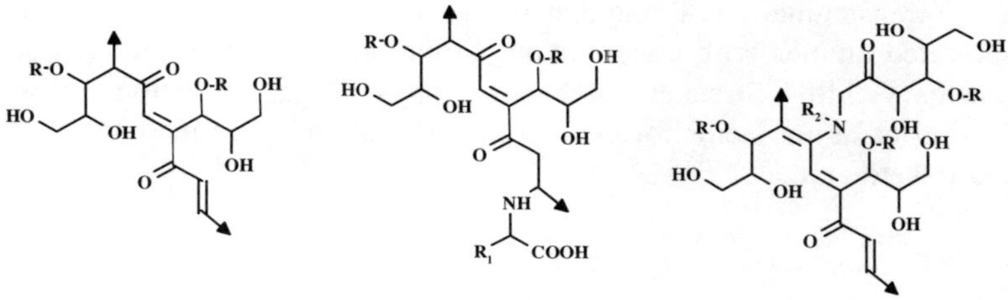

Figure 18.18

Based on extensive work concerning low-molecular-weight colored Maillard products and based on the fact that hydrolysis of food melanoidins affords amino acids and peptides, Hofmann proposed that food melanoidins could be formed by cross-linking of a low-molecular-weight chromophore and a noncolored high-molecular-weight biopolymer.[71] This hypothesis was confirmed by the isolation and characterization of a brown-orange melanoprotein bearing a lysine-bound 3(2H)-pyrrolinone chromophore (18.75) from the reaction system casein/2-furaldehyde.[72] The identification of chromophore (18.76) derived from two arginine residues, glyoxal, and 2-furaldehyde indicates that the oligomerization of the protein molecule via colored cross-linking structures can also contribute to the formation of melanoidins.[73] This is further supported by characterization of the protein-bound 1,4-*bis*(5-amino-5-carboxy-1-pentyl)pyrazinium radical (18.77) in orange-brown melanoidins (>100,000 Da) isolated from an aqueous model system containing bovine serum albumin and glycolaldehyde.[74] As already mentioned (see Section 18.3), 1,4-disubstituted pyrazinium radicals are very reactive intermediates involved in color formation. Thus, it is probable that melanoproteins contain a range of chromophores

18.75

18.76

18.77

(formed by ionic condensation as well as radical reactions), which are attached to lysine or arginine residues. Some chromophores may cause cross-linking of protein oligomers. Instead of being covalently bound to protein, the low-molecular-weight chromophores may be only physically entrapped by protein polymer as shown for a glucose/gluten model system.[75]

It can be assumed that food melanoidins represent a mixture of the above-mentioned structures. Depending on the starting materials as well as reactions conditions, certain reaction mechanisms will dominate and consequently various melanoidin structures will be obtained.

18.10 CLOSING REMARKS

This article dealt mainly with new developments on Maillard reaction mechanisms in food and related systems. In general, similar mechanisms occur also under physiological conditions. Indeed, a considerable part of Maillard research is devoted to medical chemistry focusing on health and disease.[2–8,76]

Although a great deal of work has been devoted to volatile reaction products, there is still additional effort to be made in structure elucidation of nonvolatile Maillard reaction products. These compounds may have color, taste, or antioxidative properties but can also interact with aroma compounds. Therefore, chemical characterization of principle patterns and characteristic units of these nonvolatile compounds will yield additional data to elucidate structural elements of chromophors, reductones, or compounds possessing antioxidative and other interesting properties. This may be achieved by applying modern analytical tools (LC-MS and NMR) and labeling experiments. More research is also required to extend our knowledge on Maillard reaction mechanisms in low-moisture systems and high reaction temperatures (pyrolysis conditions). Finally, the effect of structured reaction media on guiding reaction pathways is an area to be considered for future research. Ninety years after its discovery, the Maillard reaction will remain a challenging topic for scientists from various research disciplines.

REFERENCES

1. Maillard, L.C., Action des acides amines sur les sucres: formation des melanoidines par voie methodique, *C.R. Hebd. Seances Acad. Sci.*, 154, 66, 1912.
2. Eriksson, C., Ed., *Progress in Food and Nutrition Science*, Vol.5: *Maillard Reaction in Food: Chemical, Physiological and Technological Aspects*, Pergamon Press, Oxford, 1981.
3. Walter, G.R. and Feather, M.S., Eds., *The Maillard Reaction in Foods and Nutrition*, ACS Symposium Series 215, American Chemical Society, Washington, DC, 1983.
4. Fujimaki, M., Namiki, M., and Kato, H., Eds., *Developments in Food Science*, Vol.13: *Amino-Carbonyl Reactions in Food and Biological Systems*, Elsevier, Amsterdam, 1986.
5. Finot, P.A. et al., Eds., *The Maillard Reaction in Food Processing, Human Nutrition and Physiology*, Birkhäuser Verlag, Basel, 1990.

6. Labuza, T.P. et al., Eds., *Maillard Reaction in Chemistry, Food and Health*, The Royal Society of Chemistry, Cambridge, 1994.

7. O'Brien, J. et al., Eds., *The Maillard Reaction in Foods and Medicine*, The Royal Society of Chemistry, Cambridge, 1998.

8. Horiuchi, S. et al.., Eds., The Maillard reaction in food chemistry and medical science: update for the postgenomic era, *7th International Symposium on the Maillard Reaction*, Kumamoto, October 29–November 1, Elsevier Science B.V., Amsterdam, 2002.

9. Hodge, J.E., Dehydrated foods: chemistry of browning reactions in model systems, *J. Agric. Food Chem.,* 1,928, 1953.

10. Danehy, J.P., Maillard reaction: nonenzymatic browning in food systems with special reference to the development of flavour, *Adv. Food Res.,* 30, 77, 1986.

11. Namiki, M., Chemistry of Maillard reaction: recent studies on the browning reaction mechanism and the development of antioxidants and mutagens, *Adv. Food Res.,* 32, 115, 1988.

12. Ledl, F. and Schleicher, E., New aspects of the Maillard reaction in foods and in the human body, *Angew. Chem. Int. Ed. Engl.,* 29, 565, 1990.

13. Yaylayan, V.A. and Huyghues-Despointes, A., Chemistry of Amadori rearrangement products: analysis, synthesis, kinetics, reactions, and spectroscopic properties, *Crit. Rev. Food Sci. Nutr.,* 34, 321, 1994.

14. Wittmann, R. and Eichner, K., Detection of Maillard products in malts, beers, and brewing colorants, *Z. Lebensm. Unters. Forsch.,* 188, 212, 1989.

15. Hashiba, H., Isolation and identification of Amadori compounds from soy sauce, *Agric. Biol. Chem.,* 42, 763, 1989.

16. Heinzler, M. and Eichner, K., Behaviour of Amadori compounds during cocoa processing, Part I. Formation and decomposition, *Z. Lebensm. Unters. Forsch.,* 192, 24, 1991.

17. Reutter, M. and Eichner, K., Separation and determination of Amadori compounds by HPLC and post-column reaction, *Z. Lebensm. Unters. Forsch.,* 188, 28, 1989.

18. Holtermand, A., The Browning reactions, *Die Stärke,* 18, 319, 1966.

19. Cremer, D. R., Vollenbroeker, M., and Eichner, K., Investigation of the formation of Strecker aldehydes from the reaction of Amadori rearrangement products with α-amino acids in low moisture model system. *Eur. Food Res. Technol.,* 211, 400, 2000.

20. Hayashi, T. and Namiki, M., Formation of two-carbon sugar fragment at an early stage of the browning of sugar with amine, *Agric. Biol. Chem.,* 44, 2575, 1980.

21. Namiki, M. and Hayashi, T., Formation of novel free radical products in early stage of Maillard reaction, *Prog. Food Nutr. Sci.,* 5, 81, 1981.

22. Namiki, M. and Hayashi, T., A new mechanism of the Maillard reaction involving sugar fragmentation and free radical formation, in *The Maillard Reaction in Foods and Nutrition*, Waller, G. R. and Feather, M. S., Eds., ACS Symposium Series 215, American Chemical Society, Washington, DC, 1983, p. 21.

23. Hofmann, T., Bors, W., and Stettmaier, K., Studies on radical intermediates in the early stage of the nonenzymatic browning reaction of carbohydrates and amino acids, *J. Agric. Food Chem.,* 47, 379, 1999.

24. Herraiz, T. and Ough, C., Chemical and technological factors determining tetrahydro-β-carboline-3-carboxylic acid content in fermented alcoholic beverages, *J. Agric. Food Chem.,* 41, 959, 1993.

25. Rönner, B. et al., Formation of tetrahydro-β-carbolines and β-carbolines during the reaction of tryptophan with D-glucose, *J. Agric. Food Chem.,* 48, 2111, 2000.

26. Molero-Vilchez, M.D. and Wedzicha, L.B., A new approach to study the significance of Amadori compounds in the Maillard reaction, *Food Chem.,* 58, 249, 1997.

27. Glomb, M.A., Pfahler, C., and Hiller, R., Reductons participate in redox-reactions during amine catalysed sugar degradation, in *7th International Symposium on the Maillard Reaction*, Horiuchi, S., et al., Eds., Kumamoto, October 29–November 1, Elsevier, Amsterdam, 2002, 201.

28. Pischetsrieder, M., Schoetter, C., and Severin, T., Formation of aminoreductone during the Maillard reaction of lactose with N^α-acetyllysine or proteins, *J. Agric. Food Chem.*, 46, 928, 1998.

29. Kroh, L.W. and Schulz, A., News on the Maillard reaction of oligomeric carbohydrates: a survey, *Nahrung*, 45, 160, 2001.

30. Kawakishi, S. and Ushida, K. Autoxidation of Amadori compounds in the presence of copper ion and its effect on the oxidative damage of protein, in *Maillard Reaction in Food Processing, Human Nutrition and Physiology*, Finot, P. A. et al., Eds., Birkhäuser Verlag, Basel, 1990, p. 475.

31. Hofmann, T. and Schieberle, P., Formation of aroma active Strecker-aldehydes by a direct oxidative degradation of Amadori compounds, *J. Agric. Food Chem.*, 48, 4301, 2000.

32. Anet, E.L.F.J., Chemistry of non-enzymatic browning. VII. Crystalline di-D-fructoseglycine and some related compounds, *Aust.. J. Chem.*, 12, 280, 1959.

33. Hayashi, T. and Namiki, M., Role of sugar fragmentation in the Maillard reaction, in *Amino Carbonyl Reactions in Foods and Biological Systems*, Fujimaki, M., Namiki, M., and Kato, H., Eds., Elsevier, Amsterdam, 1986, p. 29.

34. Yaylayan, V. A. and Sporn, P. Novel mechanism for the decomposition of 1-(amino acid)-1-deoxy-D-fructoses (Amadori compounds): a mass spectrometric approach, *Food Chem.*, 26, 283, 1987.

35. Weenen, H. et al., C_4, C_5 and C_6 3-deoxyglycosones: structures and reactivity, in *The Maillard Reaction in Foods and Medicine*, O'Brien, J. et al., Eds., The Royal Society of Chemistry, Cambridge, 1998, p. 57

36. Tressl, R. and Kersten, E., Formation of pyrroles, 2-pyrrolidones and pyridones hy heating 4-aminobutyric acid and reducing sugars, *J. Agric. Food Chem.*, 41, 2125, 1993.

37. Tressl, R. and Wondrak, G.T., New melanoidin-like Maillard polymers from 2-deoxypentoses, *J. Agric. Food Chem.*, 46, 104, 1998

38. Tressl, R. and Kersten, E., Formation of 4-aminobutyric acid specific Maillard products from 1-13C-D-glucose, 1-13C-D-arabinose and 1-13C-D-fructose, *J. Agric. Food Chem.*, 41, 2278, 1993.

39. Bailey, R.G., Ames, J.M., and Mann, J., Identification of new heterocyclic nitrogen compounds from glucose-lysine and xylose-lysine Maillard model systems, *J. Agric. Food Chem.*, 48, 6240, 2000.

40. Tsuchida, H. et al., Identification of novel nonvolatile pyrazines in commercial caramel colours, in *Developments in Food Science*, Vol.13: *Amino-Carbonyl Reactions in Food and Biological Systems*, Fujimaki, M., Namiki, M., and Kato, H., Eds., Elsevier, Amsterdam, 1986, p. 85.

41. Hofmann, T., Identification of novel colored compounds containing pyrrole and pyrrolidine structures formed by Maillard reactions of pentoses and primary amino acids, *J. Agric. Food Chem.*, 46, 3902, 1998.

42. Hofmann, T. and Heuberger, S., The contribution of coloured Maillard reaction products to the total colour of browned glucose/L-alanine solutions and studies on their formation, *Z. Lebensm. Unters. Forsch.*, 208, 17, 1999.

43. Frank, O., Heuberger, S., and Hofmann, T., Structure determination of novel 3(6H)-pyranone chromophore and clarification of its formation from carbohydrates and primary amino acids, *J. Agric. Food Chem.*, 49, 1595, 2001.

44. Tressl, R., Helak, B., and Kersten, E., Formation of proline- and hydroxyproline-specific Maillard products from 1-13C-D-glucose, *J. Agric. Food Chem.*, 41, 547, 1993.

45. Hayase, F., Koyama, T., and Konishi, Y., Novel dehydrofuroimidazole compounds formed by advanced Maillard reaction of 3-deoxy-D-hexos-2-ulose and arginine residues in proteins, *J. Agric. Food Chem.*, 45, 1137, 1997.

46. Glomb, M.A. and Pfahler, Ch., Synthesis of 1-deoxy-D-erythro-hexo-2,3-diulose, a major hexose Maillard intermediate, *Carbohydr. Res.*, 329, 515, 2000.

47. Severin, T. and Seilmeier, W., Maillard reaction. II. Reaction of pentoses in the presence of amine acetates (in German), *Z. Lebensm. Unters. Forsch.*, 134, 230, 1967; *Chem. Abstr.* 1968, 68, 13276y.

48. Knerr, T. and Severin, T., Investigation of the glucose propylamine reaction by HPLC, *Z. Lebensm. Unters. Forsch.*, 196, 366, 1993.

49. Van den Ouweland, G.A.M. and Peer, H.G., Components contributing to beef flavour: Volatile compounds produced by the reaction of 4-hydroxy-5-methyl-3(2H)-furanone and its thio analog with hydrogen sulphide, *J. Agric. Food Chem.*, 23, 501, 1975.

50. Nursten, H.E. and O'Reilly, R., Coloured compounds formed by the interaction of glycine and xylose, *Food Chem.*, 20, 45, 1986.

51. Hofmann, T., Studies on the influence of the solvent on the contribution of single Maillard reaction products to the total color of browned pentose/alanine solutions—a quantitative correlation using the color activity concept, *J. Agric. Food Chem.*, 46, 3912, 1998.

52. Kerr, T., Pischetsrieder, M., and Severin, T., 5-hydroxy-2-methyl-4-(alkylamino)-2H-pyran-3(6H)-one: a new sugar-derived aminoreductone, *J. Agric. Food Chem.*, 42, 1657, 1994.

53. Blank, I., Devaud, S., and Fay, L.B., Study on the formation and decomposition of the Amadori compound N-(1deoxy-D-fructos-1-yl)-glycine in Maillard model systems, in *The Maillard Reaction in Foods and Medicine*, O'Brien, J. et al., Eds., The Royal Society of Chemistry, Cambridge, 1998, p. 43.

54. Hofmann, T., Acetylformoin: a chemical switch in the formation of colored Maillard reaction products from hexoses and primary and secondary amino acids, *J. Agric. Food Chem.*, 46, 3918, 1998.

55. Brands, C.M.J. and van Boekel, M.A.J.S., Reaction of monosaccharides during heating of sugar-casein system: building of reaction network model, *J. Agric. Food Chem.*, 49, 4667, 2001.

56. Davidek, T. et al., Degradation of the Amadori compound, N-(1-deoxy-D-fructos-1-yl)glycine in aqueous model systems, *J. Agric. Food Chem.*, 50, 5472, 2002.

57. Kramhöller, B., Pischetsrieder, M., and Severin, T., Maillard reactions of lactose and maltose, *J. Agric. Food Chem.*, 41, 347, 1993.

58. Pischetsrieder, M. and Severin, T., Maillard reaction of maltose-isolation of 4-(glcopyranosuloxy)-5-(hydroxymethyl)-2-methyl-3(2H)-furanone, *J. Agric. Food Chem.*, 42, 890, 1994.

59. Pischetsrieder, M., Gross, U., and Schoetter, C., Detection of Maillard products of lactose in heated or processed milk by HPLC/DAD, *Z. Lebensm. Unters. Forsch.*, 208, 172, 1999.

60. Schoetter, C. et al., Formation of aminoreductones from maltose, *Tetrahedron Lett.*, 35, 7369, 1994.

61. Hollnagel, A. and Kroh, W., Degradation of oligosaccharides in nonenzymatic browning by formation of α-dicarbonyl compounds via "peeling off" mechanism, *J. Agric. Food Chem.*, 48, 6219, 2000.

62. Schönberg, A. and Moubacher, R., The Strecker degradation of α-amino acids, *Chem. Rev.*, 50, 261, 1952.

63. Hofmann, T., Münch, P., and Schieberle, P., Quantitative model studies on aroma-active aldehydes and acids by Strecker-type reactions, *J. Agric. Food Chem.*, 48, 434, 2000.

64. Velisek, J., Davidek, T., and Davidek, J., Reaction products of glyoxal with glycine, in *The Maillard Reaction in Foods and Medicine*, O'Brien, J. et al., Eds., The Royal Society of Chemistry, Cambridge, 1998, p. 204.

65. Rizzi, G.P., Non-enzymatic transamination of unsaturated carbonyls: A general source of nitrogenous flavor compounds in foods, in *Phenolic, Sulfur, and Nitrogen Compounds in Food Flavors,* Charalambous, G. and Katz, I., Eds, ACS Symposium Series 26, American Chemical Society, Washington, DC, 1976, p. 122.

66. Shu, C.K., Pyrazine formation from amino acids and reducing sugars, a pathway other than Strecker degradation, *J. Agric. Food Chem.*, 46, 1515, 1998.

67. Tressl R. et al., Pentoses and hexoses as sources of new melanoidin-like Maillard polymers, *J. Agric. Food Chem.*, 48, 1765, 2000.

68. Yaylayan, V.A. and Kaminsky, E., Isolation and structural analysis of Maillard polymers: caramel and melanoidin formation in glycine/glucose model system, *Food Chem.*, 63, 25, 1998.

69. Cämmerer, B. and Kroh, L.W., Investigation on the influence of reaction conditions on the elementary composition of melanoidins, *Food Chem.*, 53, 55, 1995.

70. Cämmerer, B., Jalyschko, W., and Kroh, L.W., Intact carbohydrate structures as part of melanoidin skeleton, *J. Agric. Food Chem.*, 50, 2083, 2002.

71. Hofmann, T., Determination of the chemical structure of novel coloured compounds generated during Maillard-type reactions, in *The Maillard Reaction in Foods and Medicine*, O'Brien, J. et al., Eds., The Royal Society of Chemistry, Cambridge, 1998, p. 82.

72. Hofmann, T., Studies on melanoidin-type colorants generated from the Maillard reaction of protein-bound lysine and furan-2-carboxaldehyde: chemical characterisation of a red coloured domaine, *Z. Lebensm. Unters. Forsch.*, 206, 251, 1998.

73. Hofmann, T., 4-Alkylidene-2-imino-5-4-alkylidene-5-oxo-1,3-imidazol-2-inylazamethylidene-1,3-imidazoline: a novel colored substructure in melanoidins formed by Maillard reactions of bound arginine with glyoxal and furan-2-carboxaldehyde, *J. Agric. Food Chem.*, 46, 3896, 1998.

74. Hofmann, T., Bors, W., and Stettmaier, K., Radical assisted melanoidin formation during thermal processing of foods as well as under physiological conditions, *J. Agric. Food Chem.*, 47, 391, 1999.

75. Fogliano, V. et al., Formation of coloured Maillard reaction products in a gluten-glucose model system, *Food Chem.*, 66, 293, 1999.

76. Baynes, J.W. and Monnier, V.M., Eds., *The Maillard Reaction in Aging, Diabetes and Nutrition*, Alan R. Liss, New York, 1989.

19 Nonnutritional Applications of Saccharides and Polysaccharides

Piotr Tomasik

CONTENTS

19.1 General Remarks 315
19.2 Application of Nonmodified Starch 316
19.3 Applications of Chemically Modified Starches 316
      19.3.1   Alcoholyzed and Phenolyzed Starches and Sugar
               Alcohols 317
      19.3.2   Oxidized Starches 317
      19.3.3   Metal Salts and Complexes 318
      19.3.4   Ethers 318
      19.3.5   Acetals 319
      19.3.6   Esters 320
      19.3.7   Halostarches 321
      19.3.8   Aminostarches and Carbamates 322
      19.3.9   Copolymers 322
References 323

19.1 GENERAL REMARKS

Nonnutritional applications of saccharides are discussed in Chapter 3. Polysaccharides have attracted attention for their availability, versatility, and renewability. Among other functions, they are considered as a source of energy. Applications of chitin and chitosan deal mainly with nonnutritive area (see Chapter 14); any other polysaccharides, particularly when blended with other polysaccharides or saccharides, exhibit interesting rheological properties, making them suitable as potential adhesives. This area of polysaccharide chemistry is now intensively studied in many laboratories worldwide. It should be noted that dextrins and flours have also been commonly used for centuries as adhesives and glues. Beyond the food industry, dextrins are commonly used as adhesives, fillers, and binders; in textiles; and in paper sizing. Recently, dextrins have been used as blends with synthetic polymers.[1] A series of papers was published on the application of dextrins and maltodextrins as binders in ceramics based on micrometric metal oxide powders.[2] Clay ceramics use granular starch as an additive for manufacture of porous ceramic materials. Starch granules suspended in ceramic cake and then burnt out on sintering provide the required size of pores.

Among the considerable number of polysaccharides, only cellulose and starch, with their availability and price, justify their massive use for nonnutritive applications. These factors also necessitate wide development in physical and chemical modifications of these polysaccharides for such purposes.

19.2   APPLICATION OF NONMODIFIED STARCH

Starch is a common and traditional size for textiles. A series of papers has been published on the application of starch and cellulose as chromatographic sorbents. Cellulose has also found its place on the list of sorbents in common use. Gelatinized starch has been frequently proposed as a material for microcapsules for dyes, colorants, fragrances, and air-sensitive compounds. Usually, such microcapsules are formed by coacervation, freezing, or drying. Microencapsulation of fragrant compounds in processed starch granules has been recently presented. Potato starch as the most readily swelling substance appears superior for this purpose.[3,4] Biodegradable materials constitute another field of application of these polysaccharides. There are two approaches to this application. In the first, starch or cellulose is admixed to synthetic plastics based on vinyl monomers, for instance, polyethylene. To meet suitable mechanical properties and biodegradability, ca. 20% w/w of these polysaccharides is admixed to polyethylene. These polysaccharides act as biodegradable fillers in nonbiodegradable plastics.[5] In the second approach, complexes of polysaccharides and proteins are formed. Here, the anionic character of the polysaccharide appears to be an indispensable condition. It limits the number of applicable polysaccharides to pectins and potato starch with its phosphoric acid moieties esteryfying amylopectin.[6-10]

Thermolysis of starch and cellulose blended with proteogenic amino acids results in generation of secondary food aromas. Starch generates secondary food aromas, whereas cellulose generates aromas potentially useful for the cosmetics industry.[11]

19.3   APPLICATIONS OF CHEMICALLY MODIFIED
STARCHES

Nonnutritional applications of starch and cellulose are incentives for their numerous chemical modifications. Physical and enzymatic modifications (see Chapter 10) have been performed to make products of depolymerization, that is, dextrins. Modifications also include formation of sorption and inclusion complexes.[3,4]

The purposes of chemically modifying starch[12] (see Chapter 9) and cellulose are similar, but the corresponding derivatives of both polysaccharides perform differently. For instance, nitrocelluoses perform better than nitrostarches as explosives and thin films. In contrast to starch xanthate, cellulose xanthate finds wide application in production of artificial silk. These are the earliest and most common nonnutritive applications of cellulose and starch. Starch xanthates, and, particularly, starch xanthides have been proposed as heavy-metal traps for sewage and water basins.

19.3.1 ALCOHOLYZED AND PHENOLYZED STARCHES AND SUGAR ALCOHOLS

Alcoholysis (and phenolysis) of starch provides D-glucopyranosides useful in production of polyethers for rigid polyurethane foams and surfactants. Foams made from these materials are flame resistant. Alkyl pyranosides can serve as biodegradable surfactants and emulsifiers. Reduction of both polysaccharides generates polyols, which are promising as binders in metal oxide nanopowder-based ceramics.

19.3.2 OXIDIZED STARCHES

Applications of oxidized cellulose and starches include preparation of wood veneer glue, a joint cement consisting of cement mixed with gelatinized oxidized starch and asbestos; gypsum board; and textile dressings and sizes. Sizes is perhaps the largest application for oxidized starches and cellulose. Oxidized starches can be components of phosphate-free detergents.

Starch dialdehyde, the product of oxidation of starch with periodate, acetalated with glyoxal when admixed to pulp, increases the wet tensile strength of paper by 250%. It is also used to make paper sheets for thermal recording and in photolithography. Polysaccharide dialdehyde acetals, hydrazones, oximes, hydrogen sulfite compounds, and corresponding Schiff bases are known to increase the wet and dry strength of additives to paper.

Several copolymers and condensates of oxidized polysaccharides with polymers have been developed to form resins for coatings, molding powders, and materials for immobilization of enzymes, for instance, alpha amylase adhesives for labels and corrugated paper. Thin films are prepared in copolymerization of polysaccharide dialdehydes with partially hydrolyzed poly(vinyl acetate), O-(carboxymethyl)cellulose, and ethylene-vinyl alcohol.

Acidic oxidized polysaccharides, that is, polysaccharides with carboxylic groups, can trap metal ions, and heavy metal salts of oxidized polysaccharides have been suggested as dye mordants.

Oxidized polysaccharides possess binding properties with particles of artificial fertilizers. They have potential applications also as detergent builders, scale-preventing agents, and microencapsulating agents for aromas.

There exist a wide variety of medical applications for oxidized starches and their derivatives. Powdered oxidized starch decreases gauze adherence to wounds. Calcium derivatives have been proposed for treating hypocalcemia. The use of carboxystarch polymers as a component of toothpaste and mouthwashes is reported. These additives do not stain teeth and they inhibit plaque formation. Polysulfates of oxidized starches are heparin-like compounds because they prevent blood coagulation.

19.3.3 METAL SALTS AND COMPLEXES

Metal starchates carrying carboxyl, xanthate, sulfate, phosphate, and similar groups readily form corresponding salts by methatetical reactions between sodium salts of the aforementioned derivatives and other water-soluble metal salts. In this manner, they become ion exchangers used to remove heavy-metal contaminants from water. *cis*-Platinum derivatives of starch in tests with leukemia tumors in mice were found more effective and less toxic than *cis*-platin. Polysaccharide xanthates can act as flocculants for metal oxides and hydroxides. Related metal derivatives have been recommended as wood adhesives and also as binders in textile, paper, and paint industries.

19.3.4 ETHERS

Many reagents make etherification of starch and cellulose an important method for producing many useful derivatives with a wide range of applications, such as adhesives for paper and cardboard, paper coatings, and food coatings; additives to beverages, films, threads, dressings, sizes, and absorbents; viscosity-reducing agents for coal slurries; carriers for proteins and drugs; and additives for plastic shaping, emulsion paints, and protective colloids. Such ethers derived from starch retain their ability to gelatinize. Their uses with polysaccharide esters, reactive thermosetting resins, and acrylic resins are beneficial.

Ungelatinized products obtained on starch alkylation with dihaloalkanes or epichlorohydrin are simultaneously cross-linked. They form smooth, slightly cohesive pastes for textile printing. They are also used as textile sizes; thickeners; detergent builders; adjuvants in pharmaceutical preparations; coating for silver emulsion films; filling material for the anode compartment of small-size alkaline batteries; curing agent in epichlorohydrin–starch copolymers, in dextrin-containing adhesives, and in styrene–butadiene resin (SBR) latex; and printing pastes. In combination with poly(vinyl alcohol) and poly(methylacrylate), they have also been used as a fabric size. Swelling films are obtained by using a blend of (carboxymethyl)starch and chitosan dissolved in formic acid. An absorbent has been produced by reacting starch with bis(acrylamide)acetic acid, followed by treatment with chloroacetic acid. A wood adhesive has also been produced.

O-(Hydroxyethyl)starch resulting from hydrolysis of starch etherified with epichlorohydrin is one of the most intriguing starch modifications. Applications involve use in dermatological lotions; suspending media for photosensitive silver halides; ZnO elelctroconductive layers in electrographic devices; desensitizers of lithographic printing plates; components of paper coatings; sizing components for glass fibers, textiles, and yarns; antifoaming components in antifreeze media; and capsule shells. (Hydroxyethyl)starch accelerates disintegration of tablets prepared by direct compression. It can also be used as a drug carrier. In wood pulp, (hydroxyethyl)starch decreases the amount of surface lint on paper. It is also a potential soil stabilizer. Alkylation of (hydroxyethyl)starch provides a biodegradable surfactant. (Hydroxyethyl)starch films have been reported to show negligible transmission of oxygen.

(Hydroxyethyl)starch has been extensively studied as a blood expander, sometimes in a partially hydrolyzed form. Bleeding volume index of hydroxyethyl starch is found to be similar to that of whole blood. A product of DS between 0.43 and 0.55 and an average molecular weight of 216,000 was found to be superior. (Hydroxyethyl) starch appears to be a superior anticoagulant. It is also used for cryopreservation of mammalian cells, including those of the bone marrow. Various useful application of *O*-(hydroxyethyl)starch blends with low- and high-molecular-weight compounds are reported. In combination with urea, (hydroxyethyl)starch is used as a size for hydrophobic synthetic yarns; with alkylaminopropylamines or acetylated starch and poly(vinylalcohol), (hydroxyethyl)starch is used as a corrosion inhibitor and a microgel precursor. Blends with borax, urea, and ethylene glycol constitute an aerosol for ironing garments. Blends of hydroxyethyl starch and (hydroxyethyl)cellulose are effective thickeners for textile printing and developing pastes for motion-picture films, x-ray films, and photographic paper. Starch-derived ethylene glycol and glycerol glycosides react with alkylene oxides and have been tested as biodegradable detergents and components of polyurethane foams. Water-resistant short-tack adhesives are formed by blending (hydroxyethyl)starch with formaldehyde–urea copolymers and polyethylene glycol dodecanoate.

19.3.5 ACETALS

The chemical and physical stability of formaldehyde-cross-linked starch made these materials one of the first starch-based plastics. Such products were fairly stable and responded favorably to heat and pressure. As a result, they were soon applied as paper coatings and as thickeners for textile printing with dyes. Compounds from co-crosslinking with ketones were patented as adhesives for corrugated paperboard and water-stable sizing materials. Starch acetals were also used as plastics to impregnate fibrous substances, as a coating for heat-resistant fibers, as artificial lumber when mixed with sawdust, and as molding-compound binders, surfactants, and further cross-linking agents. Starch–formaldehyde acetal was also used as a cotton fabric size to improve the whiteness of household cloth after washing.

Adhesiveness of starch acetals can be improved by reacting starch with formaldehyde and ammonia or ammonium salts. The involvement of urea and amines in the acetalation offers a group of novel materials.

Protein–acetal compounds prepared with gelatin form sponges, which can be used instead of gauze and adhesive bandages. Polycondensates of aldehydes with amides of unsaturated carboxylic acids after reaction with starch produce agents for impregnating textiles, leather, wood, and paper. Strengthening agents for paper are also prepared by grafting oxidized starch with acrylamides and converting them into amines, followed by cross-linking with formaldehyde.

Materials with interesting properties became available when aldehydes other than formaldehyde were taken for acetalation. Pastes made of starch and chloral hydrate had strong adhesive and antiseptic properties. Starch cross-linked with such unsaturated aldehydes as acrolein has a wide range of applications in binding of nonwoven textile fibers and filter elements, and as thickeners or extenders for laminating fabrics.

19.3.6 ESTERS

Nitrated starch finds applications as a propellant, nitrogen fertilizer used together with ammonium salts, a component of ignition preparations, and a lacquer component. The main applications of phosphated starches are perhaps as food and cosmetics modifiers. Starch phosphates and phosphonates are sizes for textile, paper, and glass fiber sheets (phosphonium cationic starches have also been proposed for the last case), and adhesives for tobacco sheets and cigarette filters, corrugated board, and water-loss inhibitors. A granular distarch phosphate has been reported as a paper adhesive and also as a sand binder for foundry molds, and as a flocculant for coal wash tailings.

Starch phosphates have also been used as soil conditioners to increase water retention. Starch phosphates improve the dispersion and dye reception of synthetic fibers and are used to stabilize water, finger paints, and white coating colors. Phosphonoamidated starches are useful in paper sizing. Biodegradable films are produced by combining starch phosphate with poly(vinyl alcohol). Phosphonoacetyl starch added to polyacrylonitrile improves the affinity of the latter for dyes. The reaction of starch phosphate with such proteins as soybean protein and casein is used to produce protective colloids for insolubilization of pigments in coatings and paper.

The functional properties of starch sulfates depend on the degree of polymerization and the method used for sulfation. Starch sulfates are readily water soluble, and they typically form solutions having viscosities higher than those of soluble starch solutions of the same concentrations. In contrast to soluble starch, starch sulfates do not revert to solid gels. Quick-setting pastes of high thickness and good adhesiveness are available in this manner. These properties cease, however, when the solutions are neutral or made alkaline. Starch sulfates have been used as protective colloids; adhesives; thermosetters; thickeners for drilling muds and water-color paints; additives to hydraulic binders; components of electrostatic papers; additives for clay–ceramic masses; and components of water-insoluble coatings for paper, cardboard, wood, glass, and textiles. Starch sulfate is mixed with gelatin to form a wall material for microcapsules produced by coacervation. Microcapsules are also produced from sodium starch acetate or sulfate blended with gum arabic, polyacrylamide, and other compounds. Because of its polarity, starch sulfate interacts with protein. Spun-protein fibers containing starch sulfate exhibit increased thermostability.

The structure and some biological properties of sulfated starch and, separately, amylose resemble those of heparin, the well-known blood anticoagulant. Alkylsulfonyl starches with long-chain alkyl groups (C8, C10, C18) give effective surfactants even at low DS values (0.023 to 0.099). Some of these inhibit growth of *Escherichia coli, Bacillus subtilis,* and *Aspergillus niger.* Long-chain and aromatic-ring sulfones are available by reacting starch with sultones. Sulfoethylaminated starches have also been described. Thiosulfates of starch reported in the literature include benzoylthiosulfate and benzoyloxy thiosulfate, produced by treating starch with the corresponding thiosulfate in an aqueous suspension at room temperature.

Silylated starches are designed as glues, binders, coatings, water-repellent sizes, and glass-fiber sizes. Polysiloxanes react with starch in the presence of alkali aluminate, alkali hydroxide, or urea, and also with thiourea. Several modifications of silylated starches are achieved by the use of silanes having the general structure $R(CH_2)_n Si(Me)_m(OR^1)_{3-m}$, where $R = NH_2$, H, Me, Cl, $CH=CH_2$, SH, O, glycidoxy, $CH_2=CHMeCOO$; $R^1 = C_1-C_6$ alkyl; $n = 10–20$; and $m = 0–2$. Such derivatives are biodegradable.

Acylation of starch improves several of its functional properties, particularly with respect to food cooking applications. Starch acylation is used to reduce setback viscosity, and control gelatinization time, stability at low pH, freeze–thaw stability, storage characteristics, and other characteristics.

Starch esters are used as emulsifying agents, thickeners, and sizes; binders for coal briquets; and additives for hydraulic binders, paper coatings, and adhesives. Starch acetate having a degree of esterification above 0.03 is blended with diluted alkali and urea to produce an adhesive. Starch acetate has been used in wastewater treatment, especially in applications involving food processing. Starch dodecanoate having a degree of substitution above 2 has been patented as a base for chewing gum, and starch hexadecanoate is a component of shampoos. Esters of higher fatty acids are used to produce special optical effects in photographic films. Acylation of hydrolyzed starch with acid anhydrides in pyridine gives products of interest for medical and cosmetic use, and also has applications similar to those of acylated starches.

Esters from unsaturated fatty acid chlorides are usually viscous or limpid oils soluble in hydrocarbons and turpentine, whose primary applications are as varnishes, films, artificial threads, aqueous emulsions, and rubber-like plastics. Heating them in an inert gas produces insoluble products caused by polymerization involving double bonds in the acyl moieites. As mentioned in the section on nitrates, acetates of amylose are less compact than amylopectin acetate. Esters of unsaturated acids have also been proposed as remoistenable adhesives.

Mixed esters are proposed for use as petrochemical dispersants and cardboard adhesives. They are also compounded with starch and borax for use in processing corrugated board.

19.3.7 HALOSTARCHES

6-Iodoethylated starch resulting from the reaction of (hydroxypropyl)starch with *N*-iodosuccinimide in the presence of triphenylphosphine is cited as a potential contrast material for medical imaging by computed tomography.

Preparation of a chlorostarch for improved pigment retention in paper has been reported. Acetalation of chlorinated starches leads to a high-strength paper size.

19.3.8 AMINOSTARCHES AND CARBAMATES

Amino ethers (either free or preferably cationized) have a broad range of current and potential applications in paper manufacturing as coatings and sizes that improve paper strength, in retention of pigments and dyes, production of conductive coatings, textile finishes, and other uses. In the textile industry, cationic starches have been used as fiber sizes. Use of cationic starches in the processing of poly(vinyl alcohol) is also reported. Cationic starches are used as components of various latexes, as additives that decrease fluid loss (filtration rate) during aqueous drilling of soils, and as coatings for glass fibers.

Starch anthranilates have been proposed as optical brighteners. They improve the retention of inorganic fillers and fiber fines in paper processing.

Soluble carbamates have been considered for applications such as adhesives, pigmented coatings, paper coatings, and surfactants. Starch carbamates having double bonds in side chains are UV-curable and have been reported as coatings.

Starch carbamate block copolymers are biodegradable. During production, the reaction mixture can be supplemented with vegetable derivatives, molasses, polysaccharide agricultural waste, and vegetable oil fractions. Biodegradability of such materials reaches 6.7% during 12 weeks of exposure to soil. In the late 1950s, starch was used as filler for polyurethane polymers in order to reduce the cost of final products. Polyesters that were modified by reacting with diisocyanates were also molded with starch. Starch carbamate was also used for blending aminoplast or phenolic resins in order to obtain wood adhesives.

Starch polyurethane foams can be converted into sorbents for water when starch–acrylonitrile or starch–methacrylonitrile graft copolymers are cross-linked with diisocyanates. Such polyurethanes have been proposed as occlusive wound dressings and as sizes for cotton yarns. Reinforced paper starch polyurethanes are produced in a reaction of N-chlorocarbamoylethyl starch with cresol or thiolignin. Addition of mineral salts to polyurethanes prepared from starch and urea improve their performance as adhesives.

Starch and urea are prepared under pressure in order to produce an adhesive for corrugated paperboard and a thickener for frozen food. Carbamoylated starch can also be further reacted with formaldehyde and amines or with acrylamide and accompanying dimethylaminomethylation, giving flocculants for sewage disposal applications. Reactions of starch urethanes with alkylene oxides and aldehydes are used to produce hardenable compounds in waterproof films, coatings, and adhesives.

Thiosemicarbazones of starch dialdehyde and thiosemicarbazones and condensation products of starch dialdehyde with isoniazid were evaluated for their antituberculostatic activity and showed positive effects with the mice that were intravenously infected with *Mycobacterium tuberculosis*.

19.3.9 COPOLYMERS

After more than a decade of intensive studies on graft copolymerization of vinyl monomers onto starch, the first applications for these materials appeared. Applications include coatings; reinforcing materials for tires, rubbers, and plastics; anticorrosion additives for water; flocculants; superabsorbents; retention aids for pigments; sizing agents for paper and textiles; water-dispersed latexes; components of electrostatographic toners; gas chromatographic supports; and components of deodorants and various biodegradable materials.

REFERENCES

1. Tomasik, P., Wiejak, S., and Pałasiński, M., The thermal decomposition of carbohydrates. Part II: The thermolysis of starch. *Adv. Carbohydr. Chem. Biochem.*, 47, 279, 1989.
2. Schilling, C.H. et al., Application of polysaccharides in ceramics. *Zywn. Technol. Jakość*, 4(17) Suppl., 217, 1998.
3. Tomasik, P. and Schilling, C.H. Starch complexes. Part I: Complexes with inorganic guests, *Adv. Carbohydr. Chem. Biochem.*, 53, 263, 1998.
4. Tomasik, P. and Schilling, C.H., Starch complexes. Part II: Complexes with organic guests *Adv. Carbohydr. Chem. Biochem.*, 53, 345, 1998.
5. Roeper, H. and Koch, H., The role of starch in biodegradable thermoplastic materials, *Starch/Staerke*, 42, 123, 1990.
6. Tolstoguzov, V.B., *Functional Properties of Food Macromolecules,* Mitchel, J.R. and D.A. Ledward, A.D., Eds., Elsevier, London, 1986, chap. 9.
7. Tolstoguzov, V.B., *Food Properties and their Applications,* Damodaran, S. and Paraf, A., Eds., Marcel Dekker, New York, 1997, pp. 171.
8. Schmitt, C. et al., Structure and technofunctional properties of protein–polysaccharide complexes, a review, *Crit. Rev. Food Sci. Nutr.,* 38, 689, 1998.
9. Nishinari, K., Zhang, H., and Ikeda, S., Hydrocolloid gels of polysaccharides and proteins, *Curr. Opin. Colloid Interface Sci.*, 5, 195, 2000.
10. deKruif, C.G. and Tuinier, R., Polysaccharide–protein interactions, *Food Hydrocoll.,* 15, 555, 2001.
11. Bączkowicz, M. et al., On reaction of some polysaccharides with biogenic amino acids, *Starch/Stearke*, 43, 294, 1991.
12. Tomasik, P. and Schilling, C.H., Chemical modification of starch, *Adv. Carbohydr. Chem. Biochem.*, 53, 2003.

20 Carbohydrates: Nutritional Value and Health Problems

Przemysław Jan Tomasik

CONTENTS

20.1 Introduction 326
20.2 Saccharides as Structural Material 326
20.3 Saccharides as Metabolism Regulators 326
    20.3.1 Direct Regulation of Hunger and Satiety 326
    20.3.2 Absorption of Minerals, Bile Acids, and Cholesterol 327
20.4 Saccharides as Energy Source for the Organism 327
    20.4.1 Nutritional Value of Saccharides 327
    20.4.2 Saccharide Daily Allowances 329
20.5 Carbohydrates and Health 329
    20.5.1 Carbohydrate Intolerance 329
        20.5.1.1 Glucose Intolerance and Diabetes 329
        20.5.1.2 Fructose Intolerance 331
        20.5.1.3 Oligosaccharide Intolerance 331
    20.5.2 Obesity 331
    20.5.3 Caries 331
    20.5.4 Arteriosclerosis 332
    20.5.5 Effects of Carbohydrate Malnutrition 332
20.6 Fiber 332
    20.6.1 Nondigestible Saccharides as Energy Sources for Colonic
        Bacteria 332
    20.6.2 Other Effects of Fiber 333
20.7 Organoleptic Properties of Saccharides 333
20.8 Nonnutritional Saccharides 333
References..334

0-8493-1486-0/04/$0.00+$1.50
© 2004 by CRC Press LLC

20.1 INTRODUCTION

As key nutrients in the delivery of energy, carbohydrates are essential components of the human diet. Moreover, carbohydrates play a substantial role as structural and energy-storing materials of the human body. They regulate physiological and metabolic functions such as appetite, intake of minerals, and level of bile acids and cholesterol. Selected carbohydrates (prebiotics) are also used as nutrients for human commensals, that is, colonic bacteria (probiotics).

20.2 SACCHARIDES AS STRUCTURAL MATERIAL

The most abundant of saccharides are glucose dissolved in body fluids and glycogen, the storage form of glucose, present in the liver and skeletal muscles. Other carbohydrates such as ribose and deoxyribose are incorporated as essential structural components of nucleic acids. Glycoproteins and glycolipids coconstitute cell membranes. Glycation and glycosylation modify almost 50% of proteins, some lipids, and some 300 genes.[1]

Galactose formed from lactose is indispensable in developing infant nervous tissue constituted of galactocerebrosides. Mucopolysaccharides play a significant role as components of connective tissue. Endogenous heparin, another mucopolysaccharide, is an important anticoagulant. Polysaccharide chains of glycoproteins are necessary for controlling the immune system. They might be antigens of ABO system in erythrocytes.

Lactose is the sole disaccharide synthesized in the human metabolic processes. This nutrient for suckling infants is generated only in the mother's mammary glands.

20.3 SACCHARIDES AS METABOLISM REGULATORS

20.3.1 DIRECT REGULATION OF HUNGER AND SATIETY

The system regulating satiety and hunger is still a matter of controversy. Changes in the concentration of some metabolites such as blood glucose or free fatty acids are most commonly accepted as key factors. The glucostatic hypothesis accepts existence of chemoreceptor cells sensitive to glucose level in the blood in the hypothalamic region of the brain. They are considered as glucodetectors.[2] According to this hypothesis, a decrease in the glucose level causes the sensation of hunger and an increase in the blood glucose concentration produces the feeling of satiety. The most recent data show that the status of glucosensitive cells might be modified by insulin.[3] Insulin is released to blood prior to glucose absorption from the intestine. Therefore, the satiety signal generated by insulin level might well be faster than that generated by glucose. Satiety signals are also generated by the stimulation of peripheral glucodetectors localized in the liver and duodenum. These signals control neural impulsation of the vagus nerve.

20.3.2 ABSORPTION OF MINERALS, BILE ACIDS, AND CHOLESTEROL

Levels of calcium, bile acids, cholesterol, and other compounds in the human organism are dependent on the intake of digestible and nondigestible saccharides. Intestinal absorption of calcium increases with the ingestion of several carbohydrates. In some people, glucose or galactose administered orally fractionally increases calcium absorption. In healthy adults, such increase may reach 30%. The effect of lactose on calcium intake is controversial. In experimental animals, a diet containing either inulin or resistant starch (5 to 20%) improves uptake of calcium, iron, and zinc. In humans, inulin stimulates the absorption of calcium, but has no such effect on the absorption of zinc, iron, or magnesium. In this instance, however, experimental doses of inulin were below 15 to 40 g/day.

These observations can be rationalized in terms of osmotic action of oligosaccharides, acidification of intestine content by bacterial fermentation, formation of readily soluble magnesium and calcium salts of lower fatty acids generated on fermentation, or secondary hypertrophia of mucous membrane of the colon. There are also suggestions that fiber (some carbohydrates) can complex minerals such as calcium, iron, copper, magnesium, phosphorus, and zinc. The resulting complexes are insoluble and reduce bioavailability of these essential minerals.[4]

Fiber in the diet, especially pectins, complexes bile acids and reduces cholesterol intake. Tests with rats fed with a diet containing 10% w/w inulin showed a decrease in the level of triglycerides and cholesterol void as well as after a meal. About 12 to 36 g of pectins in the human daily diet reduces the total serum cholesterol by 8 to 30%. However, relevant data in healthy people are not univocal.

20.4 SACCHARIDES AS ENERGY SOURCE FOR THE ORGANISM

20.4.1 NUTRITIONAL VALUE OF SACCHARIDES

Throughout their life, humans ingest about 10,000 kg of carbohydrates. Digestible carbohydrates are the main source of energy. In developed countries, carbohydrates supply 45 to 60% of the energy requirement, but in the Third World such requirement is fulfilled in less than 85% of the population. Adults in Europe daily consume about 240 g of glucose, 65 g of fructose, and 15 g of galactose (only monosaccharides are absorbed from the gut); however, polysaccharides, mainly starch, predominate. Only thermally processed and pasted starch is digested in the gut. Such starch undergoes enzymatic hydrolysis to monosaccharides by salivary and pancreatic amylase, glucoamylase, amyloglucosidase, and disaccharidases from intestinal mucosa. Monosaccharides are absorbed from the intestine into the portal circulation and transported with blood to the liver, where fructose and galactose are converted into glucose. In the systemic circulation only glucose is present. The concentration of glucose in blood increases in the postprandial period. At that time, tissues utilize glucose absorbed directly from the gut. An excess of glucose is stored as glycogen, and excessive glucose, which cannot be transformed into glycogen, is metabolized to triglycerides.

Oxidation of glucose is exergonic. Hypothetically, oxidation of 1 mole glucose provides 38 moles ATP. Hydrolysis of 38 moles ATP to ADP releases 543.4 kcal (2276.2 kJ) of energy. Digestion and absorption of saccharides take up to 6% of energy released on the oxidation of glucose. Further, 5% of the total energy released during the glucose oxidation is required for the incorporation of glucose to the Krebs cycle (Figure 20.1). About 20% of glucose is metabolized in the less effective lactic acid and glucose-alanine cycles. A total energy efficiency of glucose reaches 60-76% of its hypothetical value. Moreover, the storage of glucose as glycogen consumes additionally 5% of the energy value of glucose.

Glucose is released from glycogen in the intraprandial period. Gluconeogenesis from amino acids and glycerol also contributes to the level of up to 130 g of glucose that can be synthesized daily. To avoid ketosis, a complete oxidation of lipids requires 180 g of glucose daily. Therefore, the minimum daily glucose intake reaches 50 g. To calculate a minimum daily allowance for other saccharides in the diet, relevant digestion indices and energy equivalents have to be considered.

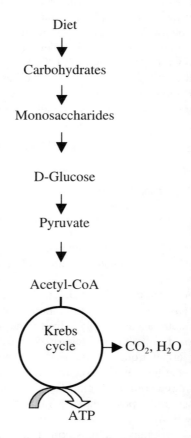

FIGURE 20.1 Incorporation of glucose in the Krebs cycle.

20.4.2 SACCHARIDE DAILY ALLOWANCES

Saccharides make up 25 to 52% of energy requirements of newborns whether fed naturally or by formula. Mother's milk includes lactose, and formula contains glucose, lactose, or maltodextrins, or all three. These components are the most suitable for infants because of the low activity of their amylolytic enzymes. A residual activity of amylase can be found in pancreatic juice in the third month of an infant's, life which slowly increases in the following months.[5] In the adult diet, carbohydrates should make up over 50 to 65% of the energy requirements, which corresponds to 320 to 460 g of saccharides daily. In humans, a diet consisting mainly of polysaccharides can provide up to 70% of energy needs. The main rule of dietetics is to limit intake of mono- and disaccharides and to increase the level of complex carbohydrates in the diet. A daily fiber intake of 27 to 40 g is recommended. In Europe, grain products are the most common source of carbohydrates. A considerable amount of carbohydrates is ingested with sweets, cakes, and soft drinks. Carbohydrates from fiber-rich fruits and vegetables constitute a minor part of ingested saccharides. Honey, candied fruits, and products of processed potato starch are the foods richest in saccharides (Table 20.1).

20.5 CARBOHYDRATES AND HEALTH

20.5.1 CARBOHYDRATE INTOLERANCE

Some pathogenetic and symptomatic diseases cause carbohydrate intolerance. Glucose intolerance resulting from the shortage of insulin is the most common. Intolerance to disaccharides is activated by the absence of specific enzymes in the gut.

20.5.1.1 Glucose Intolerance and Diabetes

Glucose intolerance relates to diabetes and impaired glucose tolerance. The latter is a stage between normal glucose homeostasis and diabetes. Frequently, impaired glucose tolerance is a prediabetic stage.

Diabetes can be an insulin-dependent diabetes mellitus (IDDM), insulin-independent diabetes mellitus (NIDDM), and a few other specific types. Insulin facilitates the transfer of glucose from blood to the cells. Shortage of insulin results in high concentration of glucose in blood and low level of this energy-producing material in the cells. A high glucose level in blood changes its osmolality, producing some clinical symptoms. More important, the low level of glucose in the cells impairs their metabolism. IDDM is caused by immunodestruction of pancreatic islets. In NIDDM, the shortage of insulin is relatively insignificant. It results from resistance of cells to insulin, impaired insulin secretion, and excessive glucagon secretion. The relationship between NIDDM and diet is not clear. Some papers demonstrate that a diet rich in simple carbohydrates and animal fat is more diabetogenic than a diet rich in complex carbohydrates; however, several papers contradict this point of view.[6]

TABLE 20.1
Nutritional Value of Selected Foodstuffs

Foodstuff	Approximate Content in 100 g		
		Carbohydrate	
	Energy (kcal)	s	Fibers
Beer	41	3.7	0
Wine	120	8	0
Tea	0	0	0
Coffee	1	0.3	0
Cola drink	49	12.5	0
Bread (mixed)	203	43.7	2.7
Graham	198	40	5
Pasta	364	72	5.3
Flour (wheat)	347	75	2.6
Rice	349	78.9	2.4
Honey	320	89	0
Jams	270	70	0.5
Sugar	390	100	0
Cottage cheese	100	2	0
Gouda cheese	380	2.3	0
Ice cream	202	23.8	0
Whole milk	62	4.8	0
Yogurt	61	4.6	0
Egg	158	1.2	0
Fats and oils	888	0	0
Apple	42	8.8	1.3
Apricot	51	27.5	0.64
Banana	62	15.7	0.33
Orange	34.3	8.5	0.3
Peach	34	9.2	0.5
Pear	47	10.9	1.6
Plum	51	12	0.5
Beef steak	283	0	0
Pork ham	182	3	0
Chicken leg	135	0	0
Fish	80–200	0	0
Broccoli	28	5	1
Corn	88	19.3	0.7
Onion	33	6.1	1.5
Peas (dry)	346	61.6	15.7
Carrots	25	5.4	1.5
Potato	59	13.5	0.8
Tomato	28	5	1.2
Cucumber	11	2.1	0.4

20.5.1.2 Fructose Intolerance

Fructose is metabolized without insulin. Therefore, it is a safe sweetener and energy source for diabetics. Fructose intolerance, however, also occurs. It is an inherited metabolic disorder resulting from the lack of fructose-1-phosphate aldolase. When the disorder is diagnosed, a fructose-free diet is recommended.[7]

20.5.1.3 Oligosaccharide Intolerance

Low activity of disaccharidases in the gut causes suppressed hydrolysis of disaccharides in its anterior part. Nonhydrolyzed disaccharides serve as nutrients for bacteria in the posterior part of the gut. Excessive fermentation in the large intestine produces abdominal bloating, flatulence, and borborygmi and is caused by gaseous products of bacterial fermentation. Unabsorbed disaccharides increase the osmotic tension of intestinal content, causing a water shift to the lumen of the gut. Osmotic tension also accelerates transit time, producing diarrhea and other abdominal complaints.

Primary and secondary lactose intolerances are serious medical problems. In primary hypolactasia, children are born with normal lactase activity, which then starts declining at the age of three. Approximately 70% of adults in non-White populations are susceptible to lactase deficiency; among the Whites, however, the percentage drops to below 40%. Secondary hypolactasia is usually a transient phenomenon. It occurs as a consequence of damage to the small intestine mucosa. With restoration of the mucosa, lactase activity improves.[8] Sucrose intolerance has been reported in Eskimos.

20.5.2 Obesity

With the exception of certain single-gene disorders, the etiology of obesity depends on a number of factors or causes and results from overloading with energy. Storage of glucose as glycogen is limited. Usually, in adult persons, reserves of glycogen do not exceed 100 g in the liver and 150 to 400 g in the muscles; as a consequence, excess of consumed sugars is converted into triglycerides. Accumulation of lipids in the human body seems to be unlimited. A number of obese people carry over 100 kg of fat in their body. Eating meals with unrestricted levels of carbohydrates, especially simple carbohydrates, produces simple obesity. Epidemiological research points to NIDDM occurring more frequently in obese people than in lean people.[9]

20.5.3 Caries

Epidemiological data show that the risk of caries is lower in countries where the annual sugar consumption is below 10 kg per person. Particularly, sucrose in food with texture permitting adhesion to the teeth produces high risk of dental caries. Therefore, some other carbohydrates and carbohydrate-derived sweeteners, such as polyols (sorbitol, maltitol), which are not metabolized by oral microorganisms, are more desirable.

20.5.4 ARTERIOSCLEROSIS

There is no clear evidence that a high-carbohydrate diet leads directly to coronary heart disease; nevertheless, the progression from overeating, through obesity and diabetes mellitus, to coronary heart disease, is well known.

20.5.5 EFFECTS OF CARBOHYDRATE MALNUTRITION

Carbohydrate deficiency caused by fasting, improperly balanced diet, or diabetes mellitus leads to ketosis: the consequence of unequal rates of β-oxidation and oxidation of acetyl CoA in the tricarboxylic acid cycle. Except in the case of diabetes mellitus, carbohydrate content in a diet that satisfies daily allowances maintains the rates of both processes and prevents ketosis.

20.6 FIBER

20.6.1 NONDIGESTIBLE SACCHARIDES AS ENERGY SOURCES FOR COLONIC BACTERIA

Prebiotics are defined as nondigestible substances, the dietary fiber, maintaining biological activity within the human organism by selective stimulation of growth or bioactivity of beneficial microorganisms either naturally present or therapeutically introduced to the intestine. They are nondigestible saccharides such as resistant starch; nonstarchy polysaccharides such as pectins, guar gum, or oat gum; oligosaccharides; sugar alcohols; and endogenic carbohydrates such as mucin and chondroitin sulfate. In the intestine, prebiotics undergo fermentation by microflora. Currently, particular attention is paid to saccharides belonging to the inulin group: Glc$p$Fru$f$ [α-D-glucopyranosyl-(β-D-fructofuranosyl)n-1-D-fructofuranose]; Fru$p$-Fru$f$ [β-D-fructopyranosyl-(β-D-fructofuranosyl)n-1-D-fructofuranose] where n = 10 to 60; fructooligosaccharides, FOS (apart from oligofructose with n = 2 to 9, and, eventually, with D-xylose, D-galactose, D-glucose, and mannose residues); and nondigestible oligosaccharides, NDO. Oligofructoses are present in wheat, onion, garlic, artichoke tubers, endive, leek, asparagus, topinambur, and bananas. Mother's milk contains relatively high levels, 3 to 6 g/l, of oligosaccharides. FOS and NDO are selectively digested by bifidobacteria and stimulate growth of their colonies. Therefore, in the gastric tracts of infants fed with the mother's milk, the population of bifidobacteria is ten times higher than in infants fed with formulae.

Intake of prebiotics promotes the growth of probiotic bacteria. It results in inhibition of the growth of pathogenic bacteria, bacterial digestion of nutrients, particularly proteins, and in reduction of their allergogenicity. Probiotics also synthesize vitamins of the B-group and K as well as cytoprotective short-chain fatty acids and polyamines such as putrescine, spermine, and spermidine. Probiotics also metabolize some drugs. Probiotics reduce symptoms of lactase deficiency. Moreover, probiotics degrade fiber, which osmotically increases the volume of stool and improves action of the colon.[10]

20.6.2   OTHER EFFECTS OF FIBER

Fiber has other effects besides adsorbing minerals. A number of papers have reported the positive and negative action of fiber. FOS influence homeostasis of cells in the intestine walls and protect them. Immunomodulation and bacteriostatic activity should also be considered. The activity might result from blocking receptors capable of interactions with pathogenic bacteria. Oligosaccharides with mannose side-chains obstruct adhesion of *Escherichia coli* to the intestinal walls. Cellobiose, the plant disaccharide, reduces infectivity of *Listeria monocytogenes*. FOS added to animal fodder significantly decrease frequency of the colon cancer. Recommended daily uptake of FOS for humans reaches 4 g. In rats, inulin also inhibits production of carcinogenic sialomucins and stimulates synthesis of anticarcenogenic sulfomucins.[11]

A fiber-rich diet increases fecal bulk, making the stools softer and more frequent. A high volume of unabsorbed food shortens the transit time and lessens the concentration of carcinogens, reducing their contact with intestinal walls.

Other less beneficial side-effects have been reported. Pectins in fiber added to naturally occurring food pectins can exceed the natural level and cause diarrhea.

Nonfiber saccharides such as raffinose and stachyose present in legume plants are not digested by the human enzymatic system. Their bacterial digestion, similar to the effect of excessive fiber, produces flatulence.

20.7   ORGANOLEPTIC PROPERTIES OF SACCHARIDES

Carbohydrates are commonly appreciated for their organoleptic properties. They add sweet taste, flavor, color, and texture to foodstuffs. The organoleptic properties resulting from the presence of carbohydrates strongly affect cephalic phase of food intake.

There are also adverse effects of modification of carbohydrates on food processing and digestion. These problems are important in case of thermally processed milk monodiet for suckling children. Carbohydrates react with lysine into dihydroalanine, provoking the degradation of cystine and forming lysinoalanine. This type of reaction, called Maillard reaction (Chapter 18), decreases nutritional value of food by reducing essential amino acids and vitamins. Thermal processing of milk leads also to epimerization of lactose to nondigestible lactulose. A decrease in lactose concentration not only changes palatability and flavor of milk, influencing the volume of ingested food, but also decreases the available energy, declines calcium absorption, reduces intestinal pH, and lowers bioavailability of galactose, essential for proper brain development.

20.8   NONNUTRITIONAL SACCHARIDES

Some saccharides are important constituents of medicines, such as cardiac glycosides, aminoglycosides (antibiotics), and artificial heparin widely used as an anticoagulant.

REFERENCES

1. Freeze, H., Congenital disorders of glycosylation: an emerging group of inherited diseases with multisystemic clinical presentation, *Clin. Chem. Lab. Med.* 40, S34, 2002.

2. Mayer, J., *Overweight: Causes, Costs, and Control*, Prentice-Hall, Englewood Cliffs, NJ, 1968, p. 35.

3. Oomura, Y., Activity of chemosensitive neurons related to the neurophysiological mechanisms of feeding, In *Recent Advances in Obesity Research,* Vol. 2, Bray, G.A., Ed., Newman Publishing, London, 1978, p. 123.

4. Coudray, C. et al., Effects of soluble dietary fibers supplementation on absorption and balance of calcium, magnesium, iron and zinc in healthy young men, *Eur. J. Clin. Nutr.,* 51, 375, 1997.

5. Klish, W.J., Carbohydrates in infant formula: future trends, in *New Perspectives in Infant Nutrition*, Ghraf, R., Falkner, F., Kleinman, R., Koletzko, B., and Moran, J., Eds., Ergon, Madrid, 1995, p. 129.

6. Nathan, D.M., Clinical practice: initial management of glycemia in type 2 diabetes mellitus, *New. Engl. J. Med.*, 347, 1342, 2002

7. Ali, M., Rellos, P., and Cox, T.M., Hereditary fructose intolerance, *J. Med. Genet.,* 35, 353, 1998.

8. de Vrese, M. and Schrezenmeir, J., Nutritional aspects of lactose, *Carbohydr. Eur.,* 25, 18, 1999.

9. Hill, J.O. and Peters, J.C., Biomarkers and functional foods for obesity and diabetes, *Br. J. Nutr.,* 88, S213, 2002.

10. Tomasik, P.J. and Tomasik, P., Prebiotics and probiotics, *Cereal. Chem.,* 80, 113, 2003.

11. Gibson, G., Dietary modulation of the human gut microflora using the prebiotics oligofructose and inulin, *J. Nutr.,* 129, 1438S, 1999.

21 Glucose — Our Lasting Source of Energy

Heinz Th. K. Ruck

CONTENTS

21.1 Introduction 335
21.2 Elements of Fuel Cells 336
21.3 Can Biomass Replace Oil? 337
21.4 Methane — An Abundant Biogas 339
21.5 Birth of the Green Biorefinery 339
21.6 Green Biorefineries for Ethanol Production 340
21.7 The Search for Supercellulases 341
21.8 How to Boost Ethanol Supply? 341
21.9 The Ethanosolv Process by T.N. Kleinert 342
21.10 ReShuffle of the Pulp and Paper Industry 342
21.11 Concluding Remarks 346
References 346

21.1 INTRODUCTION

Several estimations have been made by various authors to assess global biomass annually synthesized by autotrophic plant cells, sun, and water.[1,2] Frequently, yearly biomasses are figured close to 170 to 200×10^9 t, containing approximately 75% carbohydrates (primarily glucose), 20% lignin, and 5% protein and lipids, besides traces of terpenes, alkaloids, hormones, nucleic acids, enzymes, and chlorophyll, tailor-made to support various functions of living cells. The corresponding energy content for a 200×10^9 t biomass is 100×10^9 coal units, which is tenfold the annual global energy consumption. The cited biomass nourishes ca. 6 billion people consuming ca. 6 billion tonnes (bt) (3.5%), out of which only 300 mio (million) tonnes are channeled to the industry. It is perhaps fair to assume that ca. 2 bt each are provided as wood, cereals grown on arable land, and vegetables. Added to this, meat is bred either in stables or by stock raising on pastures covering 4.5% of the globe surface; arable land accounts for 4.8%. Fisheries provide only an insignificant quantity of proteins compared with ca. 5 billion cattle and even more pigs and sheep raised preferably on permanent grasslands.

The consumption of 3.7 bt of food is contrasted by fossil fuels, summing up in 2000 to 7.3 billion oil equivalents (OE), made up of 3.3 bt of oil, 2.3 bt of coal, and 1.8 bt OE as gas (138 trillion cubic meters), besides less than 10 million tonnes (mt) of uranium (probably to be replaced around 2030 by plutonium-free [232] thorium). Of the 7.3 billion OE, only 7% is claimed by the industry and the remaining 93% consumed for energy production in households and by traffic. In this context, a full substitution of fossil raw materials by renewable masses seems unlikely.[2] The advent of fuel cells (FCs), however, has changed the situation considerably, because the operation of space vehicles caused the development of perfluorinated proton exchange membranes (PEMs). It is now possible to supply electric power at working temperatures from 70 to 90°C just by "cold"-burning hydrogen, at present motoring even submarines. The reaction perfects the arrangement of Grove (1842) for production of electric tension in dilute sulfuric acid between platinum electrodes contacting hydrogen and oxygen,[3] and reversing electrolysis. Herein, water "replaces" coal.[4]

21.2   ELEMENTS OF FUEL CELLS

The electric energy conversion of fuel cells between 40 and 70% depends on the working temperature. The Carnot factor governing the conversion of steam into power is not valid here. Instead, fuel cells follow the relation:

$$\eta_{el} = \Delta G / \Delta H \qquad (21.1)$$

where ΔH is enthalpy of reaction and ΔG the Gibbs free enthalpy, with $\eta_{el} + \eta_{th}$ approaching 90%. Proton exchange membrane fuel cells (PEMFCs) operating at 77°C yield η_{el} of 45% whereas molten potassium-lithium-carbonate fuel cells (MCFCs) working at 627°C attain an η_{el} of 65% due to their cathode exhaust temperature around 550°C driving a final turbine.

The operation of any FC requires cheap suitable fuels such as methane, alcohols, or hydrocarbons convertible to hydrogen by tailor-made "reformers," for example:

$CH_4 + H_2O \rightarrow CO + 3H_2$ with $\Delta H_{1000K} = 226$ kJ/mol supporting the conversion

$CO + H_2O \rightarrow CO_2 + H_2$ with $\Delta H_{1000K} = -35$ kJ/mol yielding simultaneously

$CH_4 + 2H_2O \rightarrow CO_2 + 4H_2$ with $\Delta H_{1000K} = 191$ kJ/mol

calling for a rather low working temperature (250°C) in presence of noble metal catalysts (e.g., Pt or Ru). Ethanol combustion occurs above 400°C, requiring better heat insulation and other catalysts.

The MCFC operates beyond 500°C without any catalyst and can burn liquids too:

$CH_4 + 2H_2O + Heat \rightarrow CO_2 + 4H_2$ (reforming reaction utilizing hot FC thermal energy)

Also, 1 mole of CH_4 produces 4 moles of H_2, yielding more primary energy (10 to 12 kWh). At the anode, the hydrogen molecule reacts with the carbonate anion supplied by the electrolyte as a eutectic mixture melting at 480°C:

$$H_2 + CO_3^{2-} \rightarrow H_2O + CO_2 + 2e^- \text{ (free energy)}$$

The CO_2 of the reforming reaction is channeled to the cathode together with O_2 from air to restore CO_3^{2-}:

$$CO_2 + \rightarrow O_2 + 2e^- \text{ (consumed energy)} \rightarrow CO_3^{2-}$$

Due to the electrolyte temperature of ca. 650°C, the emerging air of 550°C can support the reforming reaction and also drive a turbine at 450°C, boosting η_{el} beyond 65%. Residual steam can serve for heating and refrigeration systems (NaBr absorption).[5] A superior reforming can be achieved by ethanol:

$$C_2H_5OH + 3H_2O + \text{thermal energy} \rightarrow 2CO_2 + 6H_2$$

with 1 mole of ethanol providing 6 moles of H_2. Furthermore, ethanol is readily stored like oil, warranting service for longer periods. The lifetime of an MCFC exceeds 40,000 h (4.5 years) before carbonate corrosion attacks the nickel frames at total efficiencies beyond 85%. Self-supply of farmers is feasible because traces of water do not disturb MCFCs. Self-supply holds true for PAFC also (based on phosphoric acid at 200°C) and even more for the ceramic oxide FC at 950°C with η_{el} up to 70%. The solid oxide fuel cell (SOFC) electrolyte is ZrO_2 doped with 8% YO_2 (yttrium oxide) transporting O_2^{2-} from anode to cathode. Again, catalysts of the Pt group are not required. Their immobility predestines them for installation in house blocks, etc., feeding their surplus kilowatt-hour-power into the grid. Likewise, battery-driven cars can be loaded on premises, reserving PEMFC-equipped vehicles for longer distances.

21.3 CAN BIOMASS REPLACE OIL?

In this context, one should remember that global oil reserves are estimated close to 2 to 3 billion barrels,[6] forecasting the end of the oil age within this century. The surface of our planet is made up of 28% land (148×10^6 km^2) and 72% sea (362×10^6 km^2), furnishing the rain to enable photosynthesis on land, primarily on arable regions (4.8% of 510×10^6 km^2 = 24.5×10^6 km^2) and forested areas with more than 30 in of rainfall per annum whereas pastures of varying qualities receive 10 to 30 in of rainfall only (4.5%, i.e., 23×10^6 km^2).

The residual land surface of 100×10^6 km^2 is either mountainous or desert, ca. 20% (30×10^6 km^2), with below 10 in of precipitation. Intensive agriculture requires 20–40 in per annum, but extensive grasslands require only 10–20 in., yielding shorter blades. For growth of plants, the coincidence of sunshine and rain is of paramount importance, and also the kind of precipitation, for example, convectional (tropical), orographic (mountains cause windward rains), or cyclonic rains. The last type prevails in temperate latitudes, yielding water steadily pouring out of dull, overcast skies, and in winter as snow, sleet, or hail. Cyclonics reach their main precipitation between latitudes of 45 and 55°North, primarily in the Northern Hemisphere, with a preponderance of summer rainfall.[7]

Forests and grasslands represent the two major vegetations. The former stabilize the climate, including moisture or dust content of air and the water table. The role of grasslands is less appreciated. For example, most authors credit 90% of the planetary biomass to forests — allegedly 50×10^9 t[1] — of which only 2.3×10^6 t (4.5%) is consumed by humans[2], primarily as fuel in the Third World besides timber and pulp for paper in the First World.

Regarding wood as a possible energy source, one has to consider that only 14 to 18% of forests are accessible. About 1 to 2×10^5 km² of tropical forests are eradicated yearly. The annual yield per hectare declines from the tropical 20 t wood/ha to merely 3 t/ha in Central Europe (timber quality) and still less in Northern Europe (1 to 2 t/ha). However, improved forest management and reforestation could possibly double the timber harvest. Complementing such endeavors could be the perfection of the burgeoning technique to utilize whole trees, rendering their lignin and carbohydrates into saleable products to boost the revenues per hectare to be reinvested in forestry.

As regards grasslands, one has to discern drier (steppe) and moister (prairies) areas allowing trees to grow turning to a savannah, whereas vanishing precipitation diverts the steppe to desert. The total grass-bearing surface is estimated to cover a fifth of all continents close to 30×10^6 km², including the mentioned 23×10^6 km² attended by farmers or stock-breeders for animals, etc. For extensively cultivated grasslands, a biomass of 4 t/ha of dry matter (DM) can be interpreted as modest. Nevertheless, even by such an assumption, nearly 12×10^9 t of DM/ha would contain at least 50% of cellulosics capable of yeilding 3×10^9 t of ethanol, energetically virtually equivalent to oil if upgraded by burning in fuel cells, furnishing electric and thermal power approaching almost the 7.3 billion OE spent in 2001.[2] The supervised, and, if intensively cultivated, grasslands in temperate zones would be capable of producing between 7 and 12 t of DM/ha, possibly amounting to a global 23×10^6 km², bearing ca. 23×10^6 km² $\times 10^2$ ha $\times 10$ t/ha = 20–23×10^9 t of DM/ha made up of approximately 60% polysaccharides, 15% lignin, 10–15% ashes, and 10% protein. Such polysaccharides include a soluble fraction and solid fibers, both convertible to ethanol, whereas proteins are suitable for forage and human food (especially the rubisco-fraction rich in magnesium). Lignin and ash are returned to the soil, furthering humus of enhanced nitrogen assimilation.

The actual share of grass-bearing lands is closer to 25% (37×10^6 km²) of the terrestrial surface because large sections of arable land bear grasses such as cereals, corn, sugarcane, or bamboo. Cereals in temperate zones yield 3 to 4 t/ha of starch crops compared to sugar from cane of 1 to almost 9 t/ha (in Hawaii), depending on soil, age of crop, sunshine, and rain. Carbohydrates stored below the surface of soil furnish even more starch, for example, potatoes about 6 to 7 t/ha starch or sugar beets up to 12 t/ha pure sugar. Obviously, intensive agriculture yields impressively more carbohydrates than extensive forest management in northern latitudes. Wood cellulosics can be upgraded to bioethanol via the Scholler process, yielding 275 l/t wood at 250°C in 0.5% sulfuric acid, still used for large-scale production in Russia (BIOL process).[1]

After the 1973 oil crisis, food-surplus crops in selected areas were converted to biofuel. In Brazil, 7.5% of the land produced cane sugar, yielding ca. $13 \times 10^6 \, m^3$ ethanol per annum, emerging from about 400 factories of capacities between 30 and 650 m^3/day based on energy autarky by burning the bagasse (about 15-fold the mass of obtained alcohol) with 20 to 22% lignin. Almost 12 t of liquid waste per cubic meter ethanol with 10 to 15% DM has to be disposed. In part, it is used as forage for hogs and cattle. Such a volume causes excessive sewage problems, which can be overcome by methanization of bagasse if the necessary peripheral investments are not deterrents.

The intensive efforts in recent decades to keep rivers clean particularly in industrialized areas have led to a well-established technique regarding the anaerobic fouling of organic wastes, thereby converted to methane, also called biogas. Pertinent research has been done by Sahm[8] and Braun.[9]

21.4 METHANE — AN ABUNDANT BIOGAS

The discovery of flammability of marsh gas in Volta's gun (1774) carrying an explosive mixture with air fired by an electric spark (the first machine to create static electricity by induction) did not evoke any attention. Laboratory production of methane (CH_4) was first waged after 1900. Since 1973, wastewater treatment and municipal sewage plants managed to utilize the biogas made on premises to cover their power needs. In the 1980s, farmers discovered that their organic waste yielded CH_4 in volumes sufficient to operate heat and electricity generators, serving the public grid with rewarding revenues when appropriate prices per kilowatt-hour are provided by legislation.

21.5 BIRTH OF THE GREEN BIOREFINERY

Due to the pioneering research of Carlsson[10] and Kiel[11] that advanced the idea of a green biorefinery, the farming business became aware of the size of annual crops of grass or related energy plants (alfalfa, turnips, potatoes, etc.) ranging from 7 to 12 of DM/ha for extensive and intensive land cultivation. Preservation of the crops is warranted by silage, enabling the operation of power stations for 8000 h per annum, allowing revenues ca. 3.0 €/ha. In comparison, forage turnips provide only 1.8 €/ha. Hence, investments can be redeemed within a few years, preferably for larger cooperations of farmers pooling 100 to 250 ha. The combined crop also lowers the cost of the required fermenter from 500 to less then 200 €/m^3 if the fouling installations increase from 200 to 600 m^3. Graf[12] offers a detailed discussion of pertinent aspects of this process. As a rule of thumb, 10.0 m^3 biogas can be expected annually per hectare as demonstrated by the first grass-power plant started in 1995 at Triesdorf, Germany.[13] Moreover, energy grass also furnishes an interesting biochemistry emerging from long-term attention focused on leaf protein concentrates (LPCs). Grass and lucerne (alfalfa) exhibit protein contents from 12 to 18% (alfalfa only) DM, specifically coming from ribulose 1,5 biphosphate-carboxylase/oxygenase (rubisco, E.C. 4.1.1.399), which is easily recovered by pressing fresh grasses to yield a press cake and the so-called green juice suspending about 20% of diverse matter, turning to brown juice after heating to 80°C and extraction of proteins. These serve at present as forage but might be also a human nutrient when former pastures for cattle are restructured as sources of raw materials for green biorefineries. The reduction of cattle will also reduce the emission of CH_4 (which as a greenhouse gas is ca. 30 times more effective than CO_2) to 200×10^6 t/a. About 800×10^6 t CH_4 stems from rice fields and marshes or lakes, besides anthropogenic contributions.

Green/brown juices contain 20 to 40% water-soluble carbohydrates (WSCs) fermentable by lactobacilli to lactic acid. One or two percent suffices for perfect silage of grass bales to maintain green refineries as CH_4 producers in the course of one year. Exhaustive fermentation yields 70 to 120 kg/t dry grass upgradable to aminium dilactide.[14] The finally resulting biodegradable polymer can replace films or foils made so far from polyethylene, polypropylene, or related plastics, and help combat many environmental problems.

21.6 GREEN BIOREFINERIES FOR ETHANOL PRODUCTION

Besides glucose, other sugars of WSCs are also fermetable[15]; for example, galactose, xylose, mannose, fructose, rhamnose, maltose, or sucrose, present either as WSC or as oligomers or polymers in the press cake (PC) as hemicelluloses or cellulose (fibers/lignocelluloses). PC-hemicelluloses are suspendable in dilute NaOH and subsequently depolymerized by weak acids. Such hydrolyzates yield ethanol by fermentation. PC celluloses require a more severe treatment:

1. Hydrolysis in weak acids (0.3% sulfuric acid) at 175°C for 10 h, neutralization by lime, and fermentation. The BIOL-version at 250°C needs only seconds in 0.5% H_2SO_4 in a horizontal double-wall reactor equipped with transporting screws.[1] The remaining lignin is apt to regreen deserts or furnishes an aromatic fraction on pyrolysis, replacing benzene in the "postoil" age. The lignin content of grasses ranges from 6 to 21%, impeding the hydrolysis of fibres. Consequently, sugarcane (bagasse) and bamboo (20–22% lignin) require severe digesting, almost comparable with that of wood (20–30% lignin).
2. Application of thermophile enzymes adapted for the industry to operate between 70 and 80°C to attack specific substrates, for example, pentosanes making up 7–10% of softwood and 17–25% of hardwood.[1] The resulting ethanol distills at 78°C, that is, under enzyme-active conditions (AGROPOL process).

Ethanol has been produced from starchy sources by civilized people for millennia. The world shortage of oil in 1973 provoked its use as a fuel with unexpected success.

The efforts of Brazil deserve special mention, where more than 2 million Volkswagen engines use ethanol as fuel, pure or as additive. The U.S. followed the Brazilian example, replacing cane sugar by cornstarch.

It was noticed in the U.S. that the E-15 fuel with 15% ethanol improved the octane number of petrol and allowed the renunciation of MTBE (tertiary butyl-methyl-ether), which was known to be carcinogenic. At present, annually ca. 6×10^6 t ethanol from cornstarch are provided by several distillers, offered as variable fuel with 85% ethanol and 15% petrol additive. New goals have been set after the attack on the World Trade Center on September 11, 2001. The U.S. administration has asked for 45×10^6 t in 2012, requiring 90×10^6 t of glucose left in the corn crop for pig breeding. Only the biomass of maize plants can possibly afford to extract the said amount of glucose for the 176×10^6 t of corn and straw biomass containing about 60% or ca. 100×10^6 t glucose equivalents in form of (hemi)celluloses apart from 18% lignin and 12 to 15% ash.[16] However, the presently available armamentarium stirs up no hope to convert this source of carbohydrates into ethanol at a price level accepted by the market.

21.7 THE SEARCH FOR SUPERCELLULASES

Modern biotechnology can furnish an answer to the problem of how to obtain an enzyme capable to cut the cost for hydrolysis of cellulose of actually 15 cents /kg ethanol by a factor ten to equalize the prices for ethanol and petrol. The answer can probably be expected before 2012.[16]

Ethanol is a superior fuel because it does not contribute to the atmospheric CO_2 level, is easily stored in tanks, excels the energy of petrol when burnt in fuel cells, furnishes home-made surplus electric power for battery-driven cars, and enables decentralization of power supply for the grid. What remains is actually the problem of burning ethanol in PEM fuel cells at elevated reformer temperatures. Researchers are engaged in finding suitable novel catalysts working below 400 to 500°C comparable to methanol reforming around 250°C.

21.8 HOW TO BOOST ETHANOL SUPPLY?

Focusing on quick returns on investment, a Swiss research group designed a green biorefinery concentrating on grass-based protein, fiber, and ethanol (or biogas, if preferred) by using the simultaneous saccharification and fermentation (SSF) process.[17] Pilot studies per ton dry matter of clover and grass yielded 150 to 300 kg protein concentrate as fodder (for pigs and poultry), 250 kg fiber, and 100 kg ethanol (from 200 kg WSC), apart from ca. 250 kg lignin and ash returned to the soil. With the forthcoming cellulases, another 125 kg ethanol will be available (225 kg of DM/t), further cutting down the period of payback of 6 to 8 years at an input of 5 to 10 t of DM/ha. The first refinery of this capacity went on stream in Schaffhausen, Switzerland, in 2001 as a biogas version.[17]

In 2002, Novozymes, Denmark, presented their Spirizyme® Fuel enzyme,[16] a versatile glucoamylase active between 30 and 70°C to liquefy dry dissolvable granular substances (DDGS) of corn. This technique will enable China to replace 10% of her present fuel consumption of ca. 22×10^6 t per annum by ethanol. As shown in Table 21.1, even an industrialized country such as the FRG can rely on domestic resources to produce ca. 25×10^6 t of ethanol, furnishing fuel cell energy equivalent to 50×10^6 t of naphtha actually spent for traffic and house-heating, presupposing that the administration is capable to finance a start-up of green biorefineries by the dozen. Additionally, green crops from grasslands have to be complemented by novel cooking techniques for whole-tree harvesting, the first attempts of which date back to the 1930s.

21.9 THE ETHANOSOLV PROCESS BY T.N. KLEINERT

A pilot application of the Ethanosolv process started much later in the 1990s in Newcastle, NB, in Eastern Canada, where a pulp producer operated a 15-t/d plant financed by a government grant of over $75 million Canadian. Although the result was refreshing because of satisfying pulp qualities, a saleable lignin powder as extender for plastics, and the hemicellulose fraction for final fermentation to ethanol, a long-term price slump for market pulp rendered the manufacturer bankrupt. Shortly thereafter, Karstens[20] had the idea to replace ethanol by aminoethanol, lowering the digester temperature from 160°C to ca. 100°C. A pilot plant is under construction, hopefully paving the way for potential investors. Market expectations are brilliant: in addition to the produced pulp of sulfate quality, an equal amount of lignin and fermentable hemicelluloses can be sold for at least half the pulp revenue.

The Kleinert–Karstens Ethanosolv process saves the humus-preserving lignin for amelioration of poor soils but requires electric power from the grid or a fuel cell power station on premises. The energy multiplier of Carlo Rubbia[19] based on spallation of ^{232}Th is possibly the final answer to obtain a kilowatt-hour for 2 cents free from CO_2 and plutonium. ^{232}Th is accessible in huge quantities within monazite sand, warranting an energy 100-fold that of the present total global consumption.[18]

21.10 RESHUFFLE OF THE PULP AND PAPER INDUSTRY

Table 21.1 assumes an accretion of 3t dry wood/ha, implying 2.2 t/ha polysaccharides fermentable to 1.1 t of ethanol, if the pulp too could serve that purpose. However, the printing industry calls for paper, for example, in the FRG for ca. 15×10^6 t/ha produced from pulp fibers (70 to 80% including waste paper), fillers (clay, carbonates, etc.), and additives. Paper production would benefit much if made exclusively from virgin cellulose pulp free from lignin. Such fibers would also be shortened after running through several life cycles (including deinking), but still be pure cellulose with fiber lengths over 50% of original but valuing only 70% of the virgin material after recovery of fillers. The fermentation would yield finally FC-ethanol competitive in price to petrol.

The extra profit on pulp should be invested in reforestation, as the forests in the FRG can supply 8 mha $\times$ 1.1 t/ha $\approx$ 9 mt of fuel (Table 21.1) as soon as the scheme shown in Figure 21.1 is in operation for some time. Besides, enough construction

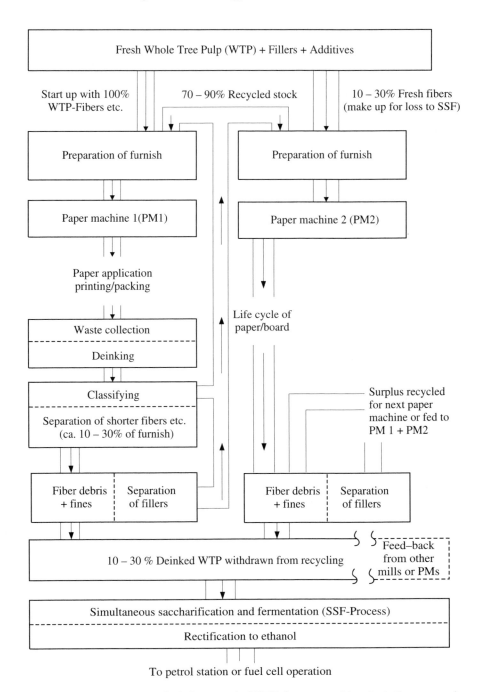

FIGURE 21.1 Application of whole tree pulp (WTP) for papermaking, including conversion to ethanol as a replacement of petrol (or additive) and for fuel cells within the mill.

TABLE 21.1

Glucose Sources for Fermentation to Ethanol in the FRG[a] in 2000

Crop	DM/ha/annum (t)	FRG Acreage (mha)	Obtainable Average Crops[b] (t/ha • /annum)			Attainable Harvest of Products (mt/annum)	
			Protein	Ethanol	Lignin/A sh	Ethanol[c]	Lignin/A sh
Grass (Gramineae)							
Couch/twist		2.4 (Out of 5)					
Rye (*Lolium perenne*)	7–12		1.5	3	3	7.2	2–3
Timothy and others[d]							
Sudan grass[e]	15–20	—	—	5	7	—	—
Alfalfa/lucerne (drought-resistant *Legumenosae*)	9–10	—	1.7	2 (?)	—	—	—
Cereals (Gramineae)							
Wheat	4–6	6.9 (Only 20% for ethanol)	0.6	2	1	2.8	1.5
Corn	8–11	1.5 (Only 20% for ethanol)	—	3	2	0.9	2.0
Straw wheat	4	6.9 (Only 20% for ethanol)	—	1.5	1.4	2.1	1.9
Maize chaff	11–15	1.5 (Only 20% for ethanol)	—	3.5	4–6	2.7	5.0
Tubers/Beets							
Potatoes (o.d.m.)	7	0.3 (Only 50% for ethanol)	—	3.2	0.5	0.5	—
Beets forage	15	0.5 (Only 50% for ethanol)	0.5	5	4.5	1.2	1.2
Beets herbage	6	0.5 (Only 50% for ethanol) for forage	—	—	—	—	—

Wood from Trees

Deciduous (beech)	3.9						
Softwood (spruce)	2.5[f]	8.2	—	1.1	0.8	9 ca. 27 for fossil oil substitutes	6.5 ca. 20 for fertilizer

[a] The FRG considered representative for central Europe (Benelux, Poland, Austria, Slovenia or Baltic States).

[b] When processed in green refineries (2B AG-type).

[c] Assuming 60% saccharides in DM (including water-soluble sugars) and about 30% lignin and ash.

[d] Other varieties *Poa pratensis*, *Aleopecurus sp.*, *Lolium hybridicum*, *Dactylis glomerata*, *Trisetum flavescens* L., etc., altogether 10,000 species ordered into 600 genera including all cereals, sugar cane (perennial) and bamboo(growing up to 80 feet high in two or three months!)

[e] Sudan grass presently emerges as a powerful energy crop; acreage unknown.

[f] Usable wood (biomass much larger); personal communication Patt, R., University of Hamburg, July 24, 2002.

wood would be at disposal after a period of reforestation. Finally, whole-tree utilization combined with a reorganization of the paper cycle would supply abundant ethanol from grasslands, restricting, however, their use as pastures for grazing animals. Entirely new horizons will be opened for lignin,[21] for example, for self-crimping fiber[22] or as an extender for biodegradable polylactides.[14]

21.11 CONCLUDING REMARKS

The grass-covered global surface with about 23×10^8 ha photosynthesizes a gargantuan amount of polysaccharides (PS), proteins, and lignins. If extensively cultivated, probably half of this area would be sufficient to produce enough ethanol to serve as the world's fuel source, exceeding the efficiency of oil by ca. 85% (electric plus thermal) when feeding fuel cells, almost doubling the output of a Carnot cycle. It is also astonishing that temperate regions yield much higher PS harvests than the tropics do. For example, sugar harvest in more south located Cuba yields ca. 2 t/ha whereas that in more north located Bavaria ca. 12 t/ha despite more intense sunshine in the South. However, maximal rainfall in the North concentrates closer to the harvest (20 to 30 in. for agriculture and beyond 30 in. for forests) of grasses and energy crops such as Sudan grass.

Another surprising feature is the amount of DM in crops of forests (3 t/ha · annum) and grasslands (7 to 20 t/ha · annum) caused by interaction of sunlight and chlorophyll but governed by water supply at the right time. Hence, green biorefineries can replace the present oil flow of nearly 3.5 bt/annum by ethanol if the explosive growth of population can be controlled.

A beneficial side effect is the supply of proteins up to 15% of DM. This indispensable food component is lost if fermentation produces biogas instead of ethanol. At any rate, lignin plus ash return to the soil as a humus preserver, increasing nitrogen assimilation fourfold. The practice to burn lignin appears in hindsight an offence against nature.

REFERENCES

1. Osterroth, D., *From Coal to Biomass*, Springer-Verlag, Berlin, 1989, p. 175.
2. Roeper, H., Renewable raw materials in Europe, *Starch/Staerke*, 54, 89, 2002.
3. Grove, W., in *Elektrochemie*, Ostwald, W., Ed., Leipzig, 1896, reprint: H. Deutsch Verlag, Frankfurt/Main.
4. Verne, J., *The Island of Secrets*, 1869. Reprinted as *Die geheimnisvolle Insel*, W. Heuschen-Villaret, Arena Verlag, Würzburg, 1974.
5. Huppmann, G. and Kraus, P., Hot module fuel cells, German Patent DE 4425186 1996.
6. Klemm, P., Discrepancy in discovery and consumption of oil, *Energie-Depesche* 16 (3), 37, 2002.
7. *New Caxton Encyclopedia*, Caxton Publishers, London, 1969, 4995 pp.
8. Sahm, H., Anaerobic degratation of halogenated aromatic compunds, *Microbiol. Ecol.*, 12, 147, 1r986.

9. Braun, R. and Danner, H., Anaerobic digestion for closing material cycles producing energy, in *Proceedings of the 2nd International Symposium on the Green Biorefinery,* Schloss Kornberg Institute, Feldbach, 1999, p. 70.

10. Carlsson, R., Status quo of the utilization of green biomass, in *Die Grüne Bioraffinerie,* Soyez, K. and Kamm, B., Eds., Verlag Gesellschaft für ökologische Technologie, Berlin 1998, p. 39.

11. Kiel, P. and Anderson, M., Agricultural residues as fermentation media, in *Cereals: Novel Uses,* Campbell, G. M., Ed., Plenum Press, New York, 1997, p. 229.

12. Graf, W., *Kraftwerk Wiese* (Book on Demand), 2nd ed., Self-Publisher, Norderstedt, 2001.

13. Krieg, H., Braun, R., and Bugar, A., *Research Project Graskraft of the Landwirtschaftliche Lehranstalt in Triesdorf,* LLA, Triesdorf (Bavaria), 1995.

14. Kamm, B. and Kamm, M., The green biorefinery: principles, technologies and products, in *Proceedings of the 2nd International Symposium on the Green Biorefinery,* Schloss Kornberg Institute, Feldbach, 1999, p. 46.

15. Kromus, S., Narodoslawsky, M., and Krotschek, C., The green biorefinery: utilization of grasslands for renewable raw materials, *Ländl. Raum* 4, 21, 2002.

16. Nedwin, G., Plefebvre, G., and Cherry, J., Search for super-cellulases, *Bio Times,* 9, 7, 2001; 3, 6, 2002.

17. Hansen, G. et al., Biorefining of grass: it's industrial now!, in *Proceedings of the European Congress,* Amsterdam, 2002. (Proceedings paged separately; obtain from Grass, St., 2B AG, Neugutstr. 66, CH-8600 Dübendorf).

18. Dietze, P. and Stadlbauer, E. A., Discussion on climate change, *GIT-Lab.-J.,* 46, 483, 2002.

19. Leuschner, U., Energy-multiplication by spallation, *IZE-Stromthemen,* 7, 4, 1997.

20. Karstens, T., Ergebnisse zur Herstellung von Chemiezellstoff nach einem schwefelfreien Verfahren ohne Verwendung von Natronlauge, in *Proceedings of the 4th International Symposium on Alternative Cellulose,* Rudolstadt, TITK-Proceedings, 2000, p. 1.

21. Liebner, F. et al., *N*-Modified lignin: high grade artificial humus and long-lasting fertilizer, in *Proceedings of the 7th European Workshop on Lignocellulosics and Pulp,* Turku, 2001, p. 475.

22. Meister, F. et al., Stärkemodifizierte ALCERU-Fasern, in *Proceedings of the 5th International Symposium on Alternative Cellulose,* Rudolstad, TITK-Proceedings, 2002, p. 1.

22 Analysis of Molecular Characteristics of Starch Polysaccharides

Anton Huber and Werner Praznik

CONTENTS

22.1 Introduction 349
22.2 Analytical Strategy 353
22.3 Granular Level 355
22.4 Molecular Level 356
    22.4.1 Molecular Dimensions 359
    22.4.2 Molecular Conformation — Branching Analysis 362
    22.4.3 Supermolecular Structures 366
References 368

22.1 INTRODUCTION

Starch is the most important form of energy reserve of crop, is formed annually worldwide in huge amounts, and is of significant commercial importance. The most important application of starch was and is for food and animal feedstock.

Whereas the main source of starch in Europe is potatoes, in the U.S. most starch comes from maize. In tropical countries, the major starch source is cassava and most starch in the Far East is from rice. Quality of starch always was adjusted, historically by breeding techniques, now additionally by gene technology, in particular, to obtain highly short-chain branched (scb) waxy-type glucans.[1-4] Today, waxy-mutants are available for maize, wheat, barley, rice (monocots), and pea and amaranth (dicots) (1) with modified physicochemical and technological (functional) properties for single mutants (waxy, amylose, sugary); (2) new functionalities, however, additionally poor yield and poor germination for double mutants (waxy/amylose, waxy/dull); (3) no modified starch structure/functionality for gene-dose of single mutants (aeae+ or ae++); and (4) novel functionalities and high yields from intermutants (wxwx+/++ae).

0-8493-1486-0/04/$0.00+$1.50

However, modified enzyme patterns in starch-containing crops modify starch glucan composition on a molecular level and result in new or adjusted quality profiles of starch raw materials. This leads to many and mutual influences (Figure 22.1) on the formation process of starch, any efficient technological utilization of these materials requires comprehensive analysis, and initial selection of raw materials and quality control during processing become important factors.

Physicochemical characteristics such as disintegration temperature and swelling behavior of starch granules in aqueous media provide basic information for classification of technological starch properties. Depending on source and harvesting period, dimension of starch granules range between 2 and 140 μm. They swell continuously if suspended in water and even more at elevated temperatures. However, at a certain temperature, the gelatinization temperature, the swelling process becomes irreversible and further increase of temperature disintegrates the granules on a molecular level. The process can readily be monitored with increasing optical transparency and decreasing viscosity of heated starch suspensions.

Temperature–time profile of Brabender viscosity of a wheat glucan–water suspension, for instance, initially shows a broad disintegration window and relatively low increase of viscosity; however, at a final period of decreasing temperatures, a pronounced reorganization capacity is indicated by a significant increase of viscosity. At identical conditions, waxy maize runs through a narrow disintegration phase between 65 and 75°C correlated with initially pronounced increase of viscosity. After disintegration, viscosity decreases, and even in the cooling period, waxy maize glucans keep their status and show no significant reaggregation (Figure 22.2). The consequences of these differences become obvious in technological properties such as freeze–thaw stability: comparably high stability of paste, high viscous waxy

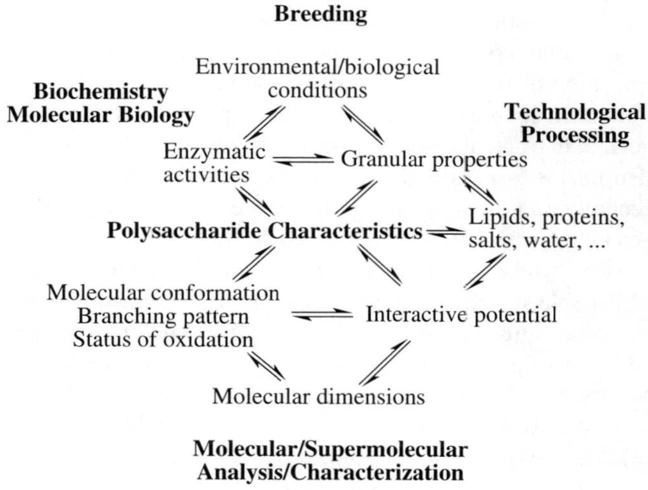

FIGURE 22.1 Controlling influences on starch polysaccharide characteristics.

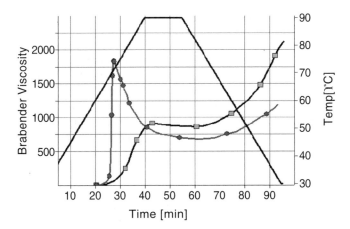

FIGURE 22.2 Stability against applied thermal and mechanical stress. Brabender viscosity for wheat glucans (—■—): broad disintegration window leads to parallel reorganization of supermolecular structures, pronounced reorganization on cooling; and waxy maize glucans (—●—): narrow disintegration window leads to minor reorganization or supermolecular structures, constant shear deformation. Applied temperature program — heating: from 30 to 90°C within 40 min; holding: 90°C for 15 min; cooling: from 90 to 30°C within 40 min.

maize, and quick collapsing of gel-forming wheat starch (Figure 22.3). Stability tests of waxy maize and wheat starch glucans against applied thermal stress provide information about gelatinization temperature, start of disintegration of supermolecular glucan structures, and the temperature window of the disintegration process.

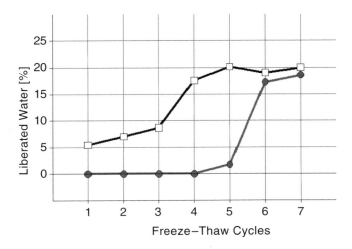

FIGURE 22.3 Freeze–thaw stability (water binding capacity of aqueous suspensions) of waxy maize glucans (—●—) and wheat glucans (—■—).

Obviously, the critical temperature and disintegration temperature range depend on molecular-level characteristics: short-chain branched (scb) waxy maize glucans are intramolecular stabilized and thus stand applied thermal stress better than do wheat starch glucans with pronounced long-chain branched (lcb) contributions (Figure 22.4). Increasing thermal energy disintegrates wheat starch glucans at significantly lower temperatures and a smaller temperature range (56–62°C; T_{max} = 58°C) than for waxy maize glucans (65–85°C; T_{max} = 78°C). Based on these data, it is obvious that for a proper classification, efficient and comprehensive analytical strategies have to be applied to complement the list of technological parameters with molecular-level characteristics.

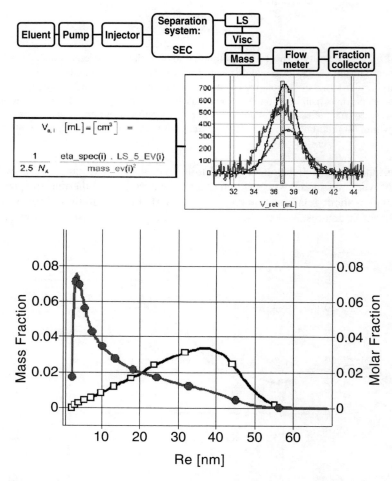

FIGURE 22.4 Stability of starch glucans on applied thermal stress. Viscosity at increasing temperature. Waxy maize (——●——): high initial stability, disintegration peak starting at 65°C. Wheat (——■——): minor initial stability, disintegration peak at 56–62°C.

22.2 ANALYTICAL STRATEGY

Analytical strategies have to be adapted as well to different raw material classes as to the set of parameters to be achieved (Table 22.1). Table 22.2 gives a flowchart of

TABLE 22.1
Analytical Strategy for Starch-Containing Crop Material[a]

Starch-containing crop material tubers/roots (e.g., potato) or seeds (e.g., cereals)		
	1.1 Pretreatment	1.1.1 Smashing and suspending in water 1.1.2 Separation from fibers by sedimentaion 1.1.3 Drying 1.1.4 Storage
	2.1 Homogenized dry material	2.1.2 Spectroscopic identification and analysis of solid body/surface
	3.1 Apolar extraction (defatting)	3.1.1 Dissolution in propanol: *lipids;* elimination of residual moisture
	4.1 Dissolution in polar media	4.1.1 Insoluble residues
5. Technological treatment: suspending ion pure water steam processing	5.1 Dissolved or suspended material in polar medium	5.1.1 Dissolution for analysis: aqueous solution at elevated temp.; alkaline: e.g., NaOH; aprotic polar: dimethylsulfoxide (DMSO)
Analysis and quantification: short-chain branched (scb) long-chain-branched (lcb) nonbranched (nb)			
1. Acidic or enzymatic hydrolysis	1. Physicochemical analysis of technological characteristics	1. Semipreparative fractionation	1. Analytical fractionation: SEC
2. Enzymatic + chemical Glc - analysis	2. Viscoelasticity: rheology Disintegration or reorganization: DSC Thermal resistance: swelling, gelatinization Order and crystallinity: X-ray	2. Fractionation of aqueous soluble glucans according differences in excluded volume	
3. Quantification: Glc + glucans		3. Destructive or nondestructive branching analysis: chain-lenghts distribution	3. Nondestructive molecular analysis: dimension, conformation, interactive properties

[a] Follow the sequence of numbering

TABLE 22.2
Experimental Approaches to Obtain Molecular and Supramolecular
Characteristics of Starch Polysaccharides

Experimental Approach	Information
Thin layer chromatography (TLC) Enzymatic assays Reversed phase high performance liquid chromatography (rpHPLC) High performance anion-exchange chromatography – pulsed amperometric detection (HPAEC-PAD)	Identification and quantification of soluble components, monosaccharides, disaccharides, oligosaccharides obtained from direct extraction or enzymatic and acidic hydrolysis
Enzymatically controlled step-by-step fragmentation + fragment analysis by rpHPLC or size-exclusion chromatography (SEC)	Branching characteristics: chain-lengths distribution, branching percentage
Semipreparative SEC: staining and complexing + spectroscopy	Mass fractions separated according excluded volume for subsequent processing or analysis; fraction-specific branching characteristics;
SEC: mass + standards	Calibrated molecular weight distribution, degree of polymerization distribution, molecular weight averages
SEC: mass, light scattering, and viscosity	Absolute molecular weight distribution, absolute degree of polymerization distribution, excluded volume distribution, polysaccharide-coil packing density distribution
Quantitative terminal labeling of polysaccharides + SEC: mass/molar	Absolute molecular weight distribution of molar fractions
Intrinsic viscosity	Excluded volume, overlapping concentration
Viscosity at varying shear stress	Viscous + elastic contributions (visoelasticity); gel properties
Viscosity at increasing thermal stress	Gelatinization temperature (range); disintegration temperature (range)
Viscosity at varying thermal + constant mechanical stress (Brabender viscograph)	Stability or resistance toward applied energy (temperature, shearing, periods); disintegration or reorganization capacities
Differential scanning calorimetry (DSC)	Modifications in conformation; phase transitions
Photon correlation spectroscopy (PCL)	Diffusion mobility of molecular and supermolecular polysaccharide structures

the appropriate sequences of preparation and analyses for starch-containing crop materials: sampling at the location of harvesting followed by comprehensive and time-consuming fractional preparations and analyses. In case of particular interest in individual parameters or a limited parameter group only, this scheme may be simplified; however, comprehensive information requires the full set.

22.3 GRANULAR LEVEL

Plant-specific and environmental-conditions-induced activities of starch synthases and branching enzymes result in individual distributions of degree of polymerization and branching characteristics for any kind of starch glucans. For storage, the crystallized insoluble form is preferred, and thus formation of starch granules starts. Formation and growth of granules is a complex process and ends up with each granule as an individual object. Nevertheless, glucans with different percentages of proteins, lipids, water and charges, primarily phosphates, are the major components of all types of starch granules. In terms of order, granules typically are 20–40% crystalline, of irregular, however, plant- and variety-specific shape with diameters of 1–120 µm and density of 1.5–1.6 g/cm^3; white to creamy color; and internally organized in layers, having dense layers [120–400 nm formed by ~16 alternating crystalline (5–6 nm) and semicrystalline (2–5 nm) rings][6] and less dense layers with higher content of water.[7] Undoubtedly, water needs to be considered as a fundamental structural feature in the formation of starch granules. Although starch contains no crystal water, H_2O is not just another bulk material. All variations of crystallinity contain more or less water and represent more or less compact order on a dominant amorphous background. Different classes of starch granule crystallinity are typically discriminated by x-ray diffraction patterns: (1) A-type: left-handed, parallel-stranded double helices crystallized in a monocline space group B2; compact packing of glucan-chains and low water content: 12 H_2O molecules with 12 anydroglucose units (AGUs); 6 AGUs per helix turn, 1.04-nm height for each turn; (2) B-type: double helices crystallized in the hexagonal (space group P6; less compact packed, higher water content: 36 H_2O molecules with 12 anhydroglucose units (AGU); 6 AGUs per helix turn, 1.04-nm height for each turn; (3) C-type: a mix of A- and B-type; however, listed as distinct type; and (4) V-type: formed by 6 AGUs in a helical structure with 0.8-nm height per helix turn.

Enzymatically supported fragmentation analysis reveals that A-type starch glucans additionally differ from B-type in their branching pattern, in particular in their ratio of terminal (A-chains) and internal (B-chains) glucan segments.[9–12]

Whereas short-chain branched (scb) glucans (amylopectin) are assumed to form crystalline lamellae by parallel double helices with branching positions in amorphous regions, nonbranched (nb) and long-chain branched (lcb) glucans (amylose) are preferably located in the amorphous layers[13] and are subject for complex formation with lipids. Additionally, limited cocrystallization of scb- and lcb-glucans forming small (~25 nm) and large (80 to 120 nm) blocks was observed by scanning electron microscopy (SEM) and atomic force microscopy (AFM).[14, 25]

[13]C CP/MAS spectra support the idea of amorphous single-chain and ordered double-helix glucans.[16] Thermal stress on B-type results in loss of water and transformation into A-type; swelling of A-type in aqueous media and destruction of crystalline structure yields B-type when recrystallyzing.

Starch granules are composed of a crystalline short-chain branched (scb) glucan (amylopectin) framework and an amorphous long-chain branched (lcb) glucan (amylose) fraction. Glucans of scb- and lcb-type are more or less incompatible: scb-glucans form compact layers of high order, whereas lcb-glucans form amorphous precipitates in less dense domains. Within the granules, the tendency for separation increases with increasing percentage of lcb-glucans in a mixture of both types. No instance of a homogeneous scb-/lcb-glucan blend in a starch granule is known.

In the native environment (plant cell), starch granules are hydrated, and thus in a swollen state, similar to the conditions of industrially hydrated starch granules: in both cases, the granules show increased order (birefringence) and reduced hilum opening. In the swollen state, lcb-glucans diffuse out of amorphous regions, and transfer of ionic compounds into and out of the granules is enabled. Extraction procedures (leaching) and drying result in the formation of compact granules with a cracked hilum.

Based on differences in swelling behavior in aqueous environment, there are reasonable suggestions for preferred localizations of scb- (amylopectin) and lcb-glucans (amylose) within starch granules: (1) Waxy maize starch granules represent an arrangement of more or less 100% scb-glucans (amylopectin) and no lcb-glucans (amylose). The scb-glucans are closely packed in concentric layers in dry waxy maize granules and expand on swelling in aqueous media. These concentric layers of scb-glucans represent the framework for the majority of starch granules. (2) Potato starch granules represent a mix of a major fraction of ~80% scb-glucans (amylopectin) and a minor fraction of ~20% lcb-glucans (amylose). The lcb-glucans are localized in distinct concentric layers alternating with scb-glucan layers. On hydration, these granules swell due to the expanding layers and simultaneously reduce the volume for amorphous lcb-glucans by encapsulation. From such granules, even lcb-glucans (amylose) may be extracted by leaching processes, which similarly reduce the amorphous layer fraction. (3) Amylomaize starch granules are composed of a minor fraction of ~30% of scb-glucans (amylopectin) and a major fraction of ~70% of lcb-glucans (amylose), with a pronounced separation of scb- and lcb-glucans. On hydration, lcb-glucans of such granules are kind of diluted but granules do not swell. From such granules lcb-glucans may be extracted by leaching processes.

22.4 MOLECULAR LEVEL

Investigations on the molecular level in most cases require dissolution of the materials of interest. Although starch polysaccharides are rather polar, due to pronounced H-bond-binding they can hardly be dissolved in neutral aqueous media. For most technical purposes, they get dissolved or suspended in H-bond-"breaker" alkaline media; however, for analytical purposes dissolution in aprotic polar medium such as dimethylsulfoxide (DMSO) is preferable as problems with ionic strength variations can be avoided. A number of chemical, enzymatically supported, and physical techniques are available to investigate molecular-level characteristics of starch glucans. However, due to the multiple and superimposed heterogeneities of these materials, the most important are fractionation and separation techniques, in particular precipitation–complexation and chromatography (Table 22.3). As the major quality of starch polysaccharides, like any polysaccharides, is to fill up volume in a more or less structurized way, any analytical approach has to qualify and quantify this performance. Size-exclusion chromatography (SEC) is a perfect tool to obtain such

TABLE 22.3
Branching Characteristics of a Wheat Starch Polysaccharide from Destructive Step-by-Step Fragmentation and Subsequent Fragment Analysis

	%	DP_n	Seg	Glc_n	Full turns	Br %
		Waxy Maize				
Molecule		940		960		5.3
C-chains	5	50	1	50	1×2–7	
B-chains	23	50	4	200	4×7	
A-chains	72	15	46	690	58×2	
A- + B-chains	95					
		Wheat: Initially n-BuOH Precipitated Fraction (22%)				
Molecule		255		255		3.6
C-chains	46	112	1	112	1×2–15	
B-chains	14	39	1	39	1×6	
A-chains	40	13	8	104	8×2	
A- + B-chains	54					
		Wheat: Subsequent MeOH Precipitation from Supernatant (78%)				
Molecule		966		966		6.3
C-chains	5	50	1	50	1×2-7	
B-chains	25	40	6	240	40×7	
A-chains	70	13	52	676	52×2	
A- + B-chains	95					

Note: %: percentage; DP_n: number average degree of polymerization; Glc_n: number of anhydro glucose units; seg: number of segments (branches); full turns: number of helical full turns for complexation; br %: percent branching.

kind of information as the SEC concept, in particular SEC combined with multiple detection, provides comprehensive information about molecular and supermolecular dimensions, molecular conformations, and interactive potentials of separated polymers and oligomers.

Simplified, the filling of volume (V_e; Equation 22.1a) by starch polysaccharides is controlled by a fine-tuning mechanism at the molecular level, which modifies polymer/oligomer domains due to changing external and internal conditions. At least three major controlling classes of how volume (V_e) is structurized can be distinguished:

- *Conformation* (mc; Equation 22.1a): Molecular symmetries in terms of helices, branching pattern (short-chain, long-chain branches, number of branching points), cross-links, oxidation status, compatibility structures, and packing density.

- *Dimension* (md; Equation 22.1a): Molecular weight, degree of polymerization, excluded volume, and transition states between geometric molecular dimensions and lengths of coherent supermolecular structures.
- *Interactive properties* (ip; Equation 22.1a): Water content, aggregation/association, supermolecular dimensions (gel-formation), viscoelasticity, and stress management:

$$V_e = ip \cdot md^{mc} \tag{22.1a}$$

where V_e is the excluded volume, ip the interactive potential, md the molecular dimension, and mc the molecular conformation.

Although variations of md, mc, and ip already provide countless structural options, diversity is even more increased by distributions of these features. In particular, starch glucans are a superimposed heterogeneous mix of regular and irregular modules of:

- highly symmetrical helices (multiple helices), primarily by lcb-glucans
- irregular "fractal structures," primarily by scb-glucans
- compact and internally H-bond stabilized structures, dominantly by scb-glucans (crystallinity)
- less compact "amorphous" domains with pronounced reorganization capability, dominantly by lcb-glucans

which easily may be customized either with respect to mass (Equation 22.1b) or molar (Equation 22.1c) fractions for the major native purpose of guaranteeing optimum energy management:

$$m\_V_eD = ipD \cdot m\_mdD^{mcD} \tag{22.1b}$$

where m_V_eD is mass fraction of excluded volumes distribution; ipD the distribution of interactive potentials, m_mdD the mass fraction of molecular dimensions distribution, and mcD the distribution of molecular conformation.

$$n\_V_eD = ipD \cdot n\_mdD^{mcD} \tag{22.1c}$$

where n_V_eD is molar fraction of excluded volumes distribution, ipD the distribution of interactive potentials, n_mdD the molar fraction of molecular dimensions distribution, and mcD the distribution of molecular conformation.

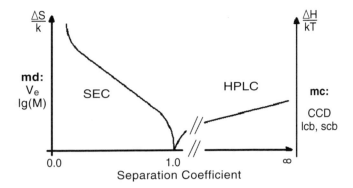

FIGURE 22.5 Separation criteria in liquid chromatography (LC). Entropy-controlled separation ($\Delta S/k$) according differences between individual components in excluded volume (V_e) by size-exclusion chromatography (SEC). Enthalpy-controlled high performance liquid chromatography separation ($\Delta H/kT$) according to differences of individual components in interaction potential with liquid chromatography matrix. S: entropy, H: enthalpy, T: temperature, k: Boltzmann constant, mc: molecular conformation, CCD: chemical composition distribution, lcb: long-chain branched, scb: short-chain branched; md: molecular dimension; V_e: excluded volume; $\lg(M)$: logarithm of molecular weight.

22.4.1 MOLECULAR DIMENSIONS

SEC is categorized in the entropy-controlled section of liquid chromatography (LC) techniques (Figure 22.5). As the separation criterion in SEC is occupied or excluded volume (V_e), application of SEC on starch glucans provides information about excluded volume heterogeneity ($V_e D$: excluded volume distribution). If SEC-separated fractions are monitored from different but complementary points of view, such as refractive index (universal mass) and light scattering and viscosity (Figure 22.6), any V_e fraction can then be qualified in detail on the contributions of md, mc, and ip to V_e, and parameters become available for (1) molecular weight distribution, (2) branching characteristics, and (3) supermolecular dimensions and coherent segment dimensions.

SEC with triple detection of mass, light scattering (LS), viscosity (Figure 22.6) provides for each fraction information about V_e (by the SEC separation criterion; Figure 22.5); absolute molecular weight (by mass-to-scattering ratio; Equation 22.2a); and Staudinger–Mark–Howink coefficient K and exponent a (by mass-to-specific viscosity ratio including molecular weight information; Equation 22.2b and Equation 22.2c). Combining the fraction data, several characteristic distributions become available from SEC mass, LS, and viscosity experiments: absolute molecular weight calibration (*raw_MWV*; Equation 22.2d), intrinsic viscosity calibration (*eta_int*; Equation 22.2e), and universal calibration (*raw_univ*; Equation 22.2h and Equation 22.2i) for the SEC system used:

$$M_i \left[\frac{g}{mol} \right] \propto \frac{mass\_ev(i)}{LS\_EV(i)} \tag{22.2a}$$

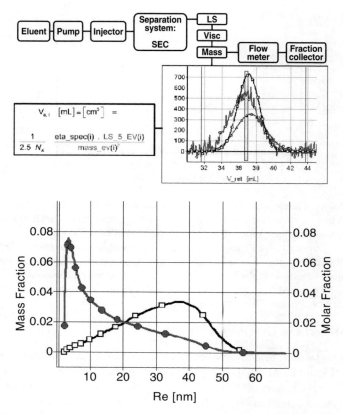

FIGURE 22.6 Determination of excluded volume distribution (m_VeD) of starch polysaccharides by SEC mass, light scattering, and viscosity triple detection, mass by refractive index DRI (—△—), scattering intensity as Rayleigh factor at $\Theta = 5°$ (—o—), specific viscosity via differential pressure (*ip*) and inlet pressure (*dp*) (—☐—). Normalized (area = 1.0) sphere equivalent radii distribution (R_eD) of SEC-separated wheat glucans, differential mass fraction (—■—), differential molar fractions (—●—).

$$[\eta]_i \left[\frac{mL}{g} \right] \quad \propto \quad \frac{mass\_ev(i)}{eta\_spec(i)} \tag{22.2b}$$

$$[\eta]_i \quad = \quad K \cdot M_i^a \tag{22.2c}$$

$$[M]_i \rightarrow lg([M]_i) \;^{i=int\_start...int\_stop} raw\_MWV: lg(M_i) \text{ vs } V_{ret} \tag{22.2d}$$

$$[\eta]_i \rightarrow lg([\eta]_i) \;^{i=int\_start...int\_stop} eta\_int \rightarrow eta\_int|lg: \; lg([\eta]_i) \text{ vs } V_{ret} \tag{22.2e}$$

$$V_e = \frac{[?] \cdot M}{2.5\ N_A} \qquad (22.2f)$$

$$V_{e,\,i}\ [mL] \equiv [cm^3] = \frac{1}{2.5\ N_A}\ \frac{eta\_spec(i) \cdot LS\_5\_EV(i)}{mass\_ev(i)^2} \qquad (22.2g)$$

$$raw\_univ(i) = lg(V_{e,\,i}) = eta\_int|lg(i) \cdot raw\_MWV(i) \qquad (22.2h)$$

$$fit\_univ \leftarrow \left(y_{fit\_univ}\right)_i = \sum_{n=0}^{1} a_n \cdot \left(x^n_{raw\_univ}\right)_i \qquad (22.2i)$$

where M (g/mol) is the molecular weight, $[\eta]$ (ml/g) the intrinisic viscosity, V_{ret} (ml) the SEC retention volume, N_A (l/mol) the Avogadro number, V_e, (cm³) the excluded volume (occupied sphere equivalent volume), *raw_MWV* the SEC elution profile of detected differential pressure, *eta_spec* the SEC profile of specific viscosity, *eta_int* the SEC profile of intrinsic viscosity, *eta_int|lg* the SEC profile of logarithmic intrinsic viscosity, *LS_5_EV* the SEC profile of scattering intensity at $\Theta = 5°$, *mass_ev* the SEC profile of mass fractions, *raw_univ* the SEC profile of logarithmic V_e distribution, *fit_univ* the SEC profile of (linear) fit to *raw_univ*, and *int_start, int_stop* are limits of selective SEC separation range (integration limits).

Such experimental stations are typically managed electronically by operator-customized software packages such as ASTRA (Wyatt Technology, CA), OmniSEC™ (Viscotek) CODAwin32 and CPCwin32 (a.h group, Graz, Austria) for data acquisition, (remote) control of connected instruments, and subsequent modular data processing. Universal calibration obtained from SEC mass, LS, and viscosity experiments then provides distribution of sphere equivalent radii of occupied volume in a twofold form (Figure 22.6): distribution of mass fractions and distribution of molar fractions, both ranging between 2 and 55 nm, however, of very different shape and significantly different maxima. Molecular weight distribution can be obtained in several ways and, depending on the procedure, yields quite different results. Figure 22.7(a) illustrates calibration functions obtained one time from light scattering and another time from universal calibration diminished for viscosity contributions. Whereas light scattering suggests ca. $17 \cdot 10^6$ g/mol, the universal calibration approach yields ca. $5.5 \cdot 10^6$ g/mol for wheat. Additionally, light scattering indicates a rather uniform sample, whereas universal calibration results a polydispersity of ca. 7. Similar results will be found for waxy maize and other starch samples. Figure 22.7(b) shows the molecular weight distributions obtained for waxy maize and wheat from viscosity-diminished universal calibration. Obviously, more information is required to check which approach gives true, or at least plausible, molecular weights. Details on molecular conformation, in particular on branching patterns should provide information about preferred interaction potentials.

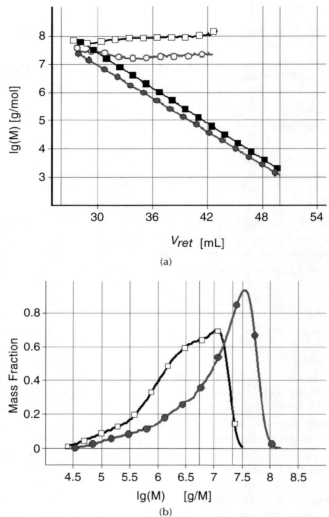

(a)

(b)

FIGURE 22.7 (a) Molecular weight calibration for SEC-separated starch glucans: absolute molecular weight calibration from mass/LS data (Equation 22.2a) for waxy maize (—■—) and wheat (—O—); universal calibration diminished for viscosity contributions (Equation 22.2i and Equation 22.2e) for waxy maize (—■—) and wheat (—●—). (b) Molecular weight distributions (differential mass fraction profiles) of SEC-separted starch glucans: wheat (—■—); weight average molecular weight, $M_w = 5.5 \cdot 10^6$ g/mol; number average molecular weight, $M_n = 0.78 \cdot 10^6$ g/mol; waxy maize (—●—) weight average molecular weight, $M_w = 21.5 \cdot 10^6$ g/mol; number average molecular weight $M_n = 1.95 \cdot 10^6$ g/mol.

22.4.2 MOLECULAR CONFORMATION — BRANCHING ANALYSIS

The influence of variations of molecular conformation, such as type of branching pattern, tendency to form supermolecular structures, and how volume is occupied by these materials, of starch polysaccharides on material quality is often underestimated.[17] Most probably, variation of branching pattern is the key tool by which adjustments are achieved at native conditions. Intensity of branching (branching percentage and location of branching positions) and the ratio of lcb- to scb-glucans is the dominant control parameter for macroscopic starch qualities.[18,19] nb/lcb-Glucans are less or poorly soluble in aqueous media, highly tend to retrograde, and easily gelatinize and form gels and films. scb-Glucans, on the other hand, show much better solubility in aqueous media, are capable of fixing a high amount of water, and are less sensitive toward varying environmental conditions.[20-25] Consequently, characterization of starch qualities includes comprehensive analysis of branching characteristics: (1) kind of branching pattern; composition of branching pattern for starch glucan fractions sampled with respect to identical excluded volume, identical molecular weight (degree of polymerization), identical internal stabilization, and identical potential for formation of supermolecular characteristics; (2) number and percentage of branching position within individual glucan molecules and kind of distribution of number of branching positions in supermolecular domains; (3) heterogeneity or homogeneity of branching positions within individual glucan molecules and distinct supermolecular domains; (4) degree of local symmetry (crystallinity) due to certain kind of branching characteristics, and influence of increased branching to symmetry and interactive properties.

Experimental approaches to determine branching characteristics are rather laborious and quite often are estimations rather than quantitative data. Basically, branching analysis can be achieved by destructive or nondestructive techniques. Destructive techniques include pure chemical-directed or enzymatically catalyzed step-by-step fragmentation, reversed-phase HPLC analysis, or quantitative derivatization and subsequent fragment analysis, for instance, by GC-MS, in both cases followed by recalculation of mean molecules as a puzzle from fragment data. Complexing and staining of starch glucans (native glucans, glucan fractions, glucan fragments) with polyiodide anions in hydrophobic caves of terminal helical starch glucan branches, and spectroscopy in terms of extinction ratio E_{640}/E_{525} provides relative information about lcb and scb characteristics of investigated samples. Application to fractions from semipreparative SEC [Figure 22.8(a) and Figure 22.8(b)] yields profiles of lcb-to-scb ratio with respect to decreasing excluded volume.

Experimentally, 125 mg freshly sublimated iodine is dissolved in the presence of 400 mg kI in 1000 ml demineralized water and diluted 1:1 with 0.1 M acidic acid to a final pH of 4.5 to 5.0 when mixed with the alkaline eluate from SEC. Polyiodide anions complexed in the helical starch glucan segments shift extinction maximum from E_{max} at 525 nm of free aqueous iodine to higher wavelengths.[26,27] Actually, the shift is controlled by both the length of helical segments and the number of available helical segments; however, correlation of E_{640} values are an appropriate indication for lcb-glucans. Correlation of scb-glucans with E_{525} values is supported by corresponding maxima in ORD/CD spectra for $\alpha(1\rightarrow4)$-glucans with DP < 40.[28,29] The ratio of E_{640} to E_{525} finally indicates branching characteristics of complexed glucans, glucan fractions, or glucan fragments as ratio of lcb- to scb-glucans.

Results for semipreparative SEC separation of a wheat and a waxy maize sample, subsequent complexing and staining of obtained fractions with polyiodide anions, and monitoring of extinction ratio of complexed (E_{640}) and noncomplexed (E_{525})

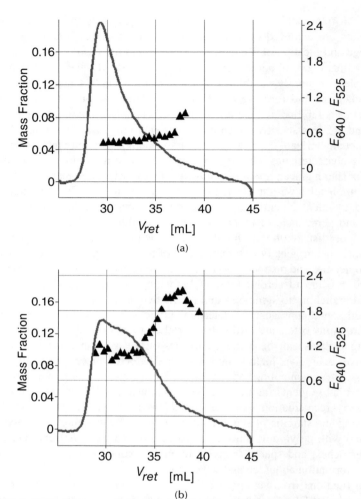

FIGURE 22.8 SEC elution profiles of (a) waxy maize starch glucans and (b) wheat starch glucans with E_{640}/E_{525} (lcb/scb) ratio. ▲ indicates kind and homogeneity of branching pattern. SEC-system: TosoHaas guard PWH + GMPWM + GMPW6000 + 5000 + 4000 + 3000 (150 mm × 7.5 mm), eluent: 0.005 M NaOH; sample volume: 0.4 ml (5 mg/ml), flow rate: 0.80 ml/min.

starch glucans are shown in Figure 22.8(a) (waxy maize) and Figure 22.8(b) (wheat). Whereas waxy maize gives rather uniform E_{640}/E_{525} values in the range of 0.5, obviously indicating scb characteristics, wheat glucans represent a mix of high-molecular-weight scb-glucans with E_{640}/E_{525} values of 0.9–1.0 and a midrange molecular weight lcb-population with E_{640}/E_{525} values of 1.5–2.0.

Table 22.4 lists the results of destructive branching analysis for waxy maize and wheat. The data are in good agreement with results from nondestructive analysis: highly short-chain branched waxy maize with mean DP = 15 for the terminal A-chains, which represent 72% of the molar mass of a mean molecule with DP = 940.

TABLE 22.4
Technological and Molecular-Level Characteristics of Starch Glucans

Characteristics	Wheat	Waxy Maize
Source and specifications	Chamtor, France; moisture: 9.8%; glucan content: 98.3%	Agrana, Austria; moisture: 11.4%; glucan content: 97.4%
Reorganization capacity after disintegration	+++	+
Disintegration of 5% paste at 95°C: η (mPa·s)	107	340
Shear stress stability	High	Medium
Acid resistance	None	None
Status of starch suspensions after first freeze–thaw cycle	Soft gel	Pasteous, high-viscous
Freeze–thaw stability	None	Medium
Type of x-ray diffraction pattern	A	A
Branching Characteristics		
General classification	scb and lcb	scb
E_{640}/E_{525} high V_e range	1.0–1.2	0.4–0.55
E_{640}/E_{525} low V_e range	1.8–2.1	0.6–1.0
scb-Fraction (mass %)	78	100
lcb-Fraction (mass %)	22	—
lcb-Fraction Fragmentation Analysis		
Primary C-chains (DP/%)	112/46	—/—
Secondary B-chains (DP/%)	39/14	—/—
Terminal A-chains (DP/%)	13/40	—/—
scb-Fraction Fragmentation Analysis		
Primary C-chains (DP/%)	50/5	50/5
Secondary B-chains (DP/%)	40/25	50/23
Terminal A-chains (DP/%)	13/70	15/72
Excluded Sphere Equivalent Radii $V_e \rightarrow R_e$ from SEC Separation		
Range (nm)	2–55	2–55
Maximum of molar fractions distribution (nm)	5	5
Maximum of mass fractions distribution (nm)	38	38
Molecular Weight: SEC Mass + Univ. Cal - Structure Viscosity		
Range (g/mol)	30,000–30,000,000	30,000–100,000,000
Weight average molecular weight (M_w) (g/mol)	5,500,000	21,000,000
Number average molecular weight (M_n) (g/mol)	780,000	1,950,000
Molecular Weight: Apparent Absolute from SEC Mass/LS		
Range (g/mol)	$20–40 \cdot 10^6$	$80–100 \cdot 10^6$
Disintegration on thermal stress	56–62°C	65–80°C
Reorganization capacity after disintegration	+++	+
Populations in Terms of Coherent Mobility		
Molecular dissolved glucans l_{coh} (nm)	10–30; max = 18	10–45; max = 25
Glucan aggregates l_{coh} (nm)	10–800; max = 250	10–800; max = 370

Wheat consists of an lcb population (ca. 22%), which may be precipitate with *n*-butanol from aqueous solution and contains significantly less A-chains, and an scb population (ca. 78%), which is very similar to that of waxy maize starch.

22.4.3 SUPERMOLECULAR STRUCTURES

As no obvious differences of molecular dimensions in terms of occupied volume ($V_e \rightarrow R_e$) will be found for different starch glucans such as wheat and waxy maize, and molecular weight at least appears somewhat ambiguous, strongly depending on applied approach, branching pattern characteristics are strongly suspected to control macroscopic material qualities by promoting formation of more or less stable supramolecular structures. Investigation of segment mobility in starch glucan solutions by noninvasive techniques such as photon correlation spectroscopy is a perfect tool to check such supramolecular structures. Dynamic light scattering (DLS) and photon correlation spectroscopy (PCS) experiments provide data on diffusive mobility of individual glucan molecules and mobility of supramolecular segments, in particular, on translational diffusion of glucans and glucan aggregates. If laser light is applied to starch glucan solutions, mobility of the segments in such solutions cause Doppler shifts, which may be monitored as the autocorrelation function, $G_2(\tau)$, (Equation 22.3a):

$$G_2(?) = \frac{1}{T} \int_0^? I(t) \cdot I(t + ?) dt \qquad (22.3a)$$

where T is the temperature (K), t the time, I the intensity of scattered laser light, and τ the correlation period.

Indirect Laplace transformation of $G_2(\tau)$ yields $G_2(t)$, which contains the translational diffusion coefficient, D_T,[44] (Equation 22.3b):

$$G_2(t) = A \cdot \left[1 + C_i \left(\int_{\tau_{min}}^{\tau_{max}} \frac{D_T(?)}{\tau^2} \cdot e^{\frac{-t}{\tau}} d\tau \right)^2 \right] \quad \text{with} \quad \tau = \frac{1}{D_T h^2} \qquad (22.3b)$$

where h is the scattering vector, and A and C are coefficients.

According to Stokes and Einstein (Equation 22.3c), D_T of observed glucans and glucan aggregates can be correlated with radius, R_H, or diameter, d, of moving equivalent spheres. As in particular supramolecular structures are dynamically formed by individual glucan molecules, d is rather the length of coherently moving segments, and thus, for practical purposes, coherence length, l_{coh}, of supramolecular segments replaces the equivalent sphere diameter d:

$$D_T = \frac{k_B T}{6\pi \eta R_H} \rightarrow l_{coh} \qquad (22.3c)$$

where k_B is the Boltzmann constant, T the temperature (K), and η the viscosity of solution.

Monitoring of mobilities within starch glucan solutions and transforming the obtained results into dimensions of moving segments results in the situation illustrated in Figures 22.9(a) and Figure 22.9(b) for wheat, and, when simplified, shows two populations: (1) a major mass fraction [Figure 22.9(a)] with dimensions (l_{coh}) of 10–30 nm, representing individual dissolved starch glucan molecules; and (2) a minor, but not negligible, fraction [Figure 22.9(b)] with dimensions (l_{coh}) of 100–800 nm representing glucan aggregates.

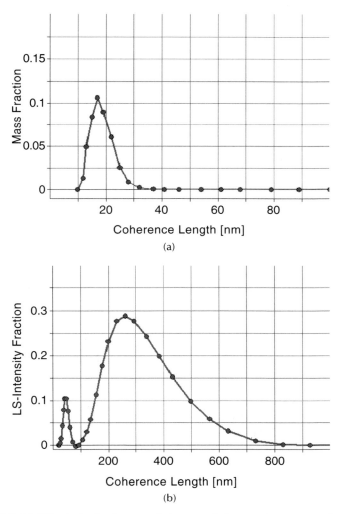

FIGURE 22.9 (a) Wheat starch glucans: photon correlation spectroscopy analysis of mass fractions of dissolved coherently moving (D_T) glucans transformed into coherence lengths (l_{coh}). (b) Wheat starch glucans: photon correlation spectroscopy analysis of intensity fractions (equivalent to information about molecular weight obtained from light scattering) of dissolved coherently moving (D_T) glucans transformed into coherence lengths (l_{coh}).

Segment dimensions approximately a magnitude higher than individual molecule dimensions can be observed for any starch glucan in aqueous environment even after days of dissolving. However, there are variations in quantity, stability against applied stress, and reorganization capacity for starches from different sources. There is good reason to assume variations in branching characteristics as the driving force for varying capabilities to form more or less stable supramolecular structures, which finally control macroscopic starch material properties.

REFERENCES

1. Weatherwax, P., A rare carbohydrate in waxy maize, *Genetics*, 7, 568, 1922.
2. Nelson, O.E. and Rines, H.W., The enzymatic deficiency in the waxy mutant of maize, *Biochem. Biophys. Res. Commun.,* 9, 297, 1962.
3. Boyer, C.D. and Hannah, L.C., Kernel mutants of corn, in S*peciality Corns,* Hallauer, A., Ed., CRC Press, Boca Raton, 1994, p. 2.
4. Obanni, M. and. BeMiller, J.N., Identification of starch from various maize endosperm mutants via ghost structures, *Cereal Chem.,* 72, 436, 1995.
5. Zobel, H.F., Molecules to granules: a comprehensive review, *Starch/Stärke*, 40, 1, 1988.
6. Cameron, R.E. and Donald, A.M., A small-angle X-ray scattering study of the annealing and gelatinization of starch, *Polymer*, 33, 2628, 1992.
7. French, D., Organisation of starch granules, in *Starch Chemistry and Technology*, 2nd ed., Whistler, R.L., BeMiller, J.N., and Paschal, E.F., Eds., Academic Press, London, 1984.
8. Perez S. and Imberty A., Structural features of starch, *Carbohydr. Eur.*, 15, 17, 1996.
9. Hizukuri, S., Kaneko, T., and Takeda, Y., Measurement of the chain length of amylopectin and its relevance to the origin of crystalline polymorphism of starch granules, *Biochim. Biophys. Acta*, 760, 188, 1983.
10. Hizukuri, S., Polymodal distribution of the chain length of amylopectin and its significance, *Carbohydr. Res.*, 147, 342, 1986.
11. Fredriksson, H. et al., The influence of amylose and amylopectin characteristics on gelatinization and retrogradation properties of different starches, *Carbohydr. Polym.*, 35, 119, 1998.
12. Gidley, M.J. and Bulpin, P.V., Crystallisation of malto-oligosaccharides as models of the crystalline forms of starch: minimum chain-length requirement for the formation of double helices, *Carbohydr. Res.*, 161, 291, 1987.
13. Jenkins, P.J. et al., In situ simultaneous small and wide-angle X-ray scattering: a new technique to study starch gelatinization, *J. Polym. Sci., Polym. Phys. Ed.*, 32, 1579, 1994.
14. Gallant, D.J., Bouchet, B., and Baldwin, P.M., Microscopy of starch: evidence of a new level of granule organization, *Carbohydr. Polym.*, 32, 177, 1997.
15. Baldwin, P.M. et al., Surface imaging of thermally-sensitive particulate and fibrous materials with the atomic force microscope: a novel sample preparation method, *J. Microscopy,* 107, 7040, 1985.
16. Gidley, M.J. and Bociek, S.M., Molecular organization in starches: a ^{13}C CP/MAS NMR study, *J. Am. Chem. Soc.*, 107(7), 7040, 1985.

17. Huber, A. and Praznik, W., Molecular characteristics of glucans: high amylose corn starch, *ACS Symposium Series No. 635: Strategies in Size-Exclusion Chromatography,* 19, 351, 1996.
18. Praznik W., et al., Molecular characteristics of high amylose starches, *Starch/Stärke* 46, 82, 1994.
19. Praznik, W. and Huber, A., Modification of branching pattern of potato maltodextrin with Q-enzyme *Zywnosc. Technologia. Jakosc. (Food Technology Quality)*, 4, 202, 1998.
20. Pfannemüller, B., Stärke, in *Polysaccharide*, Burchard, W., Ed., Springer-Verlag, Berlin, 1985, p. 25.
21. Leloup, V.M. et al., Microstructure of amylose gels, *Carbohydr. Polym.*, 18, 189, 1992.
22. Yun, S.H. and Matheson, K., Structural changes during development in the amylose and anylopectin fractions (separated by precipitation with concovaline-A) of starches from maize genotypes, *Carbohydr. Res.*, 270, 85, 1992.
23. Prakunz, W. et al., Characterization of the (l-h)-α-D-gluconbranding 6-glycosyltranferose by *in vitro* synthesis of branched starch polysaccharides, *Carbohydr. Res.*, 227, 71, 1992.
24. Schoch, T.Y., The fractionation of starch, *Adv. Carbohydr. Chem.*, 1, 247, 1995.
25. Huber, A. and Provenik, W., Dimensions and structured features of aqueous dissolved polymers, in *Carbodydrates as Organic Raw Materials,* IV, Provenik, W. and Huber, A., Eds., WUV-Universitäts Wien, 1998, Ch. 19.
26. Banks, W., Greenwood, C.T., and Khan, K.M., *Carbohydr.Res.*, 17: 25, 1971.
27. Handa, T. et al., Deep blueing mechanism of triiodide ions in amylose being associated with its conformation, *ACS Symp Ser.*, 150, 455, 1981.
28. Pfannemüller, B. and Ziegast, G., Properties of aqueous amylose and amylose-iodine solutions, *ACS Symp.Ser.*, 150, 529, 1981.
29. Bhide, S.V., Karve, M.S., and .Kale, N.R., The interaction of sodium dodecyl sulfate, a competing ligand, with iodine complexes of amylose and amylopectin, *ACS Symp. Ser.*, 150, 491, 1981.
30 Hofer, M., Basic concepts in static and dynamic light scattering: application to colloids and polymers, in *Neutron, X-ray and Light Scattering*, Lindner, P. and Zemb, Th., Eds., North-Holland, Amsterdam, 1991.

23 Spectroscopy of Polysaccharides

Jeroen J.G. van Soest

CONTENTS

23.1 Introduction 371
23.2 Vibrational (IR and Raman) Spectroscopy 372
    23.2.1 Mid-Infrared Spectroscopy (MIR) 372
        23.2.1.1 Cellulose 372
        23.2.1.2 Starch 372
        23.2.1.3 Other Polysaccharides 373
        23.2.1.4 Complex Multicomponent Systems 374
    23.2.2 Near Infrared Spectroscopy (NIR) 374
    23.2.3 Raman and Nonconventional Vibrational Spectroscopy 375
23.3 NMR Spectroscopy 375
    23.3.1 Solution State and High-Resolution NMR 376
    23.3.2 Solid-State NMR 377
    23.3.3 Some Specific Polysaccharide Examples 377
        23.3.3.1 Cellulose 377
        23.3.3.2 Starch 377
        23.3.3.3 Other Polysaccharides 378
23.4 Dielectric Relaxation Spectroscopy (DRS) 378
23.5 Other Miscellaneous Techniques 380
    23.5.1 Mass Spectrometry (MS) 380
    23.5.2 Illustrative Examples of Other Techniques 380
References 381

23.1 INTRODUCTION

Structural characterization of carbohydrates, or particularly polysaccharides, is of paramount importance for the development of carbohydrate products. Development of new analysis techniques is a continuous process. In the area of carbohydrate research and product development, some techniques predominate. Spectroscopic methods are widely used.

Among the most commonly used spectroscopy techniques are nuclear magnetic resonance (NMR), vibrational spectroscopy, circular dichroism (CD), mass spectrometry (MS), electron spin resonance (ESR), x-ray photon-electron spectroscopy (XPS), and dielectric relaxation spectroscopy (DRS). This chapter describes some of the recent interesting new developments on polysaccharides or polysaccharide-based materials. The chapter mainly focuses on vibrational, NMR, and DRS because these are commonly used for process–structure–function research, in addition to structural analysis.

23.2 VIBRATIONAL (IR AND RAMAN) SPECTROSCOPY

Fourier-transform (FT) vibrational spectroscopy is an important rapid and sensitive research tool frequently used to elucidate the detailed structure and physical properties of polymers, such as polysaccharides, and their interactions with other components. In the mid-infrared (MID) region, the absorption of radiation in the wavelength range from 2.5 to 50 μm (4000 to 200 cm^{-1}) is measured. The near-infrared (NIR) region ranges from 780 to 2500 nm. The absorption involves transitions between vibrational energy molecular states and rotational substates. Infrared (IR) and Raman provide similar information on molecular vibrations, and many of the same bands are seen. Because of the different selection rules, the spectra are considered to be complementary. IR requires a change in dipole moment during the vibration and Raman a change in polarizability. The spectrum of a polysaccharide is unique and referred to as a molecular fingerprint. Water gives only a weak line at 1640 cm^{-1} in the Raman spectrum, making Raman ideal for the study of aqueous systems. The intensity of the absorption is proportional to the number of molecules, making quantitative analysis possible. Updates were written in the 1980s and 1990s.[1–3] The polysaccharides most intensively studied by using vibrational spectroscopy are cellulose and starch. Among the others are xylan, pectin, chitin, chitosan, alginates, and galactans.

23.2.1 MID-INFRARED SPECTROSCOPY (MIR)

23.2.1.1 Cellulose

Cellulose is a β-(1→4)-linked glucan found in plants, algae, and bacteria. MIR studies handle the identification of celluloses, determination of the degree of ordering or crystallinity, and influence of source (cotton, linen, wood, ramie, algal, bacterial) and processing (pulping, regeneration) on the crystallinity of celluloses and cellulose fibers.[4,5] Spectra are sensitive to differences in Type I (natural), II, and III celluloses. Cellulose is one of the first polysaccharides on which normal coordinate analysis was done. By using MIR, polarized radiation provided information on the direction of transition moments of functional groups with respect to the fiber axis.[4–7]

23.2.1.2 Starch

Starch is the second common glucan consisting of amylose [α-(1→4)-linked] and amylopectin [with additional α-(1→6) branch points]. By using chemometric methods, such as multivariate data, discriminant and principal component analysis, artificial neural network, and principal variables, it was shown that modified starches can be qualitatively and quantitatively differentiated.[7,8] Starch gelatinization or melting and the subsequent retrogradation processes (gelation, syneresis, helix formation, crystallization) are of great importance to the quality of products.

Studies that used sample techniques such as diffuse reflectance (DRIFT) and attenuated total reflectance (ATR) showed the usefulness of IR for the determination of conformational changes during processing and storage. But absolute determination of the conformation of starch is still not possible. The O–H stretching vibrations between 3000 and 3700 cm[-1] and the C–C and C–O stretching and C–O–H bending vibrations between 800 and 1300 cm[-1] are sensitive to conformation, short-range order, crystallinity, and hydration.[7–9] Recently, the relationship between baking behavior and chemical structure of cassava starch was studied with IR[10] (Figure 23.1).

23.2.1.3 Other Polysaccharides

Pectins have a backbone of (1→4)-linked α-D-galacturonan with side chains of α-D-galactopyranose, α-L-arabinofuranose, and α-(1→2) rhamnopyranosyl residues. Most studies focused on the carboxyl groups in pectic derivatives or the degree of esterification of carbonyl groups in pectins.[7,11,12] Characteristic band shifts of the C=O ring, and the symmetric or asymmetric carboxylate vibrations showed the interaction of divalent cations with pectate or pectinate, indicating a metal coordination in agreement with the egg-box hypothesis and complex formation.

Xylan structure depends on the nature of the xylopyranan backbone, β-(1→4) or β-(1→3), and the presence of glucuronic acid side chains and arabinose. This results in marked spectral differences, which depend on water content, sorption, and hydration.[7] It was shown that water influences the molecular orientation and degree of ordering. Spectral features of (1→4) xylans are affected by the α-L-arabinofuranose substitution of the C-3, but not significantly by arabinofuranose or glucuronic

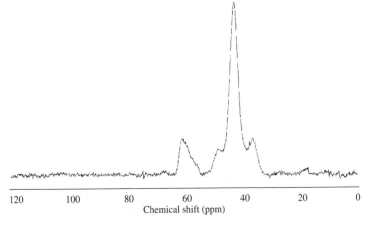

FIGURE 23.1 Solid-state ^{13}C-CP/MAS NMR spectrum of pregelatinized waxy corn starch.

substitution at C-2.[7] Most IR studies on polysaccharides are related to the identification or quantification of functional groups of the respective saccharides. In particular, the fingerprint region (400 to 1700 cm^{-1}) is useful. Examples are the analysis of amino groups of aminosugars such as chitin and chitosan, identification of carboxylate groups in carboxymethylated glucans, determination of acylation degree of chitosan, and characterization of seaweed galactans and sulfated agar-type or algal polysaccharides.[7,13]

In analogy with starch and cellulose, the crystalline structural variations of chitins were studied by IR. Other studies focused on the gelation properties or complex-forming capability of sugars with other polymers, metal ions, or small molecules such as with mucic acid.[7,14] The effect of aqueous salt solutions is clearly seen in the spectra of dextran.[7,8] The carrageenan sugar conformational regions and the counter ion groups between 1400 and 1000 cm^{-1} were assigned by using a variable-temperature ATR.[8]

23.2.1.4 Complex Multicomponent Systems

Plant cell walls consist of a complex mixture of biopolymers such as polysaccharides (cellulose, pectins), proteins, lignins, and hemicelluloses. The variations in bands and their intensities reflect compositional or conformational differences. Usually complete assignment is difficult. The primary cell wall architecture was determined by IR and Raman.[7]

Quantitative IR combined with chemometrics of complex sugar mixtures is rapid and has potential for classification, identification of components, and online monitoring in industry. Examples include estimation of the composition of degraded wood by DRIFT, screening of mutant plants by IR microspectroscopy of leaves, and screening of fruit purees.[7] Besides quantitative uses, structural studies by chemometrics were reported, such as the study on complex formation of ions and sucrose and the analysis of cellulose Iα and Iβ crystalline phases during the development of cell walls.[7,8]

23.2.2 NEAR INFRARED SPECTROSCOPY (NIR)

The most characteristic NIR wavelengths (1432, 1931, 2170, 2310, and 2477 nm) for polysaccharides are highly appropriate for quantitative studies. NIR is used in particular to determine composition and water content of various systems, such as foods, grain products and vegetables.[7,8,15–19] By using modern data analysis techniques, effects of processing on product functionality can be studied as well. NIR is applicable to the study of chemical changes involving hydrogen bonding, making it possible to study the effects on gelatinization during heating, retrogradation, hydrolysis, extrusion, and starch damage during milling.[7,8,15–20]

In carbohydrate systems, which contain polar groups such as hydroxyl and carboxylic acid groups, water molecules are adsorbed in a specific way, which varies with the type of polar group. NIR was recently used to show that adsorbed water clearly influenced the nature of the counterion pair for carboxymethylated cellulose with various degrees of substitution (DS).[21] NIR reflection analysis was employed to determine the phase-separation kinetics of aqueous starch/galactomannan systems.[22]

23.2.3 Raman and Nonconventional Vibrational Spectroscopy

Raman is often used as a technique complementary to IR. The spectral bands depend linearly on the number of molecules in a sample, making quantitative analysis a reliable and easy method in pharmaceutical, food, and polymer industrial settings.[8,23] The method was used extensively to determine the DS of modified starches.[23,24] Online monitoring starch gelatinization and hydrolysis was done with FT-Raman.[25]

IR microspectroscopy opens the door for looking at the spatial distribution of the constituents on an area as small as 10 by 10 µm. Microscopic structural variations can have a significant effect on the physical and mechanical properties of materials. During the last decade, the technique became popular to investigate materials such as plant tissues, the formation of cellulose I and II crystals in the wood cell wall, the surfaces of nutshells, flax cell wall structure and changes during growth, the location of components in strawberry, and the compositional variation of polygalactans in red algae.[6,26–28]

It is possible to look at molecular orientations of polymers by using polarized light and IR. Polysaccharide systems are complex, but polarized spectra provided information on the orientation of polysaccharides by measuring parallel or perpendicular absorptions or difference bands.[7,29] Most studies were made on cell wall showing, for instance, that cell wall polysaccharides are in an ordered state or align under mechanical stress. Other studies were made on drivable water-swollen cellulose films[30] or cellulose fiber orientation.[31]

2D-FTIR is today used frequently in polymer physics research to study the effects of applied mechanical stress on highly oriented materials.[7,32] A recent example of dynamic 2D-IR is the noninvasive study of polysaccharides in onions looking at the time-dependent response to applied mechanical stress, showing the different cellulose and pectin reorientation rates in wet cell walls. An example of using 2D spectral correlation analysis is the study of starch-based paper coatings by FTIR-photoacoustic spectroscopic depth profiling.[33]

23.3 NMR SPECTROSCOPY

NMR is of great importance in elucidating the structure, stereochemistry, and dynamics of carbohydrates. Advances in the past decades are due to the construction of high-field spectrometers (>900 MHz) and efficient computers. NMR is based on the property of nuclear spin. The atom spins of molecules can align parallel or antiparallel to a magnetic field, leading to different energy levels. Transitions between the energy levels can be stimulated and the energy absorption can be recorded as a resonance signal. The spectra contain information on the chemical environment of the atoms (chemical shift), the molecular geometry (spin–spin coupling), and the number of atoms give rise to the signal intensity (integral). Carbohydrates have two main active nuclei, ^{13}C and ^{1}H, whereas less frequently used nuclei are ^{2}H, ^{3}H, ^{11}B, ^{15}N, ^{17}O, ^{19}F, and ^{31}P. Usually, carbohydrates show a narrow spectral window, making it hard to assign the spin systems. In most complex spectra of polysaccharides, unambiguous assignment and extraction of NMR parameters, spin coupling constants, and chemical shifts are only possible if multidimensional correlation methods determining spatial proximity and connections of atoms are used. Because of the poor solubility properties of most polysaccharides, solid-state NMR is valuable.[34–37]

23.3.1 SOLUTION STATE AND HIGH-RESOLUTION NMR

Carbohydrates are often dissolved in D_2O to exchange protons of functional groups such as hydroxyl or amide to simplify the spectra. Other common solvents are water, dimethyl sulfoxide, or organic solvents such as deuterated chloroform and methanol or mixtures. Traditionally, initial estimates of the number of monosaccharide residues present are made by 1D experiments. The anomeric proton resonances are found in the shift range 4.4–5.5 ppm, and the remaining ring protons are in the range of 3.0–4.2 ppm. More details are obtained by 2D experiments such as ^{13}C-^1H HSQC, HMQC, or HMBC. (See Table 23.1 for NMR pulse sequence abbreviations.) A detailed description of the carbohydrate analysis by classical or new methods has been given recently by Duus et al.[36] The field of polysaccharide structure determination was reviewed by Mulloy and Bush and coworkers.[35,37] Therefore, in this chapter, only a few recent examples are described of using high-resolution NMR for structure–function relationship studies.

Spin lattice relaxation times (T_1, T_2) were used to describe the state of water in polysaccharide gels and even complex systems such as foods. Water ^1H relaxation was studied in methoxy pectins, showing that in solution 40% of the water might be affected by interactions with the pectins, depending on source.[38] The mechanism of water uptake in cereals and cookies was studied by combining NMR relaxometry

---

TABLE 23.1
NMR Pulse Sequence and Experiment Abbreviations

Experiment	Description
COSY	Correlation spectroscopy
DFQ-COSY	Double quantum filtered – COSY
HOHAHA/TOCSY	Homonuclear Hartmann Hahn/total correlation spectroscopy
HMBC	Heteronuclear multiple bond correlation
HMQC	Heteronuclear multiple quantum coherence
HSQC	Heteronuclear single quantum coherence
NOESY	Nuclear Overhauser enhancement spectroscopy
ROESY	Rotating frame Overhauser enhancement spectroscopy
CRS	Cross relaxation spectroscopy
CP/MAS	Cross polarization/magic angle spinning
HPDEC	High power decoupling
WISE	Wideline separation
IRCP	Inverse recovery cross polarization
SPRITE	Single-point ramped imaging with T1 enhancement

with a novel solid imaging technology (SPRITE).[39] An important physical property of polysaccharides is the glass transition temperature (T_g). Proton CRS is suitable for describing the glass-to-rubber transition, for instance, in starch plasticized with water.[40] Pulsed gradient spin echo NMR diffraction patterns were used to determine the pore size in freeze-dried starch gels without altering the sample structure with octane as a probe.[41] Table 23.1 presents the commonly used NMR pulse sequences.

23.3.2 Solid-State NMR

Since 1976, it is common practice to record high-resolution solid-state NMR spectra of polymers by using dipolar decoupling (DD), cross polarization (CP), and magic angle spinning (MAS). Solid-state NMR provides information on structural and physical properties such as crystal structure, molecular mobility, and ordering. By using DD, one radio frequency is used to observe responses while another frequency is used to irradiate the resonance of the nuclei to be decoupled. For instance, ^{13}C spectra are decoupled with ^{13}C-^1H DD. Higher resolution and more spectral details are obtained. Rapid sample spinning at the magic angle (54.7°) results in a reduction of the chemical shift anisotropy and thereby the resolution and an increase in signal intensity. Molecular anisotropic information can be regained at slower spinning rates. By using CP, polarization is transferred from abundant spins (^1H) to the dilute ^{13}C nuclei in a saccharide. CP reduces the measurement time and boosts signal intensity. Other important information is obtained by recording the spin lattice relaxation times, T_1, the spin lattice relaxation times in the rotating frame, $T_{1\rho}$, and the spin–spin relaxation times, T_2, of ^1H and ^{13}C atoms. Information is obtained on polymer compatibility and miscibility, mobility of backbone and side chains, and influence of additives such as plasticizers. 2D experiments (still time consuming) give more insight in chemical structure, molecular order, mobility, and molecular movements. The most commonly used solid-state NMR methods are CP/MAS (detecting rigid sites) and HPDEC (mobile sites).[8]

23.3.3 Some Specific Polysaccharide Examples

23.3.3.1 Cellulose

Most studies are related to structural analysis of cellulose derivatives determining the DS.[42,43] Studies on the crystalline forms of cellulose from various sources were reported by several authors.[44–47] The assignments of the noncrystalline forms of cellulose and a paracrystalline "in-fibril" form were found by using CP/MAS.[48] Other studies are related to the interaction with other polymers, such as locust bean gum, showing that cellulose crystallite surfaces involve mostly mannosyl residues of the backbone and not only the long segments lacking galactosyl residues.[49]

23.3.3.2 Starch

Early studies on starch were directed to the identification of crystal type and proportions. The C-1 region in the CP/MAS spectrum is sensitive to different starch crystals, such as A- (see Figure 23.2), B-, and V-type.[8,50] Molecular mobility of starch systems, such as bread and gels, starch retrogradation, and water diffusion were comprehensively studied. Other studies use relaxation time experiments to obtain information on, for instance, the mobility of starch in plastics. 2D wideline separation (WISE) ^{1}H-^{13}C was used to study the water organization and molecular mobility in starches and other saccharides. A recent study showed the applicability of inverse recovery CP (IRCP) for the discrimination between the mobility of the C-6 vs. the ring carbons in starch and the interaction with plasticizers.[8] Figure 23.2 presents a solid-state NMR ^{13}C-CP/MAS spectrum of native waxy maize starch.

23.3.3.3 Other Polysaccharides

The branching characteristics of glycogen were determined by 2D H–H and C–H COSY methods.[51] Interest in the use of novel NMR methods to study plant cell wall polysachharides is growing.[52–54] WISE was used to study heterogeneous cell wall polysaccharide mixtures in potato tissues and J-WISE for the study of local mobility and hydration of biopolymers in plant cell walls. NMR is used as well for biomedical polymers such as in the case of a pulsed field gradient NMR investigation on the self-diffusion of dextran as a model compound to mimic metabolites in cartilage.[55,56] Another example made use of a hyphenated method, rheo-NMR, showing that shear flow affected secondary and tertiary structures of hyaluronan.[57] The network structure of carboxylated polysaccharide hydrogels was studied by ^{13}C- and ^{15}N-CP/MAS NMR.[58]

23.4 DIELECTRIC RELAXATION SPECTROSCOPY (DRS)

DRS is used to investigate molecular motions in polymers.[8] The method is noninvasive, requires little sample preparation, and measures the response of materials to an alternating electric field after placing a sample in a capacitor. The impedance is

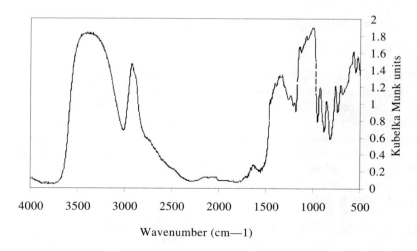

FIGURE 23.2 FTIR spectrum of potato starch.

measured and the dielectric permeability, ε', giving the polarization of the sample, and the dielectric loss, ε'', a measure of the energy loss, as a function of frequency and temperature are calculated. Dipoles and ions try to follow the direction of the applied alternating current. At low frequencies, the charges move freely and ε' remains constant and ε'' is zero. At high frequencies, the charges cannot move quickly enough with the field and are fixed in a position that does not have to be parallel to the field. During a frequency scan, a transition takes place depending on the characteristic relaxation time, τ, of the orientation process (Figure 23.3). In analogy, at low temperatures the charges are frozen in and at increased temperature they become mobile.

For polymers, several transitions are found. At temperatures below the T_g, peaks are found because certain chain segments are more mobile than other parts of the polymer. A β-transition can be seen for the mobilization of chain segments, and at even lower frequencies or temperatures γ- and δ-transitions are found for smaller segments such as the rotation of large side groups. The relaxation times of these transitions follow Arrhenius law, $\tau = \tau_0 \exp(E_a/RT)$, where τ_0 is the preexponent factor at infinite T and E_a the thermal activity energy of the dipole rotation The α-transitions at high frequencies and temperatures are due to mobility changes of whole chains (the T_g). The α-transition behaves according to Vogel–Fulcher–Tammann (VFT) law. VFT accounts for the dipole rotations, which are determined by polymer free volume changes during the T_g, $\tau = \tau_0 \exp[E_v/R(T - T_v)]$, where E_v is VFT activation energy and T_v the VFT temperature.

For dry solid polysaccharides, four relaxation processes are observed below the T_g: the local main chain motion or the β-relaxation, the side group motion in the repeating unit or the γ-relaxation, the δ-relaxation in the low frequency side for well-dried samples, and finally a β-relaxation found in wet samples around room temperature, of which the origin is not entirely clear. At higher temperatures, the α-relaxation can be measured, which is often associated with the hopping motion of ions in the disordered structure of an ionic biopolymeric material.[8,59]

Although DRS was used for synthetic polymers, only recently DRS became more popular for carbohydrates.[8,59,60] A recent review on DRS for analysis of polysaccharides was given by Einfeldt et al.[59] DRS is used to study physical ageing phenomena of polysaccharides.[61] Several studies concern complex polysaccharide systems (wood, bread, cereal).[8,62,63] Local or side-group motions in modified celluloses and starches were evaluated with the Havriliak–Negami relaxation model.[8,64] The

FIGURE 23.3 DRS main experimental results.

electrical and dielectric properties of Ba^{2+} and Ca^{2+} cross-linked alginate hydrogels were studied by single-particle electrorotation using microelectrodes.[65] Some studies were related to the effect of water on the chain mobility in polysaccharides.[8,64] Another area is the gelation, coil-helix, and sol-gel transitions occurring in gel-forming ionic polysaccharides.[66,67] The problem of the origin of the dielectric secondary relaxations of pure polysaccharides was described not long ago.[68] The activation energy of the lower temperature transition for amylose films, obtained from the frequency dependence of the relaxation, was typical of a primary α-relaxation or T_g. This indicated that although glycerol is an effective plasticizer of amylose, amylose–glycerol mixtures are only partially miscible.[69] DRS is important for studies on electroheological carbohydrate materials.[70,71] Figure 23.3 shows schematic results of the DRS analysis.

23.5 OTHER MISCELLANEOUS TECHNIQUES

23.5.1 MASS SPECTROMETRY (MS)

Structural characterization of carbohydrates with MS began in the 1960s, using electron impact ionization (EI). MS developed from a method only applicable to small organic molecules into a method handling large biomolecules.[72,73] MS is based on the separation of charged molecules by using magnetic or electric fields. After ionization, the ions are separated according to their charge and kinetic energy (electric field) or differences in momentum (magnetic field). Time of flight (TOF) is used to separate the ions with differences in velocity. Quadrupole filters separate the ions according to their direct m/z (mass–charge ratio) value. It is often a time-consuming and elaborate work to determine the chemical structure or DS of carbohydrates. Traditionally, enzymatic methods are used combined with chromatographic methods. Developments, such as soft ionization techniques [fast atom bombardment (FAB), liquid secondary ion (LSI), electrospray ionization (ESI), atmospheric pressure chemical ionization (APCI), matrix-assisted laser desorption ionization (MALDI)], have made it possible to obtain information on monosaccharide composition, DS, linkages, branching, and sequence. MS is often used as a hyphenated analysis coupled with chromatographic methods. The molar mass is limited to oligosaccharides or small polysaccharide fragments (~5000 Da) by using quadrupole detection. With TOF, detection of ions of m/z higher than several hundred thousand daltons was reported. Tandem MS (MS/MS), using for example collision-induced dissociation (CID), is an interesting option for the analysis of sugars.[72–76]

23.5.2 ILLUSTRATIVE EXAMPLES OF OTHER TECHNIQUES

Finally, because of the more limited use of other spectroscopic techniques, only some illustrative examples will be described here. CD spectroscopy measures the difference in absorption intensities for left and right circularly polarized light of an asymmetric molecule and is mainly applicable to small sugars, especially with π-electron systems.[77] Recently, a paper was published describing the origin of circular dichroism in carbohydrate polymers.[78] Other studies are related to the interaction of CD active compounds with polysaccharides.[79]

XPS is used to obtain quantitative elemental and chemical state information by irradiation of a sample surface with low energy x-rays, resulting in the emission of electrons. The electrons are separated according their kinetic energy and the distribution of the energy profile gives the photoelectron spectrum. Most studies concern polysaccharides adhered to various surfaces, chemical modifications, or coatings.[80–85]

An interesting review on ESR and its new technical developments used in polymer studies is given by Veksli et al.[86] Spin probe (using nitroxyl radicals) ESR can give information on polymer morphology, phase separation and gelation, order–disorder transitions, and interfacial behavior.[87,88] Other studies are directed to the identification of radicals in presence of (poly-)saccharides generated during enzymatic or chemical free radical reactions and x-ray or γ-irradiation.[89–91]

REFERENCES

1. Koenig, J.L., Ed., *Spectroscopy of Polymers*, ACS, Washington, 1992.
2. Wilson, R.H. and Tapp, H.S., Mid-infrared spectroscopy for food analysis: recent new applications and relevant developments in sample presentation methods, *Trends Anal. Chem.*, 18, 85, 1999.
3. Gremlich, H.U. and Yan, B., Eds., *Infrared and Raman Spectroscopy of Biological Materials,* Practical Spectroscopy Series Vol. 24, Marcel Dekker, New York, 2001.
4. Kataoka, Y. and Kondo, T., FT-IR microscopic analysis of changing cellulose crystalline structure during wood cell wall formation, *Macromolecules*, 31, 760, 1998.
5. Langkilde, F.W. and Svantesson, A., Identification of celluloses with Fourier-transform (FT) mid-infrared, FT-Raman, and near-infrared spectrometry, *J. Pharm. Biomed. Anal.*, 13, 409, 1995.
6. Rowe, R.C., McKillop, A.G., and Bray, D., The effect of batch and source variation on the crystallinity of microcrystalline cellulose, *Int. J. Pharm.*, 101, 169, 1994.
7. Kačuráková, M. and Wilson, R.H., Developments in mid-infrared FT-IR spectroscopy of selected carbohydrates, *Carbohydr. Polym.*, 44, 291, 2001.
8. Smits, A.L.M., The molecular organisation in starch based products, Ph.D. thesis, Utrecht University, Utrecht, 2001.
9. Iizuka, K. and Aishima, T., Starch gelation process observed by FT-IR/ATR spectrometry with multivariate data analysis, *J. Food Sci.*, 64, 653, 1999.
10. Demiate, I.M. et al., The relationship between baking behavior of modified cassava starches and starch chemical structure determined with FTIR spectroscopy, *Carbohydr. Polym.*, 42, 149, 2000.
11. Engelsen, S.B. and Norgaard, L., Comparative vibrational spectroscopy for determination of quality parameters in amidated pectins as evaluated by chemometrics, *Carbohydr. Polym.*, 30, 9, 1996.
12. Filippov, M.P., Practical infrared spectroscopy of pectic substances, *Food Hydrocoll.*, 6, 115, 1992.
13. Matsuhiro, B., Vibrational spectroscopy of seaweed galactans, *Hydrobiologia*, 327, 481, 1996.
14. Tian, W. et al., Sugar interaction with metal ions: crystal structure and FT-IR spectroscopic study of strontium galactarate monohydrate, *J. Inorg. Biochem.*, 78, 197, 2000.

15. Jowder, O. et al., Mid-infrared spectroscopy and chemometrics for the authentication of meat products, *J. Agric. Food Chem.*, 47, 3210, 1999.
16. Osborne, B.G., Recent developments in NIR analysis of grains and grain products, *Cereal Foods World*, 45, 11, 2000.
17. Reeves, J.B. et al., Mid infrared versus near infrared diffuse reflectance spectroscopy for quantitative and qualitative analysis of agricultural products, *Recent Res. Dev. Agric. Food Chem.*, 3, 201, 1999.
18. Campbell, M.R. et al., Prediction of starch amylose content versus total grain amylose content in corn by near-infrared transmittance spectroscopy, *Cereal Chem.*, 76, 552, 1999.
19. Blanco, M. et al., On-line monitoring of starch enzymatic hydrolysis by near-infrared spectroscopy, *Analyst*, 125, 749, 2000.
20. Evans, A.J. et al., Near infrared on-line measurement of degree of cook in extrusion processing of wheat flour, *J. Near Infrared Spectrosc.*, 7, 77, 1999.
21. Berthold, J., Olsson, R.J.O., and Salmen, L., Water sorption to hydroxyl and carboxylic acid groups in carboxymethylcellulose (CMC) studied with NIR-spectroscopy, *Cellulose*, 5, 281, 1998.
22. Closs, C.B. et al., Phase separation and rheology of aqueous starch/galactomannan systems, *Carbohydr. Polym.*, 39, 67, 1999.
23. Ma, C.Y. and Phillips, D.L., FT-Raman spectroscopy and its applications in cereal science, *Cereal Chem.*, 79, 171, 2002.
24. Chong, C. K. et al., Development of NMR and Raman spectroscopic methods for the determination of the degree of substitution of maleate in modified starches, *J. Agric. Food Chem.*, 49, 2702, 2001.
25. Schuster, K.C. et al., On-line FT-Raman spectroscopic monitoring of starch gelatinization and enzyme catalysed starch hydrolysis, *Vibr. Spectr.*, 22, 181, 2000.
26. Jarvis, M.C. and McCann, M.C., Macromolecular biophysics of the plant cell wall: concepts and methodology, *Plant Physiol. Biochem.*, 38, 1, 2000.
27. McCann, M.C. et al., Approaches to understanding the functional architecture of the plant cell wall, *Phytochemistry*, 57, 811, 2001.
28. Suutarinen, J., Anakainen, L., and Autio, K., Comparison of light microscopy and spatially resolved Fourier transform infrared (FT-IR) microscopy in the examination of cell wall components of strawberries, *Lebensm. Wiss. Technol.*, 31, 595, 1998.
29. Fenwick, K.M., Jarvis, M.C., and Apperley, D.C., Estimation of polymer rigidity in cell walls of growing and nongrowing celery collenchyma by solid-state nuclear magnetic resonance in vivo, *Plant Physiol.*, 115, 587, 1997.
30. Togawa, E. and Kondo, T., Change of morphological properties in drawing water-swollen cellulose films prepared from organic solutions. A view of molecular orientation in the drawing process, *J. Polym. Sci. B*, 37, 451, 1999.
31. Wada, M. et al., Improved structural data of cellulose IIII prepared in supercritical ammonia, *Macromolecules*, 34, 1237, 2001.
32. Noda, I., Dowrey, A.E., and Marcott, C., Two-dimensional infrared (2D IR) spectroscopy, *Mod. Polym. Spectrosc.*, 1, 1999, p. 1.
33. Halttunen, M. et al., Applicability of FTIR/PAS depth profiling for the study of coated papers, *Vibr. Spectrosc.*, 19, 261, 1999.
34. Bush, C.A., Martin-Pastor, M., and Imberty, A., Structure and conformation of complex carbohydrates of glycoproteins, glycolipids, and bacterial polysaccharides, *Annu. Rev. Biophys. Biomol. Struct.*, 28, 269, 1999.
35. Bush, C.A., Ed., Polysaccharides and complex oligosaccharides, in *Encyclopedia of Nuclear Magnetic Resonance*, Wiley, Chichester, 1996.

36. Duus, J.Ø., Gotfredsen, C.H., and Bock, K., Carbohydrate structural determination by NMR spectroscopy: modern methods and limitations, *Chem. Rev.*, 100, 4589, 2000.
37. Muloy, B., High field NMR as a technique for the determination of polysaccharide structures, *Mol. Biotechnol.*, 6, 241, 1996.
38. Kerr, W.L. and Wicker, L., NMR proton relaxation measurements of water associated with high methoxy and low methoxy pectins, *Carbohydr. Polym.*, 42, 133, 2000.
39. Cornillon, P. and Salim, L.C., Characterization of water mobility and distribution in low- and intermediate-moisture food systems, *Magn. Reson. Imag.*, 18, 335, 2000.
40. Vodovotz, Y., Dickinson, L., and Chinachoti, P., Molecular characterization around a glassy transition of starch using H-1 cross-relaxation nuclear magnetic resonance, *J. Agric. Food Chem.*, 48, 4948, 2000.
41. Cornillon, P., McCarthy, M.J., and Reid, D.S., Study of restricted diffusion by NMR in freeze dried starch gels, *J. Text. Stud.*, 28, 421, 1997.
42. Tezuka, Y., ^{13}C NMR structural study on cellulose derivatives with carbonyl gropus as a sensitive probe, in *Cellulose Derivatives, ACS Symposium Series* 688, Heinze, T.J. and Glasser, W.G., Eds., ACS, Washington, 1998, chap. 11.
43. Torri, G. et al., Novel cellulosic ethers with low degrees of substitution. II. Magic angle spinning NMR study, *Carbohydr. Polym.*, 40, 125, 1999.
44. Focher, B. et al., Structural differences between non-wood plant celluloses: evidence from solid state NMR, vibrational spectroscopy and x-ray diffractometry, *Ind. Crops Prod.*, 13, 193, 2001.
45. Larsson, P.T., Wickholm, K., and Iversen, T., A CP/MAS ^{13}C NMR investigation of molecular ordering in celluloses, *Carbohydr. Res.*, 302, 19, 1997.
46. Heux, L., Dinand, E., and Vignon, M.R., Structural aspects in ultrathin cellulose microfibrils followed by ^{13}C CP-MAS NMR, *Carbohydr. Polym.*, 40, 115, 1999.
47. Tanahashi, M. et al., Characterization of steam-exploded wood. III. Transformation of cellulose crystals and changes of crystallinity, *Mokuzai Gakkaishi*, 35, 654, 1989.
48. Wickholm, K., Larsson, P.T., and Iversen, T., Assignment of non-crystalline forms in cellulose I by CP/MAS ^{13}C NMR spectroscopy, *Carbohydr. Res.*, 312, 123, 1998.
49. Newman, R.H. and Hemmingson, J.A., Interactions between locust bean gum and cellulose characterized by ^{13}C NMR spectroscopy, *Carbohydr. Polym.*, 36, 167, 1998.
50. Karim, A.A., Norziak, M.H. and Seow, C.C., Methods for the study of starch retrogradation, *Food Chem.*, 71, 9, 2000.
51. Stanek, M., Falk, H., and Huber, A., Investigation of the branching characteristic of glycogen by means of two-dimensional ^{1}H and ^{13}C NMR spectroscopy, *Monatsh. Chem.*, 129, 355, 1998.
52. Yan, B. and Stark, R.E., A WISE NMR approach to heterogeneous biopolymer mixtures: dynamics and domains in wounded potato tissues, *Macromolecules*, 31, 2600, 1998.
53. Hediger, S., Lesage, A., and Emsley, L., A new NMR method for the study of local mobility in solids and application to hydration of biopolymers in plant cell walls, *Macromolecules*, 35, 5078, 2002.
54. Newman, R.H. and Redgwell, R.J., Cell wall changes in ripening kiwifruit: C-13 solid state NMR characterisation of relatively rigid cell wall polymers, *Carbohydr. Polym.*, 49, 121, 2002.
55. Trampel, R. et al., Self-diffusion of polymers in cartilage as studied by pulsed field gradient NMR, *Biophys. Chem.*, 97, 251, 2002.
56. Scott, J.E. and Heatley, F., Biological properties of hyaluronan in aqueous solution are controlled and sequestered by reversible tertiary structures, defined by NMR spectroscopy, *Biomacromolecules*, 3, 547, 2002.

57. Fischer, E. et al., Shear flow affects secondary and tertiary structures in hyaluronan solution as shown by rheo-NMR, *J. Mol. Struct.*, 602, 303, 2002.

58. Nooy, A.E.J. et al., Ionic polysaccharide hydrogels via the passerini and ugi multicomponent condensations: synthesis, behavior and solid-state NMR characterization, *Biomacromolecules*, 259, 1, 2000.

59. Einfeldt, J., Meissner, D., and Kwasniewski, A., Polymer dynamics of cellulose and other polysaccharides in solid state-secondary dielectric relaxation processes, *Progr. Polym. Sci.*, 26, 1419, 2001.

60. Rosa, A. de la, Heux, L. and Cavaillé, J.Y., Secondary relaxations in poly(allyl alcohol), PAA, and poly(vinyl alcohol), PVA. II. Dielectric relaxations compared with dielectric behaviour of amorphous dried and hydrated cellulose and dextran, *Polymer*, 42, 5371, 2001.

61. Borde, B. et al., Sub-Tg relaxations and physical ageing in hydrated glassy polysaccharides, in *Biopolymer Science: Food and Non Food Applications, Les Colloques 91*, Colonna, P. and Guilbert, S., Eds., INRA Service des Publ., Versailles, 1999, p. 167.

62. Yokoyama, M. et al., Mechanical and dielectric relaxations of wood in a low temperature range IV. Dielectric properties of adsorbed water at high moisture contents, *Mokuzai Gakkaishi*, 46, 523, 2000.

63. Ratkovic, S. and Pissis, P., Water binding to biopolymers in different cereals and legumes: proton NMR relaxation, dielectric and water imbition studies, *J. Mater. Sci.*, 32, 3061, 1997.

64. Einfeldt, J. et al., Influence of the p-toluene sulphonylation of cellulose on the polymer dynamics investigated by dielectric spectroscopy, *Carbohydr. Polym.*, 49, 357, 2002.

65. Esch, M. et al., Dielectric properties of alginate beads and bound water relaxation studied by electrorotation, *Biopolymers*, 50, 227, 1999.

66. Hayashi, Y. et al., Ordering in aqueous polysaccharide solutions. I. Dielectric relaxation in aqueous solutions of a triple-helical polysaccharide schizophyllan, *Biopolymers*, 63, 21, 2002.

67. Ikeda, S. and Kumagai, H., Dielectric analysis of sol-gel transition of kappa-carrageenan with scaling concept, *J. Agric. Food Chem.*, 46, 3687, 1998.

68. Meissner, D., Einfeldt, J., and Kwasniewski, A., Contributions to the molecular origin of the dielectric relaxation processes in polysaccharides — the low temperature range, *J. Non Cryst. Solids*, 275, 199, 2000.

69. Moates, G.K. et al., Dynamic mechanical and dielectric characterisation of amyloseglycerol films, *Carbohydr. Polym.*, 44, 247, 2001.

70. Rejon, L. et al., Response time and electrorheology of semidiluted gellan, xanthan and cellulose suspensions, *Carbohydr. Polym.*, 48, 413, 2002.

71. Zhao, X.P. and Duan, X., A new organic/inorganic hybrid with high electrorheological activity, *Mat. Lett.*, 54, 348, 2002.

72. Settineri, C.A. and Burlingame, A.L., Mass spectrometry of carbohydrates and glycoconjugates, in *Carbohydrate Analysis*, Rassi, Z. el, Ed., Elsevier, Amsterdam, 1995, p. 447.

73. Stahl, B. et al., Analysis of fructans from higher plants by matrix-assisted laser desorption/ionization mass spectrometry, *Anal. Biochem.*, 246, 195, 1997.

74. Garrozzo, D. et al., Matrix-assisted laser desorption/ionization mass spectrometry of polysaccharides, *Rapid Commun. Mass. Spectrom.*, 9, 809, 1995.

75. Asam, M.R. and Glish, G.L., Tandem mass spectrometry of alkali cationized polysaccharides in a quadrupole ion trap, *J. Am. Chem. Soc. Mass Spectrom.*, 9, 987, 1997.

76. Burgt, Y.E.M. et al., FAB CIDMS/MS analysis of partially methylated maltotrioses derived from methylated amylose: a study of the substituent distribution, *Carbohydr. Polym.*, 329, 341, 2000.

77. Harada, N. and Nakanishi, K., Eds., in *Circular Dichroic Spectroscopy, Exciton Coupling in Organic Stereochemistry*, University Science Books, Mill Valley, CA 1983.

78. Parra, A. and Stevens, E.S., The origin of circular dichroism in unsubstituted carbohydrate polymers, *Carbohydr. Polym.*, 41, 111, 2000.

79. Synytsya, A. et al., Interaction of metallotexaphyrins with mono- and polysaccharides, *J. Chem. Soc. Perkin Trans.*, 2, 1876, 2000.

80. Boyd, R.D. et al., Surface characterization of glass and poly(methyl methacrylate) soiled with a mixture of fat, oil, and starch, *J. Adhes. Sci. Technol.*, 14, 1195, 2000.

81. Granja, P.L. et al., Cellulose phosphates as biomaterials. II. Surface chemical modification of regenerated cellulose hydrogels, *J. Appl. Polym. Sci.*, 82, 3354, 2001.

82. Morra, M. and Cassinelli, C., Simple model for the XPS analysis of polysaccharide-coated surfaces, *J. Surf. Interf. Anal.*, 26, 742, 1998.

83. Morra, M. and Cassinelli, C., Non-fouling properties of polysaccharide-coated surfaces, *J. Biomater. Sci. Polym. Ed.*,10, 1107, 1999.

84. Dai, L.M. et al., Biomedical coatings by the covalent immobilization of polysaccharides onto gas-plasma-activated polymer surfaces, *J. Surf. Interf. Anal.*, 29, 46, 2000.

85. Nakayama, Y., Takatsuka, M., and Matsuda, T., Surface hydrogelation using photolysis of dithiocarbamate or xanthate: hydrogelation, surface fixation, and bioactive substance immobilization, *Langmuir*, 15, 1667, 1999.

86. Veksli, Z., Andreis, M. and Rakvin, B., ESR spectroscopy for the study of polymer heterogeneity, *Progr. Polym. Sci.*, 25, 949, 2000.

87. Wasserman, L.A. and LeMeste, M., Influence of water on potato starch-lipid interactions. An electron spin resonance (ESR) probe study, *J. Sci. Food Agric.*, 80, 1608, 2000.

88. Williams, P.A. and Langdon, M.J., The influence of locust bean gum and dextran on gelation of kappa-carrageenan, *Biopolymers*, 38, 655, 1996.

89. Marcazzan, M. et al., An ESR assay for alpha-amylase activity towards succinylated starch, amylose and amylopectin, *J. Biochem. Biophys. Meth.*, 38, 191, 1999.

90. King, K. and Gray, R., The effect of gamma-irradiation on guar gum, locust bean gum, gum tragacanth and gum karaya, *Food Hydrocoll.*, 6, 559, 1993.

91. Sirendi, M., Gohtani, S., and Yamano, Y., Alkoxyl and methyl radical formation during cleavage of tert-butyl hydroperoxide in the presence of polysaccharides, *J. Disp. Sci. Technol.*, 20, 699, 1999.

24 Natural and Synthetic Nonsaccharide Sweeteners

Michal Uher

CONTENTS

24.1 Natural Sweet Compounds 387
    24.1.1 Proteins 387
    24.1.2 Terpenoids 392
    24.1.3 Steroidal Saponins 394
    24.1.4 Dihydroisocoumarins (1H-2-Benzopyran-1-ones) 394
    24.1.5 Flavanoids 396
    24.1.6 Other Natural Sweet Compounds 398
24.2 Synthetic Nonnutritive Sweeteners 399
References 402

24.1 NATURAL SWEET COMPOUNDS

A number of sweet natural organic compounds have been found. They belong to families of proteins, terpenoids, steroids, dihydroisocoumarins, and flavanoids. Some of them are being currently used in the human diet as sweetening or flavoring agents.

24.1.1 PROTEINS

Except for a limited number of sweet proteins, high-molecular-weight substances do not stimulate taste cells. Up to this point, six sweet proteins have been known: thaumatin, composed of 207 amino acid chains (~22 kDa) (24.1); its aluminium salt, talin (24.2); monellin, with two 45 to 50 amino acid chains (~11 kDa) (24.3); mabinlin, having two 33 and 72 amino acid chains, respectively (~12.4 kDa) (24.4); pentadin, consisting of a 54 amino acid chain (~12 kDa) (24.5); and brazzein, with the same number of amino acids in the chain but of ~6 kDa. (24.6) (Table 24.1). Some proteins such as miraculin and curculin, although tasteless, are used as taste-modifiers. All these proteins can be used as harmless low-calorie sweeteners and are helpful in reducing the sugar intake. Table 24.1 lists selected physical and functional properties of sweet proteins as well as their occurrence.

TABLE 24.1
Natural Sweet Compounds

Compound and Structure	Origin and Occurrence	Sweetness[a]	Remarks	Applications	Stability pH	Stability Temperature (°C)	Reference
Proteins							
Thaumatins I and II (24.1)	*Taumatococcus danielli* (Maranthacaea), West Africa	1600	Licorice aftertaste	Food additive	2.7–10	<75	1–3
Talin (24.2)	Synthetic Al salt of thaumatins	3000	Extremely water soluble	Food additive	2–10	pH, air, and polysaccharide sensitive	1–3
Monelin (24.3)	*Dioscoreophyllum cuminsii* (Menispermaceae)	2000–2500	Also synthetically available	Multiple sweetener	2	<70	1–3
Mabinlin (24.4)	*Capparis masaikai* L. (Capparidaceae), Yunan, China	10	Also synthetically available	Low-calorie sweetener		<80	4
Pentadin (24.5)	*Pentadiplandra brazzena* B., Africa	500	Tastes more like sucrose than thaumatin			Heat-induced brazzein dimer	5
Brazzein (24.6)	*Pentadiplandra brazzena*	2000				<80 (4 h)	6, 7

Terpenoids

Perillaldehyde (24.7)	Perilla frutensces L. (Labiateae), China	12	Bitter aftertaste, low water solubility	Tobacco additive in Japan, substitute of maple syrup and licorice	3		2, 8
Perillaldoxime (24.8)	Synthetic	450	Bitter aftertaste, low water solubility, menthol–licorice aftertaste	Tobacco additive in Japan, substitute of maple syrup and licorice	3		2, 9
Hermandulcin (24.10)	Lippia dulcis Trev., (Verbenaceae), Mexico	1000	Bitter, also synthetically available	Application restricted by taste		Thermolabile	2, 9
Stevioside (24.11)	Stevia rebaudiana B (Asteraceae), mainly Paraguay and Brazil	30–80 (300)	Pleasant sweet taste	Sweetener for seasonings, pickles, and salted food in Japan	3–10	100 (1h)	2, 9
Rebaudioside (24.12)–(24.19)	Rubus suavissimus S. (Rosaceae), China	114	Bitter aftertaste	Chinese sweet tea			2
Baiynoside (24.20)	Phlonis betoncoides D. (Labiateae), China	250	Also synthetically available	Therapy of women's disease			10, 11
Gaudichaudioside (24.21)	Baccharis gaudichaudians, Paraguay	55	Pleasant taste with a low concomitant perception of bitterness	An auxilia tb ry food additive	>4.5	Thermally stable	12
Glycyrrhizin (24.22)	Glycyrrhiza glabra (Leguminosae), roots of licorice, Asia and Europe	170	Occurs as K, Ca, and NH_4 salts			Thermally stable	2, 13, 14

-- continued

TABLE 24.1 (continued)
Natural Sweet Compounds

Compound and Structure	Origin and Occurrence	Sweetness[a]	Remarks	Applications	Stability pH	Temperature (°C)	Reference
Periandrines (24.26)–(24.29)	*Periandra dulcis*, Mart. (Leguminosae), Brazil	85–95	More rapid taste sensation than glycyrrhizin	Sweetener			2
Mogroside V (24.30)	*Thlandiantha grosvenorii* (Cucurbitaceae), China	250	Licorice aftertaste				15
Abrusosides (24.33)–(24.36)	*Abrus precatorins* (Fabaceae), Florida	30–100	Ammonium salts are water soluble	Commercial development currently pursued		Thermally stable	15
Steroidal Saponins							
Osladin (24.37)	*Polypodium vulgare* (Polypodiaceae), Europe, Asia, America	500	Also synthetically available				2
Polypodoside A (24.38)	*Polypodium glycerrhiza* (Polypodiaceae), North America	600					2
Pterocaryosides (24.39)	*Pterocarya paliurus* (Juglandaceae), China	50–100	Ammonium salts are water soluble	Leaves of the plant are locally used for food sweetening			16

Dihydroisocoumarins

Phyllodulcin (24.40)	*Hydrangea macrophylla* (Saxifragaceae), Japan	400	Also synthetically available, low water solubility	Preparation of ritual sweet teas	2
Flavanoids					
Semisynthetic dihydrochalcones (24.41)	From naringin and neohesperidin	300–1800	Menthol-like refreshing sweet taste	Food, soft drinks, confectionaries, and drugs	2–6 stable at room temperature, 14 days at 50°C	2
Dihydroflavanols (24.42)–(24.49)	*Tessaria dodoneifolia* (Asteraceae), Paraguay	50	Also synthetically available		2
Other natural sweet substances					
Hematoxylin (24.54)	*Haematoxylon campechianum* (Leguminosae), Mexico, India	120	Also synthetically available		18
Telosmosides (24.55)	*Telosma procumteus* (Ascelepiadaceae), Vietnam, China, Philippines	1000		Vietnamese folk medicine	19
Monatin	*Schlerochilon ilicifolins*, Africa	1200–1400	Also synthetically available		20

a Relative sweetness with respect to that of a 10% aqueous solution of sucrose.

24.1.2 TERPENOIDS

The majority of intensely sweet naturally occurring compounds belong to the family of terpenoids (mono-, sesqui-, di-, and triterpenoids; see Table 24.1). Many members of this group are glycosides with one or more saccharide units, which improve their solubility in water. Perilartine, a chemically modified sweetener being the (E)-oxime of perilaldehyde (24.8), perialdehyde, and (E)-4-(methoxymethyl)-1,4-cyclohexadiene-1-carbaldoxime, another semisynthetic sweetener (24.9), are sweet monoterpens.

a (24.7)   b (24.8)   c (24.9)

The group of sesquiterpenes is represented by the colorless, oily hernandulcin (24.10). The toxicity of hernandulcin is low. Unpleasant aftertaste and natural bitterness limits its applications in spite of its high sweet potency.

(24.10)

The group of sweet diterpenes is composed of sweet *ent*-kaurene glycosides (24.11) to (24.19) isolated from the sweet herb *Stevia rebaudiana* Bertoni (Compositae). Stevioside (24.11) is the major *ent*-kaurene glycoside of the leaves and extracts of this plant apart from other glycosides (Table 24.2). The total yield of the isolated sweet glycosides is 10% (6.6% stevioside, 3.7% rebaudioside A, 2.1% rebaudioside C, and 0.53% dulcoside A).

The group of sweet diterpenes includes also baiynoside (24.20), a sweet labdane glycoside from the roots of a Chinese officinal herb Bai-Yun-Shen used in the therapy of gynecological diseases and gaudichaudioside A (24.21), a sweet diterpene glycoside isolated from the Paraguayan medical plant.[2] This fairly water-soluble compound has a pleasant taste with a low concomitant perception of bitterness.[12]

A group of triterpenoids includes the sweet oleanane glycoside glycyrrhizin (24.22). It forms colorless crystals poorly soluble in cold water but well soluble in hot water or ethanol. Glycyrrhizin, the sweet principle of the licorice root, is a

TABLE 24.2
Structure of Sweet *ent*-Kauren Glycosides

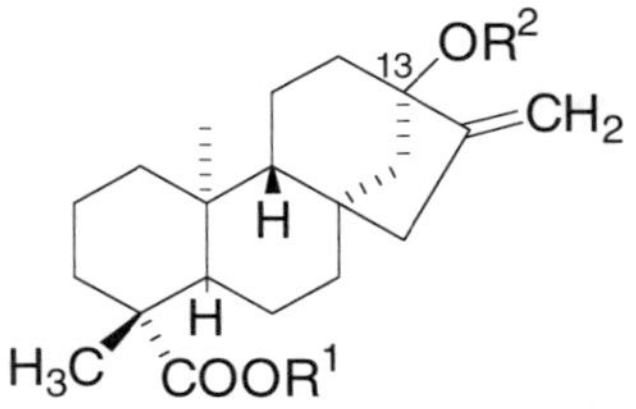

		(24.11) to (24.19)
Trivial Name	**R¹**	**R²**
Stevioside (24.11)	_-Glc	_-Glc²-_-Glc
Steviobioside	H	_-Glc²-_-Glc
Steviol (24.13)	H	H
Rebaudioside A (24.14)	_-Glc	_-Glc²-_-Glc
		3-_-Glc
Rebaudioside B (24.15)	H	As (24.14)
Rebaudioside C (24.16)	_-Glc	_-Glc²-α-Rha
		3-_-Glc
Rebaudioside D (24.17)	_-Glc²-_-Glc	As (24.14)
Rebaudioside E (24.18)	_-Glc²-_-Glc	_-Glc²-_-Glc
Dulcoside A (24.19)	_-Glc	_-Glc²-α-Rha

Note: Glc = D-glucopyranosyl; Rha = L-rhamnopyranosyl.

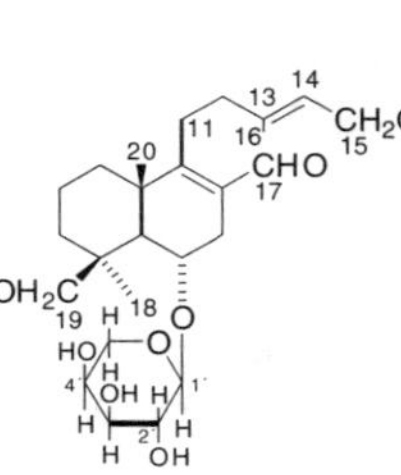

R = 2-O-β-xylopyranosyl-D-glucopyranosyl

(24.20)

(24.21)

nonsweet diglucuronide of the aglycone glycyrrhetinic acid (24.23). It can be isolated from the roots (6 to 14%) and rhizomes of licorice (23%). It is usually isolated as the calcium, potassium, or magnesium salt.

The most common commercial derivative of glycyrrhetinic acid is the ammonium salt. In glycyrrhetic acid glysosides, monoglycosides were sweeter than diglycosides. Among them, monoglucuronide of glycyrrhetic acid was the sweetest compounds (941 times sweeter than sucrose).[11,13]

The licorice extract hydrolyzed with β-glucuronidase has been used in Japan as a natural sweetener.[2] Main constituents of the extract have been investigated for evaluating its quality and safety as a food additive.[2]

Other novel sweet triterpenoid oligoglycosides, apioglycyrrhizin (24.24) and araboglycyrrhizin (24.25) have been isolated from the air-dried roots of *Glycyrhiza inflata*. Apart from glycyrrhizine, roots of the sweet-tasting wood *Periandra dulcis* Mart. contain a mixture of four triterpenoids, oleane-type glycosides, and periandrines (24.26) to (24.29).

Quantitative analysis of 30 kg of dried roots yielded 300 mg of periandrine I, 170 mg of II, 23 mg of III, and 9 mg of IV. The relative sweetness is almost equal to that of glycyrrhizin and, in contrast to the latter, the sensation sets much faster. Periandrines are not mutagenic.[8]

The sweet cucurbitane glycosides mogrosides are further representatives of the group of sweet triterpenoids. One of them, mogroside V (24.30), is used in the traditional Chinese medicine to treat colds, sore throats, as well as stomach and intestinal disorders. The composition of this group of sweeteners is complemented by sweet substances isolated from plants of the Cucurbitacae family and the sweet cycloartane glycosides abrucosides.

Roots of the plant *Bryonia dioica* Jacq contain a sweet matter of as yet unknown structure named bruyodulcoside.[8] Novel sweet cucurbitane-type glycosides carnosifloside V (24.31) and VI (24.32)[2] were isolated from the Chinese officinal plant *Hemsleya carnosiflora*. Quantitative data on their sweetness have not yet been published. Leaves of Floridean *Abrus precatorius* extracted with 1-butanol gave a group of four cycloartanne-type triterpene glycosides, the abrusosides A to D (24.33) to (24.36).

24.1.3 Steroidal Saponins

Among sweet steroidal saponins (Table 24.1), osladin (24.37), polyposide A (24.38), and pterocaryoside B (24.39) have received particular attention as compounds several hundred times sweeter than sucrose. Osladin have also been prepared synthetically.[2]

24.1.4 Dihydroisocoumarins (1*H*-2-Benzopyran-1-ones)

Phyllodulcin (24.40) (Table 24.1) is a natural dihydroisocoumarin derivative, which is used in Japan for the preparation of ritual sweet teas. It has several drawbacks as a sweetener: its water solubility is low, and the sweet taste of dilute solutions sets in very slowly and is blemished by a persistent aftertaste.[8]

Glycyrrhizine (**24.22**): R = β-GlcCOOH-β-GlcCOOH (2→1); glycyrrhetinic acid (**24.23**), R = H,
GlcCOOH-β-D-glucuronopyranosyl; apioglycyrrhizine (**24.24**), R = β-GlcCOOH-β-
apioFur(2→1); araboglycyrrhizine (**24.25**), R = β-GlcCOOH-α-L-Ara(2→1), β-GlcCOOH - β-D-
glucuronopyranosyl, β-apio-Fur - β-D-apiofuranosyl, α-L-Ara - α-L-arabinopyranosyl

Periandrine I (a) (**24.26**): R1 = β-GlcCOOH-β-GlcCOOH(2→1), R2 = CH=O;
Periandrine III (a) (**24.27**): R1 = β-GlcCOOH-β-GlcCOOH(2→1), R2 = CH2OH;
Periandrine II (b) (**24.28**): R1 = β-GlcCOOH-β-GlcCOOH(2→1), R2 = CH=O;
Periandrine IV (b) (**24.29**): R1 = β-GlcCOOH-β-GlcCOOH(2→1), R2 = CH2OH,
β-GlcCOOH-β-D-glucuronopyranosyl.

R1 = β-Glc-β-Glc(6→1), R2 = β-Glc-β-Glc(6→1)-β-Glc-β-Glc(2→1)

(**24.30**)

Carnosifloside V (**24.31**) and carnosifloside VI (**24.32**); a - R1 = β-Glc, R2 = β-Glc-β-Glc(2→1);
b - R1 = β-Glc, R2 = β-Glc-β-Glc(6→1); Glc - D-glucopyranosyl

The synthetic phyllodulcin analog 2-(3-hydroxy-4-methoxyphenyl)-1,3-benzo-
dioxane (24.41)[2] is the sweetest of all synthetic derivatives. Its very high sweetness
is marred by its instability: the molecule slowly hydrolyzes. Its aqueous solutions
stored at 25°C lose sweetness completely within 7 days.[2]

Abrusoside A (**24.33**)   β-D-Glc
Abrusoside B (**24.34**)   β-D-GlcA-6-CH32- β-D-Glc
Abrusodide C (**24.35**)   β-D-Glc2- β-D-Glc
Abrusoside D (**24.36**)   β-D-GlcA- β-D-Glc

Glc = glucopyranosyl; GlcA = glucuronopyranosyl

Osladine (**24.37**): (R1= β-Glc2-α-Rha, R2= Rha) and polypodoside A (**24.38**):
(R1=β-Glc2-α-Rha, R2=d-L-Rha, Δ7,8)
Glc = D-glucopyranosyl; Rha = L-rhamnopyranosyl

(L-Arab = L-arabinosyl-)
(**24.39**)

Phyllodulcin (**24.40**), R1 = H and phyllodulcin 8-*O*-β-D-glucoside
(**24.41**) R1 = β-D-glucopyranosyl

24.1.5 FLAVANOIDS

Bitter and bitter-sweet flavanoids (24.42) to (24.49) (Table 24.1) are readily and abundantly accessible from the lemon fruit. Flavanone-7-glycosides are a major group of citrus flavanoids (Table 24.3). Some of them are tasteless.

TABLE 24.3
Flavanoid Glycosides from Citrus Fruits

(24.42) to (24.49)

Name	R¹	R²	R³	Taste
Neohesperidin (24.42)	OCH₃	OH	β-Neohesperidosyl	Bitter
Hesperidin (24.43)	OCH₃	OH	β-Rutinosyl	Tasteless
Naringin (24.44)	OH	H	β-Neohesperidosyl	Bitter
Naringinrutinoside (24.45)	OH	H	β-Rutinosyl	Tasteless
Poncirin (24.46)	OCH₃	H	β-Neohesperidosyl	Bitter
Isosakuranetin-rutinoside (24.47)	OCH₃	H	β-Rutinosyl	Tasteless
Neoriocitrin (24.48)	OH	OH	β-Neohesperidosyl	Bitter
Eriocitrin (24.49)	OH	OH	β-Rutinosyl	Tasteless

Members of this group differ from one another in their flavanone functional groups in positions 3′ and 4′, or in their respective constituent sugar (disaccharide) unit in positions.[7] Either β-rutinose (6-O-α-L-rhamnopyranosyl-β-D-glucopyranose) or β-neospheridose (2-O-α-L-rhamnopyranosyl-β-D-glucopyranose) are the sugar units in these flavanoids. The flavanone rutinosides are tasteless and the flavanone neohesperidosides are bitter.

Hesperidin is a major flavanoidal component of oranges (*Citrus sinensis*) and lemons (*Ctrus limon*), whereas naringin is found in grapefruits (*Citrus paradisi* Macfad), neohesperidin in Spanish oranges (*Citrus aurantilum* L.), and poncirin in lemons (*Poncirus trifoliata*).[2]

Semisynthetic dihydrochalcones are available from a catalytic hydrogenation of corresponding chalcones derived from naturally occurring flavanones. Flavanones used in mixed sweeteners spread into two types: β-neohesperidose containing glycosides of naringin and neohesperidin, and β-rutinose containing hesperidin.

Flavanones and flavanone glycosides have certain applications in the pharmaceutical and food industries, respectively. Chalcones and dihydrochalcones are open-ring derivatives of flavanones. Naturally occurring dihydrochalcones are phlorizin (24.50) (the glycoside occurring in the roots of apple, plum, and cherry trees).

Dihydrochalcones (DHC) may be produced from flavanones. Semisynthetic dihydrochalcones are characterized by a menthol-like refreshing sweet taste. The sweet sensation persists longer (10 min) than in sucrose. Furthermore, chalcones

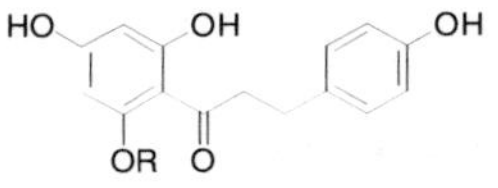

R = β-D-glucopyranosyl-

(**24.50**)

successfully suppress the unpleasant bitter taste in some drinks and pharmaceuticals. Neohesperidin DHC is used in much smaller doses than other sweeteners because of its very high sweetness. It can be used either alone or jointly with other sweeteners.

A novel sweet compound, dihydroquercetin-3-acetate (24.51), has recently been isolated from the buds of the Paraguayan medicinal herb "herba dulce."[2] An analogous artificial sweetener, dihydroquercetin-3-acetate-4′-methyl ether (24.53), was

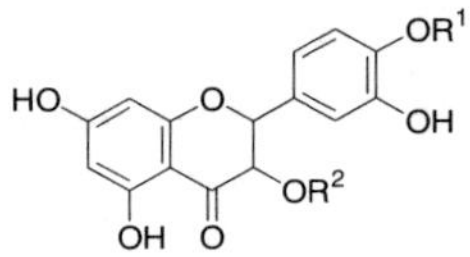

R1	R2	Compound
H	CH3CO	dihydroquercetin-3-acetate (**24.51**)
CH3	CH3CO	dihydroquercetin-3-acetate-4′-methylether (**24.52**)
H	α-L-Rha	3-O- α-L-rhamnosyltaxifoline (**24.53**)

synthesized from the commercially available 2,4,6-trihydroxyacetophenone and isovanillin. Its sweetness was about eight times higher than that of the natural product. A sweet compound, dihydroflavanol 3-O-α-L-rhamnosyl(2S,3S)taxifoline (24.52) was isolated in the 0.01% yield from the dry leaves of a common Chinese medicinal herb huang-qi (*Engelhardtia chrosolepis*).[2] Its relative sweetness or the results of the toxicological tests have not been published.

24.1.6 OTHER NATURAL SWEET COMPOUNDS

Haematoxylin (24.54),[17] a sweet polyoxypregnane glycoside, telosomoside A15

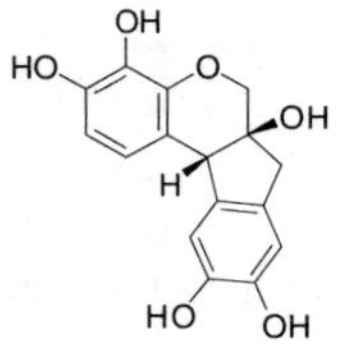

(**24.54**)

(24.55),[18] and monatin (24.56)[19]constitute this group of sweeteners (Table 24.1). Telosomoside is used in Vietnamese folk medicine.

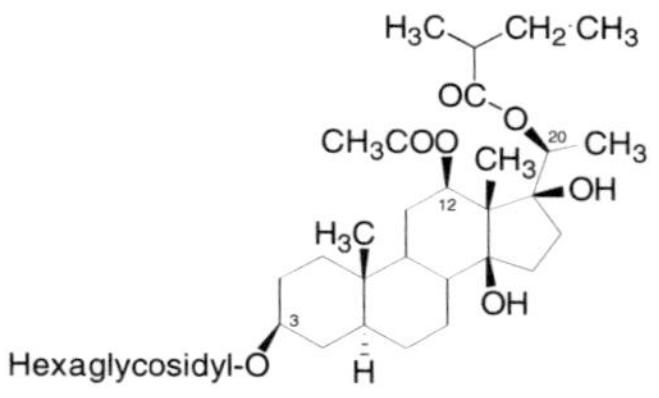

(**24.55**)

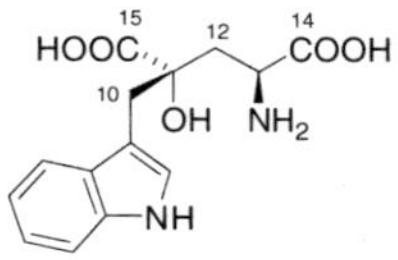

(**24.56**)

24.2 SYNTHETIC NONNUTRITIVE SWEETENERS

The detrimental effects of high consumption of sucrose- and glucose-based sugars, such as obesity and tooth decay in humans, have resulted in extensive research on synthetic nonnutritive sweeteners. The main objective was to develop a safe, high-intensity noncaloric sweetener stable at a wide range of pH and temperatures and with no aftertaste. None of the present products in the market, for example, saccharin (24.57),[2] cyclamate (24.58),[2] acesulfame-K (24.59),[2] aspartame (24.60),[2] alitame

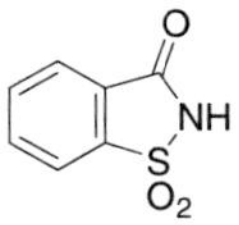

(**24.57**)

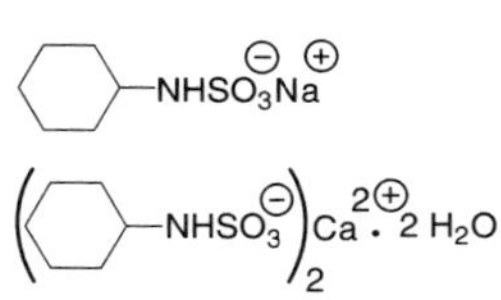

(**24.58**)

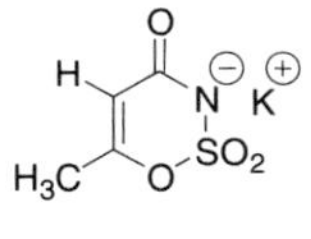

(**24.59**)

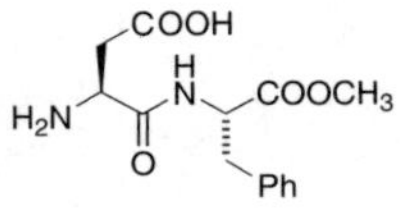

(24.60)

(24.61),[21,22] and sucralose (24.62)[23] fulfill a majority of these requirements (Table 24.4).

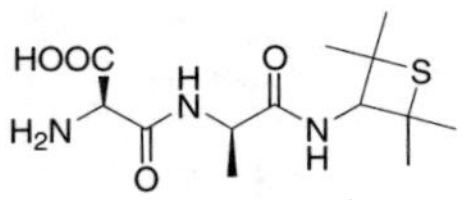

(24.61)

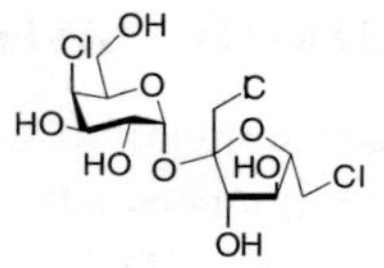

(24.62)

Other artificial sweeteners have also been discovered, but because of their toxic or carcinogenic effects they have not been approved for use. Included in this group are derivatives of urea (dulcin, suosan),[2] derivatives of benzene (5-nitro-2-propoxy-aniline), tribromobenzamides,[2] and D-tryptophan derivatives.[2]

Several research groups are still actively pursuing the search for new nonnutritive synthetic sweeteners. Potential next-generation sweeteners contain a diverse array of functionality ranging from dipeptide derivatives to guanidines, tetrazoles,[25] and ureas.[2] Among the next-generation sweeteners is neotame (24.63), a strong commercial sweetener candidate.[26]

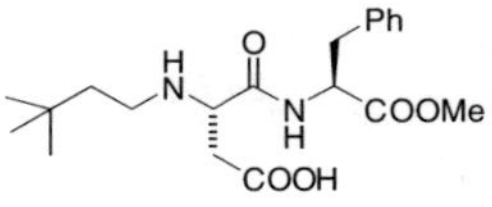

(24.63)

TABLE 24.4
Synthetic Nonnutritive Sweeteners[2,24-26]

Sweetener and Structure	Sweetness[a]	Remarks	Applications	Stability	
				pH	Temperature
Saccharine (24.57)	300–500	Bitter or metallic aftertaste	Used as Na or Ca salts in food and beverages; should not be used daily throughout the lifetime	3–10	< 150
Cyclamates (sulfamates)	30	Bitter aftertaste; their use must be clearly declared on the packaging of the products	Used as Na or Ca salts	2–10	<260
Acesulfame-K (24.59)	130	Pure sweet taste and high water solubility	For sweetening of soft drinks, table-top sweetener, masking function in pharmaceuticals	3–7	<225
Aspartame (24.60)	160–180	Clean sweet taste	Mainly in beverages	3–5	Unstable
Alitame (24.61)	2000	Pleasant sweet taste without aftertaste	Used in Australia, China, and Mexico	6–8	Good
Sucralose (24.62)	600	Pleasant sweet taste	Used in Canada	3–8	Very good
Neotame (24.63)	8000	Clean sweet taste	Next-generation sweetener		

[a]Sweetness with respect to the 10% aqueous sucrose solution as the standard.

Source: From Krutošíková, A. and Uher, M., Natural and Synthetic Sweet Substances, Ellis Horwood, Chichester, 1992, p. 224; Jenner, M.R., in Sweeteners: Discovery, Molecular Design, and Chemoreception, Walters, D. E., Orthoefer, F. T., and Du Bois, G. E., Eds., ACS, Washington, DC, 1991, chap. 6; Uher, M., Krutošíková, A., and Ková_, M., Bull. PV (Bratislava) 31, 341, 1992; Prakash, I., Bishay, I., and Schroeder, S., Synth. Commun., 29, 4461, 1999.

REFERENCES

1. Caldwell, J. and Markley, J. L. Structure-function relationships in sweet tasting proteins, *J. Chem. Soc. Pak.*, 2, 268, 1999.
2. Krutošíková, A. and Uher, M., *Natural and Synthetic Sweet Substances,* Ellis Horwood, Chichester, 1992, p . 224.
3. Kurihara, Y. Characteristics of antisweet substances, sweet proteins, and sweetness-inducing proteins, *Crit. Rev. Food Sci. Nutr.*, 32, 231, 1992.
4. Kohmura, M. and Ariyoshi, Y, Chemical synthesis and characterization of the sweet protein mabinlin II, *Biopolymers*, 46, 215, 1998.
5. van der Wel, H. et al., Isolation and characterization of pentadin, the sweet principle of *Pentadiplandra brazzeana* B., *Chem. Sens.*, 14, 75, 1989.
6. Ming, D. and Hellekant, G., Brazzein, a new high-potency thermostable sweet protein from *Pentadiplandra brazzeana* B., *FEBS Lett.*, 355, 106, 1994.
7. Izawa, H. et al., Synthesis and characterization of the sweet protein brazzein, *Biopolymers*, 39, 95, 1996.
8. Kinghorn, A. D. and Kennelly, E. J., Discovery of highly sweet compounds from natural sources, *J. Chem. Educ.*, 72, 676, 1995.
9. Tanaka, O., Improvement of taste of natural sweeteners, *Pure Appl. Chem.*, 69, 675, 1997.
10. Tanaka, T. et al., Sweet and bitter glycosides of the Chinese drug, Bai-Yun-Shen, *Chem. Pharm. Bull.*, 31, 780, 1983.
11. Nishizawa, M. and Yamada, H., Novel synthetic approaches into intensely sweet glycosides: baiynoside and osladin, *Synlett.*, 785, 1995.
12. Fullas, F. et al., Gaudichandiosides A-E., five novel diterpene glycoside constituents from the sweet-tasting plant *Baccharis gaudichaudiana*, *Tetrahedron*, 4, 8515, 1991.
13. Mizutani, K. et al., Sweetness of glycyrrhetic acid 3-O-_-monoglucuronide and the related glycosides, *Biosci. Biotech. Biochem.*, 58, 554, 1994.
14. Lin, H. M. et al., Isolation and identification of main constituents in an enzymatically hydrolysed licorice extract sweetener, *Food Addit. Contam.*, 18, 281, 2001.
15. Choi, Y. H. et al., Abrusoside A: a new type of highly sweet triterpene glycoside, *J. Chem. Soc. Chem. Commun.*, 8 87, 1988.
16. Kennelly, E. J. et al., Novel highly sweet secodammarane glycosides from *Pterocarya paliurus*, *J. Agric. Food Chem.*, 43, 2602, 1995.
17. Masuda, H. et al., Chemical study on *Haematoxylon campechianum*: a sweet principle and new dibenzb,doxocin derivatives, *Chem. Pharm. Bull.*, 39, 1382, 1995.
18. Vo Duy Huan et al., Sweet pregnane glycosides from *Telosma procumbens*, *Chem. Pharm. Bull.*, 49, 453, 2001.
19. Vleggaar, R., Ackerman, L. G. J., and Steyn, P. S., Structure elucidation of monatin, a high-intensity sweetener isolated from the plant *Schlerochiton ilicifolins*, *J. Chem. Soc. Perkin Trans.* 1, 3095, 1992.
20. Ager, D. J. et al., Commercial, synthetic nonnutritive sweeteners, *Angew. Chem. Int. Ed.*, 37, 1802, 1998.
21. Pfizer, U.S. Patent 4,411925, 1983; *Chem. Abstr.*, 101, 22238, 1984.
22. Pfizer, U.S. Patent 4,375430, 1983; *Chem. Abstr.*, 99, 6056, 1983.
23. Knight, I., The development and applications of sucralose, a new high-intensity sweetener, *Can. J. Physiol. Pharmacol.*, 72, 435, 1994.
24. Jenner, M.R., Sucrulose, in *Sweeteners: Discovery, Molecular Design, and Chemoreception* Walters, D. E., Orthoefer, F. T., and Du Bois, G. E., Eds., ACS, Washington, DC, 1991, chap. 6.

25. Uher, M., Krutošíková, A., and Kováč, M., Compounds generating sweet taste. VII. Extremelly sweet guanidine derivatives (in Slovak), *Bull. PV (Bratislava)* 31, 341, 1992.

26. Prakash, I., Bishay, I., and Schroeder, S., Neotame: synthesis, stereochemistry and sweetness, *Synth. Commun.*, 29, 4461, 1999.

Index

A

Abrusosides, 390, 394, 396
Abrus precatorius, 394
Abzymes, 140
Acacia senegal, 243
Acarbose analogs, transglycosylating activity of
    CGTases, 148
Acesulfame-K, 399, 401
Acetalation, polysaccharides, 129
Acetals
    glycosylation, 28
    hydrolysis, 29
    starch, 319–320
Acetan, 10
Acetic acid, starch esterification, 128
Acetylated starches, 165
Acetylesterases, 151, 152
Acid- and base-catalyzed degradation, 127–128
Acid-catalyzed hydrolysis
    acylation reactions, 27
    dextrinization, 124
Acids, *see also* Organic acids
    acylation reactions, 27
    esterification, 2
    honey, 76, 77
Actilight®, 206
Acylation, 27–28, 40, 321
Add H & Build Model, 64
Adhesives, 171, 319
Adipate, starch, 128, 166
Adulteration, honey, 77–78
Agars, 163, 247
    food industry applications, 161, 162, 166; *see
        also specific applications*
    glucomannans and, 241
    monosaccharide components, 235
    structures, 10
    viscosity, 236
Agavaceae, 208, 212–213
Agave, 213
Aglycones, hydrolysis, 29
Alcohol-insoluble residue (AIR), pectins, 183
Alcoholized and phenolyzed starches and sugar
        alcohols, nonnutritional
        applications, 317
Alcohols, 183
    acylation reactions, 27

amino compounds, 31
applications and physicochemical properties, 41
ethers, 39
glycosylation, 28
honey, 77
modifications, 36–38
nonnutritional applications, 317
oxidation reactions, 23
starches
    irradiation products, 126
    and retrogradation, 95
sugar, *see* Sugar alcohols
Aldaric acids, oxidation reactions, 23, 25
Aldehydes, 2, 3
    glycosylation, 28
    honey, 77
    oxidation reactions, 23
    starch irradiation, products of, 126
Aldehydrocarboxylic acids, oxidation reactions,
    24
Aldolases, 139, 140
Aldonic acids
    coordination complex formation, 22
    oxidation reactions, 23–24
Aldopyranoses
    acylation reactions, 27
    glycophore in, 65–66
Aldose (xylose) reductase, 139
Aldoses, oxidation reactions, 23
Aldotriose, 2
Aldurinic acid, coordination complex formation,
    22
Algae, 235, 246–248; *see also specific products*
    agars, 247
    alginates, 246–247
    carrageenans, 247–248
    furcellerans, 249
    monosaccharide components, 235
Alginates, 246–247
    biodelivery hydrogel preparation, 162
    food industry applications, 162, 163, 165, 166;
        see also specific applications
    monosaccharide components, 235
    structures, 10
    viscosity, 236
Alitame, 399–400, 401
Alkali modification of starch, 128
Alkanoic acids, acylation, 27–28

Alkoxylamines, 31
Alkylation, 2, 25–27
Alkyl glycosides, 29, 40
Alkyl pyranoside glycosylation, 28
Alkylsulfonyl starches, 320–321
Allium, 209
Allolactose, 139, 140
α-Hydroxy acids, coordination complex
    formation, 22
Alternan, 135
Alternansucrase, 136
Amination, 2
Amines, starch esterification, 128
Amino acids
    honey, 75–76
    nonnutritive sweeteners, 400
    Strecker degradation of, 305–307
    sugar beet and cane composition, 48
Amino compounds, chemical reactions of
    saccharides, 31–32
Aminodeoxysugars (Maillard reaction)
    degradation, 295–305
        1-amino-1,4-didoxydiketoses, 304–305
        1-deoxy-2,3-diuloses, 299–304
        3-deoxy-2-osuloses, 296–299
Amino ethers, 322
Amino- substituted compounds and carbamates
    polysaccharide modification, 129
    starch applications, nonnutritional, 322
Ammonium salts, sugar beet composition, 48
Amorphophallus konjac, 241
α-Amylase, 134, 146–147
    common properties, 145–146
    isomaltooligosaccharides, 142
    starch conversions, 143, 144
    ultrasound, yield improvement with, 133
β-Amylase, exo-acting enzymes, 147–148
Amylases
    honey, 75–76
    processed starch food products, 166
    starch conversions, 144
Amylogram peak viscosity, wheat flour, 111
Amylomaize, 84
Amylopectin, 2
    electrodialysis, 126
    phytoglycogen, 88–89
    starch granule organization, 89–90
    structure, 83–87
    structures, 17
Amylopullulanase, 145
Amylose, 94
    chemical modification of starch, 127
    electrodialysis, 126
    starch granule organization, 89
    structure, 17, 82–83

sucrose ester–starch complexes, 43
Amylose-lipid complexes, enzymatic
    conversions, 143
Amylosucrase, 136
Amyrllidaceae, fructan crops, 208, 212
Analytical chemistry, commercial products
    alkylation reactions, 26
    cyclodextrin-based, 287–288
Anhydrogalactose
    carrageenans, 247–248
    gum components, 235
Animal tissue carbohydrates, 255–269
    glycogen, 257–264
        glycogenesis, 258–259
        glycogenolysis and glycolysis, 259–264
    glycoproteins, 267–269
    glycosaminoglycans, 264–267
        metabolism, 265–267
        proteoglycans, 265
        utilization, 267
Anogeissus latifolia, 245
Anomeric carbon atom, 4
Anomeric mixtures, amine compounds, 32
Anomerization
    acylated aldopyranoses, 27
    furanosides, 29
Anthocyanines, honey, 76, 77
Anthranilates, 322
Antifoaming agents, 318
Antifreeze, sugar esters, 41
Antimicrobial properties, 333
    chitosan, 226–227
    fatty acid esters of sucrose, 43–44
Antistaling activity, 160
Apioglycyrrhizin, 394
Applications
    fatty acid esters, 42–44
    fatty acid polyesters, 44
    polydextrose, 41–42

    saccharide ethers, 42
    sugar alcohols, 41
Arabic gum, 243–244, 320
    food industry applications, 162, 163, 165; *see
        also specific applications*
    reserve polysaccharides, 235
    structures, 10
    viscosity, 236
Arabinases, 153
Arabinofuranosidases, 154
Arabinofuranosyl residues, pectin, 183–184
Arabinogalactan proteoglycans, 243–246
Arabinogalactans (Larch), 237
Arabinogalactan structures, 13
Arabinose

gum components, 235
    pectins, 183, 184
Arabinoxylans, 154, 237
    cell wall polysaccharides, 235
    viscosity, 236
Araboglycyrrhizin, 394
-aric acid, 25
Arteriosclerosis, 332
Artificial sweeteners, 69–70; *see also* Sweeteners
Ascophyllum nodosum, 246
Ascorbic acid, 1, 139
Asparagaceae, fructan crops, 208, 212
Aspartame, 399, 401
A-starch purification, 113–114
ASTRA, 361
Astragalus gummifer, 244
Aureobasidium, 36
Autoclaving, chitosan, 225
Autolytic enzymes, antimicrobial action, 43–44
Aziridine, 129

B

Baccharis gaudichaudians, 389
Bacteria, *see also* Microbial systems
    intestinal, 332
    sucrose fatty acid ester antimicrobial
            properties, 43–44
Bagasse, 51, 55
Baiynoside, 392
Bai-Yn-Shen, 392
Banana, 84, 212
Bandages, 319
Barium, 21
Barley
    amylopectin, 84, 87
    fructans, 211
    global markets, 104
Bassarin, 17
Beet sugar, 48–54
    by-products and waste, 54
    chemical composition of beets, 48
    commercial sugar varieties, 53–54
    crystallization, 52–53
    evaporation, 51–52
    extraction, 49–50
    juice purification, 50–51
    preparation of beets, 49
1-H-2-Benzopyran-1-ones, 391, 394–396
Benzoylation, acylation reactions, 27–28
Betaine, sugar beet, 48
Beverages, 166–169
Binders, sugar alcohols, 38
Biochemistry of taste transduction, 69–70
Biodegradable starch copolymers, 322

Biodelivery adhesive systems, 171
Biodelivery hydrogels, 162
Biomass
    Ethanolsolv process of Kleinert, 342
    ethanol supply, search for, 341–342
    fuel cell elements, 336–337
    green biorefineries
        birth of, 339–340
        ethanol production, 340–341
    methane, 339
    and oil, 337–339
    pulp and paper industry reshuffle, 342–346
    supercellulases, search for, 341–342
Biopolymers
    food industry applications, 162
    starch glass transition temperature, 95–96
Biosynthesis, fructans, 200–204
Biotechnology, cyclodextrins, 287
Birefringence, starch granules, 89, 95
Block copolymers, starch carbamate, 322
Blood expanders, 319
Branch structures, starch, 89–90
    amylopectin, 83, 85–86
    analysis, 357, 365
    and retrogradation, 95
Brassica napus (rape) honey, 74
Brazzelin, 388
Browning, *see* Maillard reaction
Bruyodulcoside, 394
Bulking agents
    food industry applications, 161
    functions of low-molecular weight saccharide
            blends, 160
    polydextrose, 42
Bulk sweeteners, sugar esters, 41
Butyric acid, honey, 76, 77

C

Calcium, 22, 213
    absorption of, carbohydrates and, 327, 333
    and α-amylase, 146
    biodelivery hydrogel preparation, 162
    coordination complex formation, 21
    guluronic acid and, 246
    oxidized starches, 317–318
    pectin properties, 187, 188–189, 188
    sweetness, taste transduction, 69, 70
Calcium chelators, pectin extraction, 186
Calcium D-gluconates, 24
Calcium oxide, 54
Calcium stabilizers, 169
Caloric value, sugar alcohols, 42
Cane sugar, 54–55
    by-products and waste, 55

chemical composition of cane, 48
Capparis masaikai, 388
Caramels, 38–39
Carbamates, 129, 322
Carbamoylated starch, 322
Carbohydrates
    as class of compounds, 2–3
    enzymatic conversions, *see* Enzymatic
        conversions of carbohydrates
    fiber, 332–333
    food, 5
    and health, 329, 331–332
    nonnutritional saccharides, 333
    nutritional value, 326–329, 330
    occurrence and significance, 1–2
    organoleptic properties of saccharides, 333
    structure, 3–5, 6–17
Carbonizate, starch pyrolysis products, 124
Carboxylic acids
    honey, 77
    starch esterification, 128
    starch irradiation, products of, 126
    starch modification, chemical, 128
    sugar cane composition, 48
Carboxymethyl cellulose
    alkylation reactions, 26
    food industry applications, 162, 163, 162, 165,
        166; *see also specific applications*
Cardiac glycosides, 29
Cardiovascular disease, 332
Caries, 155, 331, 399
Carnosifloside, 394, 395
Carob tree, 237
Carotenoids, honey, 76, 77
Carrageenans, 2, 163, 247–248
    food industry applications, 162, 164, 165, 166;
        see also specific applications
    glucomannans and, 241
    monosaccharide components, 235
    structures, 11
    viscosity, 236
Casein
    cyclodextrin complexes, 284
    furcellerans and, 249
Cassava
    global markets, 104, 105
    starch production, 118–120
Cassia gum, 235, 241
Catalase, honey, 75
Cations, *see* Calcium; Metal salts and complexes;
        Salts and minerals
Cattail millet, 84
Celite, 40
Cellobiohydrolase, 150
Cellobiose, 8

Cellobiose dehydrogenase, 139
Cellodextrins, 139
Cellulases, 150–151
    fungal, 150
    supercellulases, search for, 341–342
Cellulose
    energy sources, environmental
        ethanol production, 339–342
        green biorefineries, 339–341
        pulp and paper industry reshuffle,
            342–346
        supercellulases, search for, 341–342
    enzymatic conversions, 149–151
    modified starch additives, 129
    spectroscopy
        mid-IR, 372
        NMR, 377
    structures, 11
    sugar beet composition, 48
Cellulose derivatives
    alkylation reactions, 26
    food industry applications, 162, 163, 164, 165,
        166; *see also specific applications*
Cellulose xanthate, 317
Cell walls
    chitosan, 218
    pectins, 184–185, 186, 192
    polysaccharide components, 234, 235,
        236–238
Ceratonia siliqua, 237
Cereals and grains
    cell wall polysaccharides, 235
    fructan crops, 208, 211–212
    global markets, 104
    starches, 87, 94
Cesalpinia spinosa, 237
CGTase, starch conversions, 145, 148–149
Chalcones, 397
Chemical composition of honey, 74–77
    dyes and other components, 76, 77
    minerals, 76, 77
    organic acids, 76, 77
    proteins, 75–76
    saccharides, 74–75
Chemically modified starches, 94
Chemical reactions/reactivity, *see* Modifications;
        Reactivity
Chemical stability, chitosan, 223–225
Chemical structure, *see* Structure
Chicory (*Cichorium intybus*), 204–206, 208
Chinese taro, 84
Chirality, 2–3
Chitinase, 134
Chitin/chitosans, 2, 217–227
    chemical stability, 223–225

composition and molecular mass, 218–221
  chemical structure, 218–220
  molecular mass, 220–221
  sequence, 220
enzymatic conversions, 149–151
enzymatic degradation, 225–226
food industry applications, 162, 163; *see also*
    specific applications
pectin interactions, 192
physical properties, 222–223
  ion binding, 222
  solubility and charge density, 222–223
structures, 11
technical properties, 226–227
Chloral hydrate, 320
Chloride
  and α-amylase, 146
  honey, 76
Chlorinated starches, 128, 129, 321
Chlorine
  starch oxidation, 128
  wheat flour processing, 92
Cholesterols, cyclodextrin and, 277, 278–279, 285
Chondroitin sulfates, 11
Chondrus crispus (Irish moss), 248
Chromatography
  chitosan, 221

  fructans, 200
  pectins, 186
  starches, 354, 357, 359
Cichorium intybus (chicory), fructans, 204–206,
    208
Citrates, sugar beet juice purification, 50
Citrus flavanoids, 396
Ciucanu-Kerek procedure, 26
Clarity, starch pastes, 94
Coatings, low-molecular-weight saccharides, 160
Cocrystallization, starches, 94
CODAwin32, 361
Coffee, 284–285
Colloid titration, chitosan, 222
Colonic bacteria, 332
Combretaceae, 245
Complexation, starches, *see* Inclusion compounds
Complexes, coordination, 21–22
Compositae, fructans, 204–209
  chicory (*Cichorium intybus*), 204–206, 208
  globe artichoke, 208, 209
  Jerusalem artichoke (*Helianthus tuberosum*),
    206–209
Compost, cane sugar byproducts, 55
Compression, starch, 125
Conformation
  pectin, 188–189

starches, 357–358
Cooling effect, 41, 42
Coordination complex formation, 21–22
Copolymers, starch, 322–323
Corn (maize), 105–109
  analysis of molecular characteristics, 364, 365
  global markets, 104
  starches, 88
    amylopectin, 84, 85
    polysaccharide blends in foods, 164
    sucrose ester–starch–protein complexes, 43
    starch granules in native state, 356
  starch production
    processes, 107–109
    substrates, 105–106
Cosmetics and toiletry, cyclodextrin applications,
    286–287
CPCwin32, 361
Cryoprotective agents
  food industry applications, 160, 161; *see also*
    specific applications
  sugar esters, 41
Crystallinity
  amylopectin, 85
  pectin, 188
  starches, 92
Crystallization
  honey, 74
  starch retrogradation, 94–95
Curculin, 387
Curcurbitane glycosides, 394
Curdlan gel, 169
Cyamopsis tetragonoloba, 237
Cyanobacteria, amylopectin, 84
Cyano compounds, sweeteners, 63
Cyclamates, 399, 401
Cyclic saccharides, 2; *see also specific glycosides*
Cyclization, 3–5
  amine compounds, 32
  starch irradiation, products of, 126
Cycloartane glycosides, 394
Cyclodextrin glycosyltransferases (CGTases),
    145, 148–149
Cyclodextrins and cycloamylases, 148–149,
    271–288
  applications in foods, 282–285
    complexation and sequestration, 284–285
    flavors and vitamins, 282–283
    miscellaneous, 285
    physical property modifiers, 283–284
    taste modifiers, 284
  chemistry and physicochemical properties,
    272–275
  derivatives, 278–280

food industry applications, 161; *see also specific applications*
inclusion complexes, 280–282
industrial products and technologies, 285–288
  analytical separation and diagnostics, 287–288
  biotechnology, 287
  cosmetics and toiletry, 286–287
  pharmaceuticals, 285–286
  textiles, 287
metabolism and toxicology, 277–278
production of, 275–276
structures, 9
Cyclohexane diamine tetracetic acid (CDTA), 186
Cydonia oblonga, 242

D

Dairy products, 168, 169
Deacetylation, polysaccharides, 140
DEAE-Cellulose, 26
DEAE-Sephadex, 26
Debranching enzymes, 148
Demethylation
  pectin degradation, 151

  polysaccharide degradation, 140
Density, honey, 74
Deodorants, 287
Deoxyribose, 2
Deoxysugars, 2
Deoxysugars, amino (Maillard reaction), 295–305
  degradation
    1-amino-1,4-didoxydiketoses, 304–305
    1-deoxy-2,3-diuloses, 299–304
    3-deoxy-2-osuloses, 296–299
  formation and rearrangement to, from glycosylamines, 292–293
Derivatization, alkyl groups, 26
Desserts, dressings, emulsions, and spreads, 169–171
Detergents, cyclodextrins, 287
Dextran, 11, 135
Dextransucrase, 136
Dextrinization, 124–125, 125
Dextrins
  honey, 74
  oxidation, 139
Dextrose equivalent (DE), 161
Diabetes, 329
  agave, 213
  fructan-containing foods, 204
Diagnostic products, cyclodextrins, 287–289
Diastereomers, 3

Differential scanning calorimetry (DSC), starches, 87, 91, 354
Dihydrochalcones (DHC), 397–398
Dihydroisocoumarins (1-H-2Benzopyran-1-ones), 391, 394–396
Dihydroxyacetone phosphate (DHAP), 140
DIMEB, 280
Dimethylformamide, 39–40
Dipeptides
  nonnutritive sweeteners, 400
  sweetness, 61
Disaccharides, 140–141
Discoreophyllum cuminisii, 388
Dispersing agents, fatty acid polyesters, 41
DMSO, 39–40, 356
Docosanoic acid, 43
Donnan effect, 182
Dulcin, 400
Dulcoside, 392, 393
Dyes and pigments, honey, 76, 77, 76

E

Electrolysis and electrodialysis, 126
Electromagnetic radiation, starch modification, 125, 126
Electron donors and dispersion sweeteners in sucrose/galactosucrose system, 66–69
Electrophoretic light scattering, chitosans, 222
Emulsifiers
  fatty acid polyesters, 41, 42–43, 44
  functions of low-molecular-weight saccharide blends, 160
  functions of polysaccharide blends, 169, 170, 171
  green, 148–149
  sugar esters, 41
Emulsions, 163
Enantiomers, 2–3
Encapsulation, 161, 162
Endoarabinase, pectin degradation, 153
Endo-β-1,3,-glucanase, 134
Endocellulase, 150
Endodextranase, 142
Endo-glycoside hydrolases, 133, 134
Endohydrolases, 151, 152
Endopolygalacturoniases, 151, 152
Endo-xylogalacturonase, 153
Energy sources (environmental), glucose as, 335–346
  biomass, and oil, 337–339
  Ethanolsolv process of Kleinert, 342
  ethanol supply, search for, 341–342
  fuel cell elements, 336–337

green biorefineries
  birth of, 339–340
  ethanol production, 340–341
  methane, 339
  pulp and paper industry reshuffle, 342–346
  supercellulases, search for, 341–342
Energy sources (physiological), 327–329, 330, 332
Engerlhardtia chrosolepis, 398
Enterolobium gum, 245
Enzymatic conversions of carbohydrates, 132–155
  enzymes, 133–141
    glycansucrases, 135–136
    glycoside hydrolases, 133, 134
    glycosidic bond hydrolysis and transglycosylation, mechanisms of, 136–138
    glycosyltransferases, 133, 135
    glycosynthases, 138–140
    monosaccharide synthesis, 140
    oxidoreductases, 140
  oligosaccharides, 141–143
    fructooligosaccharides, 141–142
    galactooligosaccharides, 142
    isomaltooligosaccharides, 142
  polysaccharides, 143–154
    cellulose, chitin, and chitosan, 149–151
    fructans, 154–155
    pectin, 151–153
    starch, 143–149
    xylan, 153–154
  starch, 143–149
    α-amylase family, common properties, 145–146
    α-amylases, 146–147
    cyclodextrins and cycloamylases, 148–149
    debranching enzymes, 148
    exo-acting enzymes, 147–148
    miscellaneous conversion, 149
    trahalose, 149
Enzymes, honey, 75
Eriocitrin, 397
Erlose
  honey, 75, 78
  structures, 8
Erythritriol, 36, 42
Erythulose, 2
Essential oils, honey, 76
Esterases, 139
  pectin degradation, 151
  polysaccharide degradation, 140
Esterification, 2
  acylation reactions, 27

  polysaccharide modification, 128
Esters, 41
  acylation reactions, 27
  applications, 42–44
    antimicrobial properties, 43–44
    complex formation with starch and protein, 43
    emulsifying properties, 42–43
    fatty acid, 39–40, 42–44
    pectins, 183, 187
  starch
    irradiation products, 126
    nonnutritional applications, 320–321
Ethanal, 126
Ethane, starch pyrolysis products, 124
Ethanol
  Ethanolsolv process of Kleinert, 342
  green biorefineries, 339–341
  starch irradiation, products of, 126
  supercellulases, search for, 341–342
  supply, search for, 341–342
Ethanolsolv (Kleinert) process, 342
Ethers
  alkylation reactions, 26, 27
  applications and physicochemical properties, 42
  DEAE cellulose and Sephadex, 26
  modifications, 39
  polysaccharide etherification, 128–129
  saccharide, 39, 42
  starch, nonnutritional applications, 318–319, 322
Ethylene glycol, 319
Ethylene oxide, 129
Eucheuma, 248
Eugenol, 129
Exo-1,4-α-D-glucan lyase, 145
Exo-acting enzymes, 147–148
Exoglucanase, 150
Exo-glucoside hydrolases, 133, 134
Exo-N-Acetyl-α-D-galactosaminidase, 134
Exo-polygalacturonidases, 151, 152
Extraction,
Exudates, gum, 243–245

F

Fat-soluble vitamins, cyclodextrins, 283
Fat substitutes
  fatty acid polyesters, 40, 41, 44
  functions of low-molecular-weight saccharide blends, 160
Fatty acid esters
  antimicrobial properties, 43–44
  applications, 42–44

antimicrobial properties, 43–44
complex formation with starch and
    protein, 43
emulsifying properties, 42–43
applications and physicochemical properties,
    42–44
modifications, 39–40
Fatty acid methyl esters (FAME), 40, 41
Fatty acid polyesters
applications and physicochemical properties,
    44
modifications, 40-41
Fatty acids
acylation reactions, 27–28, 40
cyclodextrin complexes, 287
cyclodextrins and cycloamylases, 148
starch esters, 321
Fatty alcohols, ethers, 39
Fenugreek gum, 235, 241
Fiber
corn (maize), 106
fructans, *see* Fructans
health and nutrition, 332–333
wheat flour, 111
Fibrolose®, 206
Films
chitosan, 226
cyclodextrin complexes, 285
Fischer, Emil, 3

Fischer link, 3
Fischer synthesis, glycosylation, 28
Flavanoids
hydrolysis, 29
sweeteners, 391, 396–398
Flavones, honey, 76, 77
Flavor modifiers, cyclodextrins, 284
Flavors and vitamins, cyclodextrin complexes,
    282–283
Flaxseed gum, 241–242
reserve polysaccharides, 235
viscosity, 236
Flocculation, sugar purification, 50, 54–55
Flottweg tricanter process, 110
Foaming
cyclodextrin complexes, 287
fatty acid esters, 43
saccharide ethers, 42
Foams, starch polyurethane, 322
Food carbohydrates, 5
Food products
polydextrose uses, 42
starch gel uses, 91
sugar alcohols, 38
sweeteners, *see* Sweeteners

Food texturization and functional properties,
    159–174
low-molecular-weight saccharides, 160–161
polydextrose, 42
polysaccharide blends, 163–173
    dairy products, 168, 169
    desserts, dressings, emulsions, and spread,
        169–171
    gels and drinks, 166–169
    meat products, 172, 174
    miscellaneous, 172–173
    starch-containing, 163–165
    whipped and emulsified products, 169,
        170
polysaccharides, 162, 163
sugar esters, 41
Forget-me-not (*Myosotis*) honey, 74
Formaldehyde, 2
Formaldehyde cross-linked starch, 319, 322
Formulation aid, polydextrose, 42
Fragment analysis, starch, 354, 357, 365
Fragrances — perfumes, cyclodextrin complexes,
    286, 287, 288
Free radicals
starch graft polymerization, 129
starch modification, 126
Freezing
glass transition temperature, 96
starch modification, 124
Fructan-β-fructosidase, 154
Fructans, 135
analytical approaches, 198–200
biosynthesis and properties, 200–204
crops, 204–213
    Agavaceae, 208, 212–213
    Amyrllidaceae, 208, 212
    Asparagaceae, 208, 212
    chicory (*Cichorium intybus*), 204–206,
        208
    Compositae, 204–209
    globe artichoke, 208, 209
    Jerusalem artichoke (*Helianthus
        tuberosum*), 206–209
    Liliaceae, 208, 210–211
    monocotyledons, 208, 209–210
    Poaceae (cereals and grains), 208,
        211–212
enzymatic conversions, 154–155
occurrence, 198
Fructooligosaccharides, enzymatic conversions,
    141–142
Fructose
agave, 213
commercial sugar varieties, 53
coordination complex formation, 22

glass transition temperature, 96
glycophore localization, 59
honey, 74, 75, 76
production of, 55
sugar cane composition, 48
sweetness, 60
Fructose intolerance, 331
Fructosyl exohydrolase, 202
Fructosyl transferases, 141, 201–202
Fruits, gelled and texturized, 170
Frutafit®, 206
Frutalose®, 206
Fucose, 2, 235
Fucosidase, 134
Fucosidolactose, 8
Fucosyltransferase, 134
Fucus serratus, 246
Fuel cell elements, 336–337
Functional groups
    nonnutritive sweeteners, 400
    taste sensory physiology, 58, 59
Functional properties, *see* Food texturization and
        functional properties
Fungicides, cyclodextrin complexes, 285, 288
Furan derivatives, starch pyrolysis products, 124
Furanoses, oxidation reactions, 23
Furanosides, 28–29
Furcerellans, 249
    food industry applications, 166
    monosaccharide components, 235
    structures, 11
    viscosity, 236

G

Galactanases, 153
Galactitol, 141
Galactomannans, 169, 170
Galactooligosaccharides, enzymatic conversions,
        142
Galactopyranosyl residues, pectin, 183–184
Galactose, 140
    agars, 247
    conversion to ascorbic acid, 1
    glycophore localization, 59
    gum components, 235
    pectins, 183
    pyranose 2-oxidase conversions, 139, 141
    sweetness, 60, 66
Galactose-4-sulfate, 235
Galactose-sufated anhydro-D-galactose, 235
β-Galactosidase, 141, 153
Galactosidases, 134
Galactosucrose, 66–67
Galactosyltransferases, 134

Galacturonic acid
    gum components, 235
    oxidation reactions, 24
    pectins, 183, 185
Galacturonohydrolase, 152
Gas chromatography
    analysis of molecular characteristics, 363
    silylation for, 26
Gaudichaudioside, 389, 392
GEA Westfalia Separator Industries, 118
Gelatin, 163, 165
Gelatin form sponges, 319
Gelatinization
    physical modification, 125
    potato starch, 91–92
    starch, 89, 90, 143
Gelidium agars, 247
Gellan, 12, 170
Gelling agents, 162, 169
Gels — gel formation, 22
    biodelivery hydrogel preparation, 162
    chitosan gelling properties, 226
    food industry applications, 165
    functions of polysaccharide blends, 166–169
    pectin properties, 188, 191
    pectins, 182
Gentianose, 8
Gentobiose
    pyranose 2-oxidase conversions, 139, 140
    structures, 8
Gentose, 8
Ghatti gum, 245
    reserve polysaccharides, 235
    structures, 12
    viscosity, 236
Gigartina, 248
Glass transition temperature, starch, 95–96
Global markets, starch, 104–105
Globe artichoke, 208, 209
Glucan lyases, 145
Glucans, 135
β-Glucans, 236–237
Glucans
    analysis of molecular characteristics, 355, 365
    cell wall polysaccharides, 235
    starch granules in native state, 356
Glucaric acid, 25
Glucoamylases
    exo-acting enzymes, 148
    starch conversions, 144
    ultrasound, yield improvement with, 133
Glucomannans, 241
    reserve polysaccharides, 235
    viscosity, 236
Gluconic acid

honey, 76, 77
  oxidation reactions, 24
Glucono-1,5-lactone, 140
Glucopyranoses
  cyclodextrins, 272
  oxidation reactions, 23
  sweetness, 59
Glucopyranoside hydrolysis, 29
Glucosamine, chitin degradation, 150
Glucose
  commercial sugar varieties, 53
  glass transition temperature, 96
  glycophore localization, 59
  gum components, 235
  honey, 74, 75, 76
  oxidation reactions, 24
  production of, 55
  pyranose 2-oxidase conversions, 139, 140
  sugar cane composition, 48
  sweetness, QSAR correlation analysis, 66
Glucose intolerance, 329
Glucose oxidase, 138
  honey, 75
  isomaltooligosaccharides, 142
α-Glucosidase
  exo-acting enzymes, 147
  isomaltooligosaccharides, 142
  starch conversions, 144
β-Glucosidase, cellulose digestion, 150, 151
Glucosidases, 134, 145
Glucosyl transferases
  cyclodextrins, 275–276
  mechanisms of glycosidic bond hydrolysis,
    136–138
Glucuronic acids, oxidation reactions, 23, 24, 25
Glutamic acid, 54
Glutelins, 107
Gluten
  corn (maize), 109
  global markets, 104
  modified Martin process, 111
  sucrose ester–starch–protein complexes, 43
  Westfalia process, 113–114
  wheat flour, 111, 112
Glycansucrases (transglycosylases), 135–136; see
    also Transglycosylation
Glyceric aldehyde, 2
Glycerol, 36
Glycoconjugates, hydrolysis, 29
Glycogen
  animal tissue, 257–264
    glycogenesis, 258–259
    glycogenolysis and glycolysis, 259–264
  structures, 12
Glycolipids, 2

Glycols, taste sensory physiology, 58
Glycophores
  aldopyranoses, 65–66
  taste sensory physiology, 58; see also Taste,
    sensory chemistry
Glycoproteins, 2, 267–269
Glycosaminoglycans, animal tissue, 264–267
  metabolism, 265–267
  proteoglycans, 265
  utilization, 267
Glycosidases, lactose hydrolysis, 143
Glycoside hydrolases, 133, 134
  enzymatic conversions, 133, 134
  mechanisms of glycosidic bond hydrolysis,
    136–138
Glycosides, 2
Glycosidic bonds
  amylose, 82–83
  cyclodextrins, 272
  hydrolysis, 29
  hydrolysis and transglycosylation,
    mechanisms of, 136–138
  isomaltooligosaccharides, 142
  starch modification, chemical, 127
  structures, 5
Glycosylamines, 31
  formation and rearrangement to amino
    deoxysugars, 292–293
  other reactions, 293–295
Glycosylation, 28–29
  chemical reactions of saccharides, 28–29
  fatty acid polyesters, 41
  O-glycosylation, biological significance,
    29–31
Glycosyltransferases, see also Transglycosylation
  cyclodextrins, 148
  enzymatic conversions, 133, 135
Glycosynthases, 138–140
Glycotransferases, 134
Glycyrrhetinic acid, 393
Glycyrrhizin, 389, 392, 394, 395
Glyoxal, 2
Graft polymerization, polysaccharide
    modification, 129–130
Grain crops, fructans, 208, 211–213
Grasses, fructans, 212
Green biorefineries
  birth of, 339–340
  ethanol production, 340–341
Green emulsifiers, 148–149
Green leaf canna, 84
Guanidine, sweeteners, 400
Guar gum, 240
  food industry applications, 162, 165; see also
    specific applications

reserve polysaccharides, 235
structures, 12
viscosity, 236
Gulopyranoses, coordination complex formation, 21
Guluronic acid, 235, 246
Gum arabic, *see* Arabic gum
Gum ghatti, *see* Ghatti gum
Gums and mucilages, 232–249
agars, 247
algal gums, 246–248
alginates, 246–247
arabinogalactan proteoglycans, 243–246
arabinogalactans (Larch), 237
arabinoxylans, 237
carrageenans, 247–248
cassia gum, 241
exudates, 243–245
fenugreek gum, 241
flaxseed gum, 241–242
food industry applications, 161; *see also specific applications*
fucellarans, 249
β-glucans, 236–237
glucomannans, 241
guar gum, 240
gum arabic, 243–244
gum ghatti, 245
gum traganth, 244
Karaya gum, 244
legume seed galactomannans, 239–241
locust bean gum, 239–240
marigold flower gum, 245
microcapsules, 320
okra gum, 245
pectins (rhamnoglactouronans), 237–238
psyllium gum, 242
quince seed gum, 242
reserve polysaccharides, 239–242

rheology of gums, 234, 236
soluble mucilage polysaccharides, 238
structure, 234, 235
structures, 10, 12, 15, 16, 17
tamarind gum, 242
tara gum, 240–241
wall polysaccharides, 234, 236–238
yellow mustard seed gum, 238
Gum traganth, 244

H

Hakamori method, 26
Halogenation
polysaccharide modification, 129

starch modification, 128–129
Halostarches, nonnutritional applications, 321–322
Health
carbohydrates and, 329, 331–332
fatty acid polyesters and, 44
Helianthus tuberosum (Jerusalem artichoke), 198, 202, 206–209
Hematoxylin, 391, 398
Hemiacetals, 4
Hemicelluloses
food industry applications, 163
structures, 13
Hemiketals, 4
Hemsleya carnosiflora, 394
Heparin, 2, 15
Heparin-like compounds, 1, 318, 320, 333
Heptagonal sweetener, 61–63
Hernandulcin, 389, 392
Hesperidin, 397
Hibiscus esculentis, 245
High-amylose starches
food industry applications, 163
viscosity, 93
High-performance liquid chromatography
cyclodextrin complexes, 288
fructans, 198
starch, 363
starches, 354
Honey, 73–78
adulteration, 77–78
chemical composition, 74–77
dyes and other components, 76, 77
minerals, 76, 77
organic acids, 76, 77
proteins, 75–76
saccharides, 74–75
physical properties, 74
Honeydew honeys, 74
Huang-qi, 398
Humectants
food industry applications, 160
functions of low-molecular-weight saccharide blends, 160
polydextrose, 42
sugar esters, 41
Hyaluronic acid, 2, 15
Hydrogels, 163
Hydrolysis
aminated compounds, 31
amine compounds, 32
dextrinization, 124
glycosides, 29
mechanisms of glycosidic bond hydrolysis, 136–138

ultrasound, yield improvement with, 133
Hydrophilic–hydrophobic balance
    fatty acid esters, 42–43
    sucrose ester–starch–protein complexes, 43
Hydroxyaldehydes, 2
Hydroxycortisone, 287
Hydroxyketones, 2
Hydroxymethyl furfural, honey, 76
Hydroxypropylated starches, 165
Hydroxypropyl cyclodextrins, 278
Hydroxypropylmethyl cellulose, 162, 163, 165,
    166
Hygroscopicity, sugar alcohols, 42
HyperChem Build menu, 64
Hysteresis, agars, 247

I

Inclusion complexes
    cyclodextrins, 280–282
    physical modification, 126–127
    solvent effects, 126
Infrared spectroscopy, 372–374
Inorganic ions, *see* Metal salts and complexes;
        Salts and minerals
Interesterification, 40
Interfacial properties, fatty acid esters, 43
Intestinal flora, 332
Inulan, 135
Inulinase, 154
Inulinases, 154, 155
Inulins, 155
    food industry applications, 163, 169
    structures, 15
Inulin-type fructans, 201
Inulosucrase, 136, 142
Invertase, 134, 142
    honey, 75, 76
    ultrasound, yield improvement with, 133
Ion binding, chitosan, 222
Ionic conditions
    chitosan binding, 222
    oxidized starches, 317–318
    pectin properties, 188, 192, 193
    starch graft polymerization, 129
Ionizing radiation, starch modification, 125, 126
-ioses, 8
Iron complexes, 22
Isoamylase, 145
Isoketose, 8
Isomalt, sugar alcohol properties, 42
Isomaltooligosaccharides, enzymatic
        conversions, 142
Isomaltopentaose, 9
Isomaltose

honey, 75
    pyranose 2-oxidase conversions, 139, 141
    structures, 8
Isomaltosylglucose, 8
Isomaltotetrose, 9
Isomaltotriose, 8
Isomerases, 149
Isomerization, 2–4, 139
Isopanose, 8
Isopullulanase, 148
Isosakuranetin rutinoside, 397
Isostrain system, 165

J

Jerusalem artichoke (*Helianthus tuberosum*), 198,
        202, 206–209

K

Karaya gum, 244
    food industry applications, 165
    reserve polysaccharides, 235
    structures, 15
    viscosity, 236
Kestose, 8, 142
Ketals, 28
Ketones, 2, 3
    glycosylation, 28
    honey, 77
Ketose oxidation, 23
Kier triangle sweetener, 58–61
Kleinert (Ethanosolv) process, 342
Kojibiose, 8
Konjac flour
    blends, 170
    food industry applications, 163, 166, 170; *see
        also specific applications*
    viscosity, 236
Krebs cycle, 328

L

Lactic acid, honey, 76, 77
Lactic acid bacteria, 135
Lactitol
    food industry applications, 160
    formation of, 38
    properties of sugar alcohols, 42
Lactones, oxidation reactions, 23
Lactose, 8, 139
    galactosidase hydrolysis, 143
    glycophore localization, 59
    production of, 55

pyranose 2-oxidase conversions, 139, 140,
    141
sweetness, 60, 66
Lactulose, 8
Lamellar structure, starch–lipid complexes, 87
Laminaria, 246
Laminaribiose, 8
Lanthanum, 22
Larch arabinogalactan, 235, 236
Lauric acid esters, 43
Lauroylsucrose, 40
Legume seed galactomannans, 239–241
Leloir enzymes, 133, 135
Leucine isomers, 58
Levanases, 154
Levansucrase, 136, 142, 200
Levan-type fructans, 135, 201
Light, starch modification, 126
Light stability, cyclodextrins, 286, 288
Lignin, sugar beets, 48
Liliaceae, fructan crops, 208, 210–211
Lime milk, 51
Lime mud, 54, 55
Lime tree (*Tilia cordata*) honey, 74
Limit dextrinase, 145
Linum usitatissimum, 241
Lipases, 143
Lipid moieties, 2
Lipids
    cyclodextrins and, 277, 278–279
    starches
        amylose–lipid complexes, 143
        corn (maize), 106
        structure and properties, 87, 88, 93
        wheat, 111
    sugar beet composition, 48
Lipophilic materials,
        galactomannan–glucomannan
        blends, 170
Lippia dulcis, 389
Lithium, 22
Lithium palmitates, 40
Lithium soaps, 40
Locust bean gum, 239–240
    food industry applications, 161, 162, 166; *see
        also specific applications*
    galactomannans, 237
    reserve polysaccharides, 235
    structures, 16
    viscosity, 236
Lotus root, amylopectin, 84
Low-calorie products, 163, 169, 170; *see also*
        Sweeteners
Low-molecular-weight saccharides, and
        functional properties, 160–161

Lyases, 139, 151, 152
Lycotriose, 8
Lysophospholipases, 143
Lysozyme, 134

M

Mabinlin, 387, 388
Macrocystis pyrifera, 246
Magnesium, 22
    agave, 213
    coordination complex formation, 21
Maillard reaction, 291–310, 333
    aminodeoxysugar degradation, 295–305
        1-amino-1,4-didoxydiketoses, 304–305
        1-deoxy-2,3-diuloses, 299–304
        3-deoxy-2-oxuloses, 296–299
    glycosylamine formation and rearrangement
        to amino deoxysugars, 292–293
    glycosylamine reactions, other, 293–295
    melanoidins, 307–310
    Strecker degradation of amino acids, 305–307
    sugar esters, 41
Maize (corn), *see* Corn
Malic acid, honey, 76, 77
Malnutrition, 332
Maltese cross, 95
Maltitol, 38, 42
Maltodextrins, 161
Maltogenic amylase, 144, 146, 147
Maltohexaohydrolase, 147
Maltohexaosidase, 144
Malto-oligosaccharides, 148, 149
Malto-oligosyltrehalose hydrolase (MTH), 145,
    149
Malto-oligosyltrehalose synthase (MTS), 145,
    149
Maltopentaose, 9
Maltose, 74
    honey, 75, 76
    structure, 8
    sweetness, 60
        glycophore localization, 59
        QSAR correlation analysis, 66
Maltotetraohydrolase, 144, 147
Maltotetrose, 9
Maltotriohydrolase, 144, 147
Maltotriol, formation of, 38
Maltotriose
    honey, 75
    structure, 8
Maltulose, 8
Manihot (cassava, manioc), 104, 118–120
Mannans, structures, 13
Manninotriose, 8

Mannitol
  food industry applications, 160
  formation of, 37
  properties of sugar alcohols, 42
Mannodextrins, 139
Mannose
  glycophore localization, 59
  gum components, 235
  sweetness, 60, 66
Mannosidases, 134
Mannoside hydrolysis, 29
Mannosyltransferases, 134
Mannuronic acid
  gum components, 235
  oxidation reactions, 24
Marigold flower gum, 245
Markets, world, 104–105
Martin process, 109
Martin process, modified, 109, 110–112
Mass spectroscopy, pectins, 186
Meat products, 163, 172, 174
Mechanical agitation, starch modification, 127
Medicine and pharmacology
  biodelivery systems, 162, 171
  cyclodextrin complexes, 285–286, 287, 288
Melanoidins, 307–310
Melesitose, 8, 75
Mellibiose
  pyranose 2-oxidase conversions, 139, 141
  structure, 8
Melting temperature, starch, 87
Mesquite gum, 245
Metabolic regulators, saccharides as, 326
Metal salts and complexes, *see also* Salts and
      minerals
  chitosan binding, 222
  coordination complex formation, 21–22
  oxidized starches, 317–318
  polysaccharide modification, 130
  starch, nonnutritional applications, 318
Metanal, 2
Methane, 339
  cane sugar byproducts, 55
  energy sources, environmental, 339
  starch pyrolysis products, 124
Methanol, 126, 183
Methylated pectin, 168
Methylation
  alkylation reactions, 26
  cyclodextrins, 280
Methyl cellulose, food industry applications, 162,
      163; *see also specific applications*
Methyl esterases, pectin, 187
Methyl glucosides, hydrolysis, 29
Microbial polysaccharides, glycansucrases, 135

Microbial systems
  aldolases, 140
  chitin and chitosan, 149–151, 218
  cyclodextrins, 148
    in growth systems, 287
    production, 275–276
  fructan conversions, 155
  fructans, 200
  glycansucrases, 135
  isomaltooligosaccharides, 142
  pectin degradation, 151
  starch esters, 321
  starch metabolism
    amylases, 147
    debranching enzymes, 148
  trehalose, 149
  xylanases, 154
Microcapsules, starch esters, 320
Microcrystalline cellulose, food industry
      applications, 162, 163, 164, 165; *see
      also specific applications*
Microwave radiation, dextrins, 124, 125
Mid-infrared spectroscopy, 372–374
Milk
  carrageenans, 248
  cyclodextrin complexes, 284
  furcellerans, 249
  lactose concentration and, 333
Milk of lime, 54
Millet, 84
Minerals, *see* Metal salts and complexes; Salts
      and minerals
Miraculin, 387
Mirror images, 2–3
Mixed esters, starch, 321
Modifications, *see also* Reactivity
  caramels, 38–39
  chemical, 127–130
    acetalation, 129
    acid- and base-catalyzed degradation,
      127–128
    amination and carbamoylation, 129
    esterification, 128
    etherification, 128–129
    graft polymerization, 129–130
    halogenation, 129
    metal salts and complexes, 130
    reduction and oxidation, 128
  fatty acid esters, 39–40
  fatty acid polyesters, 40–41
  pectin, 188–189
  physicochemical properties and applications,
      41–44
  physical, 123–127
    complexation, 126–127

dextrinization, 124–125
   miscellaneous, 125–126
   pasting and gelatinization, 125
   polydextrose, 36–38
   saccharide ethers, 39
   sugar alcohols, 36–38
Modified Martin process, 109, 110–112
Mogroside, 390, 394
Moisture
   corn (maize), 106
   wheat flour, 111
Molasses, 54, 55
Molecular characteristics of starches
   analytical strategy, 353
   at granular level, 355–356
   at molecular level, 356–369
      molecular dimensions, 359–362
      molecular conformation, branching
         analysis, 362–366
      supermolecular structures, 366–368
Molecular mass, chitosan, 220–221
Monatin, 391
Monelin, 387, 388
Moniliella, 36
Monocotyledons, fructan crops, 208, 209–210
Monosaccharides, 140
   glycosylation, 28
   starch irradiation, products of, 126
   sugar beet composition, 48
   synthesis of, 140
Mucilages, *see* Gums and mucilages
Multipoint attachment theory (MPA), 61–63
Mung bean, 84, 166
Musacea, 212
Mustard seed gum, 235, 238
Mutarotation, 4, 32
Myosotis (forget-me-not) honey, 74

N

Naegeli dextrins, 85–86, 92
Naringin, 391, 397
Naringinrutinoside, 397
Near-infrared spectroscopy, 374
Nectar honeys, 74, 75
Neohesperidin, 391, 397, 398
Neohesperidose, 8
Neokestose, 8
Neopullulanase, 142, 148
Neoriocitrin, 397
Neosugar®, 206
Neotame, 400, 401
Neotrehalose, 8
Networks, pectin, 189–192
Nigerose, 8

Nitrogenous compounds, sugar beets, 48
Nofre and Tinti multipoint attachment theory,
   61–63, 65, 67
Non-Leloir enzymes, 133, 135
Nonnutritional applications, starches, 316–323
   chemically modified, 316–323
      acetals, 319–320
      alcoholized and phenolyzed starches and
         sugar alcohols, 317
      aminostarches and carbamates, 322
      copolymers, 322–323
      esters, 320–321
      ethers, 318–319
      halostarches, 321–322
      metal salts and complexes, 318
      oxidized starches, 317–318
   unmodified, 316
Nonnutritional saccharides, health and nutrition,
   333
Nonsaccharide sweeteners, commercial sugar
   varieties, 53
Nuclear magnetic resonance spectroscopy
   (NMR), 375–378
   cellulose, 377
   chitosan, 218, 219
   chitosans, 222
   cyclodextrin inclusion complexes, 281
   miscellaneous, 377–378
   pectins, 185
   solid-state NMR, 377
   solution state and high-resolution NMR,
      376–377
   starch, 377–378
   starches, 88
Nucleophilic substitution, starch etherification,
   128–129
Nutrition, 326–329, 330
   energy, 327–329, 330
   metabolic regulation, 326–327
   structural materials, 326

O

Oats, 104, 211
Obesity, 331, 399
O-glycosylation, biological significance, 29–31
Oil (petroleum), biomass and, 337–339
Oils, honey, 76
Okra gum, 235, 245
Olean, 40
Olestra, 40, 44
Oligogalacturonate, pectin, 188
Oligoglucosidase, starch conversions, 145
Oligosaccharide intolerance, 331
Oligosaccharides, sugar cane composition, 48

OmniSEC, 361
Organic acids, *see also* Amino acids; Fatty acids
  acylation reactions, 27
  esterification, 2
  honey, 76, 77
  starch esterification, 128
  starch irradiation products, 126
  starch modification, 127, 128
  sugar beet and sugar cane composition, 48
Organoleptic properties
  cyclodextrin complexation and, 283
  honey, 77
  saccharides, 333
-oses, 2, 3, 25
Osladin, 390, 394, 396
Oxalates, 50
Oxidation, 2, 23–25
  chemical reactions of saccharides, 23–25
  polysaccharide modification, 128
  starch modification, 126
Oxidized starches, nonnutritional applications,
    317–318
Oxidoreductases, 139, 140

P

Palatinose, 8
Panose, 8
Paper
  sizing and coatings, 316, 318
  starch anthranilates, 322
Pastes/pasting, 94
  physical modification, 125
  starch, 91, 92–94
  starch esters, 320
Pasting temperature, 92, 93
Pectate lyase, 152
Pectinesterases, 151
Pectins and pectic polysaccharides
    (rhamnoglactouronans), 2, 163,
    181–193, 237–238
  cell wall polysaccharides, 235
  conformation, 188–189
  coordination complex formation, 22
  enzymatic conversions, 151–153
  esterases, 140
  extraction of pectin, 187
  food industry applications, 161, 164, 165, 168;
    see also specific applications
  modification of pectin, 188–189
  networks, 189–192
  polymer chain cleavage, 139
  structures, 16, 186–187
  sugar beet composition, 48
  sugar beet juice purification, 50

Pentadin, 388
Pentadiplandra brazzena, 388
Pentosans, wheat, 112
Pentoses, alkylation reactions, 27
Peptides, sweetness, 61
Peracylation, 27
Perfumes and fragrances, cyclodextrin complexes,
    286, 287, 288
Periandrines, 390, 394, 395
Perilartine, 392
Perillaldehyde, 389
Perillaldoxime, 389
Permethylation method, 26
Pesticides, cyclodextrin complexes, 285, 287, 288
pH
  biodelivery hydrogel preparation, 162
  chitosan titration, 222–223, 224
  pectin properties, 188
  sugar beet juice purification, 51
Pharmaceuticals, *see* Medicine and pharmacology
Phenolate anion, 29
Phenols, honey, 77
Phenolyzed starches and sugar alcohols, 317
Phenylalanine, 61
Phenylglycosides, 29
Phleins, 212
Phlonis betoncoides, 389
Phosphatases, honey, 75
Phosphates
  starches, 320
    amylopectin, 91
    minor components, 87, 88
    paste properties, 94
  sugar beet juice purification, 50
Phospholipids
  cyclodextrins and, 277
  starches
    amylose and amylopectin complexes, 93
    complexes, 87, 88
    paste properties, 94
    and retrogradation, 94
Phosphoric acid esters, 2, 128
Phosphorus pentoxide, starch esterification, 128
Photographic paper, 318
Photooxidation, starch, 126
Photosensitive silver halides, 318
Photostability, cyclodextrins, 286, 288
Phyllodulcin, 391, 394–395, 396
Physical properties
  chitosan, 222–223
  food, modification of, 162
  honey, 74
Physical property modifiers, cyclodextrins,
    283–284

Physicochemical properties and applications, 41–44
    fatty acid esters, 42–44
    fatty acid polyesters, 44
    polydextrose, 41–42
    polyesters, 44
    saccharide ethers, 42
    sugar alcohols, 41
Physiological role, glycosides, 29
Phytoglycogen, 88–89
Pigments, honey, 76, 77
Plantago, 242
Plant cell starch granules, 356
Plant cell walls
    pectins, 184–185, 186, 192
    polysaccharide components, 235
Plant gums and mucilages, *see* Gums and mucilages
Plasticizers, 160
    food industry applications, 160, 161; *see also specific applications*
    starch glass transition temperature, 95–96
    sugar alcohols, 38
Plastics, beet sugar by-products, 54
Poacease (cereals and grains), fructans, 208, 211–212
Polyacrylamide, 320
Polyalcohols, *see* Sugar alcohols
Polydextroses, 38
    applications and physicochemical properties, 41–42
    food industry applications, 161
    modifications, 36–38
    uses of, 42
Polyelectrolyte gels, pectin properties, 191
Polyesters, 40–41, 44
Polygalacturonidases, pectin degradation, 151
Polyhydric alcohols, *see* Sugar alcohols
Polymer chain cleavage, 139
Polymerization, 4–5
    functions of low-molecular-weight saccharide blends, 160
    modification of polysaccharides, 129–130
    starches, glass transition, 95–96
    sugar esters, 28
Polymorphisms, starches, 95
Polyols, 41; *see also* Sugar alcohols
Polypodoside, 390, 394, 396
Polysaccharides
    and functional properties of foods, 162, 163
    modification
        chemical, 127–130
        physical, 123–127
    spectroscopy, *see* Spectroscopy
    starch, *see* Starches

sugar cane composition, 48
Polysaccharides, blends of
    and functional properties of foods, 163–164
    dairy products, 168, 169
    desserts, dressings, emulsions, and spread, 169–171
    gels and drinks, 166–169
    meat products, 172, 174
    miscellaneous, 172–173
    starch-containing, 163–165
    whipped and emulsified products, 169, 170
    spectroscopy, 374
Polysiloxanes, 321
Polysulfates, oxidized starches, 318
Polyurethane foams, starch, 322
Poncirin, 397
Portulaca oleracea gum, 245
Potassium soaps, 40
Potato starches, 94, 114–118
    amylopectin, 2, 84, 85, 87, 91
    amylose, and retrogradation, 95
    food industry applications, 165
    global markets, 104
    processed starch food products, 166
    production processes, 114–118
Potential gradient method, starch modification, 126
Precipitation
    coordination complex formation, 22
    sugar beet juice purification, 50
Production of saccharides, 47–55
    beet sugar, 48–54
        by-products and waste, 54
        chemical composition, 48
        commercial sugar varieties, 53–54
        crystallization, 52–53
        evaporation, 51–52
        extraction, 49–50
        juice purification, 50–51
        preparation of beets, 49
    cane sugar, 54–55
        by-products and waste, 55
        chemical composition, 48
Proline, honey, 75, 76, 78
Propionic acid, honey, 76, 77
Propylene glycol, 39–40
Propylene glycol alginate, 162, 163, 247
Prosopis, mesquite gum, 245
Protecting agents, low-molecular-weight saccharide blends, 160
Proteins
    animal glycoproteins, 267–269
    corn (maize), 106, 107
    food industry applications, 162, 169
    honey, 75–76

potato, 104
potato starch by-products, 118
sugar beet
    composition, 48
    juice purification, 50
sugar cane composition, 48
sweeteners, 387, 388
wheat flour, 111
Proteoglycans, animal tissues, 265
Protopectin, structures, 16
Protopectinase, 152
Psyllium gum, 235, 236, 242
Pterocaryosides, 390, 394
Pullulanase
    cyclodextrin–starch interactions, 280
    starch conversions, 145
Pullulan structures, 16
Pulp, beet sugar, 54
Pulp and paper industry, 342–346
Pyranose 2-oxidases, 139, 140
Pyranoses
    alkylation reactions, 27
    oxidation reactions, 23, 24
Pyranosides, glycosylation, 28
Pyridine, acylation reactions, 27
Pyrolysis, dextrinization, 124
Pyruvate phosphate, 140

Q

Quince seed gum, 235, 236, 242

R

Radiation, starch modification, 125, 126
Raffinose, 40
    honey, 75
    structure, 8
    sugar beet composition, 48
Raftiline®, 206
Raftulose®, 206
Raman spectroscopy, 375
RAMEB, 280
Rape (*Brassica napus*) honey, 74
Rapid Visco Amylograph, 93
Reactivity, 21–33; *see also* Modifications
    acylation, 27–28
    alkylation, 25–27
    amino compounds, 31–32
    cation complex formation, 21–22
    glycosylation, 28–29
    O-glycosylation, biological significance,
        29–31
    oxidation, 23–25
    reduction, 22–23

Rebaudioside, 389, 392, 393
Reduced calorie foods, 163, 169, 170; *see also*
        Sweeteners
Reduction, 2, 22–23
    chemical reactions of saccharides, 22–23
    polysaccharide modification, 128
Relative sweetness, 58
Reserve polysaccharides, 239–242
Retrogradation, starch, 94–95
Reversed phase HPLC, *see* High-performance
        liquid chromatography
Rhamnoglactouronans (pectins), 237–238; *see*
        also Pectins and pectic
        polysaccharides
    degradation of, 151–153
    structure, 183, 184
Rhamnose
    glycophore localization, 59
    gum components, 235
    pectins, 183
    sweetness, 60, 66
Rhamnosidase, 153
Rheology, gums, 234, 236; *see also specific gums*
Ribitol, 36
Rice
    amylopectin, 84, 87
    global markets, 104
Rice flour, 166
Root starches, 94
Rubus suavissimus, 389
Rutinose, 8
Rye fructans, 211

S

Saccharic acids, oxidation reactions, 25
Saccharide esters, acylation, 27–28
Saccharine, 401
Sago trunks, 104
Salts and minerals
    absorption of, carbohydrates and, 327
    agave, 213
    honey, 76, 77
    ionic conditions
        chitosan binding, 222
        oxidized starches, 317–318
        pectin properties, 188, 192, 193
        starch graft polymerization, 129
    starches
        gelatinization, 92
        granule organization, 89
        and retrogradation, 95
    sugar beet
        by-products, 54
        composition, 48

juice purification, 50
sugar cane composition, 48
wheat flour, 111
Saponins
    steroidal, 390, 394
    sugar beet composition, 48
Saturation of hydrogen bonds, fatty acid esters, 43
Scanning electron microscopy, starch, 89
Scent–fragrance–perfume, cyclodextrin
        complexes, 286, 287, 288
Schardinger's dextrins, 272
Schiff bases, 31–32
    chitosan, 226
    oxidized starches, 317
Schlerochilon ilcifolins, 391
Seaweeds
    agars, 247
    alginates, 246–247
    carrageenans, 247–248
    furcellerans, 249
Sensory chemistry, *see* Taste, sensory chemistry
Sephadex materials, 26
Sepharoses, alkylation, 26
Shallenberger-Acree glycophore, 58, 65
Silylated starches, 321
Silylation, 26
Silyltransferases, 134
Sizing
    paper, 316
    textiles and fibers, 318
Snell process, 39–40
Soaps, 40
Sodium, 21
Sodium hypochlorite, 128
Sodium methoxide, 40
Sodium soaps, 40
Sodium sulfate, 92
Software packages, chemical analysis, 361
Soil remediation, chitodextrin complexes, 287
Solatriose, 8
Solubility and charge density, chitosan, 222–223
Soluble mucilage polysaccharides, 238
Solvents
    fatty acid esters, 39–40
    starch modification, 126
Sonication, starch, 125
Sorbitol, 40, 139
    food industry applications, 160
    formation of, 37
    properties of sugar alcohols, 42
    sugar alcohols, 38
Sorbitol dehydrogenase, 139, 141
Sorbose, 59, 60, 140
Spectroscopy, 371–381
    cellulose, 372

dielectric relaxation (DRS), 378–380
mass spectrometry, 380
miscellaneous, examples of, 380–381
NMR, 375–378
    cellulose, 377
    miscellaneous, 377–378
    solid-state NMR, 377
    solution state and high-resolution NMR,
        376–377
    starch, 377–378
starches, 354, 365–367, 372–373
vibrational, 372–375
    cellulose, 372
    complex multicomponent systems, 374
    mid-IR, 372–374
    miscellaneous polysaccharides, 373–374
    near-IR, 374
    Raman, 375
    starch, 372–373
Spondias gum, 245
Stabilizers, fatty acid polyesters and, 44
Stachyose, 9, 40
Starch carbamate block copolymers, 322
Starches, 81–96, 104–120
    cyclodextrins and, 280
    enzymatic conversions, 143–149
        α-amylase family, common properties,
            145–146
        α-amylases, 146–147
        cyclodextrins and cycloamylases,
            148–149
        debranching enzymes, 148
        exo-acting enzymes, 147–148
        miscellaneous conversion, 149
        trahalose, 149
    food industry applications, 161
    functions of polysaccharide blends, 163–165
    isomaltooligosaccharides, 142
    lyases, 149
    molecular characteristics, 349–366
        analytical strategy, 353
        granular level, 355–356
    molecular characteristics, molecular level,
        356–369
        molecular dimensions, 359–362
        molecular conformation, branching
            analysis, 362–366
        supermolecular structures, 366–368
    nonnutritional applications of chemically
            modified products, 316–323
        acetals, 319–320
        alcoholized and phenolyzed starches and
            sugar alcohols, 317
        aminostarches and carbamates, 322
        copolymers, 322–323

esters, 320–321
ethers, 318–319
halostarches, 321–322
metal salts and complexes, 318
oxidized starches, 317–318
nonnutritional applications of unmodified
products, 316
organization of granules, 89–90
polymer chain cleavage, 139
production technologies, 105–120
cassava, 118–120
corn (maize), 105–109
potato, 114–118
wheat, 109–114
properties, 90–96
gelatinization, 91–92
glass transition temperature, 95–96
pasting, 92–94
retrogradation, 94–95
spectroscopy
mid-IR, 372–373
NMR, 377–378
structure, 82–89
amylopectin, 83–87
amylose, 82–83
minor components, 87–89
sugar cane composition, 48
wheat flour, 111
world markets, 104–105
Starch hydrolysates
food industry applications, 162, 163
formation of, 38
Starch synthase, amylopectin, 87
Sterculia, 244
Stereochemistry
monosaccharides, 19–21
monosaccharide synthesis, 140
structure, 2–4, 5, 6–17
sweetness, 58
Sterilization, chitosan, 225
Steroidal saponins, 390, 394
Steroids, cyclodextrin complexes, 287
Steviobioside, 393
Stevioside, 389, 392, 393
Strecker degradation of amino acids, 305–307
Strontium, 21
Structural materials, saccharides as, 326
Structure, 3–5, 6–17
chitosan, 218–220
gums, 234, 235
pectin, 186–187
starch, 82–89
amylopectin, 83–87
amylose, 82–83
minor components, 87–89

stereochemistry of monosaccharides, 19–21
Structure–activity relationship, sweeteners,
63–65, 66, 68
Succinic acid, honey, 76, 77
Sucralose, 400, 401
Sucrose, 8
agave, 213
honey, 74, 75
honey adulteration, 78
properties of sugar alcohols, 42
sugar beet and cane composition, 48
sweetness, 59, 60, 66
Sucrose esters
antimicrobial properties, 43–44
complex formation with starch and protein, 43
Sucrose isomerase, 139
Sucrose monolaurates, 40, 43
Sucrose polyesters, 40–41
Sugar alcohols
applications and physicochemical properties, 41
commercial sugar varieties, 53
food industry applications, 160, 161
modifications, 36–38
nonnutritional applications, 317
Sugar beet processing, 48–54
Sugar cane processing, 54–55
Sugary-2 maize, 94
Sulfamates, 401
Sulfates, starch, 128, 320–321
Sulfur dioxide
corn (maize) processing, 106
potato starch processing, 117
Sulfuric acid esters, 2
Suosan, 400
Supercellulases, search for, 341–342
Surface gelatinization, starch, 89
Surfactants, 319
fatty acid esters, 42–43
fatty acid polyesters, 41
saccharide ethers, 42
starch esters, 320–321
Sweeteners, 387–402
commercial sugar varieties, 53
natural compounds, 387–399
dihydroisocoumarins (1-H-2 Benzopyran-
1-ones), 391, 394–396
flavanoids, 391, 396–398
miscellaneous, 391, 398–399
proteins, 387, 388
steroidal saponins, 390, 394
terpenoids, 389–390, 392–394
sensory chemistry, 57–70
biochemistry of taste transduction, 69–70

electron donors and dispersion sweeteners
  in sucrose/galactosucrose system,
  66–69
glycophore in aldopyranoses, 65–66
heptagonal sweetener, 61–63
relative sweetness, 58
Shallenberger-Acree glycophore, 58
structure–activity relationship, 63–65, 66, 68
triangle sweeteners, 58–61
sugar esters, 41
sweetness, taste transduction, 69–70
synthetic compounds, 399–402
Sweet potatoes, 104
Synergistic interactions
gels, blend, 168
glucomannans, 241
Synthons, 130

T

Tagatose, 141
Takayanagi's polymer blending rule, 165
Talin, 388
Tamarind flour, 17
Tamarind gum, 235, 236, 242
Tamarind seed xyloglucan, 162, 163
Tapioca
  amylopectin, 84
  global markets, 104, 105
  starches, 94
Tapioca pudding, 91
Tara gum, 235, 236, 240–241
Taro, 84
Taste, sensory chemistry, 57–70
  biochemistry of transduction, 69–70
  electron donors and dispersion sweeteners in
    sucrose/galactosucrose system,
    66–69
  glycophore in aldopyranoses, 65–66
  heptagonal sweetener, 61–63
  relative sweetness, 58
  Shallenberger-Acree glycophore, 58
  structure–activity relationship, 63–65, 66, 68
  triangle sweeteners, 58–61
Taste modifiers
  cyclodextrins, 284
  miraculin and curculin, 387
Technical properties, chitosan, 226–227
Telsomoside, 398
Temperature, 95
  saccharide ether stability, 42
  starches
    gelatinization, 91
    glass transition temperature, 95–96
    minor components, 87

pasting, 92, 93
  and retrogradation, 95
sugar beet extraction, 50
sugar beet juice purification, 50
Terpenoid sweeteners, 389–390, 392–394
Tessaria dodoneifolia, 391
Tetrasaccharides, 9
Tetrazoles, nonnutritive sweeteners, 400
Textiles, cyclodextrins, 287
Thaumatins, 387, 388
Theanderose, 8
Thermal processing, dextrinization, 124
Thiandiantha grosvernorii, 390
Thin layer chromatography, 354
Thiosemicarbazones, 322
Thixotropic characteristics, 168
Tilia cordata (lime tree) honey, 74
Titration, chitosans, 222
TMS ethers, alkylation reactions, 26
Toxicity, cyclodextrins, 272, 277–278
Tragacantin, structures, 17
Traganth gum, 244
  food industry applications, 163, 165; *see also
    specific applications*
  reserve polysaccharides, 235
  structures, 17
  viscosity, 236
Trahalose, 149
Transduction, taste, 69–70
Transesterification, acylation reactions, 27–28
Transglycosylation, 31, 135
  chitosan, 225
  enzymatic conversions, 135–136
    glycosyltransferases, 133, 135
    mechanisms, 136–138
  glycoside hydrolases, 133
  neopullulanases, 148
  starch debranching enzymes, 148
Trehalose, 8, 40, 75
Triangle sweeteners, 58–61
Tribromobenzamides, 400
Tricanter process, Flottweg, 110
Trichosporonoides, 36
Tridentate complex formation, 21
Triethylamine, 27
Trigonella foenum-graecum, 241
Trimethylsilyl ethers, 26
Trioses, 2, 8
Trisaccharides, 8
Trityl chloride, alkylation, 25
Trityl ethers, 27
Tryptophan derivatives, sweeteners, 400
Tuber starches, 94
Turanose, 8, 75
Tyrosine groups, xylanases, 154

U

-uloses, 2
Ultrasound, glycose hydrolase yield, 133
Ultraviolet light, starch modification, 126
Umbelliferose, 8
Ureas, 31
    nonnutritive sweeteners, 400
    starch and, 322
Uronic acids, oxidation reactions, 24, 25

V

Vanillin, 129
Vascular disease, 332
Vinyl esters of fatty acids, 40
Viscosity
    chitosan degradation, 223, 224, 335
    food, modification of, 168
    honey, 74
    pectin, 187, 188
    polydextrose, 41–42
    starches, 92, 93, 125, 354
    sucrose ester–starch–protein complexes, 43
    wheat flour properties, 111
Vitamins
    agave, 213
    cyclodextrins, 282–283
    honey, 76

W

Wall polysaccharides, *see* Cell walls
Water
    cyclodextrin solubility, 280
    fatty acid esters, 42–43
    properties of sugar alcohols, 42
    sugar beet composition, 48
Water chestnut, 84
Waxy mutants, 349; *see also* Starches
    pasting profiles, 93
    starch granules in native state, 356
    starch retrogradation, 94
Westfalia process, 110, 112–114
Wetting agents, 41
Wheat
    arabinoxylans, 237
    fructans in grain, 211
Wheat starch, 109–114
    amylopectin, 84, 87
    analysis of molecular characteristics, 364,
        365, 367
    fragment analysis, 357
    pasting, 92–93
    production

    modified Martin process, 110–112
    substrates, 110, 111
    Westfalia process, 112–114
    sucrose ester–starch–protein complexes, 43
Whey protein isolate–starch mixtures, 165, 170
Whipped food products, 169, 170
World markets, starch, 104–105

X

Xanthan gum
    food industry applications, 162, 163, 166; *see
        also specific applications*
    glucomannans and, 241
    structures, 17
Xylanase, 134
Xylans
    enzymatic conversions, 153–154
    lyases and esterases, 139, 140
    structures, 14
Xylitol, 36–37
    food industry applications, 160
    properties of sugar alcohols, 42
Xylofuranosides, 28–29
Xylogalacturonan, 151, 153
Xyloglucan, tamarind, 162, 163
Xylose, 140
    glycophore localization, 59
    gum components, 235
    pectins, 183
    sweetness, 60, 66
Xylose isomerase, 149
Xylose reductase, 139
Xylosidase, 134
Xylosides, hydrolysis, 29

Y

Yeasts
    chitosan, 218
    sugar alcohols, 36
Yellow mustard seed gum, 235, 238
Yttrium, 22

Z

Zein, 107
Zinc
    acylation reactions, 27
    coordination complex formation, 21
    polysaccharide degradation, 140